Student Study Guide and Solutions Manual for

Organic Chemistry

third edition

Brown and Foote

Brent Iverson and Sheila Iverson

University of Texas at Austin

THOMSON

BROOKS/COLE

Australia • Canada • Mexico • Singapore • Spain • United Kingdom • United States

For more information about our products,
contact us at:
Thomson Learning Academic Resource Center
1-800-423-0563

For permission to use material from this text,
contact us by:
Phone: 1-800-730-2214
Fax: 1-800-731-2215
Web: www.thomsonrights.com

Asia
Thomson Learning
5 Shenton Way #01-01
UIC Building
Singapore 068808

Australia
Nelson Thomson Learning
102 Dodds Street
South Street
South Melbourne, Victoria 3205
Australia

Canada
Nelson Thomson Learning
1120 Birchmount Road
Toronto, Ontario M1K 5G4
Canada

Europe/Middle East/South Africa
Thomson Learning
High Holborn House
50/51 Bedford Row
London WC1R 4LR
United Kingdom

Latin America
Thomson Learning
Seneca, 53
Colonia Polanco
11560 Mexico D.F.
Mexico

Spain
Paraninfo Thomson Learning
Calle/Magallanes, 25
28015 Madrid, Spain

CONTENTS:

To Carina, Alexandra, Alanna, and Juliana
with love

This Student Study Guide is a companion to the third edition of *Organic Chemistry* by William Brown and Christopher Foote. All of the *problems* from the text have been reprinted in this guide, so there is no need to flip back and forth between the text and the guide. Detailed, stepwise *solutions* to all of the problems are provided.

Molecules are three-dimensional and understanding the three-dimensionality of organic chemistry is important for any student. This edition of the Student Study Guide has placed special emphasis on stereochemistry, an important aspect of three-dimensional molecular structure. Throughout the problems and answers, many of the molecules with stereocenters have the configuration explicitly denoted using wedges and dashes to indicate location in space. When the configuration is not specifically given, stereocenters are indicated by an asterisk (*). In addition, the stereochemical outcome of every reaction is stated, whether the question calls for consideration of stereochemistry or not.

The material in this volume was reviewed for accuracy by Brent and Sheila Iverson, Brian Pagenkopf of the University of Texas at Austin, and William Brown of Beloit College. If you have any comments or questions, please direct them to Professor Brent Iverson, Department of Chemistry and Biochemistry, the University of Texas at Austin, Austin, Texas 78712. E-mail: biverson@utxvms.cc.utexas.edu.

Brent and Sheila Iverson
Austin, Texas
February, 2001

CHAPTER 1
Solutions to the Problems

Problem 1.1 Write and compare the ground-state electron configurations for each pair of elements:

(a) Carbon and silicon

C (6 electrons) $1s^22s^22p^2$
Si (14 electrons) $1s^22s^22p^63s^23p^2$
Both carbon and silicon have four electrons in their outermost (valence) shells.

(b) Oxygen and sulfur

O (8 electrons) $1s^22s^22p^4$
S (16 electrons) $1s^22s^22p^63s^23p^4$
Both oxygen and sulfur have six electrons in their outermost (valence) shells.

(c) Nitrogen and phosphorus

N (7 electrons) $1s^22s^22p^3$
P (15 electrons) $1s^22s^22p^63s^23p^3$
Both nitrogen and phosphorus have five electrons in their outermost (valence) shells.

Problem 1.2 Show that the following obey the octet rule.

(a) Sulfur forms sulfide ion, S^{2-}.

S (16 electrons): $1s^22s^22p^63s^23p^4$
S^{2-} (18 electrons): $1s^22s^22p^63s^23p^6$

(b) Magnesium forms Mg^{2+}.

Mg (12 electrons): $1s^22s^22p^63s^2$
Mg^{2+} (10 electrons): $1s^22s^22p^6$

Problem 1.3 Judging from their relative positions in the Periodic Table, which element in each set is more electronegative?

(a) Lithium or potassium

In general, electronegativity increases from left to right across a row and from bottom to top of a column in the Periodic Table. This is because electronegativity increases with increasing positive charge on the nucleus and with decreasing distance of the valence electrons from the nucleus. Lithium is higher up on the Periodic Table and thus more electronegative than potassium.

(b) Nitrogen or phosphorus

Nitrogen is higher up on the Periodic Table and thus more electronegative than phosphorus.

(c) Carbon or silicon

Carbon is higher up on the Periodic Table and thus more electronegative than silicon.

Problem 1.4 Classify these bonds as nonpolar covalent, polar covalent, or ionic.
(a) S-H　　　(b) P-H　　　(c) C-F　　　(d) C-Cl

Using the rule that bonds formed from atoms with an electronegativity difference of less than 0.5 are nonpolar covalent, the following table can be constructed:

Bond	Difference in electronegativity	Type of bond
S-H	2.5 - 2.1 = 0.4	Nonpolar covalent
P-H	2.1 - 2.1 = 0	Nonpolar covalent
C-F	4.0 - 2.5 = 1.5	Polar covalent
C-Cl	3.0 - 2.5 = 0.5	Polar covalent

Problem 1.5 Using the symbols δ- and δ+, indicate the direction of polarity in these polar covalent bonds.

(a) C-N　　　　　　　　(b) N-O
δ+　 δ-　　　　　　　　δ-　 δ-
C-N　　　　　　　　　　N-O

Nitrogen is more electronegative than carbon　　　　Oxygen is more electronegative than nitrogen

(c) C–Cl

δ+ δ–
C–Cl

Chlorine is more electronegative than carbon

Problem 1.6 Draw Lewis structures, showing all valence electrons, for these molecules.
(a) C_2H_6 (b) CS_2 (c) HCN

H H
| |
H–C–C–H
| |
H H

$\ddot{S}=C=\ddot{S}$

H–C≡N:

Problem 1.7 Draw Lewis structures for these ions, and show which atom in each bears the formal charge.
(a) $CH_3NH_3^+$
Methylammonium ion

(b) CO_3^{2-}
Carbonate ion

(c) OH⁻
Hydroxide ion

H H
| |
H–C–N⁺–H
| |
H H

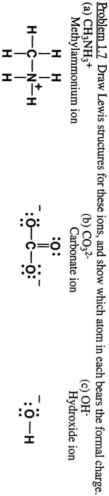

:Ö:⁻
|
:Ö–C–Ö:⁻
‖
:O:

:Ö:⁻
|
:Ö–H

Problem 1.8
Draw Lewis structures and condensed structural formulas for the four alcohols of molecular formula $C_4H_{10}O$. Classify each alcohol as primary, secondary, or tertiary.

CH_3-CH_2-CH_2-CH_2-$\ddot{O}H$

CH_3-CH-CH_2-$\ddot{O}H$
|
CH_3

:ÖH
|
CH_3-CH_2-CH-CH_3

:ÖH
|
CH_3-C-CH_3
|
CH_3

Primary   **Primary**   **Secondary**   **Tertiary**

Problem 1.9
Draw structural formulas for the three secondary amines of molecular formula $C_4H_{11}N$.

H
|
CH_3-CH_2-N-CH_2-CH_3

H
|
CH_3-N-CH_2-CH_2-CH_3

H
|
CH_3-N-CH-CH_3
|
CH_3

Problem 1.10 Draw condensed structural formulas for the three ketones of molecular formula $C_5H_{10}O$.

O
‖
CH_3-CH_2-CH_2-C-CH_3

O
‖
CH_3-CH-C-CH_3
|
CH_3

O
‖
CH_3-CH_2-C-CH_2-CH_3

Problem 1.11 Draw condensed structural formulas for the two carboxylic acids of molecular formula $C_4H_8O_2$.

O
‖
CH_3-CH_2-CH_2-C-OH

CH_3-CH-C=O
| |
CH_3 OH

Problem 1.12 Draw structural formulas for the four esters of molecular formula $C_4H_8O_2$.

O
‖
CH_3-CH_2-C-O-CH_3

O
‖
CH_3-C-O-CH_2-CH_3

O
‖
H-C-O-CH_2-CH_2-CH_3

O
‖
H-C-O-CH-CH_3
|
CH_3

Problem 1.13   Predict all bond angles for these molecules.
(a) CH_3OH

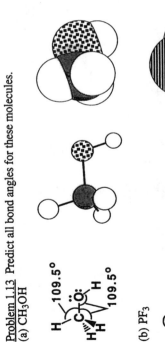

$109.5°$
$109.5°$
$109.5°$

(b) PF_3

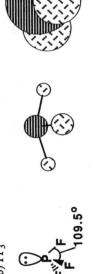

$109.5°$

(c) H_2CO_3 (Carbonic Acid)

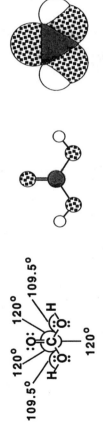

$109.5°$ $120°$ $120°$
$109.5°$
$120°$

Problem 1.14   Which molecules are polar? For each molecule that is, specify the direction of its dipole moment.
(a) CH_2Cl_2

Recall that a molecular dipole moment is determined as the vector sum of the bond dipoles in three-dimensional space. Thus, by superimposing the bond dipoles on a three-dimensional drawing, the molecular dipole moment can be determined. Note that on the following diagrams the dipole moments from the C-H bonds have been ignored because they are so small.

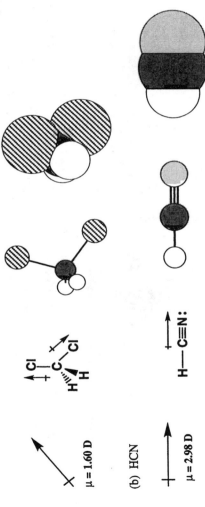

$\mu = 1.60$ D

(b) HCN

H—C≡N:

$\mu = 2.98$ D

(c) H_2O_2

The H_2O_2 molecule can rotate around the O-O single bond, so we must consider the molecular dipole moments in the various possible conformations. Conformations such as the one depicted on the left below have a net molecular dipole moment, while conformations such as the one the right below do not. The presence of at least some conformations (such as that on the left) that have a molecular dipole moment means that the entire molecule must have an overall dipole moment, in this case $\mu = 2.2$ D.

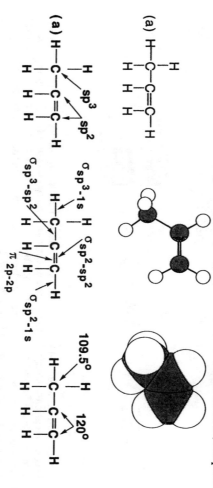

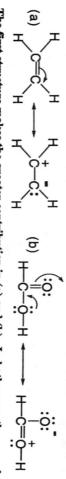

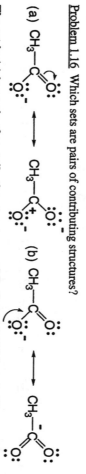

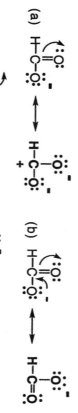

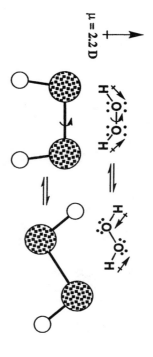

μ = 2.2 D

Problem 1.15 Draw the contributing structure indicated by curved arrows. Be certain to show all valence electrons and all formal charges.

(a)

(b)

(c) $CH_3-C-O-CH_3$ ⟷ $CH_3-C=O-CH_3$

Problem 1.16 Which sets are pairs of contributing structures?

(a) CH_3 ⟷ CH_3

(b) CH_3 ⟷ CH_3

The set in (a) is a pair of contributing structures, while the set in (b) is not. The structure on the right in set (b) is not a viable contributing structure because there are five bonds to the carbon atom.

Problem 1.17 Estimate the relative contribution of the members in each set of contributing structures.

(a)

(b)

The first structure makes the greater contribution in (a) and (b). In both cases, the second contributing structure involves the disfavored creation and separation of unlike charges.

Problem 1.18 Describe the bonding in these molecules in terms of atomic orbitals involved, and predict all bond angles.

(a)

(a) $H-C-C=C-H$

sp^3 sp^2

σ_{sp^3-1s} $\sigma_{sp^3-sp^2}$ $\sigma_{sp^2-sp^2}$ σ_{sp^2-1s}

π_{2p-2p}

109.5°   120°

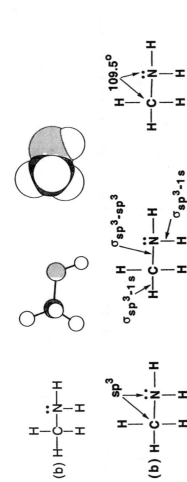

(b) H—C—N—H with H atoms, sp³

(b) H—C—N—H σ_{sp^3-1s} $\sigma_{sp^3-sp^3}$ σ_{sp^3-1s}

(b) H—C—N—H   109.5°

Electronic Structures of Atoms

Problem 1.19 Write ground-state electron configuration for each atom. After each atom is given its atomic number.
(a) Sodium (11)   Na (11 electrons) $1s^2 2s^2 2p^6 3s^1$

(b) Magnesium (12)   Mg (12 electrons) $1s^2 2s^2 2p^6 3s^2$

(c) Oxygen (8)   O (8 electrons) $1s^2 2s^2 2p^4$

(d) Nitrogen (7)   N (7 electrons) $1s^2 2s^2 2p^3$

Problem 1.20 Which atom has the ground-state electron configuration of
(a) $1s^2 2s^2 2p^6 3s^2 3p^4$

Sulfur (16) has this ground-state electron configuration

(b) $1s^2 2s^2 2p^4$

Oxygen (8) has this ground-state electron configuration

Problem 1.21 Define valence shell and valence electron.

The valence shell is the outermost occupied shell of an atom. A valence electron is an electron in the valence shell.

Problem 1.22 How many electrons are in the valence shell of each atom?
(a) Carbon

With a ground-state electron configuration of $1s^2 2s^2 2p^2$, there are four electrons in the valence shell of carbon.

(b) Nitrogen

With a ground-state electron configuration of $1s^2 2s^2 2p^3$, there are five electrons in the valence shell of nitrogen.

(c) Chlorine

With a ground-state electron configuration of $1s^2 2s^2 2p^6 3s^2 3p^5$, there are seven electrons in the valence shell of chlorine.

(d) Aluminum

With a ground-state electron configuration of $1s^2 2s^2 2p^6 3s^2 3p^1$, there are three electrons in the valence shell of aluminum.

Lewis Structures
Problem 1.23 Judging from their relative positions in the Periodic Table, which atom in each set is more electronegative?
(a) Carbon or nitrogen

In general, electronegativity increases from left to right across a row (period) and from bottom to top of a column in the Periodic Table. This is because electronegativity increases with increasing positive charge on the nucleus and with decreasing distance of the valence electrons from the nucleus. Nitrogen is farther to the right than carbon in Period 2 of the Periodic Table, thus nitrogen is more electronegative than carbon.

(b) Chlorine or bromine

Chlorine is higher up than bromine in column 7A of the Periodic Table, thus chlorine is more electronegative than bromine.

(c) Oxygen or sulfur

Oxygen is higher up than sulfur in column 6A of the Periodic Table, thus oxygen is more electronegative than sulfur.

Problem 1.24 Which of these compounds have nonpolar covalent bonds, which have polar covalent bonds, and which have ionic bonds?

(a) LiF   (b) CH_3F (c) $MgCl_2$ (d) HCl

Using the rule that an ionic bond is formed between atoms with an electronegativity difference of 1.9 or greater, the following table can be constructed:

Bond	Difference in electronegativity	Type of bond
Li-F	4.0 - 1.0 = 3.0	Ionic
C-H	2.5 - 2.1 = 0.4	Nonpolar covalent
C-F	4.0 - 2.5 = 1.5	Polar covalent
Mg-Cl	3.0 - 1.2 = 1.8	Polar covalent
H-Cl	3.0 - 2.1 = 0.9	Polar covalent

Based on these values, only LiF has an ionic bond. The other compounds have nonpolar covalent (C-H) or polar covalent (C-F, Mg-Cl, H-Cl) bonds.

Problem 1.25 Using the symbols δ- and δ+, indicate the direction of polarity, if any, in each covalent bond.

(a) C-Cl

$\overset{\delta+}{C}-\overset{\delta-}{Cl}$ Chlorine is more electronegative than carbon.

(b) S-H

$\overset{\delta-}{S}-\overset{\delta+}{H}$ Sulfur is more electronegative than hydrogen.

(c) C-S   Carbon and sulfur have the same electronegativities so there is no polarity in a C-S bond.

(d) P-H   Phosphorus and hydrogen have the same electronegativities, so there is no polarity in a P-H bond.

Problem 1.26 Write Lewis structures for these compounds. Show all valence electrons. None of them contains a ring of atoms.

(a) H_2O_2
Hydrogen peroxide

(b) N_2H_4
Hydrazine

(c) CH_3OH
Methanol

(d) CH_3SH
Methanethiol

(e) CH_3NH_2
Methylamine

(f) CH_2Cl_2
Dichloromethane

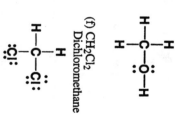

(g) CH_3OCH_3
Dimethyl ether

(h) H_2CO_3
Carbonic acid

(i) CH_2O
Formaldehyde

(j) CH_3COOH
Acetic acid

(k) CH_3COCH_3
Acetone

(l) HCN
Hydrogen cyanide

H—C≡N:

(m) HNO_3
Nitric acid

(n) HNO_2
Nitrous acid

(o) HCOOH
Formic acid

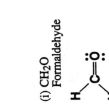
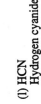

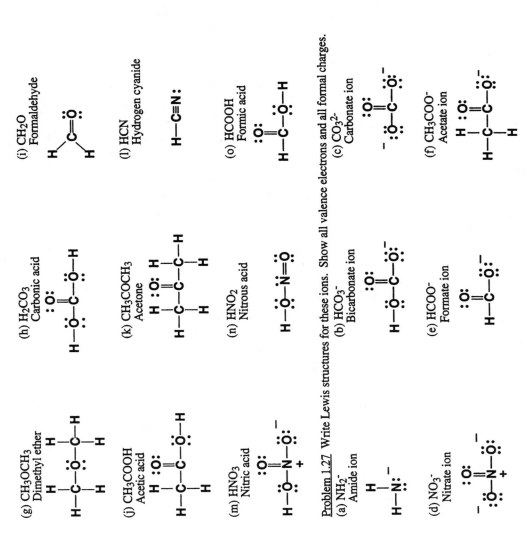

Problem 1.27 Write Lewis structures for these ions. Show all valence electrons and all formal charges.

(a) NH_2^-
Amide ion

(b) HCO_3^-
Bicarbonate ion

(c) CO_3^{2-}
Carbonate ion

(d) NO_3^-
Nitrate ion

(e) $HCOO^-$
Formate ion

(f) CH_3COO^-
Acetate ion

Problem 1.28 Complete these structural formulas by adding enough hydrogens to complete the tetravalence of each carbon. Then write the molecular formula of each compound.

(a) C_6H_{12}

(b) $C_4H_8O_2$

(c) C_4H_8O

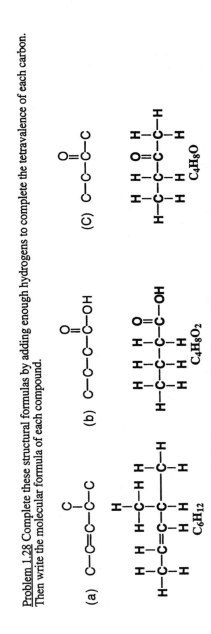

(d)

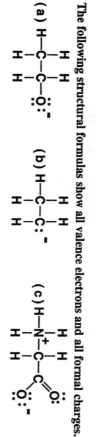

C_4H_8O

(e)

(f)

$C_3H_7NO_2$

(g)

$C_5H_{12}O$

(h)

$C_4H_8O_3$

(i)

C_3H_6O

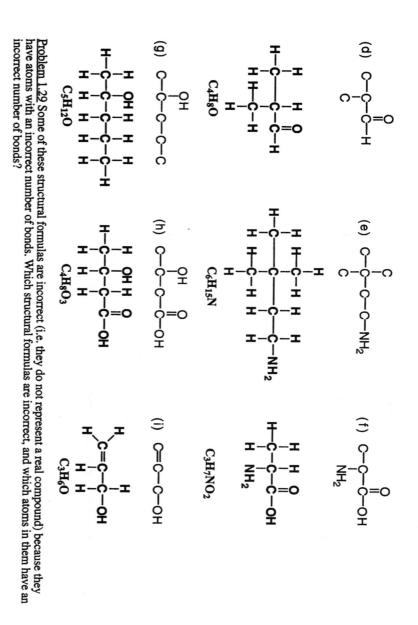

$C_6H_{15}N$

Problem 1.29 Some of these structural formulas are incorrect (i.e. they do not represent a real compound) because they have atoms with an incorrect number of bonds. Which structural formulas are incorrect, and which atoms in them have an incorrect number of bonds?

(a)

(b)

(C)

(d)

(e)

(f)

(g)

(h)

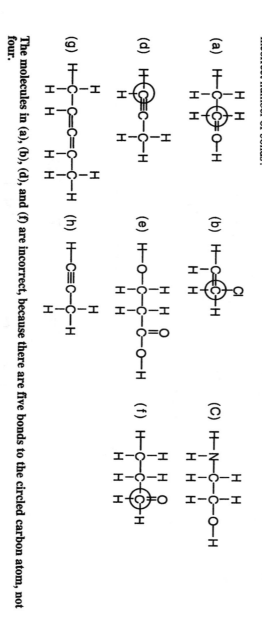

The molecules in (a), (b), (d), and (f) are incorrect, because there are five bonds to the circled carbon atom, not four.

Problem 1.30 Following the rule that each atom of carbon, oxygen, and nitrogen reacts to achieve a complete outer shell of eight valence electrons, add unshared pairs of electrons as necessary to complete the valence shell of each atom in these ions. Then assign formal charges as appropriate.

The following structural formulas show all valence electrons and all formal charges.

(a)

(b)

(c)

Problem 1.31 Following are several Lewis structures showing all valence electrons. Assign formal charges in each structure as appropriate.

There is a positive formal charge in parts (a), (e), and (f). There is a negative formal charge in parts (b), (c), and (d).

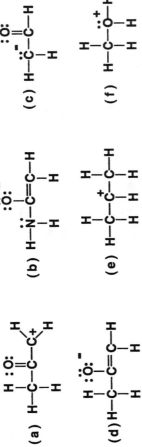

Polarity of Covalent Bonds

Problem 1.32 Which statements are true about electronegativity?
(a) Electronegativity increases from left to right in a period of the Periodic Table.
(b) Electronegativity increases from the top to the bottom in a column of the Periodic Table.
(c) Hydrogen, the element with the lowest atomic number, has the smallest electronegativity.
(d) The higher the atomic number of an element, the greater its electronegativity.

Electronegativity *increases* from left to right across a period and from bottom to top of a column in the Periodic Table. Thus, statement (a) is true, but (b), (c), and (d) are false.

Problem 1.33 Why does fluorine, the element in the upper right corner of the Periodic Table, have the largest electronegativity of any element?

Electronegativity increases with increasing positive charge on the nucleus and with decreasing distance of the valence electrons from the nucleus. Fluorine is that element for which these two parameters lead to maximum electronegativity.

Problem 1.34 Arrange the single covalent bonds within each set in order of increasing polarity.
(a) C-H, O-H, N-H   (b) C-H, B-H, O-H   (c) C-H, C-Cl, C-I
C-H < N-H < O-H   B-H < C-H < O-H   C-H < C-I < C-Cl

(d) C-S, C-O, C-N   (e) C-Li, C-B, C-Mg
C-S < C-N < C-O   C-B < C-Mg < C-Li

Problem 1.35 Using the values of electronegativity given in Table 1.5, predict which indicated bond in each set is the more polar and, using the symbols $\delta+$ and $\delta-$, show the direction of its polarity.
(a) CH_3-OH or CH_3O-H   (b) H-NH_2 or CH_3-NH_2

$$\overset{\delta-}{CH_3}\overset{\delta+}{O\text{-}H}$$

$$\overset{\delta+}{H}\text{-}\overset{\delta-}{NH_2}$$

(c) CH_3-SH or CH_3S-H   (d) CH_3-F or H-F

$$\overset{\delta-}{CH_3}\overset{\delta+}{S\text{-}H}$$

$$\overset{\delta+}{H}\text{-}\overset{\delta-}{F}$$

Problem 1.36 Identify the most polar bond in each molecule.
(a) $HSCH_2CH_2OH$ (b) $CHCl_2F$ (c) $HOCH_2CH_2NH_2$

The O-H bond   The C-F bond   The O-H bond
(1.4)   (1.5)   (1.4)

The difference in electronegativities is given in parentheses underneath each answer.

Bond Angles and Shapes of Molecules

Problem 1.37 Use the VSEPR model to predict bond angles about each highlighted atom.

Approximate bond angles as predicted by the valence-shell electron-pair repulsion model are as shown.

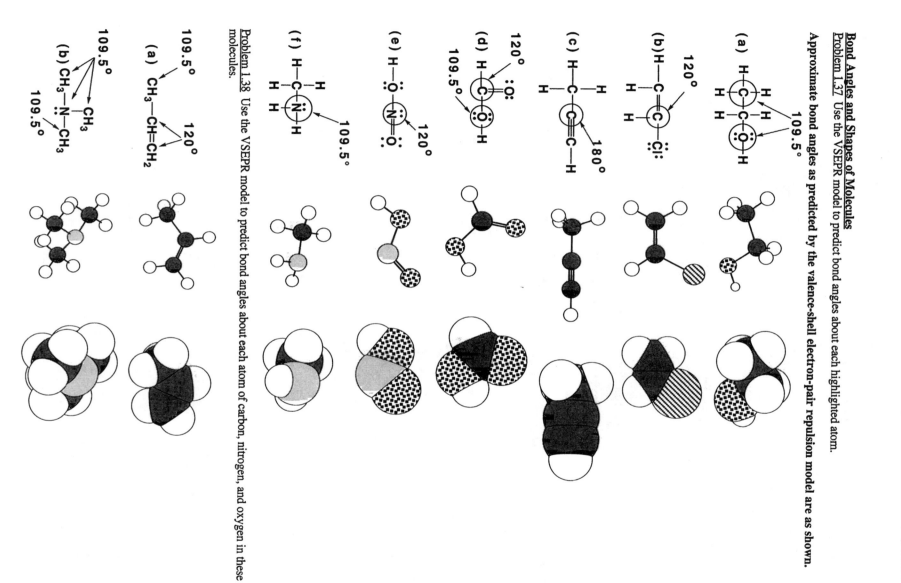

(a)
109.5°

(b) H—C—C
    120°

(c) H—C—C
    180°

(d) H—C—O
    109.5°
    120°

(e) H—O—N—O
    120°

(f) H—C—N
    109.5°

Problem 1.38 Use the VSEPR model to predict bond angles about each atom of carbon, nitrogen, and oxygen in these molecules.

(a) CH₃—CH=CH₂
    109.5°   120°

(b) CH₃—N—CH₃
       CH₃
    109.5°
    109.5°
    109.5°

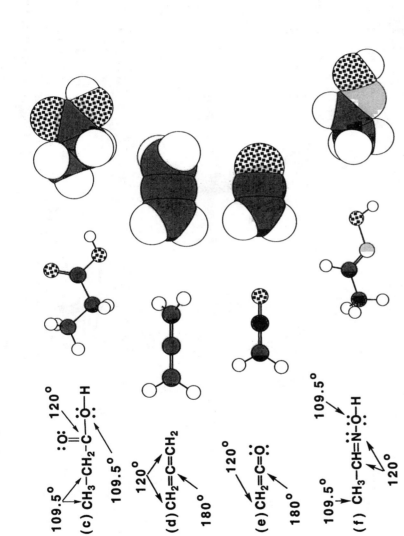

(c) 109.5°   :Ö: 120°
CH₃—CH₂—C—Ö—H
     109.5°

(d) CH₂=C=CH₂
   120°
   180°

(e) CH₂=C=Ö:
  120°
  180°

(f) 109.5°   109.5°
CH₃—CH=N—Ö—H
     120°

Problem 1.39 Use VSEPR model to predict the geometry of these ions.

(a) NH₂⁻
   109.5°
H—N:
|
H

Bent

(b) NO₂⁻
    120°
:Ö—N=Ö:

Bent

(c) NO₂⁺
    180°
:Ö=N=Ö:
     +

Linear

(d) NO₃⁻
:Ö: 120°
  N⁺
:Ö: :Ö:⁻

Trigonal planar

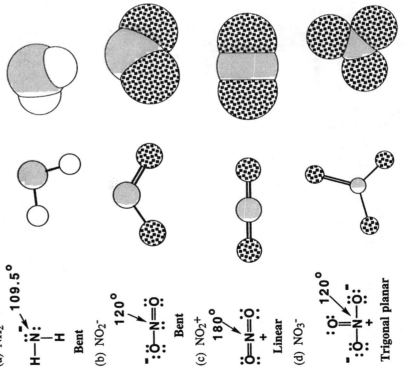

(e) CH₃COO⁻

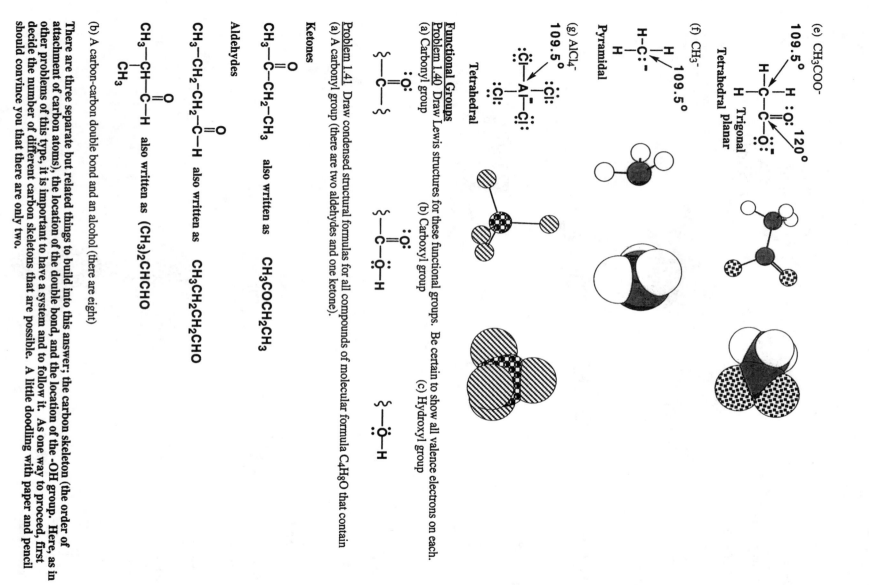

109.5° 120°
Tetrahedral planar Trigonal

(f) CH₃⁻
109.5°
Pyramidal

(g) AlCl₄⁻
109.5°
Tetrahedral

Functional Groups
Problem 1.40 Draw Lewis structures for these functional groups. Be certain to show all valence electrons on each.
(a) Carbonyl group (b) Carboxyl group (c) Hydroxyl group

Problem 1.41 Draw condensed structural formulas for all compounds of molecular formula C₄H₈O that contain
(a) A carbonyl group (there are two aldehydes and one ketone).

Ketones

$CH_3-C(=O)-CH_2-CH_3$ also written as $CH_3COCH_2CH_3$

Aldehydes

$CH_3-CH_2-CH_2-C(=O)-H$ also written as $CH_3CH_2CH_2CHO$

$CH_3-CH(CH_3)-C(=O)-H$ also written as $(CH_3)_2CHCHO$

(b) A carbon-carbon double bond and an alcohol (there are eight)

There are three separate but related things to build into this answer; the carbon skeleton (the order of attachment of carbon atoms), the location of the double bond, and the location of the -OH group. Here, as in other problems of this type, it is important to have a system and to follow it. As one way to proceed, first decide the number of different carbon skeletons that are possible. A little doodling with paper and pencil should convince you that there are only two.

Next locate the double bond on these carbon skeletons. There are three possible locations for it.

$$C-C-C \quad \text{and} \quad \overset{\displaystyle C}{\underset{\displaystyle C-C-C}{|}}$$

$$C=C-C-C \quad \text{and} \quad C-C=C-C \quad \text{and} \quad \overset{\displaystyle C}{\underset{\displaystyle C=C-C}{|}}$$

Finally, locate the -OH group and then add the remaining seven hydrogens to complete each structural formula. For the first carbon skeleton, there are four possible locations of the -OH group; for the second carbon skeleton there are two possible locations; and for the third, there are also two possible locations. Four of these compounds (marked by a # symbol) are not stable and are in equilibrium with a more stable aldehyde or ketone. You need not be concerned, however, with this now. Just concentrate on drawing the required eight structural formulas.

$HO-CH=CH-CH_2-CH_3$ # $\overset{\displaystyle OH}{\underset{\displaystyle CH_2=C-CH_2-CH_3}{|}}$ $\overset{\displaystyle OH}{\underset{\displaystyle CH_2=CH-CH-CH_3}{|}}$

$CH_2=CH-CH_2-CH_2-OH$ $HO-CH_2-CH=CH-CH_3$ # $\overset{\displaystyle OH}{\underset{\displaystyle CH_3-C=CH-CH_3}{|}}$

$\overset{\displaystyle CH_3}{\underset{\displaystyle HO-CH=C-CH_3}{|}}$ $\overset{\displaystyle CH_3}{\underset{\displaystyle CH_2=C-CH_2-OH}{|}}$

Problem 1.42 What is the meaning of the term tertiary (3°) when it is used to classify alcohols? Draw a structural formula for the one tertiary (3°) alcohol of molecular formula $C_4H_{10}O$.

A tertiary alcohol is one in which the -OH group is on a tertiary carbon atom. A tertiary carbon atom is one that is bonded to three other carbon atoms.

$$\overset{\displaystyle OH}{\underset{\displaystyle CH_3}{\overset{|}{\underset{|}{CH_3-C-CH_3}}}}$$

Problem 1.43 What is the meaning of the term tertiary (3°) when it is used to classify amines? Draw a structural formula for the one tertiary (3°) amine of molecular formula $C_4H_{11}N$.

A tertiary amine is one in which the nitrogen atom is bonded to three carbon atoms.

$$\underset{\displaystyle CH_3}{\overset{\displaystyle CH_3-N-CH_2-CH_3}{|}}$$

1.44 Draw structural formulas for
(a) The four primary (1°) amines of molecular formula $C_4H_{11}N$.

$CH_3-CH_2-CH_2-CH_2-NH_2$ $\underset{\displaystyle CH_3}{\overset{\displaystyle CH_3-CH-CH_2-NH_2}{|}}$ $\underset{\displaystyle CH_3}{\overset{\displaystyle CH_3-CH_2-CH-NH_2}{|}}$ $\underset{\displaystyle CH_3}{\overset{\displaystyle CH_3}{\overset{|}{\underset{|}{CH_3-C-NH_2}}}}$

(b) The three secondary (2°) amines of molecular formula $C_4H_{11}N$.

$\underset{\displaystyle CH_3}{\overset{\displaystyle CH_3-CH_2-CH_2-NH-CH_3}{}}$ $\underset{\displaystyle CH_3}{\overset{\displaystyle CH_3-CH-NH-CH_3}{|}}$ $CH_3-CH_2-NH-CH_2-CH_3$

(c) The one tertiary (3°) amine of molecular formula $C_4H_{11}N$.

$$\underset{\displaystyle CH_3}{\overset{\displaystyle CH_3-CH_2-N-CH_3}{|}}$$

Problem 1.45 Draw structural formulas for the three tertiary (3°) amines of molecular formula $C_5H_{13}N$.

$$CH_3-CH_2-N-CH_2-CH_3 \qquad CH_3-N-CH_2-CH_2-CH_3 \qquad CH_3-N-CH-CH_3$$

with CH_3 groups on the nitrogens (and CH_3 on the third structure's carbon)

Problem 1.46 Draw structural formulas for
(a) The eight alcohols of molecular formula $C_5H_{12}O$.

To make it easier for you to see the patterns of carbon skeletons and functional groups, only carbon atoms and hydroxyl groups are shown in the following solutions. To complete these structural formulas, you need only supply enough hydrogen atoms to complete the tetravalence of each carbon.

There are three different carbon skeletons on which the -OH group can be placed:

C—C—C—C—C   C—C—C—C   C—C—C
                         | |
                         C   C—C

Three alcohols are possible from the first carbon skeleton, four from the second carbon skeleton, and one from the third carbon skeleton.

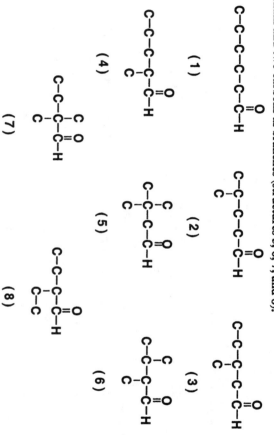

(b) The eight aldehydes of molecular formula $C_6H_{12}O$.

Following are structural formulas for the eight aldehydes of molecular formula $C_6H_{12}O$. They are drawn starting with the aldehyde group and then attaching the remaining five carbons in a chain (structure 1), then four carbons in a chain and one carbon as a branch on the chain (structures 2, 3, and 4) and finally three carbons in a chain and two carbons as branches (structures 5, 6, 7, and 8).

(c) The six ketones of molecular formula C₆H₁₂O.

Following are structural formulas for the six ketones of molecular formula $C_6H_{12}O$. They are drawn first with all combinations of one carbon to the left of the carbonyl group and four carbons to the right (structures 1, 2, 3, and 4) and then with two carbons to the left and three carbons to the right (structures 5 and 6).

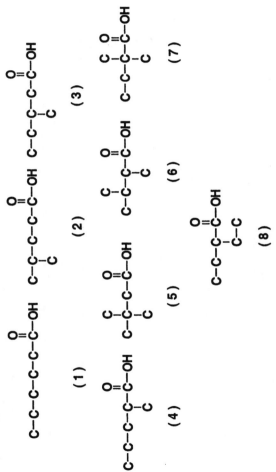

(d) The eight carboxylic acids of molecular formula $C_6H_{12}O_2$.

There are eight carboxylic acids of molecular formula $C_6H_{12}O_2$. They have the same carbon skeletons as the eight aldehydes of molecular formula $C_6H_{12}O$ shown in part (b) of this problem. In place of the aldehyde group, substitute a carboxyl group.

(e) The nine carboxylic esters of molecular formula $C_5H_{10}O_2$.

Start with unbranched carbon chains of all possible lengths, then add branching to complete the set.

Problem 1.47 Identify the functional groups in each compound.

(a)

Hydroxyl group

CH₃–CH–C–OH
with O double bond

Lactic acid   Carboxyl group

(b)

Hydroxyl group   Hydroxyl group

HO–CH₂–CH₂–OH

Ethylene glycol

(c)

CH₃–CH–C–OH
with O double bond and NH₂

Alanine

Carboxyl group

Amino group (1°)

(d)

Hydroxyl group

Hydroxyl group

HO–CH₂–CH–C–H
with O double bond and OH

Glyceraldehyde

Carbonyl group
(Aldehyde)

(e)

CH₃–C–CH₂–C–OH
with two O double bonds

Acetoacetic acid

Carbonyl group
(Ketone)   Carboxyl group

(f)

Amino group (1°)

H₂NCH₂CH₂CH₂CH₂CH₂CH₂NH₂

1,6-Hexanediamine

Amino group (1°)

Polar and Nonpolar Molecules

Problem 1.48 Draw a three-dimensional representation for each molecule. Indicate which ones have a dipole moment and in what direction it is pointing.

In the following diagrams, the C-H bond dipole moment has been left out because this is a nonpolar covalent bond.

(a) CH₃F

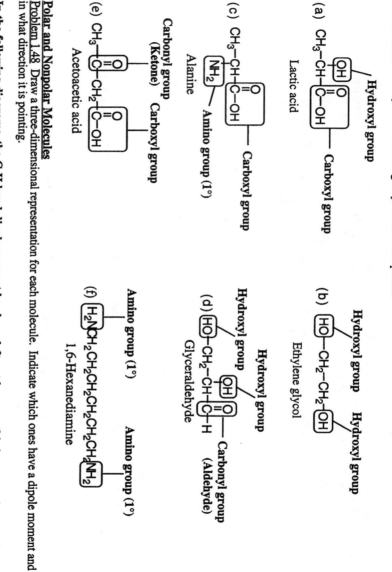

μ = 1.85 D

(b) CH₂Cl₂

μ = 1.60 D

(c) CH₂ClBr

μ = 1.50 D

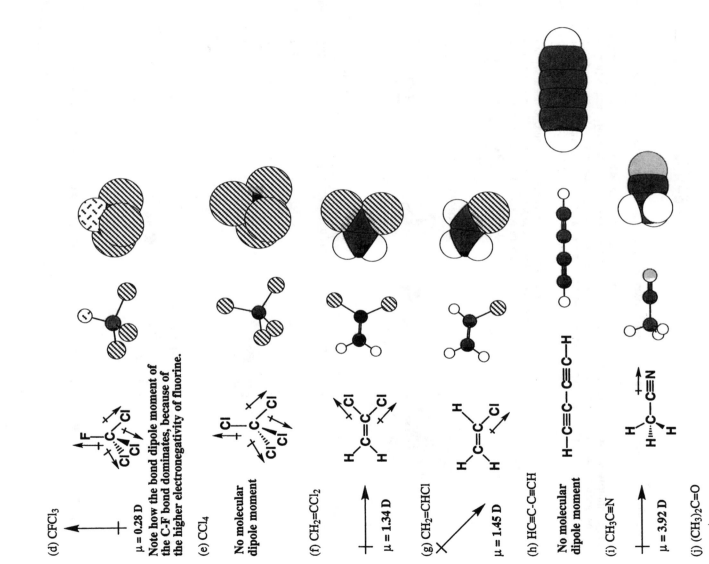

(d) CFCl₃

μ = 0.28 D

Note how the bond dipole moment of the C-F bond dominates, because of the higher electronegativity of fluorine.

(e) CCl₄

No molecular dipole moment

(f) CH₂=CCl₂

μ = 1.34 D

(g) CH₂=CHCl

μ = 1.45 D

(h) HC≡C-C≡CH

No molecular dipole moment

(i) CH₃C≡N

μ = 3.92 D

(j) (CH₃)₂C=O

μ = 2.88 D

(k) BrCH=CHBr (two answers)

The two bromine atoms can either be on opposite sides or on the same side of the double bond. Recall that double bonds do not rotate.

No molecular
dipole moment

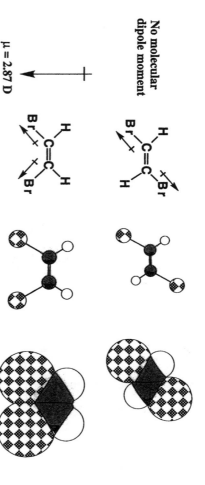

μ = 2.87 D

Problem 1.49 Tetrafluoroethylene, C_2F_4, is the starting material for the synthesis of the polymer polytetrafluoroethylene (PFTE), one form of which is known as Teflon. Tetrafluoroethylene has a zero dipole moment. Propose a structural formula for this molecule.

No molecular
dipole moment

Tetrafluoroethene

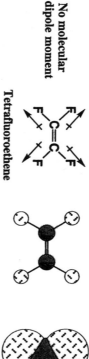

Resonance and Contributing Structures
Problem 1.50 Which statements are true about resonance contributing structures?
(a) All contributing structures must have the same number of valence electrons.
(b) All contributing structures must have the same arrangement of atoms.
(c) All atoms in a contributing structure must have complete valence shells.
(d) All bond angles in sets of contributing structures must be the same.

For sets of contributing structures, electrons (usually π electrons or lone pair electrons) move, but the atomic nuclei maintain the same arrangement in space. Thus, statements (b) and (d) are true. In addition, the total number of electrons, valence and inner shell electrons, in each contributing structure must be the same, so statement (a) is also true. However, the movement of electrons often leaves one or more atoms without a filled valence shell in a given contributing structure, so statement (c) is false.

Problem 1.51 Draw the contributing structure indicated by the curved arrow(s). Assign formal charges as appropriate.

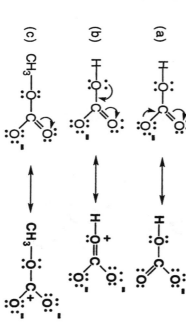

(d)

(e)

(f)

Problem 1.52 Using the VSEPR model, predict the bond angles about the carbon and nitrogen atoms in each pair of contributing structures in problem 1.51. In what way do the bond angles change from one contributing structure to the other?

As stated in the answer to Problem 1.50, bond angles do not change from one contributing structure to another.

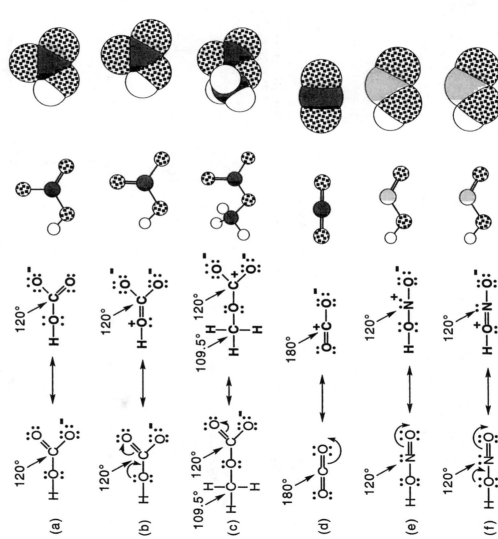

(a)   120°   ⟷   120°

(b)   120°   ⟷   120°

(c)   109.5° 120° ⟷ 109.5° 120°

(d)   180°   ⟷   180°

(e)   120°   ⟷   120°

(f)   120°   ⟷   120°

Problem 1.53 In the Problem 1.51 you were given one contributing structure and asked to draw another. Label pairs of contributing structures that are equivalent. For those sets in which the contributing structures are not equivalent, label the more important contributing structure.

(a) The two structures are equivalent because each involves a similar separation of charge.
(b, c, d, e, f) The first structure is more important, because the second involves creation and separation of unlike charges.

Problem 1.54 Are the structures in each set valid contributing structures?

(a)

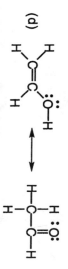

The structure on the right is not a valid contributing structure because there are 10 valence electrons around the carbon atom.

(b)   H—N̈=N̈=N̈   ⟷   H—N̈—N≡N:

Both of these are valid contributing structures.

(c)

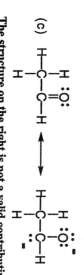

The structure on the right is not a valid contributing structure because there are two extra electrons and thus it is a completely different species.

(d)

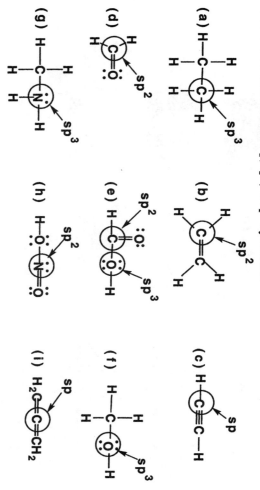

Although each is a valid Lewis structure, they are not valid contributing structures for the same resonance hybrid. An atomic nucleus, namely a hydrogen, has changed position. Later you will learn that these two molecules are related to each other, and are called tautomers.

Molecular Orbital Theory
Problem 1.55 State the orbital hybridization of each highlighted atom.

Each circled atom is either sp, sp², or sp³ hybridized.

(a) H—C—C—H   sp³

(b) C   sp²

(c) H—C≡C—H   sp

(d) C   sp²

(e) C   sp² / sp³

(f) H—C—O—H   sp³

(g) C—N   sp² / sp³

(h) O—N=O   sp²

(i) H₂C=C=CH₂   sp

<u>Problem 1.56</u> Describe each highlighted bond in terms of the overlap of atomic orbitals.

Shown is whether the bond is sigma or pi, as well as the orbitals used to form it.

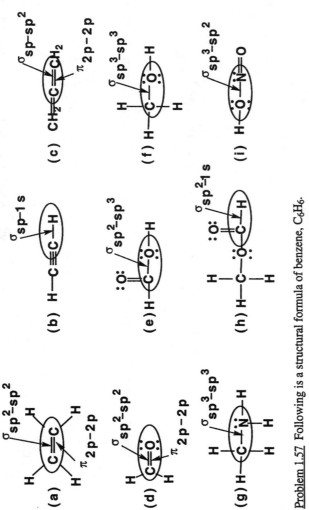

(a)

(b)

(c)

(d)

(e)

(f)

(g)

(h)

(i)

<u>Problem 1.57</u> Following is a structural formula of benzene, C_6H_6.

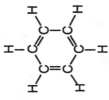

(a) Predict each H-C-C and C-C-C bond angle in benzene.

Each carbon atom in benzene has three regions of electron density around it, so according to the VSEPR model, the carbon atoms are trigonal planar. Predict each H-C-C bond angle to be 120° and each C-C-C bond angle to be 120°.

(b) State the hybridization of each carbon atom in benzene.

Each carbon atom is sp^2 hybridized because each one makes three σ bonds and one π bond.

(c) Predict the shape of a benzene molecule.

Because all of the carbon atoms in the ring are sp^2 hybridized and thus trigonal planar, predict carbon atoms in benzene to form a flat hexagon in shape, with the hydrogen atoms in the same plane as the carbon atoms.

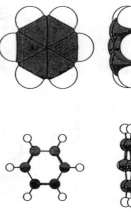

Problem 1.58 Following is the structural formula of the prescription drug famotidine, manufactured by Merck Sharpe & Dohme under the name Pepcid (see the Merck Index, 12th ed., #3972). The primary clinical use of Pepcid is for the treatment of active duodenal ulcers and benign gastric ulcers. Pepcid is a competitive inhibitor of histamine H_2 receptors and reduces both the gastric acid concentration and volume of gastric secretions.

(a) Complete the Lewis structure of famotidine showing all valence electrons and any formal positive or negative charges.

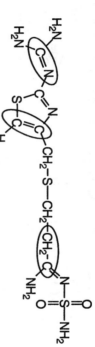

(b) Describe each circled bond in terms of the overlap of atomic orbitals.

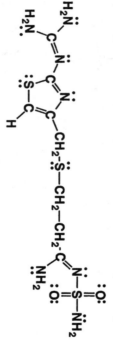

Problem 1.59 Draw a Lewis structure for methyl isocyanate, CH_3NCO, showing all valence electrons. Predict all bond angles in this molecule and the hybridization of each atom C, N, and O.

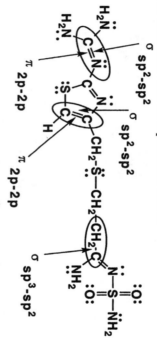

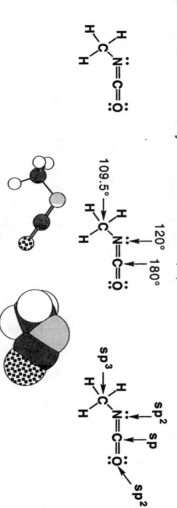

Additional Problems

Problem 1.60 Why are the following molecular formulas impossible?

(a) CH_5

Carbon atoms can only accommodate four bonds, and each hydrogen atom can only accommodate one bond. Thus, there is no way for a stable bonding arrangement to be created that utilizes one carbon atom and all five hydrogen atoms.

(b) C_2H_7

Since hydrogen atoms can only accommodate one bond each, no single hydrogen atom can make stable bonds to both carbon atoms. Thus, the two carbon atoms must be bonded to each other. This means that each of the bonded carbon atoms can accommodate only three more bonds. Therefore, only six hydrogen atoms can be bonded to the carbon atoms, not seven hydrogen atoms.

Problem 1.61 Each compound contains both ionic and covalent bonds. Draw the Lewis structure for each compound and show by dashes which are covalent bonds and by indication of charges which are ionic bonds.

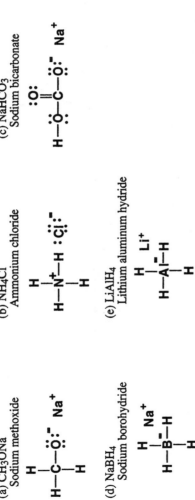

(a) CH_3ONa
Sodium methoxide

(b) NH_4Cl
Ammonium chloride

(c) $NaHCO_3$
Sodium bicarbonate

(d) $NaBH_4$
Sodium borohydride

(e) $LiAlH_4$
Lithium aluminum hydride

In naming these compounds, the cation is named first followed by the name of the anion.

Problem 1.62 Predict whether the carbon-metal bond in these organometallic compounds is nonpolar covalent, polar covalent, or ionic. For each polar covalent bonds show the direction of polarity by the symbols δ+ and δ-.

(a) (0.7) CH_3CH_2—Pb—CH_2CH_3 with CH_2CH_3 and CH_2CH_3 groups; δ- δ+
Tetraethyllead

(b) (1.3) CH_3—Mg—Cl ; δ- δ+
Methylmagnesium chloride

(c) (0.6) CH_3—Hg—CH_3 ; δ- δ+
Dimethylmercury

All of these carbon-metal bonds are polar covalent because the difference in electronegativities is between 0.5 and 1.9. In each case, carbon is the more electronegative element. The difference in electronegativities is given above the carbon-metal bond in each answer.

Problem 1.63 Silicon is immediately under carbon in the Periodic Table. Predict the geometry of silane, SiH_4.

Silicon is in Group 4 of the Periodic Table, and like carbon, has four valence electrons. In silane, SiH_4, silicon is surrounded by four regions of electron density. Therefore, predict all H-Si-H bond angles to be 109.5°, so the molecule is tetrahedral around Si.

Problem 1.64 Phosphorus is immediately under nitrogen in the Periodic Table. Predict the molecular formula for phosphine, the compound formed from phosphorus and hydrogen. Predict the H-P-H bond angle in phosphine.

Like nitrogen, phosphorus has five valence electrons, so predict that phosphine has the molecular formula of PH_3 in analogy to ammonia, NH_3. In phosphine, the phosphorus atom is surrounded by four areas of electron density; one lone pair of electrons and single bonds to three hydrogen. Therefore, predict all H-P-H bond angles to be 109.5°, so the molecule is pyramidal.

<u>Problem 1.65</u> Following are three contributing structures for diazomethane, CH_2N_2.

(a) Using curved arrows, show how each contributing structure is converted to the one on its right.

The arrows are indicated on the above structures.

(b) Which contributing structure makes the largest contribution to the hybrid?

The middle and left structures have filled valence shells, so these will make a larger contribution to the hybrid than the structure on the right, in which the terminal nitrogen atom has an unfilled valence shell. The structure in the middle has the negative charge on the more electronegative atom, N, compared with the structure on the left (negative charge on C), so the structure in the middle will make the largest contribution to the resonance hybrid.

<u>Problem 1.66</u> Draw a Lewis structure for the azide ion, N_3^-. (The order of attachment is N-N-N and they do not form a ring). How does the resonance model account for the fact that the lengths of the N-N bonds in this ion are identical.

It is not possible to draw a single contributing structures Lewis structure that adequately describes the azide anion. Rather, it can be drawn as the three contributing structures shown below.

Taken together, the three contributing structures present a symmetric picture of the bonding, thus explaining why both N-N bonds are identical.

<u>Problem 1.67</u> Draw a Lewis structure for the ozone molecule, O_3. (The order of attachment is O-O-O and they do not form a ring). How does the resonance model account for the fact that the length of each O-O bond in ozone (128 pm) is shorter than the O-O single bond in hydrogen peroxide (HOOH 147 pm), but longer than the O-O double bond in the oxygen molecule (123 pm).

It is not possible to draw a single Lewis structure that adequately describes the ozone molecule. Rather, it is better to draw ozone as a hybrid of two contributing structures.

Taken together, the two contributing structures present a symmetric picture of the bonding in which each O-O bond is intermediate between a single bond and a double bond. Recall that bonds become shorter as bond order increases. As a result, the bonds in ozone are shorter than the single O-O bond in HOOH, but longer than the OO double bond in the oxygen molecule.

<u>Problem 1.68</u> Cyanic acid, HOCN, and isocyanic acid, HNCO, dissolve in water to yield the same anion on loss of H^+.

(a) Write a Lewis structure for cyanic acid.

$$H-\ddot{O}-C\equiv N:$$

(b) Write a Lewis structure for isocyanic acid.

$$\ddot{O}=C=\ddot{N}-H$$

(c) Account for the fact that each acid gives the same anion on loss of H^+.

Loss of an H^+ from the two different acids gives the same anion that can best be described by drawing the following two contributing structures.

$$H-\ddot{\underset{..}{O}}-C\equiv N: \xrightarrow{\text{loss of } H^+} {}^-:\ddot{\underset{..}{O}}-C\overset{\frown}{\equiv}N: \longleftrightarrow :\ddot{\underset{..}{O}}=C=\ddot{N}:^-$$

$$\ddot{\underset{..}{O}}=C=\ddot{N}-H \xrightarrow{\text{loss of } H^+} :\ddot{\underset{..}{O}}=C=\ddot{N}:^-$$

tert-Butyl cation

Problem 1.69 In Chapter 6, we study a group of organic cations called carbocations. Following is the structure of one such carbocation, the *tert*-butyl cation.

(a) How many electrons are in the valence shell of the carbon bearing the positive charge?

There are six valence shell electrons on the carbon atom bearing the positive charge, two contained in each of the three single bonds.

(b) Predict the bond angles around this carbon.

According to the VSEPR model, there are three areas of electron density around the central carbon atom, so predict a trigonal planar geometry and C-C-C bond angles of 120°.

Top view

Side view

(c) Given the bond angle you predicted in (b), what hybridization do you predict for this carbon?

Given the trigonal planar geometry predicted in (b), predict sp² hybridization of this carbon atom.

Problem 1.70 Many reactions involve a change in hybridization of one or more atoms in the starting material. In each reaction, identify the atoms in the organic starting material that change hybridization and indicate what the change is. We examine these reactions in more detail later in the course.

(a)

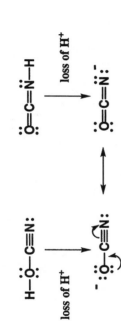

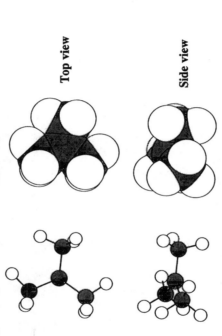

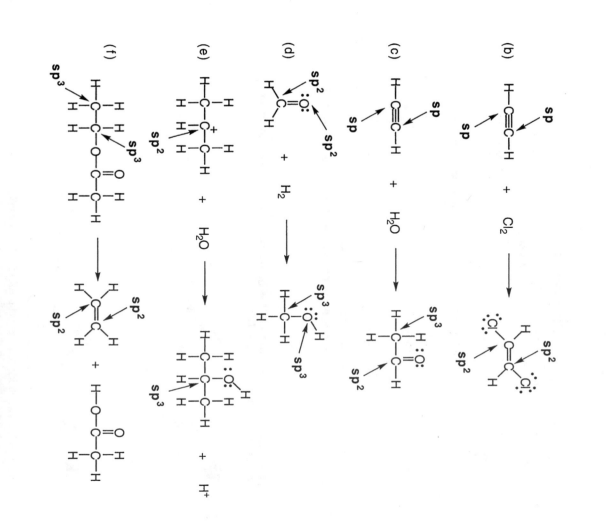

Horn shark Heterodontus francisci
Catalina Island, California

CHAPTER 2
Solutions to the Problems

Problem 2.1 Write line-angle drawings for each compound and determine if the formulas in each set represent the same compound or constitutional isomers

(a)

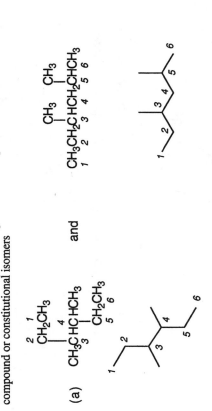

These molecules are constitutional isomers. Each has six carbons in the longest chain. The first has one-carbon branches on carbons 3 and 4 of the chain; the second has one-carbon branches on carbons 2 and 4 of the chain.

(b)

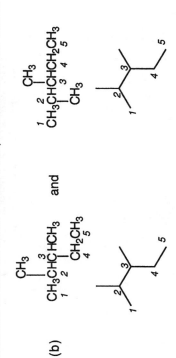

These molecules are identical. Each has five carbons in the longest chain, and one-carbon branches on carbons 2 and 3 of the chain.

Problem 2.2 Write condensed structural formulas and line-angle drawings for the three constitutional isomers of molecular formula C_5H_{12}.

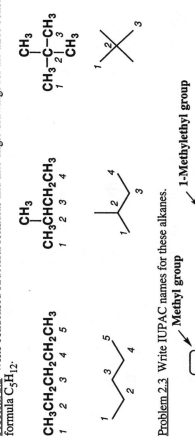

Problem 2.3 Write IUPAC names for these alkanes.

(a)

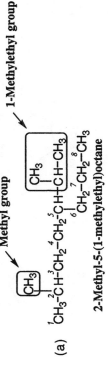

2-Methyl-5-(1-methylethyl)octane

(b) $CH_3-CH_2-CH_2-\overset{4}{C}-CH_2-CH_2-CH_3$
$\overset{1}{C}H_3-\overset{2}{C}H_2-\overset{3}{C}H_2-\overset{4}{C}-\overset{5}{C}H_2-\overset{6}{C}H_2-\overset{7}{C}H_3$

$\boxed{CH_2-CH_2-CH_3}$ ⟶ Propyl group

$\boxed{CH_3-CH-CH_3}$ ⟶ 1-Methylethyl group

4-(1-Methylethyl)-4-propylheptane

Problem 2.4 Write the common name of each alkane in Problem 2.3.

(a) **5-Isopropyl-2-methyloctane**   (b) **4-Isopropyl-4-propylheptane**

Note how the 1-methylethyl group is referred to as an isopropyl group in the common nomenclature.

Problem 2.5 Write the molecular formula, IUPAC name, and common name for each cycloalkane.

(a)
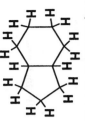

(b)
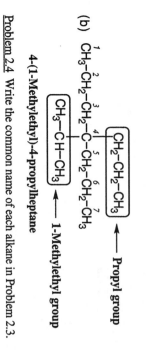

Molecular Formula C_9H_{18}
(2-Methylpropyl)cyclopentane (IUPAC)
Isobutylcyclopentane (Common)

Molecular Formula $C_{11}H_{22}$
(1-Methylpropyl)cycloheptane (IUPAC)
sec-**Butylcycloheptane (Common)**

(c)

Molecular Formula C_6H_{12}
1-Ethyl-1-methylcyclopropane (IUPAC and Common)

Problem 2.6 Write molecular formulas for each bicycloalkane, given its number of carbon atoms.
(a) Hydrindane (9 carbons)   b) Decalin (10 carbons)   (c) Norbornane (7 carbons)

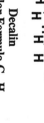

Hydrindane
Molecular Formula C_9H_{16}

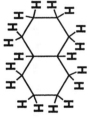

Decalin
Molecular Formula $C_{10}H_{18}$

Norbornane
Molecular Formula C_7H_{12}

Problem 2.7 Draw structural formulas for the following bicycloalkanes.
(a) Bicyclo[3.1.0]hexane   (b) Bicyclo[2.2.2]octane   (c) Bicyclo[4.2.0]octane   (d) 2,6,6-Trimethylbicyclo[3.1.1]heptane

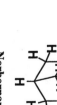

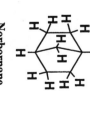

Problem 2.8 Combine the proper prefix, infix, and suffix and write the IUPAC name for each compound.

(a) $CH_3\overset{\displaystyle O}{\overset{\|}{C}}CH_3$
Propanone

(b) $CH_3CH_2CH_2CH_2\overset{\displaystyle O}{\overset{\|}{C}}H$
Pentanal

(c)
OH
Cyclopentanol

(d)
Cycloheptene

Problem 2.9 For 1,2-dichloroethane,
(a) Draw Newman projections for all eclipsed conformations formed by rotation from from 0° to 360° about the carbon-carbon single bond.

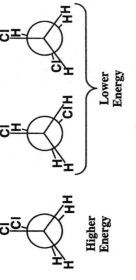

Higher
Energy

Lower
Energy

(b) Which eclipsed conformation(s) has the lowest energy. Which has the highest energy?

The chlorine atoms are the largest by far. As a result, when the chlorine atoms are eclipsed with each other (structure on the left), their steric interaction causes the higher overall energy.

(c) Which, if any, of these conformations are related by reflection?

The two lower energy conformations (structures on the right) are related by reflection as they represent "mirror images" of each other.

Problem 2.10 Following is a chair conformation of cyclohexane with carbon atoms numbered 1 through 6.

(a) Draw hydrogen atoms that are above the plane of the ring on carbons 1 and 2 and below the plane of the ring on carbon 4.
(b) Which of these hydrogens are equatorial; which are axial?
(c) Draw the alternative chair conformation. Which hydrogens are equatorial; which are axial? Which are above the ring? Which are below it?

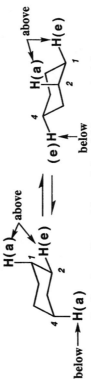

In the above figure (a) = axial and (e) = equatorial.

Problem 2.11 Draw the alternative chair conformation for the trisubstituted cyclohexane given in Example 2.11. Label all diaxial interactions in this chair conformation.

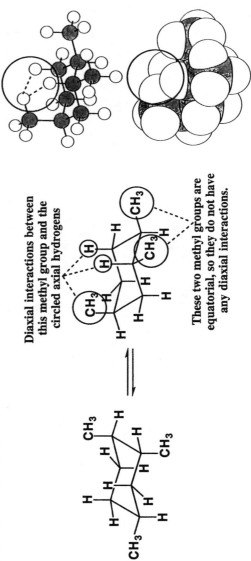

Diaxial interactions between this methyl group and the circled axial hydrogens

These two methyl groups are equatorial, so they do not have any diaxial interactions.

Note how in the above equilibrium, the new chair conformation (on the right) is the more stable due to having fewer diaxial interactions (See 2.14)

Problem 2.12 Draw a chair conformation of 1,4-dimethylcyclohexane in which one methyl group is equatorial and the other is axial. Draw the alternative chair conformation and calculate the ratio of the two conformations at 25°C.

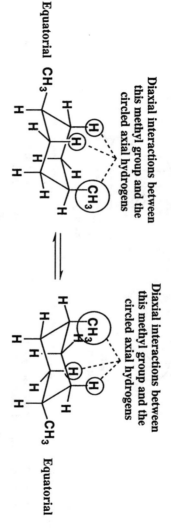

Diaxial interactions between this methyl group and the circled axial hydrogens

Equatorial

Diaxial interactions between this methyl group and the circled axial hydrogens

Equatorial

Problem 2.13 Which cycloalkanes show cis,trans isomerism? For each that does, draw both isomers.

Each chair conformation has diaxial interactions between the circled axial methyl group and circled axial hydrogen atoms. Because each chair has the same number of diaxial interactions, they are of the same energy. The ratio of these two conformations must therefore be 1:1.

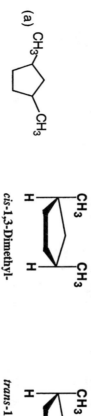

cis-1,3-Dimethyl-cyclopentane

trans-1,3-Dimethyl-cyclopentane

1,3-Dimethylcyclopentane shows cis, trans isomerism. In the following drawings, the ring is drawn as a planar pentagon with substituents above and below the plane of the pentagon.

(b)

Ethylcyclopentane does not show cis,trans isomerism.

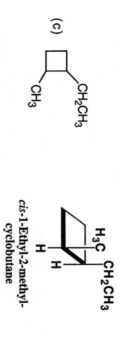

(a) CH₃

CH₃

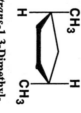

(c)

CH₂CH₃

CH₃

cis-1-Ethyl-2-methyl-cyclobutane

trans-1-Ethyl-2-methyl-cyclobutane

1-Ethyl-2-methylcyclobutane shows cis,trans isomerism.

Problem 2.14 Following is a planar hexagon representation for one isomer of 1,2,4-trimethylcyclohexane.
(a) Draw the alternative chair conformations of this compound and state which chair conformation is the more stable.

Following are alternative chair conformations for the above isomer of 1,2,4-trimethylcyclohexane. The alternative chair conformation on the right is the more stable because it has only one axial methyl group.

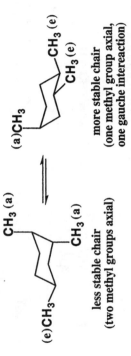

(e)CH₃

CH₃ (a)

CH₃ (a)

**less stable chair
(two methyl groups axial)**

⇌

(a)CH₃

CH₃ (e)

CH₃ (e)

**more stable chair
(one methyl group axial,
one gauche intereaction)**

(b) Calculate the energy difference between these two conformations. (hint: 1,2-trans diequatorial substituents have a gauche interaction and are destabilized by about 3.8 kJ/mol. Don't forget to include this factor in your calculation!

⟋⟍⟋ ⟶ Gauche!

The chair conformation on the left has two axial methyl groups, while the chair conformation on the right has only one. However, the chair conformation on the right has the additional guache interaction. Each axial methyl group adds 7.28 kJ/mol in energy while the gauche interaction adds 3.8 kJ/mol. The difference is (7.28 kJ/mol + 7.28 kJ/mol) - (7.28 kJ/mol + 3.8 kJ/mol) = 3.48 kJ/mol. Therefore, the chair conformation on the right is 3.48 kJ/mol more stable.

<u>Problem 2.15</u> Which of the following stereoisomer is more stable?

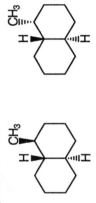

H

CH₃

H

**more stable isomer
(methyl substituent is equatorial)**

CH₃

H

H

In the first isomer of *trans*-decalin, the methyl substituent is equatorial and in the second isomer, it is axial. The equatorial methyl isomer is more stable.

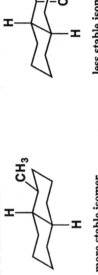

H

CH₃

H

**less stable isomer
(methyl substituent is axial)**

<u>Problem 2.16</u> Arrange the alkanes in each set in order of increasing boiling point.
(a) 2-Methylbutane, 2,2-dimethylpropane and pentane

2,2-Dimethylpropane (bp 9.5ºC)	**2-Methylbutane** (bp 29ºC)	**Pentane** (bp 36ºC)	

(b) 3,3-dimethylheptane, 2,2,4,4-tetramethylpentane, nonane

2,2,4,4-Tetramethylpentane (bp 99ºC)	**3,3-Dimethylheptane** (bp 137ºC)	**Nonane** (bp 151ºC)

Constitutional Isomerism
<u>Problem 2.17</u> Which statements are true about constitutional isomers?
(a) They have the same molecular formula.
(b) They have the same molecular weight.
(c) They have the same order of attachment of atoms.
(d) They have the same physical properties.

Statements (a) and (b) are true, statements (c) and (d) are false.

Problem 2.18 Which structural formulas represent identical compounds and which represent constitutional isomers?

(a) CH₃CH₂CH₂CHCH₃
         |
         Cl

(b) [cyclobutane with Cl]

(c) ClCH₂—[cyclopropane]

(d) CH₃CHCH₃
         |
         CH₂Cl

(e) ClCH₂CHCH₃
          |
          CH₃

(f) CH₃CH₂CH₂CH₂Cl

(g) CH₃CHCl
          |
          CH₂CH₃

(h) CH₃CCH₃
          | |
         CH₃ Cl

Following are names and molecular formulas of each

(a) 2-Chlorobutane;   C₄H₉Cl
(b) Chlorocyclobutane;   C₄H₇Cl
(c) Chloromethylcyclopropane; C₄H₇Cl
(d) 1-Chloro-2-methylpropane; C₄H₉Cl
(e) 1-Chloro-2-methylpropane; C₄H₉Cl
(f) 1-Chlorobutane;   C₄H₉Cl
(g) 2-Chlorobutane;   C₄H₉Cl
(h) 2-Chloro-2-methylpropane; C₄H₉Cl

The following are identical: (d),(e)   (a),(g)

The following compounds are constitutional isomers of molecular formula C₄H₉Cl: 2-chlorobutane (a)(g), 1-chloro-2-methylpropane (d)(e), 1-chlorobutane (f) and 2-chloro-2-methylpropane (h).

The following compounds are constitutional isomers of molecular formula C₄H₇Cl: chlorocyclobutane (b) and chloromethylcyclopropane (c).

Problem 2.19 Indicate whether the compounds in each set are constitutional isomers.

(a) CH₃–CH₂–OH   and   CH₃–O–CH₃

(b) CH₃–C–CH₃   and   CH₃–CH₂–C–H
          ‖   ‖
          O   O

(c) CH₃–C–O–CH₃   and   CH₃–CH₂–C–OH
          ‖   ‖
          O   O

(d) CH₃–CH–CH₂–CH₃   and   CH₃–C–CH₂–CH₃
            | ‖
            OH   O

(e) [cyclopentane]   and   CH₃–CH₂–CH₂–CH₂–CH₃

(f) [cyclopentane]   and   CH₂=CH—CH₂–CH₂–CH₃

Sets (a), (b), (c), and (f) are constitutional isomers; sets (d) and (e) are not.

Problem 2.20 Draw structural formulas and write IUPAC names for the nine constitutional isomers of molecular formula C₇H₁₆.

CH₃CH₂CH₂CH₂CH₂CH₂CH₃
Heptane
(bp 94.8° C)

CH₃CHCH₂CH₂CH₂CH₃
      |
      CH₃
2-Methylhexane
(bp 90.0° C)

CH₃CH₂CHCH₂CH₂CH₃
            |
            CH₃
3-Methylhexane
(bp 92.0° C)

CH₃CCH₂CH₂CH₃
      |
      CH₃
      CH₃
2,2-Dimethylpentane
(bp 79.2° C)

CH₃CHCHCH₂CH₃
      | |
      CH₃ CH₃
2,3-Dimethylpentane
(bp 89.8° C)

CH₃CHCH₂CHCH₃
      | |
      CH₃   CH₃
2,4-Dimethylpentane
(bp 80.5° C)

CH₃
|
CH₃CH₂CCH₂CH₃
|
CH₃

3,3-Dimethylpentane
(bp 86.1° C)

CH₂CH₃
|
CH₃CH₂CH₂CH CH₂CH₃

3-Ethylpentane
(bp 93.5° C)

H₃C CH₃
\ /
CH₃CHCHCH₃
|
CH₃

2,2,3-Trimethylbutane
(bp 80.9° C)

Problem 2.21 Draw structural formulas for all of the following.
(a) Alcohols of molecular formula C₄H₁₀O.

CH₃–CH₂–CH₂–CH₂–OH   CH₃–CH₂–CH–CH₃
                                      |
                                     OH

CH₃
|
CH₃–C–OH
|
CH₃

CH₃
|
CH₃–CH–CH₂–OH

(b) Aldehydes of molecular formula C₄H₈O.

      O
      ‖
CH₃–CH₂–CH₂–C–H

      O
      ‖
CH₃–CH–C–H
    |
    CH₃

(c) Ketones of molecular formula C₅H₁₀O.

      O
      ‖
CH₃–CH₂–C–CH₂–CH₃

      O
      ‖
CH₃–CH₂–CH₂–C–CH₃

      O
      ‖
CH₃–CH–C–CH₃
    |
    CH₃

(d) Carboxylic acids of molecular formula C₅H₁₀O₂.

      O
      ‖
CH₃–CH₂–CH₂–CH₂–C–OH

      O
      ‖
CH₃–CH–CH₂–C–OH
    |
    CH₃

      O
      ‖
CH₃–CH₂–CH–C–OH
          |
          CH₃

CH₃ O
| ‖
CH₃–C–C–OH
|
CH₃

Nomenclature of Alkanes and Cycloalkanes
Problem 2.22 Write IUPAC names for these alkanes and cycloalkanes. Name substituents both by IUPAC names and common names (if common names exist).

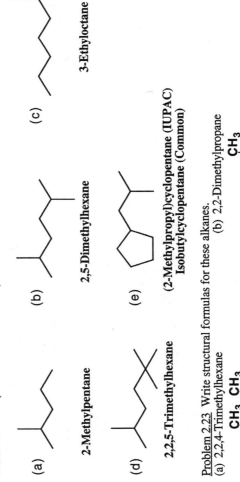

(a)

2-Methylpentane

(b)

2,5-Dimethylhexane

(c)

3-Ethyloctane

(d)

2,2,5-Trimethylhexane

(e)

(2-Methylpropyl)cyclopentane (IUPAC)
Isobutylcyclopentane (Common)

Problem 2.23 Write structural formulas for these alkanes.
(a) 2,2,4-Trimethylhexane

CH₃ CH₃
| |
CH₃CCH₂CHCH₂CH₃
|
CH₃

(b) 2,2-Dimethylpropane

CH₃
|
CH₃CCH₃
|
CH₃

(c) 3-Ethyl-2,4,5-trimethyloctane

$$CH_3CH_2 \quad CH_3$$
$$CH_3CHCHCHCHCH_2CH_2CH_3$$
$$CH_3 \quad CH_3$$

(d) 5-Butyl-2,2-dimethylnonane

$$CH_3$$
$$CH_3CCH_2CH_2CHCH_2CH_2CH_2CH_3$$
$$CH_3 \quad CH_2CH_2CH_2CH_3$$

(e) 4-(1-Methylethyl)octane

$$CH_3CH_2CH_2CHCH_2CH_2CH_2CH_3$$
$$CH_3CHCH_3$$

(f) 3,3-Dimethylpentane

$$CH_3$$
$$CH_3CH_2CCH_2CH_3$$
$$CH_3$$

(g) *trans*-1,3-Dimethylcyclopentane

(h) *cis*-1,2-Diethylcyclobutane

Problem 2.24 Explain why each is an incorrect IUPAC name. Write the correct IUPAC name for the intended compound.

(a) 1,3-Dimethylbutane

$$CH_3$$
$$CH_3CHCH_2CH_2CH_3$$

The longest chain is pentane.
Its IUPAC name is 2-methylpentane.

(b) 4-Methylpentane

$$CH_3$$
$$CH_3CHCH_2CH_2CH_3$$

The pentane is numbered incorrectly.
Its IUPAC name is 2-methylpentane.

(c) 2,2-Diethylbutane

$$CH_2CH_3$$
$$CH_3CH_2CCH_2CH_3$$
$$CH_3$$

The longest chain is pentane.
Its IUPAC name is 3-ethyl-3-methylpentane.

(d) 2-Ethyl-3-methylpentane

$$CH_3$$
$$CH_3CH_2CH CHCH_2CH_3$$
$$CH_3$$

The longest chain is hexane.
Its IUPAC name is 3,4-dimethylhexane.

(e) 2-Propylpentane

$$CH_3CH_2CH_2CHCH_2CH_2CH_3$$

(f) 2,2-Diethylheptane

$$CH_2CH_3$$
$$CH_3CH_2CCH_2CH_2CH_2CH_2CH_3$$
$$CH_3$$

The longest chain is octane.
Its IUPAC name is 3-ethyl-3-methyloctane.

(g) 2,2-Dimethylcyclopropane

The ring is numbered incorrectly.
Its IUPAC name is 1,1-dimethylcyclopropane.

(h) 1-Ethyl-5-methylcyclohexane

The ring is numbered incorrectly.
Its IUPAC name is 1-ethyl-3-methylcyclohexane.

The IUPAC System of Nomenclature
Problem 2.25 For each of the following IUPAC names, draw the corresponding structural formula for each compound.

(a) Ethanol

$$CH_3-CH_2-OH$$

(b) Butanal

$$CH_3-CH_2-CH_2-C\overset{O}{\underset{}{=}}-H$$

(c) Butanoic acid

$$CH_3-CH_2-CH_2-C\overset{O}{\underset{}{=}}-OH$$

(d) Ethanoic acid

CH_3-C-OH with =O

(e) Heptanoic acid

$CH_3(CH_2)_5-C-OH$ with =O

(f) Propanoic acid

CH_3-CH_2-C-OH with =O

(g) Octanal

$CH_3-(CH_2)_6-C-H$ with =O

(h) Cyclopentene

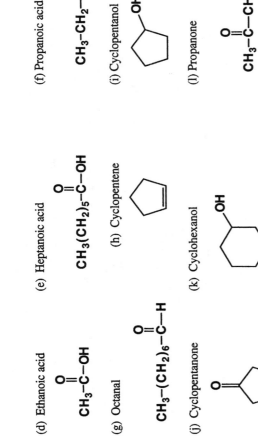

(i) Cyclopentanol

OH

(j) Cyclopentanone

O

(k) Cyclohexanol

OH

(l) Propanone

CH_3-C-CH_3 with =O

Problem 2.26 Write IUPAC names for these compounds.

(a) $CH_3-CH_2-C-CH_3$ with =O
Butanone

(b) CH_3-CH_2-C-H with =O
Propanal

(c) $CH_3-CH_2-CH_2-CH_2-CH_2-C-OH$ with =O
Hexanoic acid

(d) $CH_3-CH-CH_3$ with OH
2-Propanol

(e) Cyclohexanone
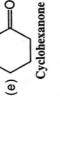

(f) Cyclopropanol

OH

(g) $CH_3-CH=CH_2$
Propene

(h) Cyclohexene
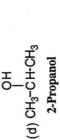

Problem 2.27 Torsional strain resulting from eclipsed C-H bonds is approximately 4.2 kJ (1.0 kcal)/mol and that for eclipsed C-H and C-CH$_3$ bonds is approximately 6.3 kJ (1.5 kcal)/mol. Given this information, sketch a graph of energy versus dihedral angle for propane.

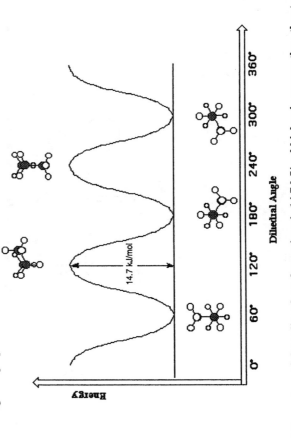

Notice that the energy of the eclipsed conformations is 14.7 kJ/mol higher in energy than the staggered conformations. This is because each eclipsed conformation has two C-H bonds eclipsed with other C-H bonds (worth 4.2 kJ/mol each) and one C-H bond eclipsed to a C-CH$_3$ bond (worth 6.3 kJ/mol).

Problem 2.28 How many different staggered conformations are there for 2-methylpropane? How many different eclipsed conformations are there?

Looking down any of the carbon-carbon bonds, there is one staggered and one eclipsed conformation of 2-methylpropane.

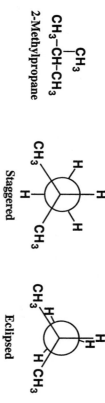

CH₃
|
CH₃–CH–CH₃

2-Methylpropane

Staggered   Eclipsed

Problem 2.29 Consider 1-bromopropane, CH₃CH₂CH₂Br.

(a) Draw a Newman projection for the conformation in which -CH₃ and -Br are anti (dihedral angle 180°).

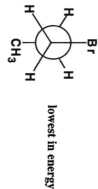

lowest in energy

(b) Draw Newman projections for the conformations in which -CH₃ and -Br are gauche (dihedral angles 60° and 300°).

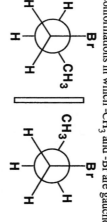

related by reflection

(c) Which of these is the lowest energy conformation.

The anti (dihedral angle 180°) is the lowest energy conformation.

(d) Which of these conformations, if any, are related by reflection?

The two gauche conformations are of equal energy, and are related by reflection.

Problem 2.30 Consider 1-bromo-2-methylpropane and draw the following.

(a) The staggered conformation(s) of lowest energy.

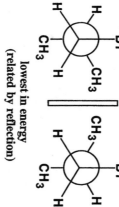
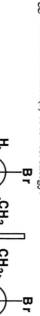

lowest in energy
(related by reflection)

(b) The staggered conformation(s) of highest energy.

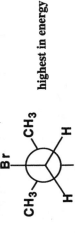

highest in energy

The lower energy staggered conformations have one methyl group anti (dihedral angle 180°) to the bromine and are related by reflection. The staggered conformation with methyl groups at dihedral angles of both 60° and 300° to the bromine have more nonbonded interaction strain and are thus higher in energy.

<u>Problem 2.31</u> In cyclohexane, an equatorial substituent is equidistant from the axial group and the equatorial group on an adjacent carbon. Show that this is so by building a molecular model and measuring these distances.

The best way to see this fact is to draw a Newman projection of one of the carbon-carbon bonds. As can be seen, the axial hydrogen from the carbon atom in the front is in between, and thus equidistant to, the axial and equatorial hydrogen atoms on the rear carbon atom.

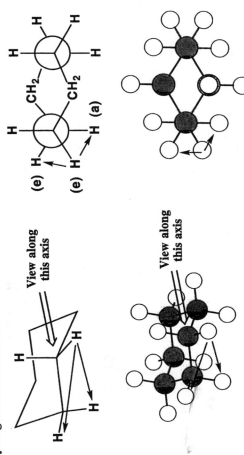

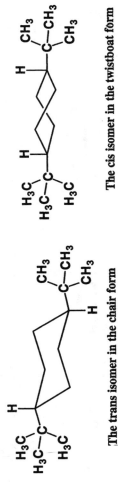

<u>Problem 2.32</u> *trans*-1,4-Di-*tert*-butylcyclohexane exists in a normal chair conformation. *cis*-1,4-Di-*tert*-butylcyclohexane, however, adopts a twist-boat conformation. Draw both isomers and explain why the cis isomer is more stable in the twist-boat conformation.

The trans isomer in the chair form   The cis isomer in the twistboat form

The *cis* isomer adopts a twist-boat conformation because each of the bulky *tert*-butyl groups can be in an pseudo-equatorial position as shown above. If the *cis* isomer existed in a normal chair conformation, then one *tert*-butyl group would be equatorial, while the other would be forced axial resulting in a large amount of nonbonded interaction strain.

Problem 2.33 From studies of the dipole moment of 1,2-dichloroethane in the gas phase at room temperature, it is estimated that the ratio of molecules in the anti conformation to gauche conformation is 7.6 to 1. Calculate the difference in energy between these two conformations.

$$\Delta G° = -RT\ln K_{eq}$$

$$K_{eq} = \frac{7.6}{1} = 7.6 \quad \text{so} \quad \ln K_{eq} = 2.0$$

Plugging in the gas constant (R = 8.314 J·K⁻¹·mol⁻¹) and temperature (T = 298 K)

$$\Delta G° = - (8.314 \text{ J·K}^{-1}\text{·mol}^{-1})(298 \text{ K})(2.0) = 5.0 \times 10^3 \text{ J/mol} = \boxed{5.0 \text{ kJ/mol}}$$

Cis, trans Isomerism in Cycloalkanes

Problem 2.34 Draw structural formulas for the cis and trans isomers of 1,2-dimethylcyclopropane.

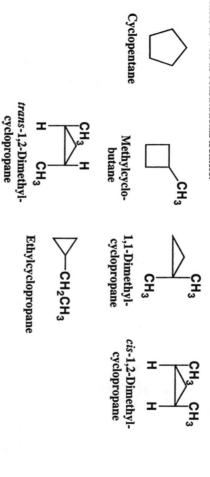

cis-1,2-Dimethyl-
cyclopropane

trans-1,2-Dimethyl-
cyclopropane

Problem 2.35 Name and draw structural formulas for all cycloalkanes of molecular formula C_5H_{10}. Be certain to include cis, trans isomers as well as constitutional isomers.

Cyclopentane

Methylcyclo-
butane

1,1-Dimethyl-
cyclopropane

cis-1,2-Dimethyl-
cyclopropane

trans-1,2-Dimethyl-
cyclopropane

Ethylcyclopropane

Problem 2.36 Using a planar pentagon representation for the cyclopentane ring, draw structural formulas for the cis and trans isomers of:
(a) 1,2-Dimethylcyclopentane
(b) 1,3-Dimethylcyclopentane.

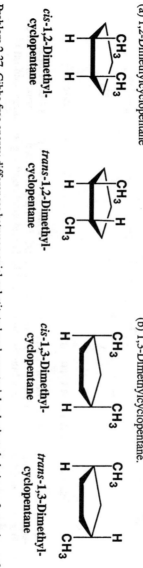

cis-1,2-Dimethyl-
cyclopentane

trans-1,2-Dimethyl-
cyclopentane

cis-1,3-Dimethyl-
cyclopentane

trans-1,3-Dimethyl-
cyclopentane

Problem 2.37 Gibbs free energy differences between axial-substituted and equatorial-substituted chair conformations of cyclohexane were given in Table 2.4.
(a) Calculate the ratio of equatorial to axial tert-butylcyclohexane at 25°C.

According to the value given in Table 2.4, the equatorial tert-butylcyclohexane is 21 kJ/mol more stable than the axial conformation.

Rearranging $\Delta G° = -RT\ln K_{eq}$ gives $\ln K_{eq} = \dfrac{-\Delta G°}{RT}$

Plugging in the gas constant ($R = 8.314$ J•K^{-1}•mol^{-1}) and temperature ($T = 298$ K) as well as the value for $\Delta G°$ converted to J/mol gives

$$\ln K_{eq} = \frac{21{,}000 \text{ J/mol}}{(8.314 \text{ J·K}^{-1}\text{·mol}^{-1})(298 \text{ K})} = 8.5$$

$$K_{eq} = e^{8.5} = 4.9 \times 10^{3}$$

So the ratio of equatorial to axial is 4.9×10^{3}:1

(b) Explain, by using molecular models why the conformational equilibria for methyl, ethyl, and isopropyl substituents are comparable but the conformational equilibrium for *tert*-butylcyclohexane lies considerably farther toward the equatorial conformation.

Rotation is possible about the single bond connecting the axial substituent to the ring. Axial methyl, ethyl, and isopropyl groups can assume a conformation where a hydrogen creates the 1,3-diaxial interactions. With a *tert*-butyl substituent, however, a bulkier -CH$_3$ group must create the 1,3-diaxial interaction. Because of the increased steric strain (nonbonded interactions) created by the axial *tert*-butyl, the energy of the axial conformation is considerably greater than that for the equatorial conformation. As seen below, an axial isopropyl group can adopt a conformation with only a minimal 1,3 diaxial interaction:

On the other hand, an axial *tert*-butyl group leads to a very severe 1,3 diaxial interaction:

Problem 2.38 When cyclohexane is substituted by an ethynyl group, -C≡CH, the energy difference between axial and equatorial conformations is only 1.7 kJ/mol (0.41 kcal/mol). Compare the conformational equilibrium for methylcyclohexane with that for ethynylcyclohexane and account for the difference between the two.

For the ethynyl case, using the same equations as in part (a) of 2.37 gives:

$$\ln K_{eq} = \frac{1{,}700 \text{ J/mol}}{(8.314 \text{ J·K}^{-1}\text{·mol}^{-1})(298 \text{ K})} = 0.69$$

$$K_{eq} = e^{0.69} = 2.0$$

So the ratio of equatorial to axial ethynyl is 2:1

Using the value of 7.28 kJ/mol given in Table 2.4 for the methyl case:

$$\ln K_{eq} = \frac{7,280 \text{ J/mol}}{(8.314 \text{ J}\cdot K^{-1}\cdot mol^{-1})(298 \text{ K})} = 2.94$$

$$K_{eq} = e^{2.94} = 18.9$$

So the ratio of equatorial to axial methyl is 18.9:1

The above ratios make sense since as can be seen with the following structures, the bulkier methyl group is expected to have more severe 1,3 diaxial interactions than the linear -C≡CH group.

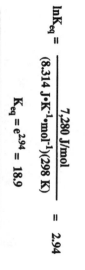

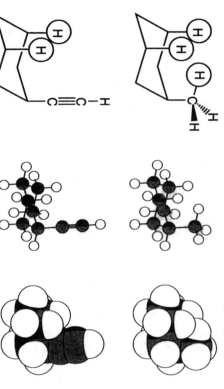

Problem 2.39 Draw the alternative chair conformations for the cis and trans isomers of 1,2-dimethylcyclohexane, 1,3-dimethylcyclohexane, and 1,4-dimethylcyclohexane.

(a) Indicate by a label whether each methyl group is axial or equatorial.
(b) For which isomer(s) are the alternative chair conformations of equal stability?
(c) For which isomer(s) is one chair conformation more stable than the other?

Cis and trans isomers are drawn as pairs. The more stable chair is labeled in cases where there is a difference.

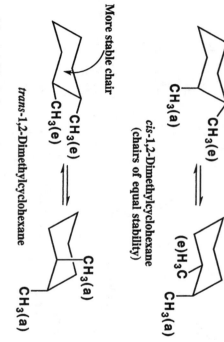

cis-1,2-Dimethylcyclohexane
(chairs of equal stability)

trans-1,2-Dimethylcyclohexane

cis-1,3-Dimethylcyclohexane

trans-1,3-Dimethylcyclohexane
(chairs of equal stability)

cis-1,4-Dimethylcyclohexane
(chairs of equal stability)

More stable chair

trans-1,4-Dimethylcyclohexane

Problem 2.40 Use your answers from problem 2.39 to complete the table showing correlations between cis, trans and axial, equatorial for the disubstituted derivatives of cyclohexane.

These relationships are summarized in the following table.

Position of Substitution	*cis*	*trans*
1,4	a,e or e,a	e,e or a,a
1,3	e,e or a,a	a,e or e,a
1,2	a,e or e,a	e,e or a,a

Problem 2.41 Calculate the difference in energy in kilojuoles per mole between the alternative chair conformations of:
(a) *trans*-4-methylcyclohexanol

More stable chair

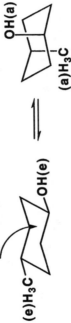

trans-4-Methylcyclohexanol

Using the values in Table 2.4, the diequatorial conformation (on the left) will be 7.28 + 3.9 = 11.2 kJ/mol more stable.

(b) *cis*-4-Methylcyclohexanol

Slightly more stable chair

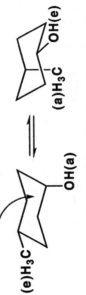

cis-4-Methylcyclohexananol

The conformation on the left is the slightly more stable because the bulkier methyl group is equatorial. The molecule on the right has the smaller -OH group equatorial. The difference in energy will be 7.28 - 3.9 = 3.4 kJ/mol

(c) *trans*-1,4-Dicyanocyclohexane

More stable chair

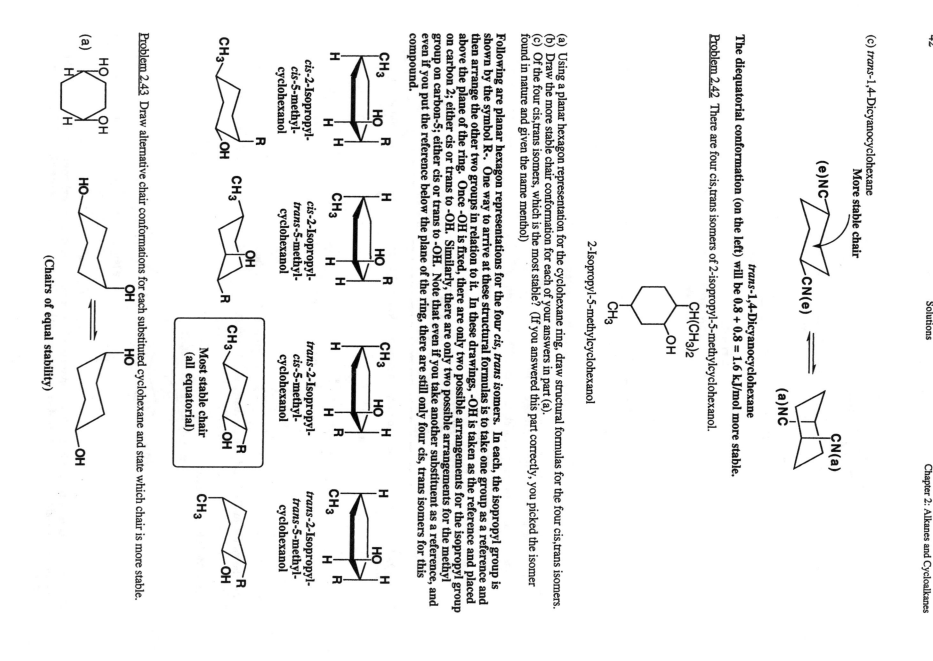

trans-1,4-Dicyanocyclohexane
The diequatorial conformation (on the left) will be 0.8 + 0.8 = 1.6 kJ/mol more stable.

Problem 2.42　There are four cis,trans isomers of 2-isopropyl-5-methylcyclohexanol.

2-Isopropyl-5-methylcyclohexanol

(a) Using a planar hexagon representation for the cyclohexane ring, draw structural formulas for the four cis, *trans* isomers.
(b) Draw the more stable chair conformation for each of your answers in part (a).
(c) Of the four cis,trans isomers, which is the most stable? (If you answered this part correctly, you picked the isomer found in nature and given the name menthol)

Following are planar hexagon representations for the *four cis, trans isomers*. In each, the isopropyl group is shown by the symbol R-. One way to arrive at these structural formulas is to take one group as a reference and then arrange the other two groups in relation to it. In these drawings, -OH is taken as the reference and placed above the plane of the ring. Once -OH is fixed, there are only two possible arrangements for the isopropyl group on carbon 2; either cis or trans to -OH. Similarly, there are only two possible arrangements for the methyl group on carbon-5; either cis or trans to -OH. Note that even if you take another substituent as a reference, and even if you put the reference below the plane of the ring, there are still only four cis, trans isomers for this compound.

cis-2-Isopropyl-
cis-5-methyl-
cyclohexanol

cis-2-Isopropyl-
trans-5-methyl-
cyclohexanol

trans-2-Isopropyl-
cis-5-methyl-
cyclohexanol

trans-2-Isopropyl-
trans-5-methyl-
cyclohexanol

**Most stable chair
(all equatorial)**

Problem 2.43　Draw alternative chair conformations for each substituted cyclohexane and state which chair is more stable.

(a)

(Chairs of equal stability)

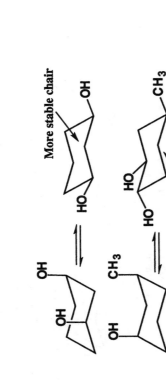

(b)

More stable chair

(c)

More stable chair

(d)

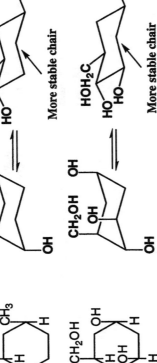

More stable chair

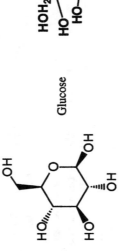

Glucose

Problem 2.44 Glucose (Section 25.1) contains a six-membered ring. In the more stable chair conformation of this molecule, all substituents on the ring are equatorial. Draw this more stable conformation.

Problem 2.45 1,2,3,4,5,6-Hexachlorocyclohexane shows cis, trans isomerism. At one time a crude mixture of these isomers was sold as an insecticide. The insecticidal properties of the mixture arise from one isomer known as the lindane, which is cis-1,2,4,5-trans-3,6-hexachlorocyclohexane (The Merck Index 12th ed, #5526).

(a) Draw a structural formula for 1,2,3,4,5,6-hexachlorocyclohexane disregarding for the moment the existence of cis, trans isomerism. What is the molecular formula of this compound?

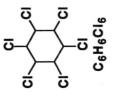

$C_6H_6Cl_6$

(b) Using a planar hexagon representation for the cyclohexane ring, draw a structural formula for lindane.

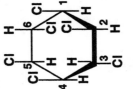

(c) Draw a chair conformation for the lindane and again labels which chlorine atoms are axial and which are equatorial.

(d) Draw the alternative chair conformation of the lindane and again label which chlorine atoms are axial and which are equatorial.

(e) Which of the alternative chair conformations of the lindane is more stable? Explain.

The two chairs are of equal stability; in each three -Cl atoms are axial and three are equatorial

Physical Properties

Problem 2.46 In Problem 2.20, you drew structural formulas for all isomeric alkanes of molecular formula C_7H_{16}. Predict which isomer has the lowest boiling point and which has the highest boiling point.

Names and boiling points of these isomers are given in the solution to Problem 2.20 The isomer with the lowest boiling point is 2,2-dimethylpentane, bp 79.2°C. The isomer with the highest boiling point is heptane, bp 94.8°C.

Problem 2.47 What generalization can you make about the densities of alkanes relative to that of water?

All alkanes that are liquid at room temperature are less dense than water. This is why alkanes such as those in gasoline and petroleum float on water.

Problem 2.48 What unbranched alkane has about the same boiling point as water? (Refer to Table 2.5 on the physical properties of alkanes.) Calculate the molecular weight of this alkane and compare it with that of water.

Heptane, C_7H_{16}, has a boiling point of 98.4°C and a molecular weight of 100. Its molecular weight is approximately 5.5 times that of water. Although considerably smaller, the water molecules are held together by hydrogen bonding while the much larger heptane molecules are held together only by relatively weak dispersion forces.

Reactions of Alkanes

Problem 2.49 Complete and balance the following combustion reactions. Assume that each hydrocarbon is converted completely to carbon dioxide and water.

(a) Propane + O_2 $\longrightarrow$ $CH_3CH_2CH_3$ + 5 O_2 $\longrightarrow$ 3 CO_2 + 4 H_2O

(b) Octane + O_2 $\longrightarrow$ 2 $CH_3(CH_2)_6CH_3$ + 25 O_2 $\longrightarrow$ 16 CO_2 + 18 H_2O

(c) Cyclohexane + O_2 $\longrightarrow$ + 9 O_2 $\longrightarrow$ 6 CO_2 + 6 H_2O

(d) 2-Methylpentane + O_2 $\longrightarrow$ 2 $CH_3CHCH_2CH_2CH_3$ + 19 O_2 $\longrightarrow$ 12 CO_2 + 14 H_2O

with CH_3 group

Problem 2.50 Following are heats of combustion per mole for methane, propane, and 2,2,4-trimethylpentane. Each is a major source of energy. On a gram-for-gram basis, which of these hydrocarbons is the best source of heat energy.

Hydrocarbon	Component of	ΔH° [kJ(kcal)/mol)]
CH_4	natural gas	-891 (-213)
$CH_3CH_2CH_3$	LPG	-2221 (-531)
$CH_3CCH_2CHCH_3$ with CH_3, CH_3, CH_3	gasoline	-5452(-1304)

On a gram-per-gram basis, methane is the best source of heat energy.

Hydrocarbon	Molecular Weight	Heat of Combustion (kJ/mol)	Heat of Combustion (kJ/gram)
Methane	16.04	-891	- 55.6
Propane	44.09	-2221	-50.4
2,2,4-Trimethylpentane	114.2	-5452	-47.7

Problem 2.51 Given below are structural formulas and heats of combustion of ethylene oxide and acetaldehyde. Which of these compounds is the more stable. Explain.

Acetaldehyde   Ethylene oxide

-1164 kJ (-278 kcal)/mol   -1264 kJ (-302.1 kcal)/mol

These molecules are constitutional isomers, so their heats of combustion are directly comparable. The molecule with the smaller (less negative) heat of combustion is the more stable, having less energy to release during the combustion process. As a result, acetaldehyde is the more stable molecule.

Problem 2.52 Without consulting tables, arrange these compounds in order of decreasing (less negative) heat of combustion: hexane, 2-methylpentane, 2,2-dimethylbutane.

Branching increases stability of an alkane. Therefore, the more highly branched the isomer, the smaller (less negative) the heat of combustion. The molecules listed in order of decreasing heat of combustion turn out to be exactly as they were listed in the question:

                    Hexane, 2-Methylpentane, 2,2-Dimethylbutane

Problem 2.53 Which would you predict to have the larger (more negative) heat of combustion, cis-1,4-dimethylcyclohexane or trans-1,4-dimethylcyclohexane.

The less stable molecule will have the larger (more negative) heat of combustion. Since these molecules are constitutional isomers with virtually identical ring strain, any difference in energy between them must be the result of differences in conformational stability. As listed in the answer to Problem 2.40, the cis isomer has two chair conformations of equal energy, each one with one axial and one equatorial methyl group. The trans isomer has chair conformations of different stability, the more stable of which is the diequatorial conformation that has no diaxial interactions. By virture of having diaxial interactions in both chair conformations, the cis isomer is higher in energy and thus will have the larger (more negative) heat of combustion.

Molecular Modeling

Problem 2.54 Using the models on the CD, measure the distance between hydrogens on adjacent carbon atoms in the staggered and eclipsed conformations of ethane and estimate the ratio of eclipsed/staggered distance. To measure distance, click on one atom (it will become darkened), and then move the pointer to any other atom.

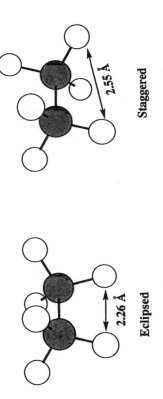

As shown in the above structures, the ratio of eclipsed/staggered distance is 2.261 Å/2.495 Å = 0.906.

Problem 2.55 Using the model of the envelope conformation of cyclopentane on the CD, measure all C-C-C bond angles, and compare them with the value of 105° given in the text. How do you account for the difference in these values?

The version of the structure optimization routine used by Chem3D is not sophisticated enough to recognize the torsional strain of the eclipsed hydrogen interactions. Thus, the structure you see is a flat pentagon, as opposed to the correct molecular structure, the so-called "envelope" conformation.

Problem 2.56 As you see from Figure 2.23, the strain energy in cyclopentane and cycloheptane are approximately equal, but there is zero strain in cyclohexane. Using the models of cyclohexane and cycloheptane on the CD, see if you can determine why the cycloheptane ring is strained. Check for angle strain, and close contacts that result in nonbonded interaction strain.

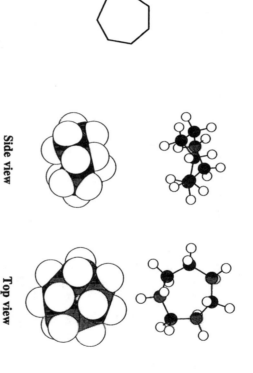

Side view   Top view

Problem 2.57 Using the model of a chair cyclohexane on the CD, rotate the model so that you view the chair from above, from the side, from the foot piece to the headpiece, and so on. As you do these rotations, convince yourself that the six axial C-H bonds are parallel and that they alternate up, down, and so on. Also convince yourself that there are six equatorial C-H bonds and that those on opposite carbons (1 and 4, 2 and 5, and so on) are parallel and trans to each other.

As can be seen in the side view, there is some torsional strain since not all the bonds are fully staggered. In addition, as can be seen in the top view, the bond angles are somewhat irregular around the ring. Thus, there is also ring strain in the molecule.

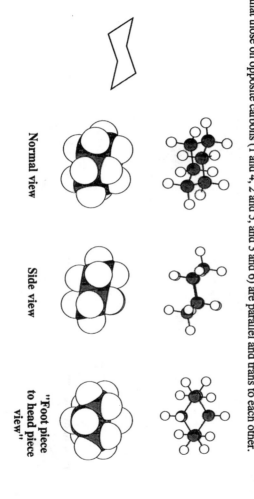

Normal view   Side view   "Foot piece to head piece view"

Problem 2.58 Build a molecular model of axial-methylcyclohexane and measure the distance between the methyl group and the ring hydrogens on carbons 2, 3, 4, 5, and 6. You should find that the axial methyl group is closer to axial hydrogens on carbons 3 and 5 than to any other ring hydrogens.

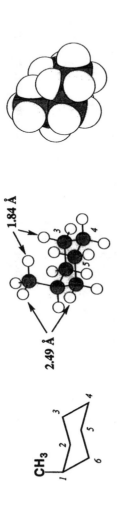

According to the Chem3D model, the closest contacts between a hydrogen atom on the axial methyl group and an axial hydrogen is 1.84 Å. This is much closer than the 2.49 Å distance between the other methyl group hydrogen atoms and the nearest equatorial ring hydrogen atom.

Problem 2.59 What kinds of conformations do the six-membered rings exhibit in adamantane and twistane? In answering this question, you will find it helpful to manipulate the models of adamantane and twistane supplied on the CD that accompanies the text.

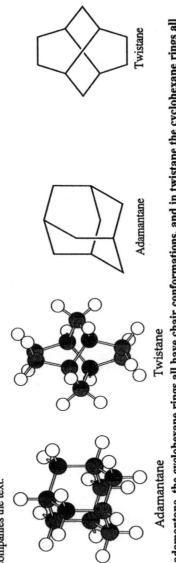

Adamantane   Twistane

In adamantane, the cyclohexane rings all have chair conformations, and in twistane the cyclohexane rings all have twist-boat conformations.

Problem 2.60 On the CD are models of these three small-ring compounds, and the strain energy of each. Which of the three is the most stable? Which is the least stable? What explanation can you offer for their relative stabilities?
(a) [1.1.1]propellane   (b) [2.1.1]propellane   (c) [2.2.1]propellane

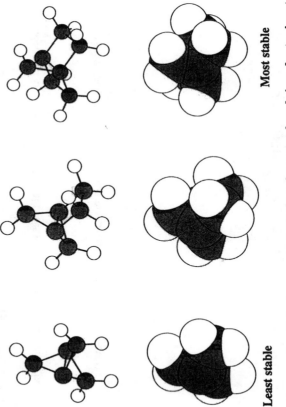

Least stable   Most stable

Four-membered rings are more stable than three-membered rings due to ring strain. Therefore, the [2.2.1]propellane is the most stable, and the [1.1.1]propellane is the least stable.

CHAPTER 3
Solutions to the Problems

Problem 3.1 Each molecule has one stereocenter. Draw stereorepresentations for the enantiomers of each.

Each part has a tetrahedral stereocenter. The stereocenters are labeled with an asterisk.

(a)

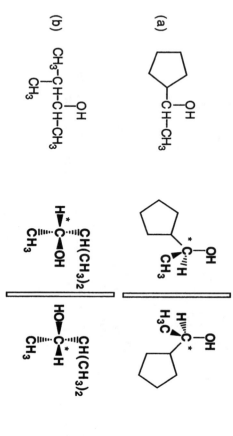

(b)

Problem 3.2 Assign priorities to the groups in each set.
(a) -CH₂OH and -CH₂CH₂OH

The -CH₂OH group has higher priority because the FIRST point of difference is the underlined O atom of -CH₂OH that takes priority over the underlined C atom of -CH₂CH₂OH.

(b) -CH₂OH and -CH=CH₂

$$-\overset{1}{C}H=\overset{2}{C}H_2 \quad \text{is treated as} \quad$$

The FIRST point of difference is the underlined O atom of -CH₂OH that takes priority over any of the atoms attached to carbon 1 of -CH=CH₂. Thus, the -CH₂OH group takes priority over the -CH=CH₂ group.

(a) -CH₂OH and -C(CH₃)₃

The FIRST point of difference is the underlined O atom of -CH₂OH that takes priority over any of the carbon atoms attached to the central carbon atom of -C(CH₃)₃.

Problem 3.3 Assign an R or S configuration to each stereocenter.

The drawings underneath each molecule show the order of priority, the perspective from which to view the molecule, and the R,S designation for the configuration.

(a)

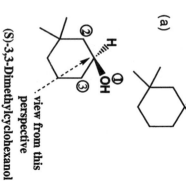

view from this
perspective

(S)-3,3-Dimethylcyclohexanol

If you view from the perspective shown, this is what you see

(b)

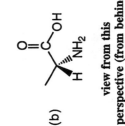

view from this perspective (from behind)

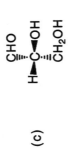

If you view from the perspective shown, this is what you see

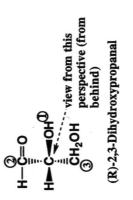

(R)-2-Aminopropanoic acid

(c)

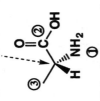

view from this perspective (from behind)

(R)-2,3-Dihydroxypropanal

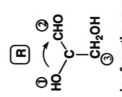

If you view from the perspective shown, this is what you see

Problem 3.4 Following are stereorepresentations for the four stereoisomers of 3-chloro-2-butanol.

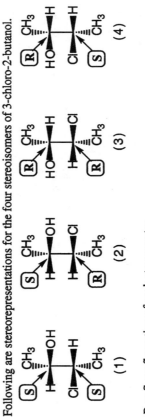

(1)   (2)   (3)   (4)

(a) Assign an R or S configuration of each stereocenter.

The configuration of each stereocenter has been labeled on the above structures. This labeling often helps when trying to establish stereochemical relationships between molecules.

(b) Which compounds are enantiomers?

Enantiomers are stereoisomers that are mirror images of each other. The pairs of enantiomers are structures (1) and (3) (S,S and R,R) as well as structures (2) and (4) (S,R and R,S).

(c) Which compounds are diastereomers?

Diastereomers are stereoisomers that are not mirror images of each other. Therefore, there are several sets of diastereomers. The following chart describes the relationship between any pair of molecules

Problem 3.5 Following are four Newman projection formulas for tartaric acid.

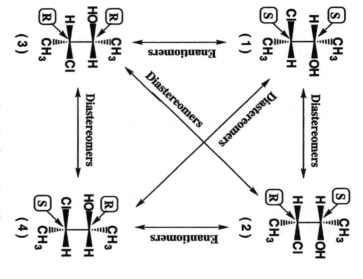

(1)

(2)

(3)

(4)

(a) Which represent the same compound?

Compounds (1) and (4) are the same compound having the configuration (2R,3R). Compounds (2) and (3) are also the same compound, having the configuration (2R,3S).

(b) Which represent enantiomers?

There are no sets of enantiomers in this set.

(c) Which represent a meso compound?

The meso compound of tartaric acid has the (2R,3S) configuration so compounds (2) and (3) are the meso compound.

(d) Which are diastereomers?

Compounds (1) and (4) are diastereomers of compounds (2) and (3).

Problem 3.6 How many stereoisomers exist for 1,3-cyclopentanediol?

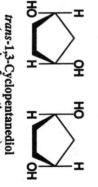

cis-1,3-Cyclopentanediol
(achiral, a meso compound)

trans-1,3-Cyclopentanediol
(a pair of enantiomers)

1,3-Cyclopentanediol has three stereoisomers. The two trans isomers are enantiomers, the cis isomer is a meso compound. *Cis*-1,3-cyclopentanediol can be recognized as a meso compound because it is superposable upon its mirror image. Alternatively, it has a plane of symmetry that bisects it into two mirror halves.

Problem 3.7 How many stereoisomers exist for 1,4-cyclohexanediol?

1,4-Cyclohexanediol can exist as a pair of cis,trans isomers. Each is achiral because of a plane of symmetry that bisects each molecule into two mirror halves. In the figure below, the plane of symmetry in each molecule is in the plane of the paper. As a result of each isomer being achiral, there are only two stereoisomers of 1,4-cyclohexanediol.

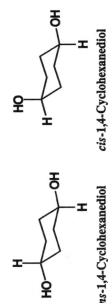

trans-1,4-Cyclohexanediol   *cis*-1,4-Cyclohexanediol

Problem 3.8 The specific rotation of progesterone, a female sex hormone, is +172, measured at 20°C. Calculate the observed rotation for a solution prepared by dissolving 300 mg of progesterone in 15.0 mL of dioxane and placing it in a sample tube 10.0 cm long.

The concentration of progesterone, expressed in grams per milliliter is:

$$300 \text{ mg} / 15 \text{ mL} = 0.020 \text{ g/mL}$$

$$\text{specific rotation} = \frac{\text{observed rotation (degrees)}}{\text{length (dm) x concentration (g/mL)}}$$

Rearranging this formula to solve for observed rotation gives:

observed rotation (degrees) = specific rotation x length (dm) x concentration (g/mL)

Plugging in the experimental values gives the final answer.

observed rotation (degrees) = +172 x 1.00 dm x 0.020 g/mL = $\boxed{+3.4}$

Problem 3.9 One commercial synthesis of naproxen (the active ingredient in Aleve and a score of other over-the-counter and prescription nonsteroidal anti-inflammatory drug preparations) gives the enantiomer shown in 97% enantiomeric excess.

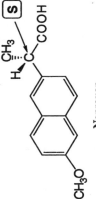

Naproxen

(A nonsteroidal antiinflammatory drug)

(a) Assign an R or S configuration to this enantiomer of naproxen.

The stereocenter has the S configuration.

(b) What are the percentages of R and S enantiomers in the mixture?

$$\text{Enantiomeric excess (ee)} = \frac{S - R}{S + R} \text{ x } 100 = \%S - \%R = 97\%$$

solving the above equation gives

$$\boxed{98.5\% \text{ S and } 1.5\% \text{ R}}$$

Chirality
Problem 3.10 Think about the helical coil of a telephone cord or a spiral binding and suppose that you view the spiral from one end and find that it is a left-handed twist. If you view the same spiral from the other end, is it a right-handed or a left-handed twist?

A helical coil has the same handedness viewed from either end.

Problem 3.11 Next time you have the opportunity to view a collection of sea shells that have a helical twist, study the chirality of their twists. Do you find an equal number of left-handed and right-handed spiral shells or mostly all of the same chirality? What about the handedness of different species of spiral shells?

This question was just meant to make you think about chirality in nature, but if you do know the answer please share it with your class.

Problem 3.12 One reason we can be sure that sp³-hybridized carbon atoms are tetrahedral is the number of stereoisomers that can exist for different organic compounds.
(a) How many stereoisomers are possible for $CHCl_3$, CH_2Cl_2, and $CHClBrF$ if the four bonds to carbon have a tetrahedral arrangement?

Both tetrahedral $CHCl_3$ and tetrahedral CH_2Cl_2 are achiral, so no stereoisomers are possible.

On the other hand, tetrahedral CHBrClF is chiral so there are two stereoisomers possible.

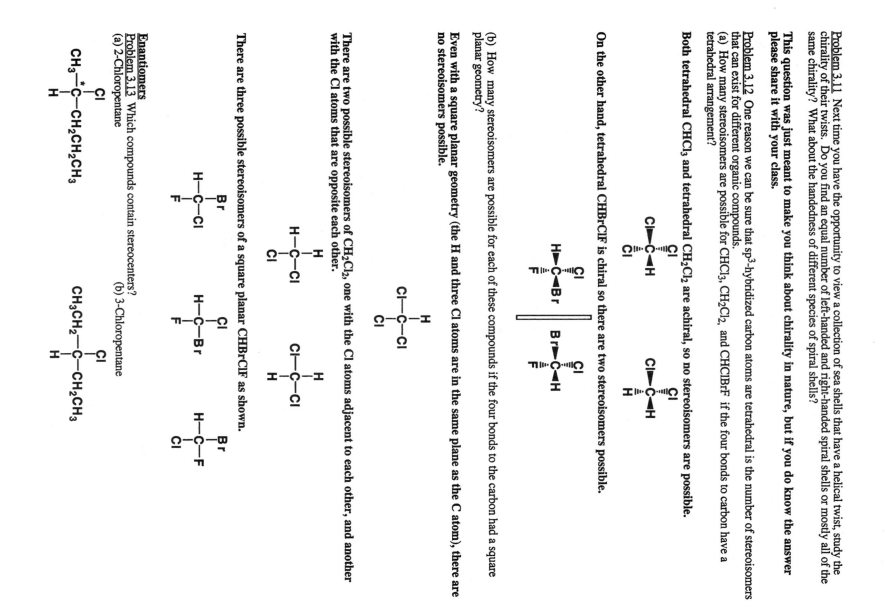

(b) How many stereoisomers are possible for each of these compounds if the four bonds to the carbon had a square planar geometry?

Even with a square planar geometry (the H and three Cl atoms are in the same plane as the C atom), there are no stereoisomers possible.

There are two possible stereoisomers of CH_2Cl_2, one with the Cl atoms adjacent to each other, and another with the Cl atoms that are opposite each other.

There are three possible stereoisomers of a square planar CHBrClF as shown.

Enantiomers
Problem 3.13 Which compounds contain stereocenters?
(a) 2-Chloropentane   (b) 3-Chloropentane

(c) 3-Chloro-1-pentene

CH₂=CH—C*—CH₂CH₃ (with Cl and H)

(d) 1,2-Dichloropropane

ClCH₂—C*—CH₃ (with Cl and H)

Stereocenters are present in (a), (c), and (d). They are marked with an asterisk (*).

Problem 3.14 Using only C, H, and O, write structural formulas for the lowest molecular weight chiral

(a) Alkane

CH₃CH₂CH₂—C*—CH₂CH₃ (with H and CH₃)

3-Methylhexane

(b) Alcohol

HO—C*—CH₂CH₃ (with H and CH₃)

2-Butanol

(c) Aldehyde

CH₃CH₂—C*—CH (with O, H, CH₃)

2-Methylbutanal

(d) Ketone

CH₃CH₂—C*—C—CH₃ (with H, O, CH₃)

3-Methyl-2-pentanone

(e) Carboxylic acid

CH₃CH₂—C*—C—OH (with H, O, CH₃)

2-Methylbutanoic acid

(f) Carboxylic ester

H—C—O—C*H—CH₂CH₃ (with O, CH₃)

2-Butyl formate

Problem 3.15 Draw mirror images for these molecules.

The mirror images are shown in bold.

(a)

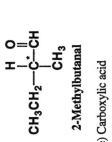

(c)

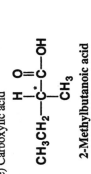

(e)

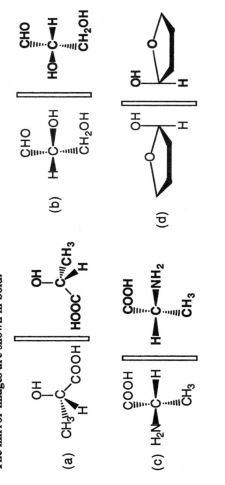

(b)

(d)

(f)

(g)

(h)

Problem 3.16 Following are several stereorepresentations for lactic acid. Take (a) as a reference structure. Which of the stereorepresentations are identical with (a) and which are mirror images of (a)?

(a)

(b)

(c)

(d)

All of the above stereorepresentations have the (S)-configuration so they are identical.

Problem 3.17 Mark each stereocenter in the following molecules with an asterisk. How many stereoisomers are possible for each molecule?

(a)

CH₃–C–CH=CH₂
        |
        OH

No stereocenters

(b)

COOH
 |
H–C–OH
 |
CH₃

No stereocenters

(c)

CH₃
 |
CH₃–CH–CH–COOH
          |
          NH₂

CH₃
 |
CH₃–CH–CH–COOH *
          |
          NH₂

2 stereoisomers
(a pair of enantiomers)

(d)

        O
        ‖
CH₃–C–CH₂–CH₃

No stereocenters

(e)

CH₂OH
 |
H–C–OH
 |
CH₂OH

No stereocenters

(f)

          OH
          |
CH₃–CH₂–CH–CH=CH₂

          OH
          |
CH₃–CH₂–CH–CH=CH₂ *

2 stereoisomers
(a pair of enantiomers)

(g)

CH₂–COOH
 |
H–C–OH
 |
CH₂–COOH

No Stereocenters

Problem 3.18 Show that butane in a gauche conformation is chiral. Do you expect that resolution of butane at room temperature is possible?

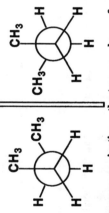

As can be seen from the above Newman projections, the two gauche conformations are non-superposable mirror images. However, these conformations rapidly interconvert at room temperature through rotation around the central C-C bond, so they cannot be resolved.

Designation of Configuration: The R-S Convention
Problem 3.19 Assign the priorities to the groups in each set.

The groups are ranked from highest to lowest under each problem. Remember that priority is assigned at the first point of difference.

(a) -H -CH₃ -OH -CH₂OH

-OH > -CH₂OH > -CH₃ > -H

(b) -CH₂CH=CH₂ -CH=CH₂ -CH₃ -CH₂COOH

-CH=CH₂ > -CH₂COOH > -CH₂CH=CH₂ > -CH₃

(c) -CH₃ -H -COO⁻ -NH₃⁺

-NH₃⁺ > -COO⁻ > -CH₃ > -H

(d) -CH₃ -CH₂SH -NH₃⁺ -COO⁻

-NH₃⁺ > -CH₂SH > -COO⁻ > -CH₃

Problem 3.20 Following are structural formulas for the enantiomers of carvone (The Merck Index, 12th ed., #1925). Each has a distinctive odor characteristic of the source from which it is isolated. Assign R and S or configuration to the single stereocenter in each enantiomer. e

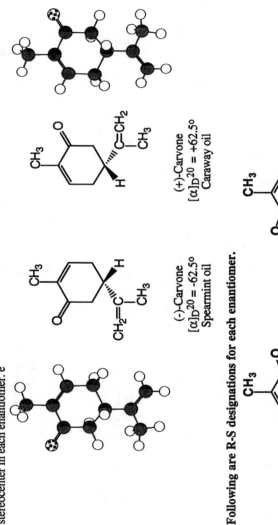

(-)-Carvone
[α]D²⁰ = -62.5°
Spearmint oil

(+)-Carvone
[α]D²⁰ = +62.5°
Caraway oil

Following are R-S designations for each enantiomer.

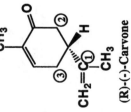

(R)-(-)-Carvone

(S)-(+)-Carvone

Problem 3.21 Following is a staggered conformation for one of the enantiomers of 2-butanol.

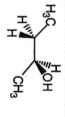

(a) Is this (R)-2-butanol or (S)-2-butanol?
The structure drawn is (S)-2-butanol.

(b) Draw a Newman projection for this staggered conformation, viewed along the bond between carbons 2 and 3.

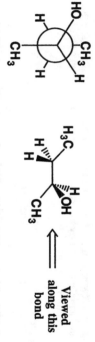

Viewed along this bond

(c) Draw a Newman projection for two more staggered conformations of this molecule. Which of your conformations is the more stable? Assume that -OH and -CH₃ are comparable in size.

More stable

Viewed along this bond

Less stable

Viewed along this bond

Assuming that -OH and -CH₃ are the same size, then the structure drawn in part (b) and the upper structure shown in part (c) are of equal stability, and these are more stable than the lower structure shown in part (c). This lower structure is less stable because both the -OH and -CH₃ groups are adjacent to the -CH₃ group that is on the rear carbon atom.

Problem 3.22 For centuries, Chinese herbal medicine has used extracts of *Ephedra sinica* to treat asthma. Phytochemical investigation of this plant resulted in isolation of ephedrine, a very potent dilator of the air passages of the lungs. The naturally occurring stereoisomer is levorotatory and has the following structure (see The Merck Index, 12th ed., #3645). Assign R or S configuration to each stereocenter.

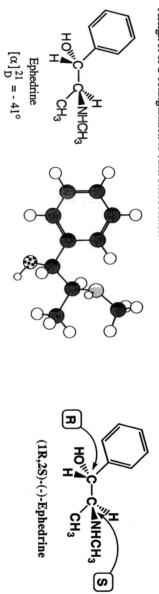

Ephedrine
$[\alpha]_D^{21} = -41°$

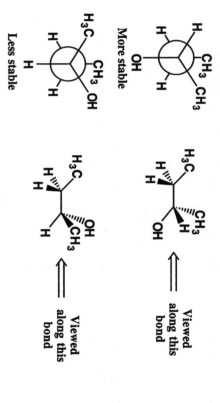

(1R,2S)-(-)-Ephedrine

Problem 3.23 When oxaloacetic acid and acetyl-coenzyme A (acetyl-CoA) labeled with radioactive carbon-14 in position 2 are incubated with citrate synthase, an enzyme of the TCA cycle, only the following enantiomer of [2-14C]-citric acid is formed stereospecifically. Note that citric acid containing only 12C is achiral. Assign an R or S configuration to this enantiomer of [2-14C] citric acid. *Note*: 14Carbon has a higher priority than 12carbon.

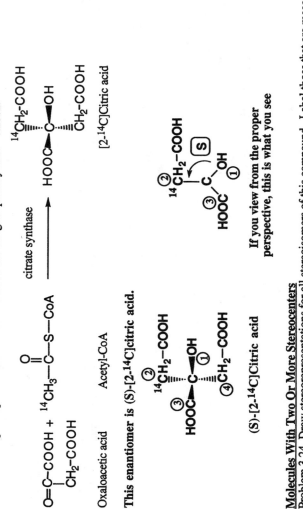

Oxaloacetic acid   Acetyl-CoA   [2-14C]Citric acid

This enantiomer is (S)-[2-14C]citric acid.

(S)-[2-14C]Citric acid

If you view from the proper perspective, this is what you see

Molecules With Two Or More Stereocenters

Problem 3.24 Draw stereorepresentations for all stereoisomers of this compound. Label those that are meso compounds and those which are pairs of enantiomers.

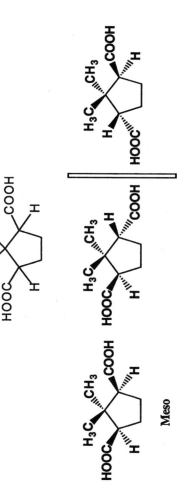

Meso

A pair of enantiomers

Problem 3.25 Mark each stereocenter in the following molecules with an asterisk. How many stereoisomers are possible for each molecule?

(a) CH3-*CH-CH-*CH-COOH
              | | |
             OH OH OH

2² = 4 Stereoisomers
(two pairs of enantiomers)

(b)
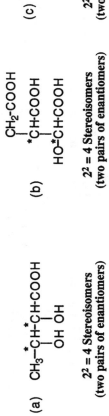

2² = 4 Stereoisomers
(two pairs of enantiomers)

(c)
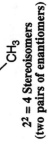

2² = 4 Stereoisomers
(two pairs of enantiomers)

(d)

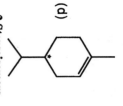

2 Stereoisomers
(one pair of enantiomers)

(e)

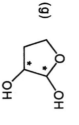

2³ = 8 Stereoisomers
(four pairs of enantiomers)

(f)

2² = 4 Stereoisomers
(two pairs of enantiomers)

(g)

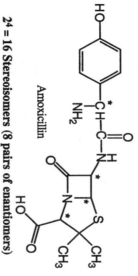

2² = 4 Stereoisomers
(two pairs of enantiomers)

(h)

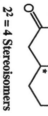

2² = 4 Stereoisomers
(two pairs of enantiomers)

Problem 3.26 Label the eight stereocenters in cholesterol. How many stereoisomers are possible for a molecule of this structural formula?

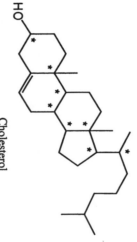

Cholesterol
2⁸ = 256 Stereoisomers (128 pairs of enantiomers)

Problem 3.27 Label the four stereocenters in amoxicillin (The Merck Index, 12ᵗʰ ed., #617), which belongs to the family of semisynthetic penicillins.

Amoxicillin

2⁴ = 16 Stereoisomers (8 pairs of enantiomers)

Problem 3.28 If the optical rotation of a new compound is measured and found to have a specific rotation of +40, how can you tell if the actual rotation is not really +40 plus some multiple of +360 (that is, the rotation is not actually +40 + n(+360), where n has only integer values). In other words, how can you tell if the rotation is not actually a value such as +400 or +760?

You could dilute the solution by a factor of two and then remeasure the rotation. If the new rotation is 20°, then the orginal rotation was 40°. If the new rotation is 200°, then the original rotation was 400°.

Problem 3.29 Are the formulas within each set identical, enantiomers, or diastereomers?

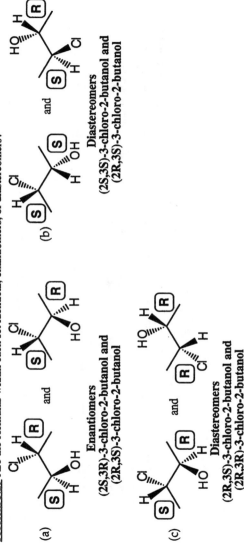

(a)

Enantiomers
(2S,3R)-3-chloro-2-butanol and
(2R,3S)-3-chloro-2-butanol

(b)

Diastereomers
(2S,3S)-3-chloro-2-butanol and
(2R,3S)-3-chloro-2-butanol

(c)

Diastereomers
(2R,3S)-3-chloro-2-butanol and
(2R,3R)-3-chloro-2-butanol

Problem 3.30 Which of the following are meso compounds?

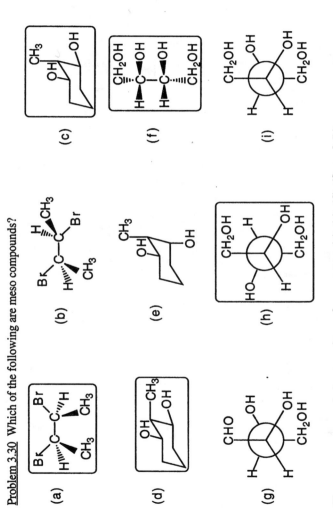

(a)

(b)

(d)

(e)

(c)

(g)

(h)

(f)

(i)

The meso compounds have a plane of symmetry: (a), (c), (d), (f), and (h).

Problem 3.31 Vigorous oxidation of the following bicycloalkene breaks the carbon-carbon double bond and converts each carbon of the double bond to a COOH group. Assume that the conditions of oxidation have no effect on the configuration of either the starting bicycloalkene or the resulting dicarboxylic acid. Is the dicarboxylic acid produced from this oxidation one enantiomer, a racemic mixture, or a meso compound?

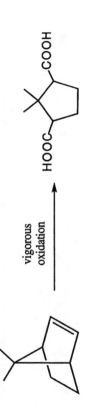

vigorous
oxidation

The two carboxyl groups, derived from oxidation of the double bond, must be cis to each other. Therefore, the compound is meso with the following configuration.

Problem 3.32 A long polymer chain, such as polyethylene (-CH₂CH₂-)ₙ, can potentially exist in solution as a chiral object. Give two examples of chiral structures that a polyethylene chain could adopt.

Although there are no stereocenters in polyethylene, a long polyethylene chain could exist in chiral conformations such as a chiral helix or some kind of chiral knot.

meso-2,2-Dimethylcyclopentane-1,3-dicarboxylic acid

Molecular Modeling

Problem 3.33 ChemDraw provides a very easy way to make mirror images. First, create a stereocenter in ChemDraw, make a copy and place it adjacent to your original. Second, select the copy and, (3) from the Object menu, select Flip Horizontal. As shown here, this procedure converts an enantiomer to its mirror image.

1. Make a copy of (R)-lactic acid

2. Select the copy

3. Select "Flip Horizontal" from the "Object" menu

(R)-lactic acid

Copy of (R)-lactic acid

(S)-lactic acid

Now try this procedure with these molecules chosen from the text.

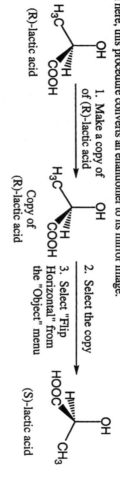

(a)

(b)

(c)

(d)

Problem 3.34 ChemDraw is able to assign an R or S configuration to a stereocenter. To do this, use ChemDraw to build a model of a chiral compound. Then select Show Stereochemistry. As practice, build structures 3.33(a-d) in ChemDraw show the configuration of each stereocenter.

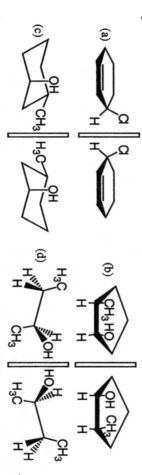

(a)

(b)

(c)

(d)

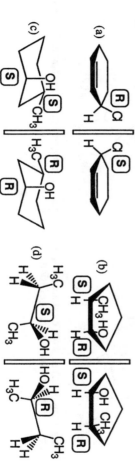

Problem 3.35 The following molecule is an attractant pheromone for the olive fly.

(a) Rotate the three-dimensional model on the CD and convince yourself that each six-membered ring has a strain-free chair conformation.

No, careful inspection verifies that they are not superposable mirror images of each other.

(b) This molecule has no stereocenter and yet it is chiral. Examine the three-dimensional model and convince yourself that it has no plane or center of symmetry and that it is, in fact, chiral.

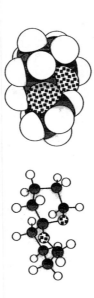

(c) "The presence of a stereocenter in an organic molecule is a sufficient condition for chirality, but it is not a necessary condition." Explain.

Some molecules are chiral because of their three dimensional structures, not because they have a stereocenter. Thus, the presence of a stereocenter is a sufficient but not a necessary condition for chirality.

Problem 3.36 TThe following molecule belongs to the class of compounds called allenes. The functional group of an allene is two adjacent carbon-carbon double bonds. Disubstituted allenes of this type are chiral. The specific rotation of the enantiomer shown is -314.

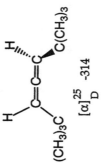

$[\alpha]_D^{25}$ -314

(a) Examine the three-dimensional model on the CD and convince yourself that it has no plane or center of symmetry.

(b) Make its mirror image and convince yourself that the original and the mirror image are nonsuperposable; that is, that they are a pair of enantiomers.

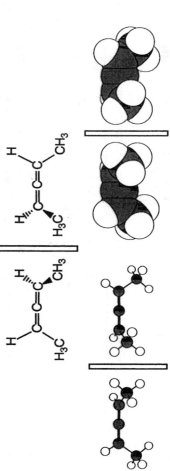

CHAPTER 4
Solutions to the Problems

Problem 4.1 For each conjugate acid-base pair, identify the first species as an acid or base and the second species as its conjugate acid or base. In addition, draw Lewis structures for each species, showing all valence electrons and any formal charges.

(a) H_2SO_4, HSO_4^-

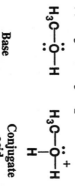

Acid   Conjugate base

(b) NH_3, NH_2^-

Acid   Conjugate base

(c) CH_3OH, $CH_3OH_2^+$

Base   Conjugate acid

Problem 4.2 Write these reactions as proton-transfer reactions. Label which reactant is the acid and which is the base; which product is the conjugate base of the original acid and which is the conjugate acid of the original base. Write Lewis structures for each reactant and product and use curved arrows to show the flow of electrons in each reaction.

(a) CH_3-S-H + OH^- ⟶ CH_3-S^- + H_2O

Acid   Base   Conjugate acid   Conjugate base

(b) $H_2C=O$ + HCl ⟶ $H_2C=OH^+$ + Cl^-

Base   Acid   Conjugate acid   Conjugate base

Problem 4.3 Following is a structural formula for guanidine, the form in which migratory birds excrete excess metabolic nitrogen. The hydrochloride salt of this compound is a white crystalline powder, freely soluble in water and ethanol.
(a) Write a Lewis structure for guanidine showing all valence electrons.

Guanidine

(b) Does proton transfer occur preferentially on one of its -NH_2 groups or on its =NH group?

The proton goes on the =NH group because this is the most basic nitrogen atom, and the resulting conjugate acid has 2 H atoms on each nitrogen. Distributing the H atoms evenly in this way is the most stable possible arrangement due to the resonance stabilization indicated below.

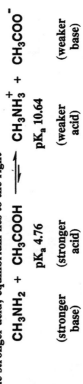

Distributing the charge over the atoms as shown is the basis for the resonance stabilization.

Problem 4.4 Write an equation to show the proton-transfer between each alkene or cycloalkene and HCl. Where isomeric carbocations are possible, show each.

(a) $CH_3CH_2CH=CHCH_3$

2-Pentene

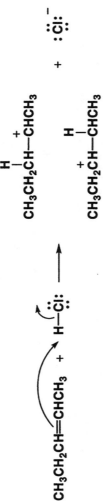

$CH_3CH_2CH=CHCH_3$ + H—Cl:

$$CH_3CH_2CH—CHCH_3 \overset{H}{\overset{|}{\underset{+}{}}}$$ + :Cl:⁻

$$CH_3CH_2CH—CHCH_3 \overset{|}{\underset{+}{\overset{H}{}}}$$

Note how there are two different cations produced upon protonation of the alkene

(b)

Cyclohexene

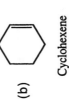

Problem 4.5 For each value of K_a, calculate the corresponding value of pK_a. Which compound is the stronger acid?
(a) Acetic acid, $K_a = 1.74 \times 10^{-5}$ (b) Chloroacetic acid $K_a = 1.38 \times 10^{-3}$

The pK_a is equal to -$\log_{10}K_a$. The pK_a of acetic acid is 4.76 and the pK_a of chloroacetic acid is 2.86. Chloroacetic acid, with the smaller pK_a value, is the stronger acid.

Problem 4.6 Predict the position of equilibrium and calculate the equilibrium constant, K_{eq}, for these acid-base reactions. The pK_a of methylammonium is 10.64.

(a) CH_3NH_2 + CH_3COOH ⇌ $CH_3NH_3^+$ + CH_3COO^-

| CH_3NH_2 | CH_3COOH | $CH_3NH_3^+$ | CH_3COO^- |
| Methylamine | Acetic acid | Methylammonium ion | Acetate ion |

Acetic acid is the stronger acid; equilibrium lies to the right

CH_3NH_2 + CH_3COOH ⇌ $CH_3NH_3^+$ + CH_3COO^-

| | pK_a 4.76 | pK_a 10.64 | |
| (stronger base) | (stronger acid) | (weaker acid) | (weaker base) |

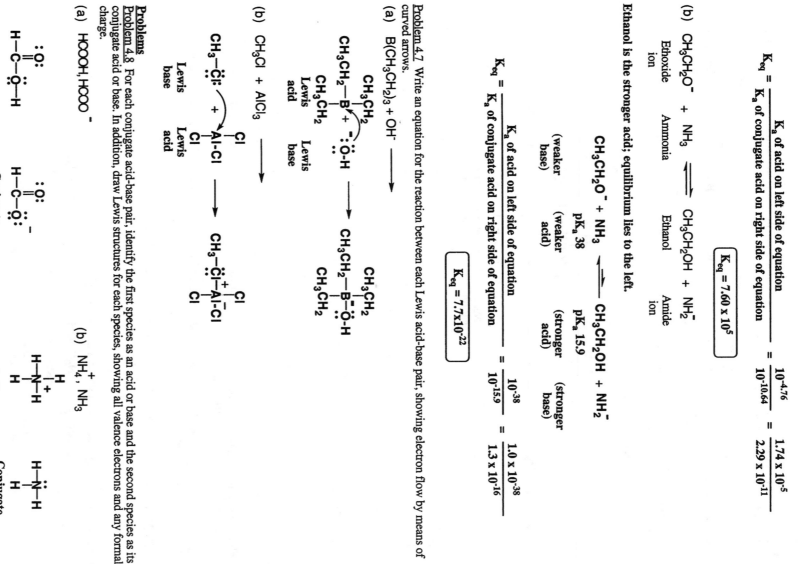

$$K_{eq} = \frac{K_a \text{ of acid on left side of equation}}{K_a \text{ of conjugate acid on right side of equation}} = \frac{10^{-4.76}}{10^{-10.64}} = \frac{1.74 \times 10^{-5}}{2.29 \times 10^{-11}}$$

$$\boxed{K_{eq} = 7.60 \times 10^5}$$

(b) $CH_3CH_2O^-$ + NH_3 $\rightleftharpoons$ CH_3CH_2OH + NH_2^-

Ethoxide   Ammonia   Ethanol   Amide
ion   ion

Ethanol is the stronger acid; equilibrium lies to the left.

$$CH_3CH_2O^- + NH_3 \rightleftharpoons CH_3CH_2OH + NH_2^-$$

            (weaker   pK_a 38   pK_a 15.9
            base)   (weaker   (stronger   (stronger
                      acid)   acid)   base)

$$K_{eq} = \frac{K_a \text{ of acid on left side of equation}}{K_a \text{ of conjugate acid on right side of equation}} = \frac{10^{-38}}{10^{-15.9}} = \frac{1.0 \times 10^{-38}}{1.3 \times 10^{-16}}$$

$$\boxed{K_{eq} = 7.7 \times 10^{-22}}$$

Problem 4.7 Write an equation for the reaction between each Lewis acid-base pair, showing electron flow by means of curved arrows.

(a) $B(CH_3CH_2)_3$ + OH^- $\longrightarrow$

(b) CH_3Cl + $AlCl_3$ $\longrightarrow$

Problems

Problem 4.8 For each conjugate acid-base pair, identify the first species as an acid or base and the second species as its conjugate acid or base. In addition, draw Lewis structures for each species, showing all valence electrons and any formal charge.

(a) $HCOOH$, $HCOO^-$

(b) NH_4^+, NH_3

(c) CH₃CH₂O⁻, CH₃CH₂OH

$CH_3CH_2O^-$, CH_3CH_2OH

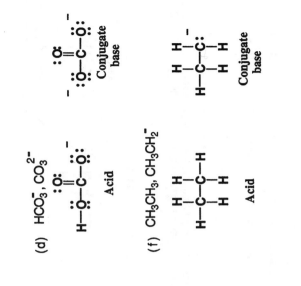

CH₃CH₂—Ö:⁻
Base

CH₃CH₂—Ö—H
Conjugate
acid

(d) HCO₃⁻, CO₃²⁻

HCO_3^-, CO_3^{2-}

Acid

Conjugate
base

(e) H₂PO₄⁻, HPO₄²⁻

$H_2PO_4^-$, HPO_4^{2-}

Acid

Conjugate
base

(f) CH₃CH₃, CH₃CH₂⁻

CH_3CH_3, $CH_3CH_2^-$

Acid

Conjugate
base

(g) CH₃S⁻, CH₃SH

CH_3S^-, CH_3SH

CH₃CH₂—S:⁻
Base

CH₃CH₂—S—H
Conjugate
acid

Problem 4.9 Complete a net ionic equation for each proton-transfer reaction using curved arrows to show the flow of electron pairs in each reaction. In addition, write Lewis structures for all starting materials and products. Label the original acid and its conjugate base; the original base and its conjugate acid. If you are uncertain about which substance in each equation is the proton donor, refer to Table 4.1 for the relative strengths of proton acids.

(a) NH₃ + HCl ⟶

NH_3 + HCl ⟶

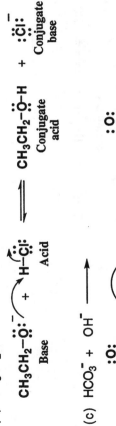

Base   Acid   Conjugate   Conjugate
                              acid   base

(b) CH₃CH₂O⁻ + HCl ⟶

$CH_3CH_2O^-$ + HCl ⟶

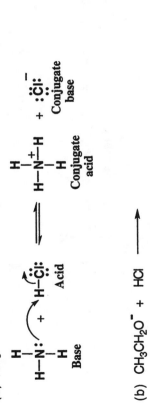

Base   Acid   Conjugate   Conjugate
                              acid   base

(c) HCO₃⁻ + OH⁻ ⟶

HCO_3^- + OH^- ⟶

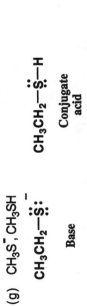

Acid   Base   Conjugate   Conjugate
                              base   acid

Problem 4.10 Complete a net ionic equation for each proton-transfer reaction using curved arrows to show the flow of electron pairs in each reaction. Label the original acid and its conjugate base; the original base and its conjugate acid.

(a) NH_4^+ + OH^- ⟶

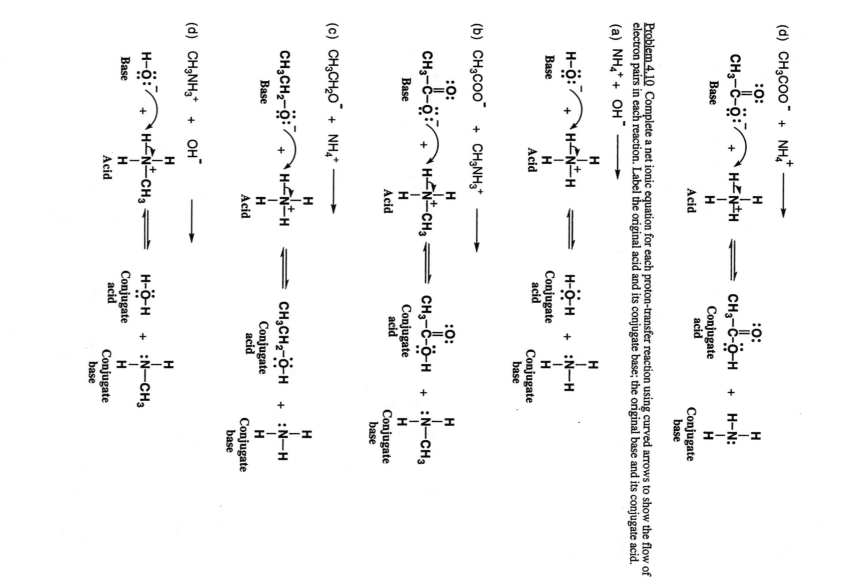

(b) CH_3COO^- + $CH_3NH_3^+$ ⟶

(c) $CH_3CH_2O^-$ + NH_4^+ ⟶

(d) $CH_3NH_3^+$ + OH^- ⟶

Problem 4.11 Each molecule or ion can function as a base. Write the structural formula of the conjugate acid formed by reaction of each with HCl.

(a) CH_3CH_2OH

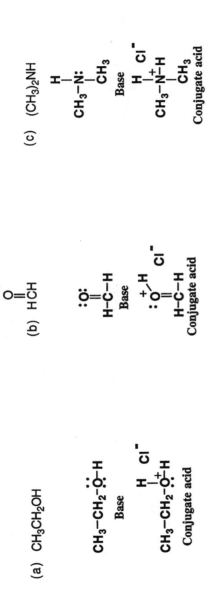

(b) HCH with =O

(c) $(CH_3)_2NH$

(d) HCO_3^-

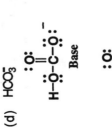

Problem 4.12 In acetic acid, the O-H proton is more acidic than the CH_3 protons. Explain.

Within a row, acidity increases with increasing electronegativity of the atom attached to hydrogen. Oxygen is more electronegative than carbon, so the hydrogen on the oxygen atom is more acidic. Note that the same argument applies when the appropriate contributing structures are considered. When the proton is removed from the oxygen atom of a carboxylic acid and a negatively-charged oxygen atom results, the negative charge can be shared with the adjacent oxygen atom of the carboxylate anion via the following contributing resonance structures:

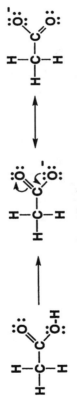

Contributing structures

In BOTH contributing structures, the negative charge is carried out on a relatively electronegative oxygen atom.

When the same type of analysis is carried out on the anion produced by deprotonation of the -CH_3 group, it can be seen that this anion is less stable because one of the important contributing structures places the negative charge on the less electronegative carbon atom.

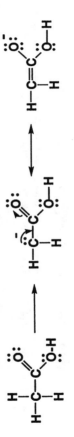

Problem 4.13 As we shall see in Chapter 19, hydrogens on a carbon adjacent to a carbonyl group are far more acidic than those not adjacent to a carbonyl group. The anion derived from acetone, for example, is more stable than is the anion derived from ethane. Account for the greater stability of the anion from acetone in terms of the resonance effect.

$$CH_3CCH_2-H \qquad\qquad CH_3CH_2-H$$

Acetone   Ethane

$pK_a = 22$ $pK_a = 51$

The anion of acetone can be stabilized by resonance with the pi bond of the adjacent carbonyl group as shown in the following contributing structures.

$$CH_3-\overset{..}{\underset{..}{O}}{:}^- \quad \longleftrightarrow \quad CH_3-C=\overset{..}{\underset{..}{O}}{:}^-$$

In this way, the negative charge is delocalized significantly, leading to a more stable anion and thus acetone is more acidic. Note how the negative charge is placed on the more electronegative oxygen atom in the structure on the right. The anion of deprotonated ethane does not have any opportunities for resonance stabilization.

Quantitative Measure of Acid and Base Strength

Problem 4.14 Which has the larger numerical value:
(a) The pK_a of a strong acid or the pK_a of a weak acid?

The weaker acid will have the pK_a with a larger numerical value.

(b) The K_a of a strong acid or the K_a of a weak acid?

The stronger acid will have the K_a with a larger numerical value.

Problem 4.15 In each pair, select the stronger acid:
(a) Pyruvic acid (pK_a 2.49) or lactic acid (pK_a 3.85)

The stronger acid is the one with the smaller pK_a and, therefore, the larger value of K_a. Pyruvic acid is the stronger acid.

(b) Citric acid (pK_a1 3.08) or phosphoric acid (pK_a1 2.10)

Phosphoric acid is the stronger acid.

Problem 4.16 Arrange the compounds in each set in order of increasing acid strength. Consult Table 4.1 for pK_a values of each acid.

(a) CH_3CH_2OH $HOCO^-$ C_6H_5COH

      Ethanol   Bicarbonate ion   Benzoic acid

pK_a: 15.9   10.33   4.19

The compounds are already in order of increasing acid strength. Ethanol is the weakest acid, benzoic acid is the strongest acid, and bicarbonate ion is in between.

(b) $HOCOH$ CH_3COH HCl

     Carbonic acid   Acetic acid   Hydrogen chloride

pK_a: 6.36   4.76   -7

Again, the compounds are already in order of increasing acid strength. Carbonic acid is the weakest acid, hydrogen chloride is the strongest acid, and acetic acid is in between.

Problem 4.17 Arrange the compounds in each set in order of increasing base strength. Consult Table 4.1 for pK_a values of the conjugate acid of each base.

The weaker the conjugate acid (higher pK_a), the stronger the base.

(a)

NH_3 $HOCO^-$ $CH_3CH_2O^-$ pK$_a$ of conjugate acid

9.24   6.36   15.9

Base strength increases in the order:

$$HOCO^- \; < \; NH_3 \; < \; CH_3CH_2O^-$$

(b)

OH^- $HOCO^-$ CH_3CO^- pK$_a$ of conjugate acid

15.7   6.36   4.76

Base strength increases in the order:

$$CH_3CO^- \; < \; HOCO^- \; < \; OH^-$$

(c)

H_2O NH_3 CH_3CO^- pK$_a$ of conjugate acid

-1.74   9.24   4.76

Base strength increases in the order:

$$H_2O \; < \; CH_3CO^- \; < \; NH_3$$

(d)

NH_2^- CH_3CO^- OH^- pK$_a$ of conjugate acid

38   4.76   15.7

Base strength increases in the order:

$$CH_3CO^- \; < \; OH^- \; < \; NH_2^-$$

Position of Equilibrium in Acid-Base Reactions

Problem 4.18 Unless under pressure, carbonic acid (Table 1.8) in aqueous solution breaks down into carbon dioxide and water, and carbon dioxide is evolved as bubbles of gas. Write an equation for the conversion of carbonic acid to carbon dioxide and water.

$$HOCOH \longrightarrow H_2O \; + \; CO_2\uparrow$$

This relationship explains why carbonated drinks evolve CO_2 gas when they are opened and the pressure is released.

Problem 4.19 Will carbon dioxide be evolved when sodium bicarbonate is added to an aqueous solution of these compounds? Explain.
(a) Sulfuric acid   (b) Ethanol   (c) Ammonium chloride

In order for carbon dioxide to be evolved, the sodium bicarbonate must be protonated to give carbonic acid (Problem 4.18). The pK$_a$ of carbonic acid is 6.36. The pK$_a$'s for sulfuric acid, ethanol and ammonium chloride are -5.2, 15.9, and 9.24, respectively. Thus, sulfuric acid is the only acid strong enough to protonate sodium bicarbonate and evolve carbon dioxide.

Problem 4.20 Acetic acid, CH_3COOH, is a weak organic acid, pK_a 4.76. Write an equation for the equilibrium reactions of acetic acid with each base. Which equilibria lie considerably toward the left? Which lie considerably toward the right?

(a) $NaHCO_3$

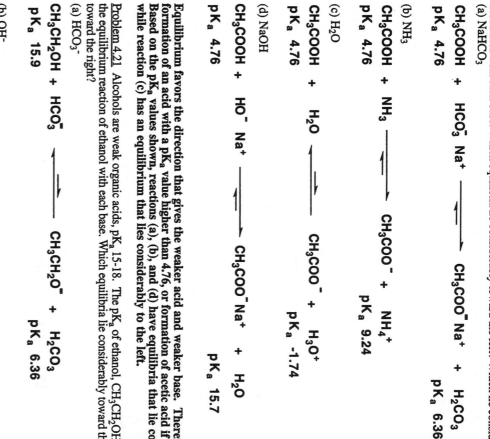

$$CH_3COOH \; + \; HCO_3^- \; Na^+ \; \rightleftharpoons \; CH_3COO^- \; Na^+ \; + \; H_2CO_3$$
$$pK_a \; 4.76 \qquad\qquad\qquad\qquad\qquad\qquad\qquad\qquad\qquad pK_a \; 6.36$$

(b) NH_3

$$CH_3COOH \; + \; NH_3 \; \rightleftharpoons \; CH_3COO^- \; + \; NH_4^+$$
$$pK_a \; 4.76 \qquad\qquad\qquad\qquad\qquad\qquad\qquad pK_a \; 9.24$$

(c) H_2O

$$CH_3COOH \; + \; H_2O \; \rightleftharpoons \; CH_3COO^- \; + \; H_3O^+$$
$$pK_a \; 4.76 \qquad\qquad\qquad\qquad\qquad\qquad\qquad pK_a \; -1.74$$

(d) $NaOH$

$$CH_3COOH \; + \; HO^- \; Na^+ \; \rightleftharpoons \; CH_3COO^- Na^+ \; + \; H_2O$$
$$pK_a \; 4.76 \qquad\qquad\qquad\qquad\qquad\qquad\qquad\qquad pK_a \; 15.7$$

Equilibrium favors the direction that gives the weaker acid and weaker base. Therefore, equilibrium will favor formation of an acid with a pK_a value higher than 4.76, or formation of acetic acid if that is the weaker acid. Based on the pK_a values shown, reactions (a), (b), and (d) have equilibrium that lie considerably to the right, while reaction (c) has an equilibrium that lies considerably to the left.

Problem 4.21 Alcohols are weak organic acids, pK_a 15-18. The pK_a of ethanol, CH_3CH_2OH, is 15.9. Write equations for the equilibrium reaction of ethanol with each base. Which equilibria lie considerably toward the left? Which lie considerably toward the right?

(a) HCO_3^-

$$CH_3CH_2OH \; + \; HCO_3^- \; \rightleftharpoons \; CH_3CH_2O^- \; + \; H_2CO_3$$
$$pK_a \; 15.9 \qquad\qquad\qquad\qquad\qquad\qquad\qquad pK_a \; 6.36$$

(b) OH^-

$$CH_3CH_2OH \; + \; OH^- \; \rightleftharpoons \; CH_3CH_2O^- \; + \; H_2O$$
$$pK_a \; 15.9 \qquad\qquad\qquad\qquad\qquad\qquad\qquad pK_a \; 15.7$$

(c) NH_2^-

$$CH_3CH_2OH \; + \; NH_2^- \; \rightleftharpoons \; CH_3CH_2O^- \; + \; NH_3$$
$$pK_a \; 15.9 \qquad\qquad\qquad\qquad\qquad\qquad\qquad pK_a \; 38$$

(d) NH_3

$$CH_3CH_2OH \; + \; NH_3 \; \rightleftharpoons \; CH_3CH_2O^- \; + \; NH_4^+$$
$$pK_a \; 15.9 \qquad\qquad\qquad\qquad\qquad\qquad\qquad pK_a \; 9.24$$

Equilibrium favors the direction that gives the weaker acid and weaker base. Therefore, equilibrium will favor formation of an acid with a pK_a value higher than 15.9, or formation of ethanol if that is the weaker acid. Based on the pK_a values shown, only reaction (c) has an equilibrium that lies considerably to the right, while reactions (a) and (d) have equilibria that lie considerably to the left. Reaction (b) has an equilibrium that lies only slightly to the left because the pK_a for ethanol is only slightly higher than that for water.

Solutions

Problem 4.22 Benzoic acid, C_6H_5COOH, is only slightly soluble in water, but its sodium salt, $C_6H_5COO^-Na^+$, is quite soluble in water. In which solutions will benzoic acid dissolve?
(a) Aqueous sodium hydroxide?　　　　　(b) Aqueous sodium bicarbonate?
(c) Aqueous sodium carbonate?

The pK_a of benzoic acid is 4.19. The pK_a values for the conjugate acids of sodium hydroxide, sodium bicarbonate ($NaHCO_3$), and sodium carbonate (Na_2CO_3) are 15.7, 6.36, and 10.33, respectively. Thus, equilibrium will favor reaction of benzoic acid with all three of these bases to give the soluble $C_6H_5COO^-Na^+$. Therefore, benzoic acid will dissolve in aqueous solutions of all three bases.

Problem 4.23 4-Methylphenol, $CH_3C_6H_4OH$ (pK_a 10.26), is only slightly soluble in water, but its sodium salt, $CH_3C_6H_4O^-Na^+$, is quite soluble in water. In which solutions will 4-methyphenol dissolve?
(a) Aqueous NaOH?　　　　　(b) Aqueous $NaHCO_3$?
(c) Aqueous Na_2CO_3?

The pK_a of 4-methylphenol is 10.26. The pK_a values for the conjugate acids of sodium hydroxide, sodium bicarbonate($NaHCO_3$), and sodium carbonate (Na_2CO_3) are 15.7, 6.36, and 10.33, respectively. Thus, equilibrium will favor reaction of 4-methylphenol with sodium hydroxide (by a large amount) and sodium carbonate (only slightly) to give the soluble $CH_3C_6H_5O^-Na^+$. 4-Methylphenol will dissolve in aqueous solutions of these two bases. Sodium bicarbonate is not a strong enough base to deprotonate 4-methylphenol, so 4-methylphenol will not dissolve in an aqueous solution of sodium bicarbonate.

Problem 4.24 For an acid-base reaction, one way to determine the predominant species at equilibrium is to say that the reaction arrow points to the acid with the higher value of pK_a. For example:

$$NH_4^+ + H_2O \longrightarrow NH_3 + H_3O^+$$
$$pK_a\ 9.24 \qquad\qquad\qquad pK_a\ -1.74$$

$$NH_4^+ + OH^- \longrightarrow NH_3 + H_2O$$
$$pK_a\ 9.24 \qquad\qquad\qquad pK_a\ 15.7$$

Explain why this rule works.

In acid-base reactions, the position of equilibrium favors reaction of the stronger acid and stronger base to give the weaker acid and weaker base. The acid with the higher pK_a is the weaker acid, so the arrow will point toward it.

Problem 4.25 Will acetylene react with sodium hydride according to the following equation to form a salt and hydrogen? Using pK_a values given in Table 4.1, calculate K_{eq} for this equilibrium.

$$HC\equiv CH\ +\ Na^+\ H^-\ \rightleftharpoons\ HC\equiv C^-\ Na^+\ +\ H_2$$

Acetylene　　　Sodium hydride　　　Sodium acetylide　　　Hydrogen
$pK_a\ 25$　　　　　　　　　　　　　　　　　　　　　　　　　$pK_a\ 35$

Since the pK_a for hydrogen is larger than the pK_a for acetylene, equilibrium will be to the right, favoring formation of the salt and hydrogen.

$$K_{eq} = \frac{K_a(\text{Acetylene})}{K_a(\text{Hydrogen})} = \frac{1 \times 10^{-25}}{1 \times 10^{-35}} = \boxed{1 \times 10^{10}}$$

Problem 4.26 Using pK_a values given in Table 4.1, predict the position of equilibrium in this acid-base reaction and calculate its K_{eq}.

$$H_3PO_4\ +\ CH_3CH_2OH\ \rightleftharpoons\ H_2PO_4^-\ +\ CH_3CH_2OH_2^+$$
$$pK_a\ 2.1 \qquad\qquad\qquad\qquad\qquad\qquad pK_a\ -2.4$$

Since the pK_a for the ethyloxonium ion ($CH_3CH_2OH_2^+$) is smaller than the pK_a for phosphoric acid (H_3PO_4), equilibrium will be to the left, favoring formation of the phosphoric acid and ethanol.

$$K_{eq} = \frac{K_a(H_3PO_4)}{K_a(CH_3CH_2OH)} = \frac{7.9 \times 10^{-3}}{2.5 \times 10^{2}} = \boxed{3.2 \times 10^{-5}}$$

Note the pKa value for the ethyloxonium ion ($CH_3CH_2OH_2^+$) is not given in Table 4.1. A good estimate would be the pKa value given for the similar species H_3O^+ (-1.71), and using this value would lead to the same conclusion that equilibrium lies to the left. Nevertheless, the actual pKa value for the ethyloxonium ion ($CH_3CH_2OH_2^+$) is -2.4 so that was used in the calculation.

Lewis Acids and Bases

Problem 4.27 For each equation, label the Lewis acid and the Lewis base. In addition, use curved arrows to show the flow of electrons in each reaction.

(a)

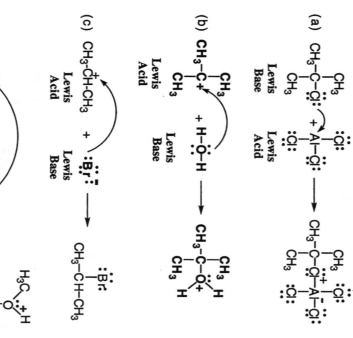

Lewis Base   Lewis Acid

(b)

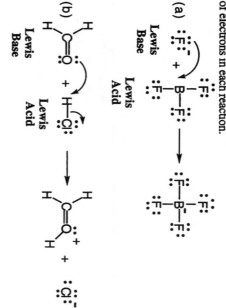

Lewis Base   Lewis Acid

Problem 4.28 Complete the equation for the reaction between each Lewis acid-base pair. In each equation, label which starting material is the Lewis acid and which the Lewis base; and use curved arrows to show the flow of electrons in each reaction. In doing this problem, it is essential that you show valence electrons for all atoms participating in each reaction.

(a)

CH_3—C—$\ddot{Cl}$: + :$\ddot{Al}$—$\ddot{Cl}$: ⟶ CH_3—C—$\ddot{Cl}$—$\overset{+}{Al}$—$\ddot{Cl}$:

Lewis Base   Lewis Acid

(b)

CH_3—C$^+$ + H—$\ddot{O}$—H ⟶ CH_3—C—$\overset{+}{O}$

Lewis Acid   Lewis Base

(c)

CH_3—$\overset{+}{CH}$—CH_3 + :$\ddot{Br}$: ⟶ :$\ddot{Br}$:

Lewis Acid   Lewis Base CH_3—CH—CH_3

(d)

CH_3—$\overset{+}{CH}$—CH_3 + CH_3—CH—$\ddot{O}$—H ⟶ CH_3—CH—$\overset{+}{O}$

Lewis Acid   Lewis Base

Problem 4.29 Each of these reactions can be written as a Lewis acid-Lewis base reaction. Label the Lewis acid and the Lewis base; use curved arrows to show the flow of electrons in each reaction. In doing this problem, it is essential that you show valence electrons for all atoms participating in each reaction.

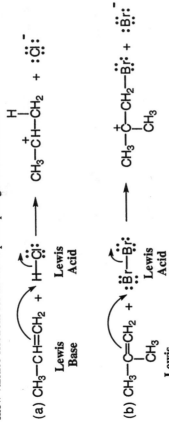

(a) $CH_3-CH=CH_2$ + $H-\ddot{\underset{..}{Cl}}:$ ⟶ $CH_3-\overset{+}{CH}-CH_2$ + $:\ddot{\underset{..}{Cl}}:^-$

          Lewis   Lewis

          Base   Acid

(b) $CH_3-\overset{+}{C}=CH_2$ + $:\ddot{\underset{..}{Br}}-\ddot{\underset{..}{Br}}:$ ⟶ $CH_3-\overset{+}{\underset{CH_3}{C}}-CH_2-\ddot{\underset{..}{Br}}:$ + $:\ddot{\underset{..}{Br}}:^-$

   CH_3   Lewis

   Lewis   Acid

   Base

Additional Problems

Problem 4.30 2,4-Pentanedione is a considerably stronger acid than propanone (acetone). Write a structural formula for the conjugate base of each acid and account for the greater stability of the conjugate base from 2,4-pentanedione.

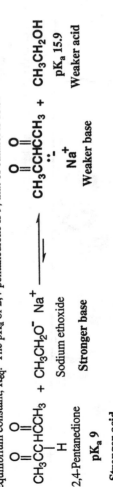

$$CH_3\overset{O}{\overset{\|}{C}}CH_2-H$$
Propanone
$pK_a = 22$

$$CH_3\overset{O}{\overset{\|}{C}}\underset{H}{\overset{O}{\overset{\|}{C}}HCCH_3}$$
2,4-Pentanedione
$pK_a = 9$

A carbonyl group adjacent to a negatively charged carbon atom, an arrangement that is referred to as an enolate ion, can lead to resonance stabilization by virtue of a resonance contributor that places the negative charge on oxygen along with formation of a carbon-carbon double bond. Thus, there are two important contributing structures that describe the resonance stabilization of the conjugate base of propanone.

$$\underset{\underset{CH_3\overset{..}{C}-\overset{..}{C}H_2}{}}{:\overset{..}{\overset{\|}{O}}:^-} \longleftrightarrow \underset{CH_3C=CH_2}{:\overset{..}{\overset{..}{O}}:^-}$$

However, there are two adjacent carbonyl groups and thus three important contributing structures that describe the resonance stabilization of the conjugate base of 2,4-pentanedione.

$$\underset{CH_3C-CH-CCH_3}{:\overset{..}{O}:\quad:\overset{..}{O}:} \longleftrightarrow \underset{CH_3C-CH=CCH_3}{:\overset{..}{O}:^-\;:\overset{..}{O}:} \longleftrightarrow \underset{CH_3C=CH-CCH_3}{:\overset{..}{O}:\quad:\overset{..}{O}:^-}$$

The added contributing structure provides additional resonance stabilization for the conjugate base of 2,4-pentanedione compared to propanone, making 2,4-pentanedione the stronger acid.

Problem 4.31 Write an equation for the acid-base reaction between 2,4-pentanedione and sodium ethoxide and calculate its equilibrium constant, K_{eq}. The pK_a of 2,4-pentanedione is 9; that of ethanol is 15.9.

$$\underset{\underset{\underset{Stronger\ acid}{pK_a\ 9}}{2,4-Pentanedione}}{CH_3\overset{O}{\overset{\|}{C}}C\underset{H}{\overset{O}{\overset{\|}{C}}HCCH_3}} + \underset{\underset{Stronger\ base}{Sodium\ ethoxide}}{CH_3CH_2O^-\ Na^+} \rightleftharpoons \underset{\underset{Weaker\ base}{Na^+}}{CH_3\overset{O}{\overset{\|}{C}}\underset{..}{\overset{..}{C}}CHCCH_3} + \underset{\underset{Weaker\ acid}{pK_a\ 15.9}}{CH_3CH_2OH}$$

Since the pK_a for 2,4-pentanedione is smaller than the pK_a for ethanol, equilibrium will be to the right, favoring formation of the 2,4-pentanedione salt and ethanol.

Problem 4.32 The *sec*-butyl carbocation can react as both a Lewis acid and a Brønsted-Lowry acid in the presence of a water-sulfuric acid mixture. In each case, however the product is different. The two reactions are:

$$K_{eq} = \frac{K_a(2,4\text{-Pentandione})}{K_a(\text{Ethanol})} = \frac{1.0 \times 10^{-9}}{1.3 \times 10^{-16}} = \boxed{7.9 \times 10^6}$$

(1)
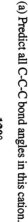

CH₃-CH-CH₂-CH₃ + H₂O ⟶ CH₃-CH-CH₂-CH₃

sec-Butyl cation

(2)

CH₃-CH-CH₂-CH₃ + H₂O ⟶ CH₃-CH=CH-CH₃ + H₃O⁺

sec-Butyl cation

(a) In which reaction(s) does this cation react as a Lewis acid? In which does it react as a Brønsted-Lowry acid?

As shown by the curved arrows that indicate the flow of electrons, in reaction (1), the *sec*-butyl cation is acting as a **Lewis acid** by reacting with the **Lewis base** (water). In reaction (2), the *sec*-butyl cation is acting as a **Brønsted-Lowry acid** by donating a proton to the water molecule.

(b) Write Lewis structures for reactants and products and show by the use of curved arrows how each reaction occurs.

(1)   CH₃-CH-CH₂-CH₃ ⟶ CH₃-CH-CH₂-CH₃

(2)   CH₃-CH-CH-CH₃ ⟶ CH₃-CH=CH-CH₃ + H-O⁺-H

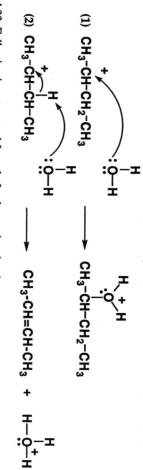

Problem 4.33 Following is a structural formula for the *tert*-butyl cation.

H₃C-C⁺-CH₃
   |
   CH₃

tert-Butyl cation
(a carbocation)

(a) Predict all C-C-C bond angles in this cation.

(b) What is the hybridization of the carbon bearing the positive charge?

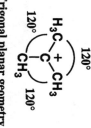

120°

H₃C-C⁺-CH₃
120° | 120°
     CH₃

Trigonal planar geometry

According to the VSEPR model, the geometry of the central carbon atom will be trigonal planar because there are three regions of electron density.

sp²
hybridization

H₃C-C⁺-CH₃
   |
   CH₃

sp²
hybridization

The positively-charged carbon atom is sp² hybridized. The three sp² hybrid orbitals take part in sigma bonding to the three carbon atoms, and the unhybridized 2p orbital is empty.

(c) Write a balanced equation to show its reaction as a Lewis acid with water; to show its reaction as a Brønsted-Lowry acid with water.

As a Lewis acid

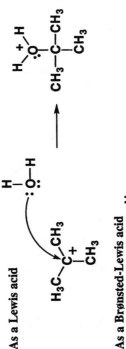

As a Brønsted-Lewis acid

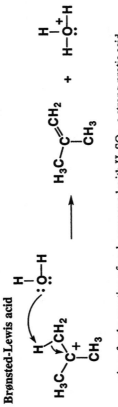

Problem 4.34 Write equations for the reaction of each compound with H_2SO_4, a strong protic acid.

(a) CH_3OCH_3

(b) $CH_3CH_2SCH_2CH_3$

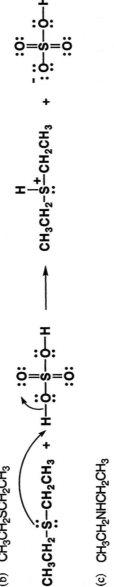

(c) $CH_3CH_2NHCH_2CH_3$

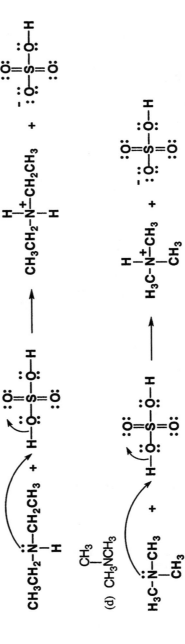

(d) CH_3NCH_3 (CH_3)

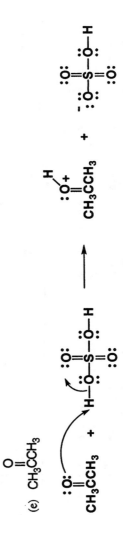

(e) CH_3CCH_3 (O)

Solutions

(f)

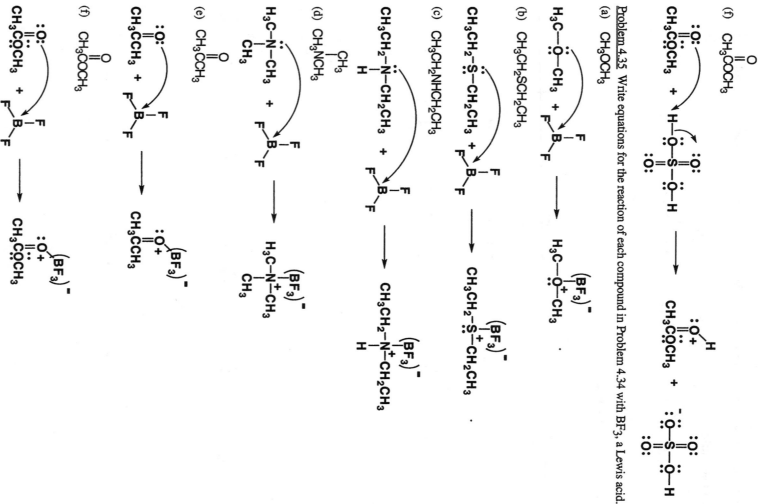

Problem 4.35 Write equations for the reaction of each compound in Problem 4.34 with BF₃, a Lewis acid.

(a) CH₃OCH₃

(b) CH₃CH₂SCH₂CH₃

(c) CH₃CH₂NHCH₂CH₃

(d) CH₃NCH₃ with CH₃

(e) CH₃CCH₃

(f) CH₃COCH₃

Problem 4.36 Write a structural formula for the conjugate base formed when each compound is treated with one mole of a base stronger than the compound's conjugate base.

This problem is difficult because the molecules contain more than one functional group that can be deprotonated. In each case, use the entries in Table 4.1 to determine which functional group is the most acidic, then draw the conjugate base that is formed from deprotonation of that most acidic functional group.

(a) $HOCH_2CH_2NH_2$

(b) $HSCH_2CH_2NH_2$

(c) $HOCH_2CH_2C{\equiv}CH$

$^-{:}\ddot{O}CH_2CH_2NH_2$

$^-{:}\ddot{S}CH_2CH_2NH_2$

$^-{:}\ddot{O}CH_2CH_2C{\equiv}CH$

(d) $HOCCH_2CH_2SH$ (with $O{=}$)

(e) CH_3CHCOH (HO, $O{=}$)

(f) $\overset{+}{H_3}NCH_2CH_2COH$ (with $O{=}$)

$^-{:}\ddot{O}CCH_2CH_2SH$ (with $O{=}$)

$CH_3CHCO^-{:}$ (HO, $O{=}$)

$\overset{+}{H_3}NCH_2CH_2CO^-{:}$ (with $O{=}$)

(g) $H_2NCH_2CH_2COH$ (with $O{=}$)

(h) $HSCH_2CH_2OH$

$H_2NCH_2CH_2CO^-{:}$ (with $O{=}$)

$^-{:}\ddot{S}CH_2CH_2OH$

Problem 4.37 Explain why the hydronium ion, H_3O^+, is the strongest acid that can exist in aqueous solution. What is the strongest base that can exist in aqueous solution?

When a strong acid (HA) is placed in aqueous solution, the following equilibrium is established:

$$HA \ + \ H_2O \ \rightleftharpoons \ A^- \ + \ H_3O^+$$

Since equilibrium favors formation of the weaker acid and weaker base, if HA is a stronger acid than H_3O^+, equilibrium will favor the right side of the equation, forming H_3O^+. As a result, the strongest acid than can exist in aqueous solution is H_3O^+.
In an analogous way, when a strong base (Base) is added to water, the following equilibrium is established:

$$Base \ + \ H_2O \ \rightleftharpoons \ Base\overset{+}{-}H \ + \ HO^-$$

Equilibrium favors formation of the weaker acid and weaker base. Any base stronger than HO⁻ will drive the equilibrium toward formation of HO⁻. As a result, the strongest base that can exist in aqueous solution is HO⁻.

Problem 4.38 What is the strongest base that can exist in liquid ammonia as a solvent?

When a strong base (Base) is added to liquid ammonia, the following equilibrium is established:

$$Base \ + \ H_3N \ \rightleftharpoons \ Base\overset{+}{-}H \ + \ H_2N^-$$

Equilibrium favors formation of the weaker acid and weaker base. Any base stronger than H_2N^- will drive the equilibrium toward formation of H_2N^-. As a result, the strongest base that can exist in liquid ammonia is H_2N^-.

Problem 4.39 For each pair of molecules or ions, select the stronger base and write its Lewis structure.

The stronger base will have the weaker conjugate acid. For each pair, choose the base having a conjugate acid with the higher pKa value in Table 4.1.

(a) CH_3S^- or $\boxed{CH_3O^-}$

(b) $\boxed{CH_3NH^-}$ or CH_3O^-

(c) CH_3COO^- or $\boxed{OH^-}$

(d) $CH_3CH_2O^-$ or $\boxed{H^-}$

(e) NH_3 or $\boxed{OH^-}$

(f) $\boxed{NH_3}$ or H_2O

(g) CH₃COO⁻ or HCO₃⁻

$$H-\overset{\cdot\cdot}{\underset{\cdot\cdot}{O}}-\overset{\overset{\cdot\cdot}{O}}{\underset{}{C}}=\overset{\cdot\cdot}{\underset{\cdot\cdot}{O}}:^-$$

(h) HSO₄⁻ or OH⁻

$$H-\overset{\cdot\cdot}{\underset{\cdot\cdot}{O}}:^-$$

(i) OH⁻ or Br⁻

$$H-\overset{\cdot\cdot}{\underset{\cdot\cdot}{O}}:^-$$

Problem 4.40 Account for the fact that nitroacetic acid, O₂NCH₂COOH (pK_a 1.68) is a considerably stronger acid than acetic acid, CH₃COOH (pK_a 4.76).

In general, the more stable the conjugate base, the stronger the acid. The nitro group is much more electronegative than a hydrogen atom, so the nitro group stabilizes the negative charge of the carboxylate anion through the inductive effect analogous to halogen atoms as described in the text.

This partial positive charge helps to stabilize the negative charge of the carboxylate through the inductive effect.

Problem 4.41 Sodium hydride, NaH, is available commercially as a gray-white powder. It melts at 800°C with decomposition. It reacts explosively with water, and ignites spontaneously on standing in moist air.
(a) Write a Lewis structure for the hydride ion and for sodium hydride. Is your Lewis structure consistent with the fact that this compound is a high-melting solid? Explain.

(b) When sodium hydride is added very slowly to water, it dissolves with the evolution of a gas. The resulting solution is basic to litmus. What is the gas evolved? Why has the solution become basic?

When NaH is placed in water, the following reaction takes place:

$$H:^- \ + \ H-\overset{\cdot\cdot}{\underset{\cdot\cdot}{O}}-H \ \rightleftharpoons \ H-H\,(gas) \ + \ H-\overset{\cdot\cdot}{\underset{\cdot\cdot}{O}}:^-$$

The Lewis structure of sodium hydride indicates an ionic bond between the hydride anion and sodium cation. Ionic compounds such as NaCl are generally high-melting solids.

$$H:^- \qquad H:^- \ Na^+$$

The hydride anion   Sodium hydride

Since the hydride ion is a stronger base than HO⁻, reaction goes to the right. Formation of H₂ gas, which bubbles from the reaction, also drives the reaction to the right. Formation of HO⁻ drives up the pH of the solution.

(c) Write an equation for the reaction between sodium hydride and 1-butyne, CH₃CH₂C≡CH. Use curved arrows to show the flow of electrons in this reaction.

$$H:^- \ + \ H-C\equiv CCH_2CH_3 \ \longrightarrow \ H-H\,(gas) \ + \ {}^-:C\equiv CCH_2CH_3$$

Problem 4.42 An ester is a derivative of a carboxylic acid in which the hydrogen of the carboxyl group is replaced by an alkyl group (Section 1.3E). Draw the structural formula of methyl acetate, which is derived from acetic acid by replacement of the H of its -OH group by a methyl group. Determine if proton transfer to this compound from HCl occurs preferentially on the oxygen of the C=O group or the oxygen of the OCH₃ group.

Remember that protonation of a neutral species occurs most readily in the location where the positive charge can be stabilized most effectively. Protonation of an ester occurs preferentially on the oxygen atom of the C=O group. Only protonation at this site leads to a product that is resonance stabilized as indicated by the two contributing structures shown. In later chapters you will learn that the C=O group is especially polarized, placing significant partial negative charge on the oxygen atom, a factor that also contributes to basicity.

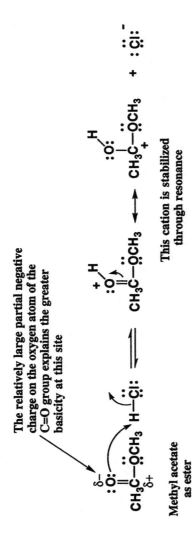

The relatively large partial negative charge on the oxygen atom of the C=O group explains the greater basicity at this site

This cation is stabilized through resonance

Methyl acetate as ester

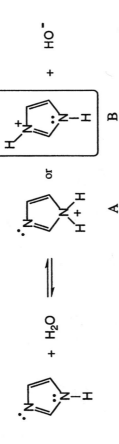

Problem 4.43 Following is a structural formula for imidazole, a building block of the essential amino histidine (Chapter 27). It is also a building block of histamine (The Merck Index, 12th ed., #4756), a compound all too familiar to persons with allergies and takers of antihistamines. When imidazole is dissolved in water, proton transfer to it gives a cation. Is this cation better represented by structure A or B?

A   B

Protonation of a neutral species occurs most readily in the location where the positive charge can be stabilized most effectively. The cation is better represented by B. Only cation B can be stabilized by resonance as indicated by the two contributing structures shown here.

Problem 4.44 Methyl isocyanate, CH₃-N=C=O, is used in the industrial synthesis of a type of pesticide and herbicide known as a carbamate. As a historical note, an industrial accident in Bhopal, India in 1984 resulted in leakage of an unknown quantity of this chemical into the air. An estimated 200,000 persons were exposed to its vapors and over 2000 of these died.

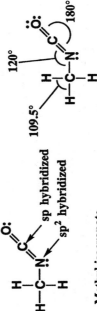

Methyl isocyanate

(a) Write a Lewis structure for methyl isocyanate and predict its bond angles. What is the hybridization of its carbonyl carbon? Of its nitrogen atom?

sp hybridized
sp² hybridized

Methyl isocyanate

109.5°   120°   180°

(b) Methyl isocyanate reacts with strong acids, such as sulfuric acid, to form a cation. Will this molecule undergo protonation more readily on its oxygen or nitrogen atom? In considering contributing structures to each hybrid, do not consider structures in which more than one atom has an incomplete octet.

Remember that protonation of a neutral species occurs most readily in the location where the positive charge can be stabilized most effectively, i.e. by resonance. As can be seen below, protonation at either the oxygen atom or the nitrogen atom will lead to cations that have three major contributing structures. The middle contributing structures in both cases contain a carbon atom with an unfilled octet, so these middle structures make a relatively small contribution to their respective resonance hybrids. Oxygen is more electronegative than nitrogen, so oxygen is less able to accomodate a full positive charge compared to nitrogen. As a result, the contributing structure on the right also makes a relatively small contribution to the overall resonance hybrid in the case of nitrogen protonation. This means that for the cation produced upon protonation of nitrogen, there is relatively little in the way of overall resonance stabilization.

Protonation on nitrogen - less resonance stabilization

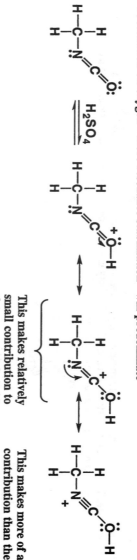

These make relatively small contributions to the resonance hybrid

Protonation on oxygen - more resonance stabilization - will predominate

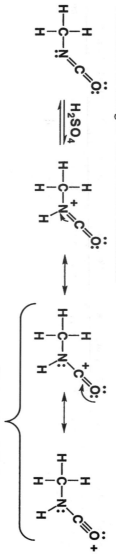

This makes relatively small contribution to the resonance hybrid

This makes more of a contribution than the analogous structure shown above for nitrogen protonation

On the other hand, the analogous contributing structure shown on the right for oxygen protonation makes more of a contribution to the resonance hybrid of that species. Greater contribution from a contributing structure leads to greater overall resonance stabilization. Therefore, protonation on oxygen will predominate for methyl isocyanate.

Problem 4.45 Offer an explanation for the following observations.

(a) H_3O^+ is a stronger acid than NH_4^+.

Oxygen is more electronegative than nitrogen, so the proton is more acidic on H_3O^+.

(b) Nitric acid, HNO_3, is a stronger acid than nitrous acid, HNO_2.

There is one more electronegative oxygen atom on nitric acid to help stabilize the deprotonated anion through a resonance effect.

(c) Ethanol and water have approximately the same acidity.

Both have the proton attached to sp^3 hybridized oxygen atoms, and neither deprotonated species can be resonance stabilized.

(d) Trifluoroacetic acid is a stronger acid than trichloroacetic acid.

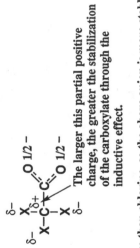

$\delta-$
$\delta- \quad X$
$\quad \quad | \quad \delta+$
$X-C-C$
$\quad \quad | \quad \quad \| \quad O \ 1/2-$
$\quad \quad X \quad C$
$\quad \quad \delta- \quad \| \quad O \ 1/2-$

The larger this partial positive charge, the greater the stabilization of the carboxylate through the inductive effect.

Fluorine is more electronegative than chlorine, so the above anion is more stable when X = F compared to when X = Cl due to increased inductive effects.

Gray reef shark Carcharhinus ambylrhynchos
Rangiroa, French Polynesia

CHAPTER 5
Solutions to the Problems

Problem 5.1 Write the IUPAC name of each alkane.

(a)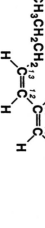

(b)

3-Methyl-1-butene

Problem 5.2 Which alkenes show cis, trans isomerism? For each alkene that does, draw the trans isomer.
(a) 2-Pentene
(b) 2-Methyl-2-pentene
(c) 3-Methyl-2-pentene

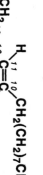

trans-2-Pentene

No cis, trans isomers since there are two methyl groups on one end of the double bond.

2,3-Dimethyl-2-butene

trans-3-Methyl-2-pentene

Problem 5.3 Name each alkene and specify its configuration by the E-Z system.

(a)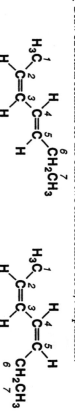

(E)-1-Chloro-2,3-dimethyl-2-pentene

(b)

(Z)-1-Bromo-1-chloropropene

(c)

(E)-2,3,4-Trimethyl-3-heptene

Problem 5.4 Write the IUPAC name of each cycloalkene

(a)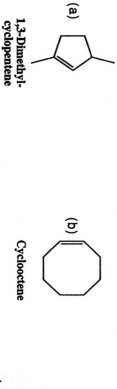

1,3-Dimethyl-cyclopentene

(b)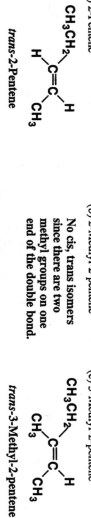

Cyclooctene

(c)

4-(1,1-Dimethylethyl)-cyclohexene (IUPAC)
4-tert-Butylcyclohexene (Common)

Problem 5.5 Draw structural formulas for the other two stereoisomers for 2,4-heptadiene.

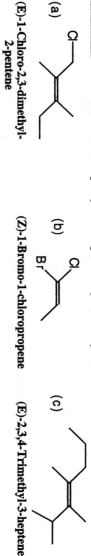

cis,trans-2,4-Heptadiene

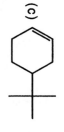

cis,cis-2,4-Heptadiene

Problem 5.6 (10E, 12Z)-10,12-hexadecadiene-1-ol is a sex pheromone of the silkworm. Draw a structural formula for this compound.

CH₃CH₂CH₂ $\overset{13}{\underset{}{\text{C}}}$ $\overset{12}{\underset{}{\text{C}}}$ CH₂(CH₂)₇CH₂OH

(10E,12Z)-10,12-Hexadecadiene-1-ol

Structure of Alkenes

Problem 5.7 Predict all bond angles about each highlighted carbon atom. To make these predictions, use the valence shell electron-pair repulsion model (Section 1.4).

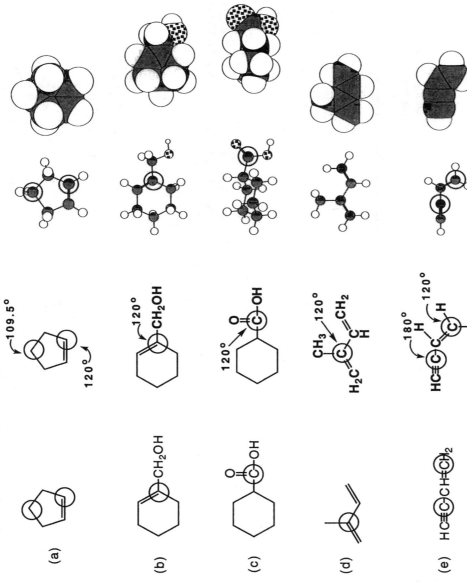

Problem 5.8 For each highlighted carbon atom in Problem 5.7, identify which atomic orbitals are used to form each sigma bond and which are used to form each pi bond.

Each bond is labeled sigma or pi and the orbitals overlapping to form each bond are shown.

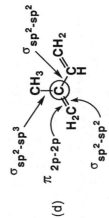

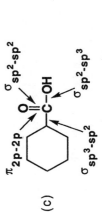

Problem 5.9 Following is the structural formula of propadiene (allene)

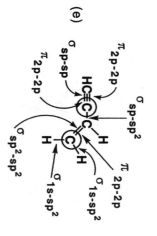

$CH_2=C=CH_2$

Propadiene
(Allene)

(a) State the orbital hybridization of each carbon atom.
(b) Describe each carbon-carbon double bond in terms of the overlap of atomic orbitals.
(c) Predict all bond angles in allene.

sp²

$CH_2=C=CH_2$

sp

sp²

π 2p-2p

σ sp²-sp

180°

120°

(d) Draw a stereorepresentation showing the shape of this molecule.

The central carbon atom of allene is sp hybridized with bond angles of 180° about it. The terminal carbons are sp² hybridized with bond angles of 120° about each. The planes created by H-C-H bonds at the ends of the molecule are perpendicular to each other.

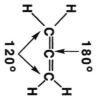

Problem 5.10 Following are lengths for a series of C-C single bonds. Propose an explanation for the differences in bond lengths.

Structure	Length of C-C single bond (pm)
CH_3-CH_3	153.7
$CH_2=CH-CH_3$	151.0
$CH_2=CH-CH=CH_2$	146.5
$HC\equiv C-CH_3$	145.9

The s electrons are on average held closer to atomic nuclei than p electrons. Thus, hybrid orbitals with higher s character have the electrons held closer to the nucleus and thus make bonds that are shorter. As shown in the table, a σ bond formed from overlap of an sp³ orbital with an sp² orbital is shorter than a σ bond formed from overlap of an sp³ orbital with another sp³ orbital. Similarly, sp³-sp overlap produces a bond that is shorter than that produced by sp³-sp² overlap.

(e)

π 2p-2p

σ sp-sp²

σ sp-sp

π 2p-2p

HC≡C-C

σ sp²-sp²

H

H

C

H

π 2p-2p

σ 1s-sp²

σ 1s-sp²

Nomenclature of Alkenes

Problem 5.11 Draw structural formulas for these alkenes.

(a) *trans*-2-Methyl-3-hexene

(b) 2-Methyl-2-hexene

(c) 2-Methyl-1-butene

(d) 3-Ethyl-3-methyl-1-pentene

(e) 2,3-Dimethyl-2-butene

(f) *cis*-2-Pentene

(g) (Z)-1-Chloropropene

(h) 3-Methylcyclohexene

(i) 1-Isopropyl-4-methylcyclohexene

(j) (E)-2,6-Dimethyl-2,6-octadiene

(k) 3-Cyclopropyl-1-propene

(l) Cyclopropylethylene

(m) 2-Chloropropene

(n) Tetrachloroethylene

(o) 1-Chlorocyclohexene

(p) Bicyclo[2.2.1]-2-heptene

or

(q) Bicyclo[4.4.0]-1-decene

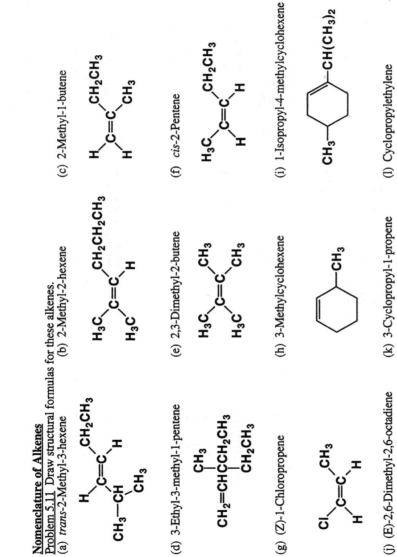
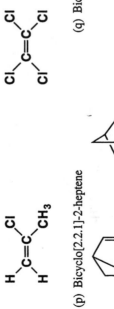
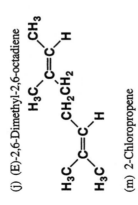

Problem 5.12 Name these alkenes and cycloalkenes.

(a)

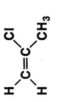

(E)-3-(2-methylpropyl)-2-Octene

(b)

4-Chloro-1,4-dimethylcyclopentene

(c)

Ethenylcyclohexane
Vinylcyclohexane

(d)

2,4-Dimethyl-2-pentene

(e)

(E)-1,4-Dichloro-2-butene
trans-1,4-Dichloro-2-butene

(f)

Tetrafluoroethene

(g)

5-Chloro-5-ethyl-
1,3-cyclopentadiene

(h)

1,4-Cyclohexadiene

(i)

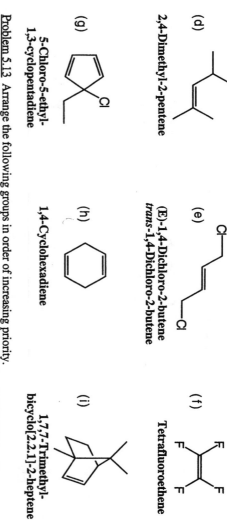

1,7,7-Trimethyl-
bicyclo[2.2.1]-2-heptene

Problem 5.13 Arrange the following groups in order of increasing priority.
(a) -CH₃ -H -Br -CH₂CH₃
-H < -CH₃ < -CH₂CH₃ < -Br

(b) -OCH₃ -CH(CH₃)₂ -B(CH₂CH₃)₂ -H
-H < B(CH₂CH₃)₂ < -CH(CH₃)₂ < -OCH₃

(c) -CH₃ -CH₂OH -CH₂NH₂ -CH₂Br
-CH₃ < -CH₂NH₂ < -CH₂OH < -CH₂Br

Problem 5.14 Assign an E or Z configuration and a cis or trans configuration to these dicarboxylic acids, each of which is
an intermediate in the tricarboxylic acid cycle. Following each is given its common name.

(a)

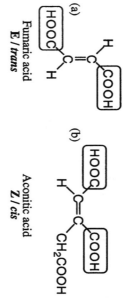

Fumaric acid
E / *trans*

(b)

Aconitic acid
Z / *cis*

The highest priority group on each sp² carbon atom is circled.

Problem 5.15 Name and draw structural formulas for all alkenes of molecular formula C₅H₁₀. As you draw these alkenes,
remember that cis and trans isomers are different compounds and must be counted separately.

Four alkenes of molecular formula C₅H₁₀ do not show cis, trans isomerism.

CH₂=CHCH₂CH₂CH₃

1-Pentene

CH₂=CCH₂CH₃
   |
  CH₃

2-Methyl-1-butene

CH₃C=CHCH₃
  |
 CH₃

2-Methyl-2-butene

CH₂=CHCHCH₃
   |
  CH₃

3-Methyl-1-butene

One alkene of molecular formula C₅H₁₀ shows cis, trans isomerism.

CH₃   H
  \\   /
   C=C
  /   \\
H   CH₂CH₃

trans-2-Pentene

CH₃   CH₂CH₃
  \\   /
   C=C
  /   \\
H   H

cis-2-Pentene

Problem 5.16 For each molecule that shows cis, trans isomerism, draw the cis isomer.

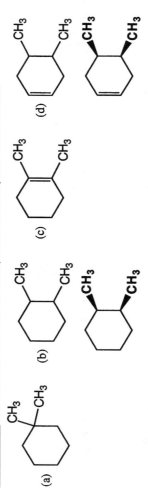

(a)

(b)

(c)

(d)

Problem 5.17 Draw structural formulas for all compounds of molecular formula C_5H_{10} that are:

(a) Alkenes that do not show cis, trans isomerism.

Four alkenes of molecular formula C_5H_{10} do not show cis, trans isomerism.

$CH_2=CHCH_2CH_2CH_3$

1-Pentene

$CH_2=CCH_2CH_3$ with CH_3

2-Methyl-1-butene

$CH_2=CHCHCH_3$ with CH_3

3-Methyl-1-butene

$CH_3C=CHCH_3$ with CH_3

2-Methyl-2-butene

(b) Alkenes that do show cis, trans isomerism.

One alkene of molecular formula C_5H_{10} shows cis, trans isomerism.

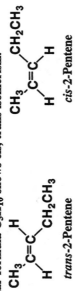

trans-2-Pentene

cis-2-Pentene

(c) Cycloalkanes that do not show cis, trans isomerism.

Four cycloalkanes of molecular formula C_5H_{10} do not show cis, trans isomerism.

Ethylcyclopropane   1,1-Dimethylcyclopropane   Methylcyclobutane   Cyclopentane

(d) Cycloalkanes that do show cis, trans isomerism.

Only one cycloalkane of molecular formula C_5H_{10} shows cis, trans isomerism.

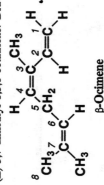

cis-1,2-Dimethyl-cyclopropane

trans-1,2-Dimethyl-cyclopropane

Problem 5.18 β-Ocimene (The Merck Index, 12th ed., #6837), a triene found in the fragrance of cotton blossoms and several other essential oils, has the IUPAC name (Z)-3,7-dimethyl-1,3,6-octatriene. Draw a structural formula for β-ocimene.

β-Ocimene

(Z)-3,7-Dimethyl-1,3,6-octatriene

Problem 5.19 Draw the structural formula for at least one bromoalkene of molecular formula C_5H_9Br that shows:

(a) Neither E,Z isomerism nor chirality.

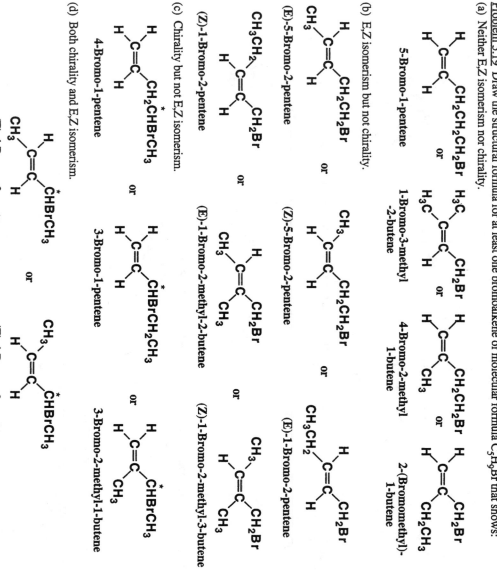

5-Bromo-1-pentene or 1-Bromo-3-methyl-2-butene or 4-Bromo-2-methyl-1-butene or 2-(Bromomethyl)-1-butene

(b) E,Z isomerism but not chirality.

(E)-5-Bromo-2-pentene or (Z)-5-Bromo-2-pentene or (E)-1-Bromo-2-pentene

(c) Chirality but not E,Z isomerism.

(Z)-1-Bromo-2-pentene or (E)-1-Bromo-2-methyl-2-butene or (Z)-1-Bromo-2-methyl-3-butene

4-Bromo-1-pentene or 3-Bromo-1-pentene or 3-Bromo-2-methyl-1-butene

(d) Both chirality and E,Z isomerism.

3-Bromo-1-pentene or (Z)-4-Bromo-2-pentene

Problem 5.20 Following are structural formulas and common names for four molecules that contain both a carbon-carbon double bond and another functional group. Give each an IUPAC name.

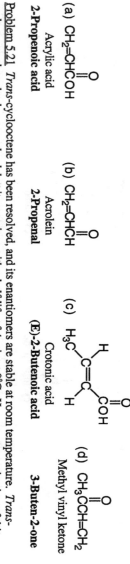

(a) $CH_2=CHCOH$
Acrylic acid
2-Propenoic acid

(b) $CH_2=CHCH$
Acrolein
2-Propenal

(c) Crotonic acid
(E)-2-Butenoic acid

(d) $CH_3CCH=CH_2$
Methyl vinyl ketone
3-Buten-2-one

Problem 5.21 *Trans*-cyclooctene has been resolved, and its enantiomers are stable at room temperature. *Trans*-cyclononene has also been resolved, but it racemizes with a half-life of 4 min at 0°C. How can racemization of this cycloalkene take place without breaking any bonds? Why does *trans*-cyclononene racemize under these conditions but *trans*-cyclooctene does not? You will find it especially helpful to build molecular models of these cycloalkenes.

Following are enantiomers of *trans*-cyclooctene and *trans*-cyclononene. It may be helpful to construct molecular models and prove to yourself that the two different configurations of the ring are in fact non-superposable mirror images of each other. For both *trans*-cyclooctene and *trans*-cyclononene, the enantiomers are configurational isomers. The enantiomers are interconverted by a change in configuration that is analogous to the chair flipping of chair cyclohexane. The *trans* double bond adds a considerable degree of rigidity to the ring, since the other atoms must exist in a slightly "stretched" configuration to accommodate the *trans* geometry. Nevertheless, the *trans*-cyclononene ring has more carbon atoms, so it is more flexible and can undergo the configurational interconversion more readily.

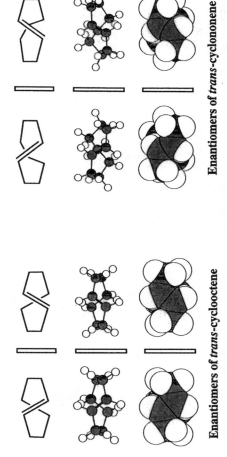

Enantiomers of *trans*-cyclooctene

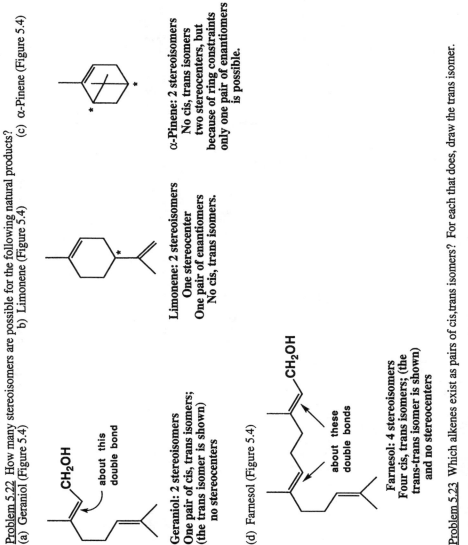

Enantiomers of *trans*-cyclononene

Problem 5.22 How many stereoisomers are possible for the following natural products?

(a) Geraniol (Figure 5.4)

CH₂OH

about this double bond

Geraniol: 2 stereoisomers
One pair of cis, trans isomers;
(the trans isomer is shown)
no stereocenters

(b) Limonene (Figure 5.4)

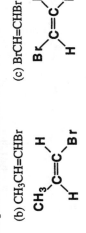

Limonene: 2 stereoisomers
One stereocenter
One pair of enantiomers
No cis, trans isomers.

(c) α-Pinene (Figure 5.4)

α-Pinene: 2 stereoisomers
No cis, trans isomers,
two stereocenters, but
because of ring constraints
only one pair of enantiomers
is possible.

(d) Farnesol (Figure 5.4)

CH₂OH

about these double bonds

Farnesol: 4 stereoisomers
Four cis, trans isomers; (the
trans-trans isomer is shown)
and no stereocenters

Problem 5.23 Which alkenes exist as pairs of cis,trans isomers? For each that does, draw the trans isomer.

For an alkene to exist as a pair of cis, trans isomers, both carbon atoms of the double bond must have two different substituents. Thus, (b), (c), and (e) exist as a pair of cis, trans isomers. The trans isomer for each alkene is drawn under its respective condensed molecular formula.

(a) $CH_2=CHBr$

(b) $CH_3CH=CHBr$

CH₃   H
    C=C
H   Br

(c) $BrCH=CHBr$

Br   H
    C=C
H   Br

(d) $(CH_3)_2C=CHCH_3$

Top left (rotated): "90" page number and "Solutions" and "Chapter 5: Alkenes 1" on right side.

(e) (CH3)2CHCH=CHCH3 with structure

Problem 5.24 Four stereoisomers exist for 3-penten-2-ol.

Structure diagram with "about this double bond" and CH3—CH=CH—CH—CH3 with OH, 3-Penten-2-ol

(a) Explain how these four stereoisomers arise.

There is one double bond that provides for cis, trans isomers, and one stereocenter in 3-penten-2-ol.

(b) Draw the stereoisomer having the E configuration about the carbon-carbon double bond and the R configuration at the stereocenter.

image with HO structure

Terpenes
Problem 5.25 Show that the structural formula of vitamin A (Section 5.2G) can be divided into four isoprene units bonded head-to-tail linkages and cross-linked at one point to form the six-membered ring.

Isoprene chain cross-linked here
image with CH2OH

Problem 5.26 Following is the structural formula of lycopene C40H56 (The Merck Index, 12th ed., #5650), a deep-red compound that is partially responsible for the red color of ripe fruits, especially tomatoes. Approximately 20 mg of lycopene can be isolated from 1 kg of ripe tomatoes. Lycopene is an important antioxidant that may help prevent oxidative damage in atherosclerosis.

Lycopene structure

(a) Show that lycopene is a terpene, that is, its carbon skeleton can be divided into two sets of four isoprene units with the units in each set joined head-to-tail.

Head-to-head bond joining two four isoprene units

Lycopene

(e) (CH₃)₂CHCH=CHCH₃

$(CH_3)_2CH$

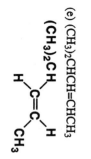

Problem 5.24 Four stereoisomers exist for 3-penten-2-ol.

about this double bond

$CH_3—CH{=}CH—\overset{*}{CH}—CH_3$ with OH

3-Penten-2-ol

(a) Explain how these four stereoisomers arise.

There is one double bond that provides for cis, trans isomers, and one stereocenter in 3-penten-2-ol.

(b) Draw the stereoisomer having the E configuration about the carbon-carbon double bond and the R configuration at the stereocenter.

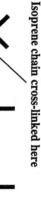

Terpenes
Problem 5.25 Show that the structural formula of vitamin A (Section 5.2G) can be divided into four isoprene units bonded head-to-tail linkages and cross-linked at one point to form the six-membered ring.

Isoprene chain cross-linked here

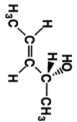

CH₂OH

Problem 5.26 Following is the structural formula of lycopene $C_{40}H_{56}$ (The Merck Index, 12th ed., #5650), a deep-red compound that is partially responsible for the red color of ripe fruits, especially tomatoes. Approximately 20 mg of lycopene can be isolated from 1 kg of ripe tomatoes. Lycopene is an important antioxidant that may help prevent oxidative damage in atherosclerosis.

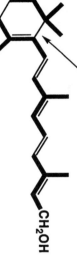

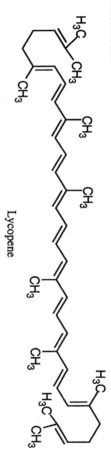

Lycopene

(a) Show that lycopene is a terpene, that is, its carbon skeleton can be divided into two sets of four isoprene units with the units in each set joined head-to-tail.

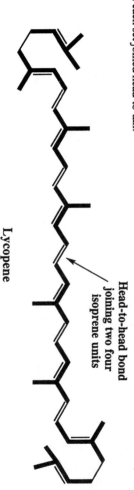

Head-to-head bond joining two four isoprene units

Lycopene

(b) How many of the carbon-carbon double bonds in lycopene have the possibility for cis,trans isomerism? Of these, which are trans and which are cis?

The double bonds on the two ends of the molecule cannot show cis, trans isomerism. The other 11 double bonds can show cis,trans isomerism, but they are all trans.

Problem 5.27 As you might suspect, β-carotene, C₄₀H₅₆, precursor to vitamin A, was first isolated from carrots. Dilute solutions of β-carotene are yellow, hence its use as a food coloring. In plants, it is almost always present in combination with chlorophyll to assist in the harvesting of the energy of sunlight. As tree leaves die in the fall, the green of their chlorophyll molecules is replaced by the yellow and reds of carotene and carotene-related molecules. Compare the carbon skeletons of β-carotene and lycopene. What are the similarities? What are the differences?

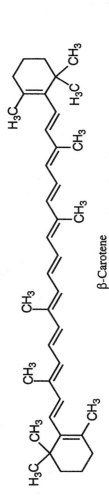

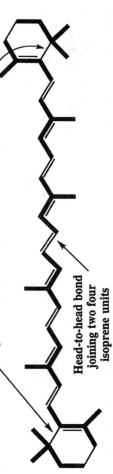

β-Carotene

The main structural difference between β-carotene and lycopene is that β-carotene has six-membered rings on the ends, not an open chain. On the other hand, both β-carotene and lycopene can be divided into two sets of four isoprene units as shown below, and all of the double bonds are E in both molecules.

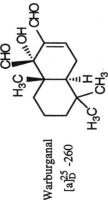

Isoprene chain cross-linked at these two points

Head-to-head bond joining two four isoprene units

Problem 5.28 Following is the structural formula of warburganal (The Merck Index, 12 ed., # 10173), a crystalline solid isolated from the plant *warburgia ugandensis*, Canellaceae. An important use of warburganal is its antifeeding activity against the African army worm. In addition, it acts as a plant growth regulator and has cytotoxic, antimicrobial, and molluscicidal properties.

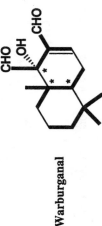

Warburganal
[a]²⁵_D -260

(a) Show that warburganal is a terpene.
(b) Label each stereocenter and specify the number of stereoisomers possible for a molecule of this structure.

Warburganal

There are three stereocenters, so there are 2 x 2 x 2 = 8 stereoisomers possible for warburgnal.

Problem 5.29 α-Santonin (The Merck Index, 12th ed., #8509), isolated from the flower heads of certain species of Artemisia, is an anthelminthic, that is, a drug used to rid the body of worms (helminths). It has been estimated that over one third of the world's population is infested with these parasites. α-Santonin in oral doses of 60 mg is used as an anthelmintic for roundworms *Ascaris lumbricoides*.

α-Santonin
[α]$_D^{25}$ -170 to -175

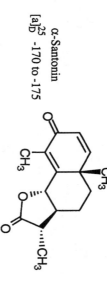

or

(a) Locate the three isoprene units in santonin, and show how the carbon skeleton of farnesol might be coiled and then cross-linked to give α-santonin. Two different coiling patterns of the carbon skeleton of farnesol can lead to α-santonin. Try to find them both.

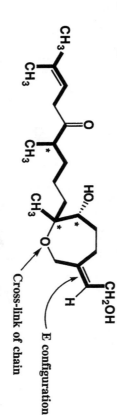

Farnesol

(b) Label all stereocenters in α-santonin. How many stereoisomers are possible for this molecule?

The four stereocenters of α-santonin are marked on the structures above. There are 2 x 2 x 2 x 2 = 16 stereoisomers possible for α-santonin.

Problem 5.30 In many parts of South America, extracts of the leaves and twigs of *Montanoa tomentosa* are brewed with water to make a "tea" used to stimulate menstruation, to facilitate labor, and as an abortifacient. Phytochemical investigations of this plant have resulted in isolation of a very potent fertility-regulating compound called zoapatanol. (See The Merck Index, 12th edition, #10318).

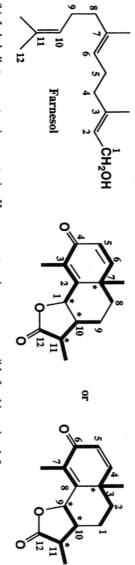

Zoapatanol

(a) Show that the carbon skeleton of zoapatanol can be divided into four isoprene units bonded head-to-tail and then cross-linked in one point along the chain.

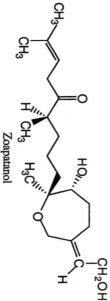

(b) Specify the configuration about the carbon-carbon double bond to the seven-membered ring according to the E,Z system.

The double bond in question has the E configuration, because the hydroxymethyl group is on the side of the double bond opposite the higher priority carbon atom that is linked to the ether oxygen.

(c) How many stereoisomers are possible for this molecule? In answering this problem, you must consider both E,Z isomerism and R,S isomerism.

There is just the one double bond capable of E,Z isomerism and three stereocenters as shown on the above structure. Thus, there are a total of 2 x 2 x 2 x 2 = 16 stereoisomers possible.

<u>Problem 5.31</u> Pyrethrin II and pyrethrosin (The Merck Index, 12 ed., #s 8148 and 8149) are two natural products isolated from plants of the chrysanthemum family. Pyrethrin II is a natural insecticide and is marketed as such.
(a) Label all stereocenters in each molecule and all carbon-carbon double bonds about which there is the possibility for cis,trans isomerism.

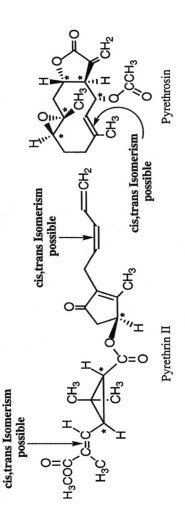

Pyrethrin II

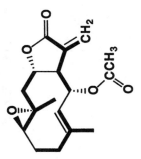

Pyrethrosin

(a) State the number of stereoisomers possible for each molecule.

For pyrethin II there are two double bonds capable of cis, trans isomerism and three stereocenters for a total of 2 x 2 x 2 x 2 x 2 = 32 possible stereoisomers. For pyrethrosin there is one double bond capable of cis, trans isomerism and five stereocenters for a total of 2 x 2 x 2 x 2 x 2 x 2 = 64 possible stereoisomers.

(b) Show that the bicyclic ring system of pyrethrosin is composed of three isoprene units.

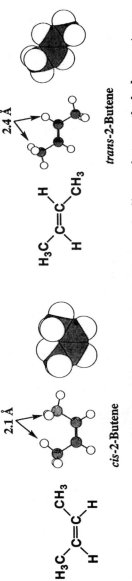

<u>Problem 5.32</u> Measure the CH₃, CH₃ distance in the energy-minimized model of *cis-2*-butene, and the CH₃, H distance in the energy-minimized model of *trans-2*-butene. In which isomer is the nonbonded interaction strain greater?

As the methyl groups rotate, the distances change. For *cis-2*-butene, the distance between the hydrogen atoms on adjacent methyl groups varies, but is 2.1 Å in the energy minimized structure. For *trans-2*-butene, the distance between the hydrogen atoms of the methyl group and the alkene hydrogen atom also varies as the methylgroup rotates, with a value ot 2.4 Å in the energy minimized structure. Clearly, there is greater nonbonded interaction strain in *cis-2*-butene.

Problem 5.33 Measure the C=C-C bond angles in the energy-minimized models of the cis and trans isomers of 2,2,5,5-tetramethyl-3-hexene. In which case is the deviation from the VSEPR model predictions greater?

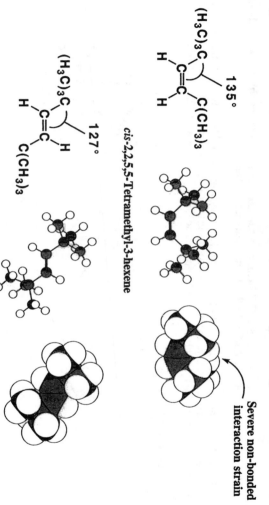

135°

cis-2,2,5,5-Tetramethyl-3-hexene

Severe non-bonded
interaction strain

127°

trans-2,2,5,5-Tetramethyl-3-hexene

The VSEPR model predicts 120° bond angles around the double bond. The minimized cis isomer has a 135° bond angle due to the severe non-bonded interaction strain present. The trans isomer has less non-bonded interaction strain and a 126° bond angle. Thus, the cis isomer is the one with significantly more deviation from the VSEPR model.

Problem 5.34 Measure the C-C-C and C-C-H bond angles in the energy-minimized model of cyclohexene and compare them with those predicted by the VSEPR model. Explain any differences.

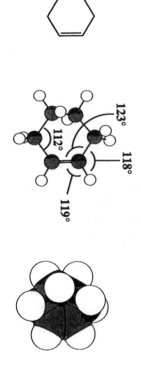

123°
112°
118°
119°

Cyclohexene

The VSEPR model predicts 120° bond angles around the double bond. Clearly, the angles predicted by Chem3D are not exactly 120° as predicted by the VSEPR model for an sp² hybridized carbon atom. This difference is the result of strain introduced by the ring of cyclohexene.

Problem 5.35 Measure the C-C-C and C-C-H bond angles in the energy-minimized models of cis and trans isomers of cyclooctene. Compare these values with those predicted by the VSEPR model. In which isomer are deviations from the VSEPR model predictions greater?

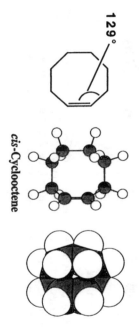

129°

cis-Cyclooctene

Solutions

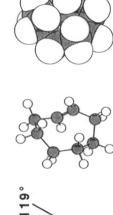

119°

trans-Cyclooctene

Only the key C-C=C bond angles are shown on the above structures. The VSEPR model predicts 120° bond angles around the double bond. The angles deviate more in the cis somer compared to the trans isomer due to increased ring strain in the cis isomer. Note how the trans isomer can more easily accommodate staggered conformations along the entire chain.

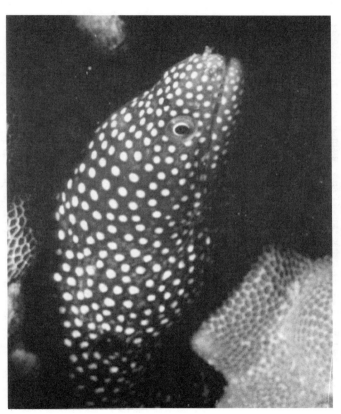

Whitemouth moray *Gymnothorax meleagris*
Kona coast, Hawaii

CHAPTER 6
Solutions to the Problems

Problem 6.1 Suppose that the activation energy for a particular chemical reaction is 105.5kJ (25.2 kcal)/mol. By what factor is the rate of reaction increased when the reaction takes place at 35°C compared with the rate at 25°C?

$$\frac{k_2}{k_1} = \frac{Ae^{-\Delta G^{\ddagger}/RT_2}}{Ae^{-\Delta G^{\ddagger}/RT_1}}$$

Taking the logarithm of each side, converting to base 10, and rearranging gives:

$$\log \frac{k_2}{k_1} = \frac{\Delta G^{\ddagger}}{2.303\ R} \left(\frac{1}{T_1} - \frac{1}{T_2} \right)$$

Plugging in the actual values gives:

$$\log \frac{k_2}{k_1} = \frac{105.5\ (kJ/mol)}{2.303\ (8.314 \times 10^{-3}\ kJ/mol\ K)} \left(\frac{1}{298\ K} - \frac{1}{308\ K} \right)$$

Solving the equation gives:

$$\log \frac{k_2}{k_1} = (5.51 \times 10^3\ K)(1.09 \times 10^{-4}\ K^{-1}) = 0.601$$

$$\boxed{\frac{k_2}{k_1} = 10^{(0.601)} = 3.99}$$

Problem 6.2 Complete the first three entries in this table for reactions taking place at 25°C. Given the pattern of these first three entries, estimate the approximate values for the remaining two entries. How many kilojuoles per mole in activation energy corresponds to a power of 10 in relative rates?

Plugging in the appropriate values gives:

$$\Delta E_a = -2.303\ RT \log \frac{k_2}{k_1} = (-5.69\ kJ/mol) \left(\log \frac{k_2}{k_1} \right)$$

$\Delta G^{\ddagger}$ (kJ/mol)	$\frac{k_2}{k_1}$
0	1
-5.69	10
-11.38	100
-17.07	1000
-22.76	10000

Problem 6.3 Name and draw the structural formula for the product of each alkene addition reaction.

(a) $CH_3-CH=CH_2$ + HI ⟶ CH_3CHCH_3
|
I

2-Iodopropane

(b)

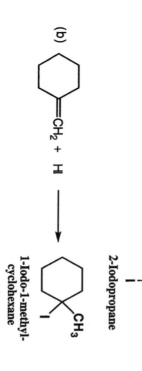

1-Iodo-1-methyl-
cyclohexane

Problem 6.4 Arrange these carbocations in order of increasing stability.

(a)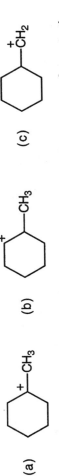

(b)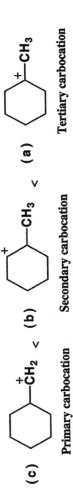

(c)

The order of increasing stability of carbocations is methyl < primary < secondary < tertiary. Thus the three carbocations can be ranked as follows:

(c)  < (b) < (a)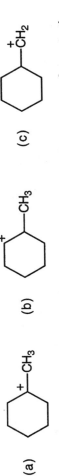

Primary carbocation   Secondary carbocation   Tertiary carbocation

Problem 6.5 Propose a mechanism for addition of HI to 1-methylcyclohexene to give 1-iodo-1-methylcyclohexane. Which step in your mechanism is rate determining?

Step 1:

   Slow, rate determining

A rate-determining proton transfer from HI to the carbon-carbon double bond to a 3° carbocation intermediate.

Step 2:

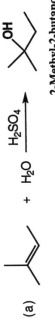   Fast

Reaction of the carbocation intermediate with iodide ion completes the valence shell of carbon and gives the product.

Problem 6.6 Draw the structural formula for the product of each alkene hydration reaction.

(a)

 + H₂O →(H₂SO₄)→

2-Methyl-2-butanol

(b)

+ H₂O →(H₂SO₄)→

2-Methyl-2-butanol

Problem 6.7 Propose a mechanism for the acid-catalyzed hydration of 1-methylcyclohexene to give 1-methylcyclohexanol. Which step in your mechanism is rate determining?

Step 1:

   Slow, rate-determining step

Proton transfer of from the acid catalyst to the alkene gives a 3° carbocation intermediate in the rate-determining step.

Step 2:

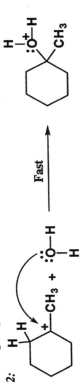   Fast

Step 3:

Problem 6.8 Acid-catalyzed hydration of 3-methyl-1-butene gives 2-methyl-2-butanol as the major product. Propose a mechanism for formation this alcohol.

Step 1:

Slow, rate-determining step

Proton transfer from the acid catalyst to the alkene gives a 2° carbocation intermediate in the rate-determining step.

Step 2:

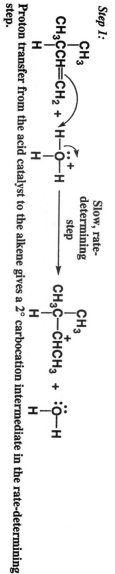

Fast

The 2° carbocation rearranges to the more stable 3° carbocation.

Step 3:

Fast

Step 4:

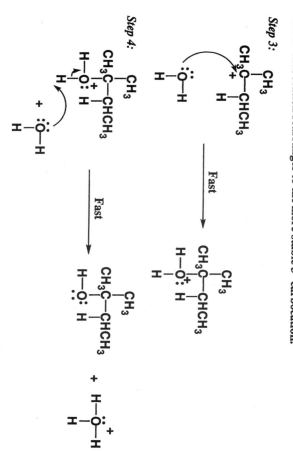

Fast

Problem 6.9 Complete these reactions.

(a) CH₃CCH=CH₂ + Br₂ $\xrightarrow{CH_2Cl_2}$ CH₃CCHBrCH₂Br

(b)

+ Cl₂ $\xrightarrow{CH_2Cl_2}$

A new stereocenter is created, so the product is a racemic mixture of two enantiomers.

Problem 6.10 Draw the structure of the chlorohydrin formed when 1-methylcyclohexene is treated with Cl_2/H_2O.

$+ \ Cl_2 \ / \ H_2O \longrightarrow$

Two new stereocenters are created and due to the trans geometry of the addition reaction, the product will be a racemic mixture of the enantiomers shown.

Problem 6.11 Draw structural formulas for the trialkylborane and alkene that give the following alcohols under the reaction conditions shown.

(a)
$$CH_3CHCH=CH_2 \xrightarrow{BH_3} \text{A trialkylborane} \xrightarrow[NaOH]{H_2O_2} CH_3-CH-CH_2-CH_2-OH$$

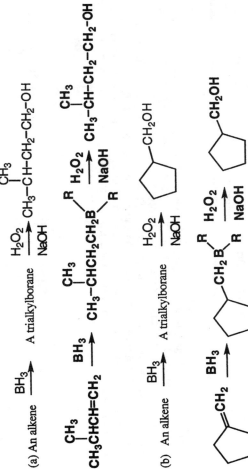

(b)   An alkene $\xrightarrow{BH_3}$ A trialkylborane $\xrightarrow[\text{NaOH}]{H_2O_2}$

Problem 6.12 Use a balanced half-reaction to show that each transformation involves a reduction.

(a)

$\longrightarrow$

Two hydrogens are required to produce the product alcohol from the ketone. Therefore, the balanced half-reaction needs two protons and two electrons (for charge balance) on the left-hand side. Since the electrons are on the left-hand side of the equation, the reaction is a two-electron reduction.

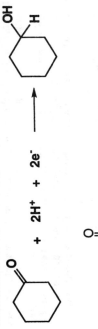

$+ \ 2H^+ \ + \ 2e^- \longrightarrow$

(b) $CH_3-CH_2-\overset{\overset{\displaystyle O}{\|}}{C}OH \longrightarrow CH_3CH_2CH_2OH$

Two hydrogens are required to produce the product alcohol from the carboxylic acid. Therefore, the balanced half-reaction needs two protons and two electrons (for charge balance) on the left-hand side. Additionally, the product alcohol has one less oxygen atom than the carboxylic acid starting material, so there must be an H_2O molecule added to the right side of the equation to balance the oxygen atoms. This H_2O molecule has two more hydrogens that must be balanced by adding two more protons and electrons to the left-hand side of the equation, giving a total of four protons and four electrons on the left-hand side. Since the electrons are on the left-hand side of the equation, the reaction is a four-electron reduction.

$$CH_3-CH_2-\overset{\overset{\displaystyle O}{\|}}{C}OH \ + \ 4H^+ \ + \ 4e^- \longrightarrow CH_3CH_2CH_2OH \ + \ H_2O$$

Problem 6.13 What alkene of molecular formula C_6H_{12}, when treated with ozone and then dimethyl sulfide, gives the following product(s)?

(a) $C_6H_{12} \xrightarrow[\text{2. (CH}_3)_2\text{S}]{\text{1. O}_3}$ CH_3CH_2CH (only product)

$$CH_3CH_2CH=CHCH_2CH_3$$
(cis or trans)

(b) $C_6H_{12} \xrightarrow[\text{2. (CH}_3)_2\text{S}]{\text{1. O}_3}$ $CH_3CH + CH_3CCH_2CH_3$ (equal moles of each)

$$\begin{array}{c} CH_3 \\ | \\ CH_3CH=CCH_2CH_3 \end{array}$$
(cis or trans)

(c) $C_6H_{12} \xrightarrow[\text{2. (CH}_3)_2\text{S}]{\text{1. O}_3}$ CH_3CCH_3 (only product)

$$\begin{array}{c} H_3C \quad\quad CH_3 \\ | \quad\quad\quad | \\ CH_3C=CCH_3 \end{array}$$

Energetics of Chemical Reactions

Problem 6.14 Most chemical reactions occur as written if they are exothermic, that is, if the bonds that are formed are stronger than the ones that are broken in the starting materials. To determine if a reaction is exothermic as written, add the bond dissociation energies of all bonds broken in the starting materials (it costs energy to break bonds). Subtract from this the total of bond dissociation energies of all bonds formed in the products (formation of bonds liberates energy). If the sum of these numbers is negative, the reaction is exothermic (energy is liberated) and the reaction proceeds to the right as written. If the sum of these numbers is positive, the reaction is endothermic (it requires energy) and it does not proceed to the right as written. Using the table of bond dissociation energies at 25°C, determine which of the following reactions are energetically favorable at room temperature, that is, if a suitable catalyst could be found, which would proceed to the right as written?

Bond	Bond dissociation energy [kJ(kcal)/mol]	Bond	Bond dissociation energy [kJ(kcal)/mol]
H-H	435(104)	C-Si	301(72)
O-H	463(111)	C=C	611(146)
C-H	413(98.7)	C=O (aldehyde)	728(174)
N-H	391(93.4)	C=O (CO_2)	803(192)
Si-H	318(76)	C=O	1075(257)
C-C	346(82.6)	N≡N	950(227)
C-N	305(73)	C≡C	837(200)
C-O	358(85.5)	O=O	498(119)
C-I	213(51)		

The following reactions can only occur to a significant extent as written if they are exothermic, that is, if the bonds that are formed are stronger than the ones that are broken in the reaction. Recall that a catalyst increases the rate, but does not change the overall thermodynamics of a reaction.

To find out if a reaction is exothermic, the dissociation energy of all the bonds in the molecules on each side of the equation are added together. If the bond dissociation energy total from the right side of the equation is higher than the total from the left side of the equation, then the reaction is exothermic (ΔH for the reaction is negative).

(a) $CH_2=CH_2 + 2H_2 + N_2 \longrightarrow H_2N-CH_2-CH_2-NH_2$

The bond dissociation energies from the left side of the equation:

611 + (4 x 413) + 435 + 950 = 3648 kJ/mol
(C=C) (4 C-H) (H-H) (N≡N)

The bond dissociation energies from the right side of the equation:

(4 x 391) + (2 x 305) + 346 + (4 x 413) = 4172 kJ/mol
(4 N-H) (2 C-N) (C-C) (4 C-H)

This reaction is exothermic because 4172 kJ/mol is larger than 3648 kJ/mol.

(b)   CH$_2$=CH$_2$ + CH$_4$ ⟶ CH$_3$-CH$_2$-CH$_3$

The bond dissociation energies from the left side of the equation:
611 + (4 x 413) + (4 x 413) = 3915 kJ/mol
(C=C)   (4 C-H)   (4 C-H)

The bond dissociation energies from the right side of the equation:
(2 x 346) + (8 x 413) = 3996 kJ/mol
(2 C-C)   (8 C-H)

This reaction is exothermic because 3996 kJ/mol is larger than 3915 kJ/mol.

(c)   CH$_2$=CH$_2$ + (CH$_3$)$_3$SiH ⟶ H-CH$_2$-CH$_2$-Si(CH$_3$)$_3$

The bond dissociation energies from the left side of the equation:
611 + (4 x 413) + (9 x 413) + (3 x 301) + 318 = 7201 kJ/mol
(C=C)   (4 C-H)   (9 C-H)   (3 C-Si) (Si-H)

The bond dissociation energies from the right side of the equation:
(346) + (5 x 413) + (9 x 413) + (4 x 301) = 7332 kJ/mol
(C-C)   (5 C-H)   (9 C-H)   (4 C-Si)

This reaction is exothermic because 7332 kJ/mol is larger than 7201 kJ/mol.

(d)   CH$_2$=CH$_2$ + CHI$_3$ ⟶ H-CH$_2$-CH$_2$-C(I)$_3$

The bond dissociation energies from the left side of the equation:
611 + (5 x 413) + (3 x 213) = 3315 kJ/mol
(C=C)   (5 C-H)   (3 C-I)

The bond dissociation energies from the right side of the equation:
(2 x 346) + (5 x 413) + (3 x 213) = 3396 kJ/mol
(2 C-C)   (5 C-H)   (3 C-I)

This reaction is exothermic because 3396 kJ/mol is larger than 3315 kJ/mol.

(e)   CH$_2$=CH$_2$ + CO + H$_2$ ⟶ H-CH$_2$-CH$_2$-CH
$$\overset{\displaystyle O}{\underset{\displaystyle ||}{}}$$

The bond dissociation energies from the left side of the equation:
611 + (4 x 413) + 1075 + 435 = 3773 kJ/mol
(C=C)   (4 C-H)   (C≡O) (H-H)

The bond dissociation energies from the right side of the equation:
(2 x 346) + (6 x 413) + 728 = 3898 kJ/mol
(2 C-C)   (6 C-H)   (C=O)

This reaction is exothermic because 3898 kJ/mol is larger than 3773 kJ/mol.

(f) 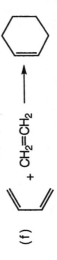 + CH$_2$=CH$_2$ ⟶

The bond dissociation energies from the left side of the equation:
(3 x 611) + 346 + (10 x 413) = 6309 kJ/mol
(3 C=C)   (C-C)   (10 C-H)

The bond dissociation energies from the right side of the equation:
611 + (5 x 346) + (10 x 413) = 6471 kJ/mol
(C=C)   (5 C-C)   (10 C-H)

This reaction is exothermic because 6471 kJ/mol is larger than 6309 kJ/mol.

(g)

The bond dissociation energies from the left side of the equation:
(2 x 611) + 346 + (2 x 803) + (6 x 413) = 5652 kJ/mol
(2 C=C)   (C-C) (2 C=O)   (6 C-H)

The bond dissociation energies from the right side of the equation:
611 + (3 x 346) + (6 x 413) + (2 x 358) + 728 = 5571 kJ/mol
(C=C)   (3 C-C)   (6 C-H)   (2 C-O)   (C=O)

This reaction is endothermic because 5571 kJ/mol is smaller than 5652 kJ/mol.

(h) $HC \equiv CH + O_2 \longrightarrow H-\overset{\overset{O}{\|}}{C}-\overset{\overset{O}{\|}}{C}-H$

The bond dissociation energies from the left side of the equation:
837 + (2 x 413) + 498 = 2161 kJ/mol
(C≡C)   (2 C-H)   (O=O)

The bond dissociation energies from the right side of the equation:
346 + (2 x 728) + (2 x 413) = 2628 kJ/mol
(C-C) (2 C=O)   (2 C-H)

This reaction is exothermic because 2628 kJ/mol is larger than 2161 kJ/mol.

(i) $2CH_4$ + O_2 $\longrightarrow$ $2CH_3OH$

The bond dissociation energies from the left side of the equation:
2(4 x 413) + 498 = 3802 kJ/mol
2(4 C-H)   (O=O)

The bond dissociation energies from the right side of the equation:
2(3 x 413) + 2(358) + 2(463) = 4120 kcal/mol
2(3 C-H)   2(C-O)   2(O-H)

This reaction is exothermic because 4120 kcal/mol is larger than 3802 kJ/mol.

Electrophilic Additions
Problem 6.15 Draw structural formulas for the isomeric carbocation intermediates formed on treatment of each alkene with HCl. Label each carbocation primary, secondary, or tertiary, and state which of the isomeric carbocations is formed more readily.

(a) $CH_3-CH_2-\overset{\overset{CH_3}{|}}{C}=CH-CH_3$

$CH_3-CH_2-\overset{\overset{CH_3}{|}}{\underset{+}{C}}-CH_2-CH_3$ + $CH_3-CH_2-\overset{\overset{CH_3}{|}}{CH}-\overset{+}{CH}-CH_3$

Tertiary   Secondary
(Formed more readily)

(b) $CH_3-CH_2-CH=CH-CH_3$

$CH_3-CH_2-\overset{+}{CH}-CH_2-CH_3$ + $CH_3-CH_2-CH_2-\overset{+}{CH}-CH_3$

Both secondary carbocations
(Formed at equal rates)

(c)

Tertiary   Secondary
(Formed more readily)

(d)

Primary   +   Tertiary
(Formed more readily)

Problem 6.16 Arrange the alkenes in each set in order of increasing rate of reaction with HI. Draw the structural formula of the major product formed in each case, and explain the basis for your ranking.

(a)   CH₃–CH=CH–CH₃   and   CH₃–C=CH–CH₃
                                                            CH₃

2-Butene

CH₃–CH=CH–CH₃ ⟶ CH₃–CH₂–CH–CH₃⁺ ⟶ CH₃–CH₂–CH–CH₃
                                 A secondary   I*
                                 carbocation   2-Iodobutane
                                                   (sec-Butyl iodide)
                                                    New stereocenter created
                                                    so racemic mixture

   CH₃
   |
CH₃–C=CH–CH₃ ⟶ CH₃–C–CH₂–CH₃⁺ ⟶ CH₃–C–CH₂–CH₃
                                  CH₃   CH₃
                                                          |

2-Methyl-2-butene   A tertiary   2-Iodo-2-methylbutane
                      carbocation   (Major product)

The reaction of 2-methyl-2-butene is the only one that can form a tertiary carbocation, so 2-methyl-2-butene is the compound that reacts faster with HI.

(b)

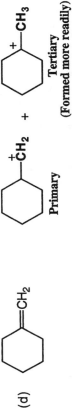

and

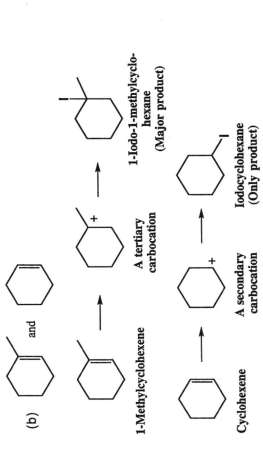

1-Methylcyclohexene

A tertiary carbocation

1-Iodo-1-methylcyclo-
hexane
(Major product)

Cyclohexene

A secondary carbocation

Iodocyclohexane
(Only product)

Only 1-methylcyclohexene can form a tertiary carbocation, so 1-methylcyclohexene reacts faster with HI.

Problem 6.17 Predict the organic product(s) of the reaction of 2-butene with each reagent.

(a) H₂O (H₂SO₄)
       *
CH₃–CH–CH₂–CH₃
        |
        OH

(b) Br₂
          *
CH₃–CH–CH–CH₃
        | |
        Br Br

(c) Cl₂
          *
CH₃–CH–CH–CH₃
        | |
        Cl Cl

(d) Br₂ in H₂O
      *   *
CH₃–CH–CH–CH₃
        | |
        Br OH

(e) HI
          *
CH₃–CH–CH₂–CH₃
        |
        I

(f) Cl₂ in H₂O
      *   *
CH₃–CH–CH–CH₃
        | |
        Cl OH

(g) Hg(OAc)₂, H₂O

CH₃—CH—CH—CH₃
    | |
    OH HgOAc

(with * over both CH carbons)

(h) the product in (g) + NaBH₄

CH₃—CH—CH₂—CH₃
    |
    OH

(with * over the CH carbon)

In parts (a), (e), and (f), a single stereocenter is created so the products are racemic mixtures. In parts (b), (c), (d), (f), and (g) two stereocenters are created. Racemic mixtures will be the result, the exact identity of which depends on whether the 2-butene starting material is cis or trans (not given in the problem).

Problem 6.18 Draw a structural formula of an alkene that undergoes acid-catalyzed hydration to give each alcohol as the major product. More than one alkene may give each alcohol as the major product.

(a) 3-Hexanol   (b) 1-Methylcyclobutanol

CH₃CH₂CH=CHCH₂CH₃

(c) 2-Methyl-2-butanol

    (cis or trans)

         CH₃
         |
H₂C=CCH₂CH₃   or   CH₃C=CHCH₃

                         CH₃
                         |
                        CH₃CH=CH₂

(d) 2-Propanol

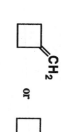

1-Methylcyclopentene with CH₂ and or CH₃ cyclobutane

Problem 6.19 Reaction of 2-methyl-2-pentene with each reagent is regiospecific. Draw a structural formula for the product of each reaction, and account for the observed regiospecificity.

In each case, the reaction mechanism involves formation of a tertiary carbocation, that then reacts with a nucleophile to give the product shown.

(a) HI

        CH₃
        |
CH₃—C—CH₂—CH₂—CH₃
        |
        I

(b) HBr

        CH₃
        |
CH₃—C—CH₂—CH₂—CH₃
        |
        Br

(c) H₂O in the presence of H₂SO₄

        CH₃
        |
CH₃—C—CH₂—CH₂—CH₃
        |
        OH

(d) Br₂ in H₂O

        CH₃
        |
CH₃—C—CH—CH₂—CH₃
        | |
        HO Br

(with * over middle carbon)

(e) Hg(OAc)₂ in H₂O

        CH₃
        |
CH₃—C—CH—CH₂—CH₃
        | |
        HO HgOAc

(with * over middle carbon)

In parts (d) and (e) one new stereocenter is created, so racemic mixtures are produced.

Problem 6.20 Reaction of 1-methylcyclopentene with each reagent is regiospecific and stereospecific. Account for the observed regiospecificity and stereospecificity.

(a) BH₃

Attack of the borane occurs in a concerted fashion, simultaneously forming both the new C-H and C-B bonds on the same face of the double bond, that is syn. Largely for steric reasons, the H atom ends up on the more hindered carbon atom (the one with more/bulkier substituents) and the B atom ends up on the less hindered carbon atom.

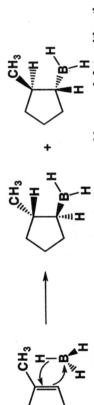

Bond forming and bond breaking is concerted. Note that the borane could approach from either the top or the bottom of the alkene, leading to a racemic mixture of the two syn addition products as shown.

(b) Br₂ in H₂O

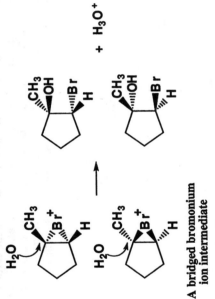

Attack by H₂O on the carbon of the bromonium ion intermediate bearing the methyl group followed by loss of a proton gives the *trans* bromohydrin. Note that the bromonium ion could form on either face of the alkene, leading to a racemic mixture of the two enantiomers shown.

(c) Hg(OAc)₂ in H₂O

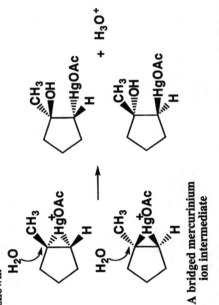

Attack by water on the bridged mercurinium ion intermediate followed by loss of a proton results in -OH *trans* to -HgOAc. Note that the mercurinium ion could form on either face of the alkene, leading to a racemic mixture of the two enantiomers shown.

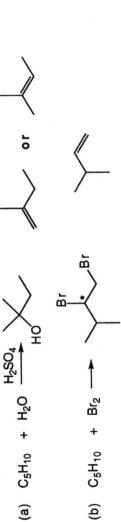

Problem 6.21 Draw a structural formula for an alkene with the indicated molecular formula that gives the compound shown as the major product. Note that more than one alkene may give the same compound as the major product.

(a) C_5H_{10} + H_2O $\xrightarrow{H_2SO_4}$

(b) C_5H_{10} + Br_2 $\longrightarrow$

Note that a stereocenter is created so the product is actually a racemic mixture.

(c) C_7H_{12} + HCl $\longrightarrow$

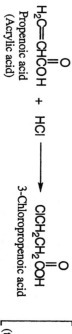

or

Problem 6.22 Account for the fact that addition of HCl to 1-bromopropene gives exclusively 1-bromo-1-chloropropane.

$$CH_3CH=CHBr \quad + \quad HCl \quad \longrightarrow \quad CH_3CH_2CHBrCl$$
1-Bromopropene   1-Bromo-1-chloropropane

The exclusive product must be derived from the significantly more stable carbocation. In this case, the significantly more stable carbocation is the one with the positive charge on the carbon atom attached to the bromine atom, despite the fact that this carbocation is primary versus the alternative secondary carbocation. Thus, the bromine atom must be able to stabilize an adjacent cationic carbon atom. It turns out that the stabilization is primarily a resonance effect, involving the lone pairs of the bromine atom as shown. Note how the resonance structure on the right illustrates how the positive charge is partially delocalized onto the bromine atom.

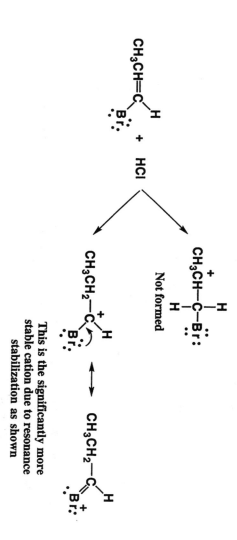

CH₃CH=C⟨H / Br̈:⟩ + HCl

CH₃CH₂⁺—C—Br̈:
                H
Not formed

CH₃CH₂—C⁺⟨H / Br̈:⟩ ⟷ CH₃CH₂—C⟨H / Br̈:⁺⟩

This is the significantly more stable cation due to resonance stabilization as shown

Problem 6.23 Propenoic acid (acrylic acid) reacts with HCl to give 3-chloropropanoic acid. It does not produce 2-chloropropanoic acid. Account for this result.

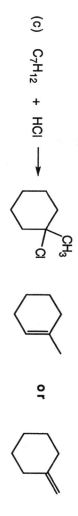

$$H_2C=CHCOH \quad + \quad HCl \quad \longrightarrow \quad ClCH_2CH_2COH$$
                O   O
Propenoic acid   3-Chloropropenoic acid
(Acrylic acid)

$$\left[\begin{array}{cc} Cl & O \\ | & \| \\ CH_3CHCOH \\ \text{2-Chloropropenoic acid} \\ \text{(this product is not formed)} \end{array} \right]$$

The exclusive product must be derived from the significantly more stable carbocation. In this case, the significantly more stable carbocation is the one with the positive charge on the terminal carbon atom, despite the fact that this carbocation is primary versus the alternative secondary carbocation. Thus, the carbonyl group attached to the internal carbon atom must be destabilizing to an adjacent cationic carbon atom. It turns out that the destabilization is primarily an inductive effect, based on the fact that a carbonyl group is electron withdrawing. An electron withdrawing group is destabilizing since removing charge density from a carbocation increases the charged character and thus the energy of the carbocation even further.

$$H_2C=CHCOH + HCl$$

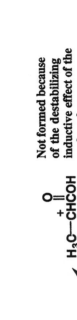

Not formed because of the destabilizing inductive effect of the carbonyl group

$$\overset{O}{\overset{\|}{H_3C-\overset{+}{C}HCOH}}$$

$$\overset{O}{\overset{\|}{H_2\overset{+}{C}-CH_2COH}}$$

This is the significantly more stable carbocation

<u>Problem 6.24</u> Draw a structural formula for the alkene of molecular formula C_5H_{10} that reacts with Br_2 to give each product.

(a)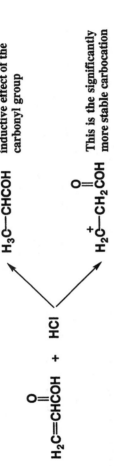

(b)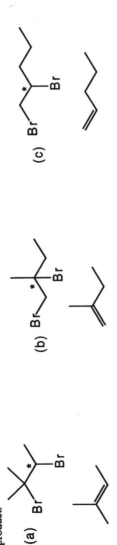

(c)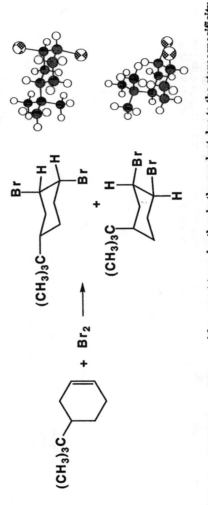

A stereocenter is created in each case, so the products of these bromine addition reactions are actually racemic mixtures.

<u>Problem 6.25</u> Draw alternative chair conformations for the product formed by addition of bromine to 4-*tert*-butylcyclohexene. The Gibbs free energy differences between equatorial and axial substituents on a cyclohexane ring are 21 kJ (4.9 kcal)/mol for *tert*-butyl and 2.0 – 2.6 kJ (0.48 - 0.62 kcal)/mol for bromine. Estimate the relative percentages of the alternative chair conformations you drew in the first part of this problem.

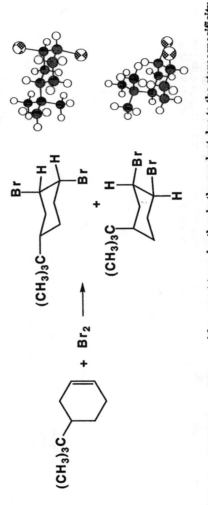

Note that the bromine atoms are *trans* with respect to each other in the product due to the stereospecificity of the reaction. Recall that large substituents are sterically disfavored in axial positions. The upper product has the large *tert*-butyl group in the strongly favored equatorial position along with both bromine atoms in the somewhat disfavored axial positions. The lower product has the large *tert*-butyl in the strongly disfavored axial position along with both bromine atoms in the somewhat more favored equatorial positions. The conformational energy difference based on the axial vs. equatorial *tert*-butyl group is 21 kJ/mol, and we will use an intermediate value of 2.3 kJ/mol for the conformational energy difference for the axial vs. equatorial bromine atoms. Thus, the relative conformation energy can be estimated as being favorable for the upper structure by an amount equal to the value that is favorable for the *tert*-butyl group (21 kJ/mol) minus the disfavorable contributions of the two bromine atoms (2 x +2.3 kJ/mol) for a total of 16.4 kJ/mol.

At equilibrium the relative amounts of each form are given by the equation:

$$\Delta G° = -RT\ln K_{eq}$$

Here K_{eq} refers to the ratio of the alternative chair conformations. Converting to base 10 and rearranging gives

$$\log K_{eq} = \frac{-\Delta G°}{(2.303)RT}$$

Taking the antilog of both sides gives:

$$K_{eq} = 10^{\left(\frac{-\Delta G^\circ}{2.303 RT}\right)}$$

Plugging in the values for ΔG°, R, and 298 K gives the final answer:

$$K_{eq} = 10^{\left(\frac{-(-16.4 \text{ kJ/mol})}{(2.303)(8.314 \times 10^{-3} \text{ kJ/mol K})(298 \text{ K})}\right)} = 10^{2.87} = 7.41 \times 10^2$$

Thus, the structure with the *tert*-butyl group equatorial will be favored by about 741 to 1 at equilibrium at room temperature.

Problem 6.26 Draw a structural formula for the cycloalkene of molecular formula C_6H_{10} that reacts with Cl_2 to give each compound.

(a)

(b)

(c)

(d)

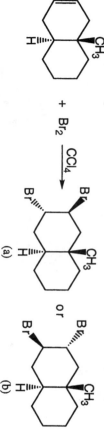

Note that the chlorine addition reactions in parts (a), (b), and (c) actually a produce racemic mixtures of enantiomers, only one of which is shown.

Problem 6.27 Reaction of this bicycloalkene with bromine in carbon tetrachloride gives a trans dibromide. In both (a) and (b), the bromine atoms are trans to each other. However, only one of these products is formed. Which trans dibromide is formed? How do you account for the fact that it is formed to the exclusion of the other trans dibromide?

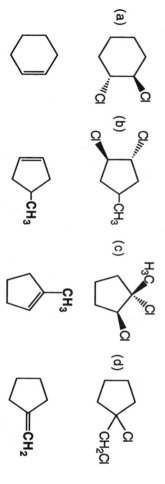

(a)   or   (b)

Product (a) is formed. Electrophilic addition of bromine to an alkene occurs via a bridged bromonium ion intermediate and anti addition of the two bromine atoms. In a cyclohexane ring, anti addition corresponds to trans and diaxial addition. Only in formula (a) are the two added bromines trans and diaxial. In (b) they are trans, but diequatorial, so this isomer cannot be formed. In this case, the starting material is a single enantiomer, so only one enantiomer is produced as the product.

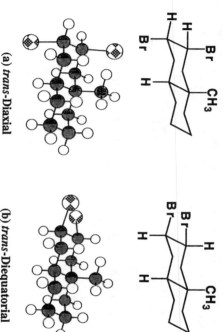

(a) *trans*-Diaxial   (b) *trans*-Diequatorial

Problem 6.28 Terpin, prepared commercially by the acid-catalyzed hydration of limonene (Figure 5.5), is used medicinally as an expectorant for coughs (see The Merck Index, 12th ed., #9314)

Limonene

$$+ \ 2H_2O \ \xrightarrow{H_2SO_4} \ C_{10}H_{20}O_2$$

Terpin

(a) Propose a structural formula for terpin hydrate and a mechanism for its formation.

Add water to each double bond by protonation to give a 3° carbocation, reaction of each carbocation with water, and loss of the protons to give terpin hydrate. Since 3° carbocations are produced in either case, it is not clear which double bond would actually react first.

Step 1:

Step 2:

Step 3:

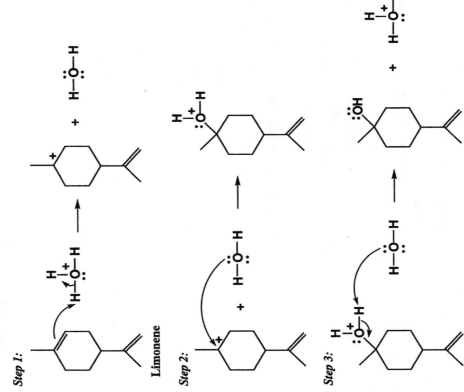

Limonene

Step 4:

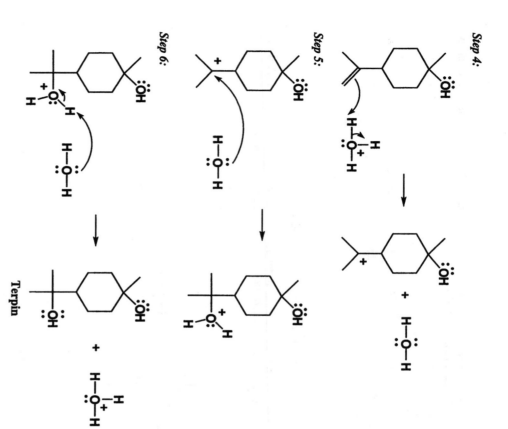

Step 5:

Step 6:

Terpin

(b) How many cis,trans isomers are possible for the structural formula you have proposed?

There are two cis,trans isomers, shown here as chair conformations with the (CH₃)₂COH- side chain equatorial.

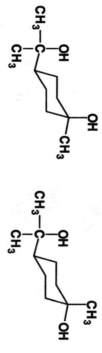

Problem 6.29 Propose a mechanism for this reaction. In so doing, account for its regioselectivity.

$$H_3C-C=CH_2 \quad + \quad ICl \quad \longrightarrow \quad H_3C-C-CH_2I$$
$$\overset{\displaystyle CH_3}{} \qquad\qquad\qquad\qquad \overset{\displaystyle CH_3}{\underset{\displaystyle Cl}{}}$$

In the addition reaction of an unsymmetrical electrophilic reagent to a double bond, the "positive" portion of the adding reagent adds to the carbon atom of the double bond so as to yield the more stable carbocation as an intermediate. Iodine is less electronegative (by 0.5 unit on the Pauling scale) than chlorine, so iodine is added first as shown.

Step 1:

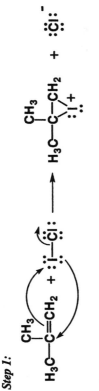

Step 2:

Problem 6.30 Treatment of 2-methylpropene with methanol in the presence of sulfuric acid gives *tert*-butyl methyl ether. Propose a mechanism for the formation of this product.

$$CH_3-C=CH_2 + CH_3OH \xrightarrow{H_2SO_4} CH_3-C-CH_3$$

Proton transfer to the alkene gives a tertiary carbocation intermediate. Reaction of this intermediate with the oxygen atom of methanol followed by transfer of a proton gives *tert*-butyl methyl ether.

Step 1:

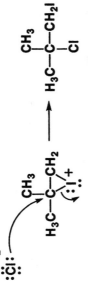

Step 2:

Step 3:

Problem 6.31 When 2-pentene is treated with Cl_2 in methanol, three products are formed. Account for the formation of each product. (You need not explain their relative percentages.)

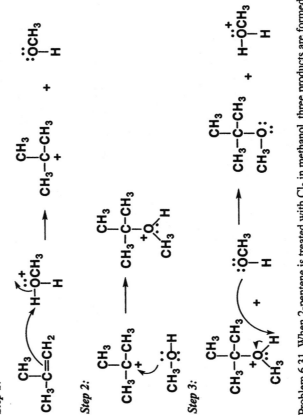

In this reaction, the chlorine reacts with the alkene to produce the chloronium ion intermediate that can then react at either carbon atom to give the four different products shown. Note that chlorine is a poor nucleophile, therefore 2,3-dichloropentane is only a minor product. The reactions produce two new stereocenters in each case. The structures shown actually represent a pair of enantiomers, the identity of which depends on whether the starting alkene is cis or trans (not specified).

Step 1:

Problem 6.32 Treatment of cyclohexene with HBr in the presence of acetic acid gives bromocyclohexane (85%) and cyclohexyl acetate (15%). Propose a mechanism for the formation of the latter product.

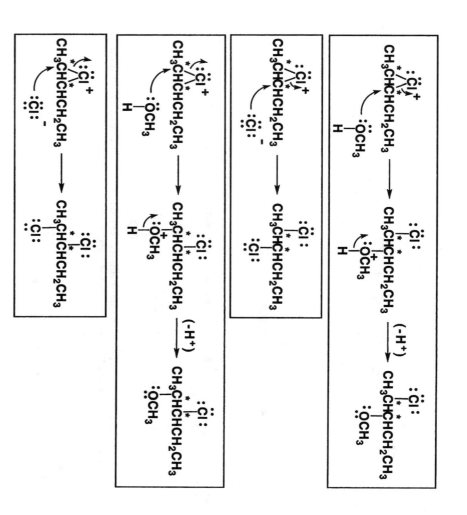

Reaction of cyclohexene with a proton gives a secondary carbocation intermediate. Reaction of this intermediate with an oxygen atom of acetic acid, followed by proton transfer gives cyclohexyl acetate.

Step 1:

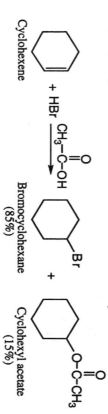

Step 2:

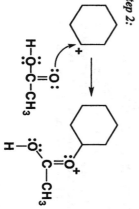

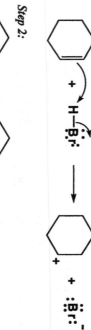

Step 3:

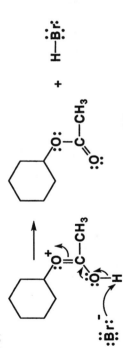

Problem 6.33 Propose a mechanism for this reaction:

$$CH_2=CHCH_2CH_2CH_3 + Br_2 + H_2O \longrightarrow$$
1-Pentene

$$\overset{Br}{\underset{}{CH_2}}-\overset{OH}{\underset{*}{CH}}CH_2CH_2CH_3 + HBr$$
1-Bromo-2-pentanol

Reaction of 1-pentene with bromine gives a bridged bromonium ion intermediate. Anti attack of water on this intermediate at the more substituted secondary carbon, followed by proton transfer gives 1-bromo-2-pentanol.

Step 1:

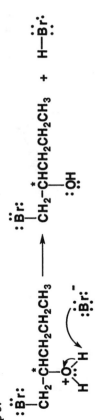

Step 2:

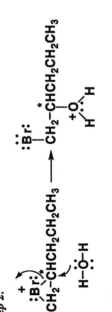

Step 3:

$$: \ddot{B}r : \\ CH_2-\overset{*}{CH}CH_2CH_2CH_3 \longrightarrow \overset{:\ddot{B}r:}{\underset{}{CH_2}}-\overset{*}{CH}CH_2CH_2CH_3 + H-\ddot{B}r:$$

A new stereocenter is formed, so the product is a racemic mixture.

Problem 6.34 Treatment of 4-penten-1-ol with bromine in water forms a cyclic bromoether. Account for the formation of this product rather than a bromohydrin as was formed in Problem 6.33.

$$\text{OH} + Br_2 \longrightarrow$$
4-Penten-1-ol

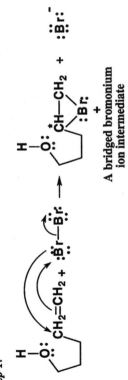

Reaction of the alkene with bromine gives a bridged bromonium ion intermediate. Reaction of this intermediate with the oxygen atom of the hydroxyl group followed by proton transfer gives the observed cyclic ether, a derivative of tetrahydrofuran.

Step 1:

Step 2:

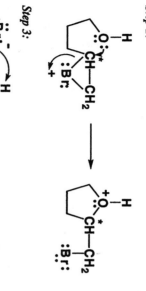

Step 3:

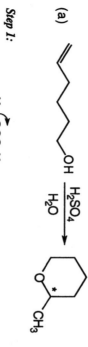

A stereocenter is created, so a racemic mixture is produced.

<u>Problem 6.35</u> Provide a mechanism for each reaction:

(a)

Step 1:

Step 2:

Step 3:

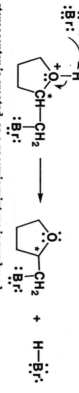

A stereocenter is created, so a racemic mixture is produced.

(b)

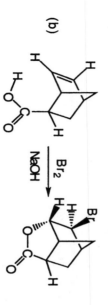

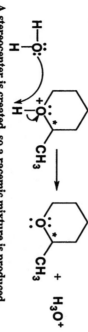

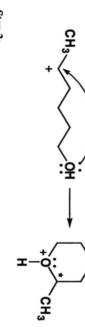

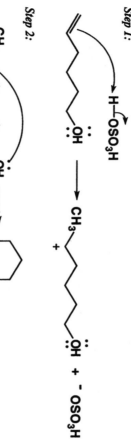

Step 1:

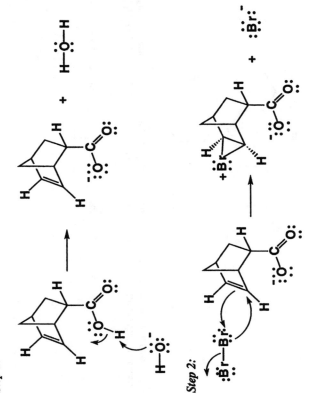

Step 2:

Step 3:

The stereochemistry of the product is set by the stereochemistry of the starting material.

<u>Problem 6.36</u> Treatment of 1-methyl-1-vinylcyclopentane with HCl gives mainly 1-chloro-1,2-dimethylcyclohexane. Propose a mechanism for the formation of this product.

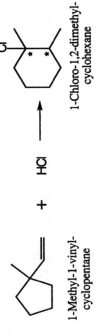

1-Methyl-1-vinyl-
cyclopentane

1-Chloro-1,2-dimethyl-
cyclohexane

The initially formed secondary carbocation can rearrange to form a more stable tertiary carbocation and a six-membered ring as shown. This new cation reacts with Cl· to give the final product.

Step 1:

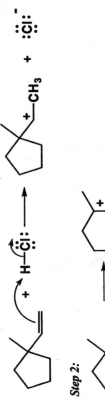

Step 2:

Step 3:

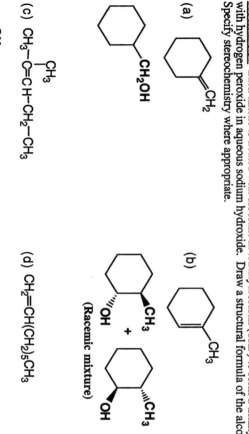

Note that two stereocenters were created, so a racemic mixture of stereoisomers is produced.

Hydroboration

Problem 6.37 Each alkene is treated with diborane in tetrahydrofuran (THF) to form a trialkylborane that is then oxidized with hydrogen peroxide in aqueous sodium hydroxide. Draw a structural formula of the alcohol formed in each case. Specify stereochemistry where appropriate.

(a)

CH₂OH

(b)

(Racemic mixture)

(c)   CH₃—C=CH—CH₂—CH₃
                |
               CH₃

(d)   CH₂=CH(CH₂)₅CH₃

CH₃
|
CH₃CHCHCH₂CH₃
|
OH

(Racemic mixture)

(e)   (CH₃)₃CCH=CH₂

CH₃
|
CH₃CCH₂CH₂OH
|
CH₃

HOCH₂CH₂(CH₂)₅CH₃

Problem 6.38 Reaction of α-pinene with diborane followed by treatment of the resulting trialkylborane with alkaline hydrogen peroxide gives an alcohol with the following structural formula. Of the four possible cis,trans isomers, one is formed in over 85% yield.

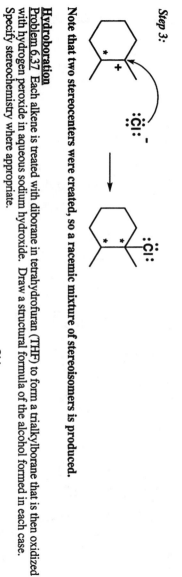

α-Pinene

(a) Draw structural formulas for the four possible cis,trans isomers of the bicyclic alcohol.

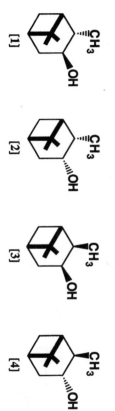

[1]   [2]   [3]   [4]

(b) Which is the structure of the isomer formed in 85% yield? How do you account for its formation? Hint: Examine the Chem3D model of α-pinene, and determine from which face of the double bond borane is more likely to approach.

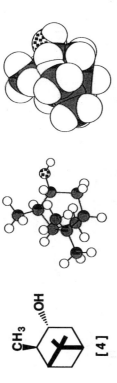

[4]

Shown in part (a) are perspective formulas for the four possible cis,trans isomers. Hydroboration followed by treatment with alkaline hydrogen peroxide results in syn (cis) addition of -H and -OH. Furthermore, boron adds to the less substituted carbon and from the less hindered side. In hydroboration of α-pinene, boron adds to the disubstituted carbon of the double bond and from the side opposite the bulky dimethyl substituted bridge. Compound [4] is the product formed in 85% yield.

Oxidation

Problem 6.39 Write structural formulas for the major organic product(s) formed by reaction of 1-methylcyclohexene with each oxidizing agent.
(a) H_2O_2/OsO_4

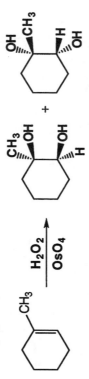

Note that even though the two OH groups are added syn with respect to each other, there are still two enantiomers produced in the reaction. The OsO$_4$ reagent can attack from either face of the double bond.

(b) O_3 followed by $(CH_3)_2S$

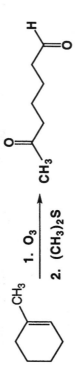

Problem 6.40 Each alkene is treated with ozone and then with dimethyl sulfide. Draw the structural formula of the organic product(s) formed from each.

(a)

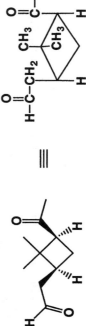

(b)

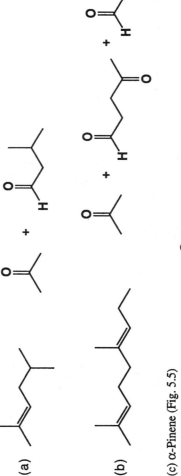

(c) α-Pinene (Fig. 5.5)

(d) Limonene (Fig. 5.5)

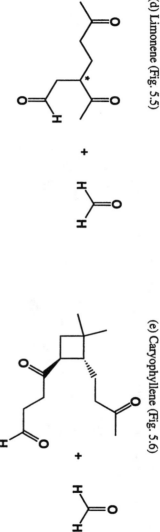

(e) Caryophyllene (Fig. 5.6)

Problem 6.41 Draw the structural formula of the alkene that reacts with ozone followed by dimethyl sulfide to give each product or set of products.

(a) C$_7$H$_{12}$ $\xrightarrow[\text{2. (CH}_3)_2\text{S}]{\text{1. O}_3}$

(b) C$_{10}$H$_{18}$ $\xrightarrow[\text{2. (CH}_3)_2\text{S}]{\text{1. O}_3}$

(c) C$_{10}$H$_{18}$ $\xrightarrow[\text{2. (CH}_3)_2\text{S}]{\text{1. O}_3}$

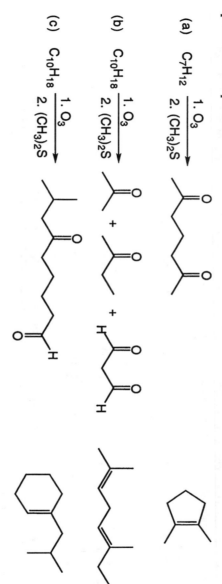

Problem 6.42 (a) Draw a structural formula for the bicycloalkene of molecular formula C$_8$H$_{12}$ that, on treatment with ozone followed by dimethyl sulfide, gives cyclohexane-1,4-dicarbaldehyde.

C$_8$H$_{12}$ $\xrightarrow[\text{2. (CH}_3)_2\text{S}]{\text{1. O}_3}$

Cyclohexane-1,4-dicarbaldehyde

Following is a representation of the bicycloalkene.

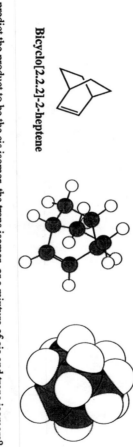

Bicyclo[2.2.2]-2-heptene

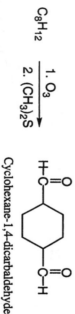

(b) Do you predict the product to be the cis isomer, the trans isomer, or a mixture of cis and trans isomers? Explain. **The product is the cis isomer, since the alkene is on one face of the starting bicyclo[2.2.2]-2-heptene.**

(c) Draw a suitable stereorepresentation for the more stable chair conformation of the dicarbaldehyde formed in this oxidation.

In either of the alternative chair conformations of the product, one carbaldehyde group is axial and the other is equatorial.

Reduction

Problem 6.43 Predict the major organic product(s) of the following reactions. Show stereochemistry where appropriate.

(a)

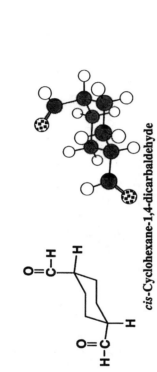

Geraniol

+ 2H₂ $\xrightarrow{\text{Pt}}$

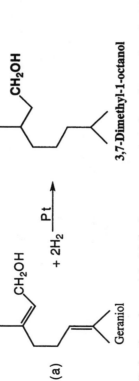

3,7-Dimethyl-1-octanol

Reduction of geraniol adds hydrogen atoms to each carbon-carbon double bond.

(b)

α-Pinene

+ H₂ $\xrightarrow{\text{Pt}}$

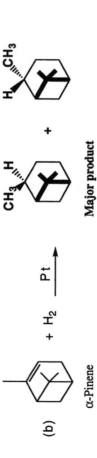

Major product

Reduction of α-pinene adds hydrogen atoms preferentially from the less hindered side of the double bond, namely the side opposite the one-carbon bridge bearing the two methyl groups. Predict, therefore, that the major isomer formed is the first one shown.

Problem 6.44 The heat of hydrogenation of *cis*-2,2,5,5-tetramethyl-3-hexene is -154 kJ (-36.7 kcal)/mol whereas that of the trans isomer is only -113 kJ (-26.9 kcal)/mol.
(a) Why is the heat of hydrogenation of the cis isomer so much larger than that of the trans isomer?

A larger value means the *cis*-2,2,5,5-tetramethyl-3-hexene is less stable than the trans isomer. This makes sense because there is so much nonbonded interaction strain due to the alkyl groups smashing into each other in the cis isomer.

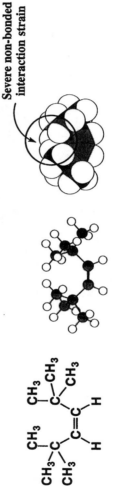

Severe non-bonded
interaction strain

cis-2,2,5,5-Tetramethyl-3-hexene

cis-Cyclohexane-1,4-dicarbaldehyde

(b) If a catalyst could be found that allowed equilibration of the cis and trans isomers at room temperature (such catalysts do exist), what would be the ratio of trans to cis isomers?

The difference in energy between the two isomers is -154 - (-113) = -41 kJ/mol. Using the equation derived in the answer to problem 6.25 gives:

$$K_{eq} = 10^{\left(\dfrac{-\Delta G^\circ}{2.303 RT}\right)} = 10^{\left(\dfrac{-(-41\ kJ/mol)}{(2.303)(8.314 \times 10^{-3}\ kcal/mol\ K)(298\ K)}\right)} = 10^{7.2} = 1.5 \times 10^7$$

Thus, at room temperature, the ratio would be 1.5 x 10⁷ to 1 in favor of the trans isomer.

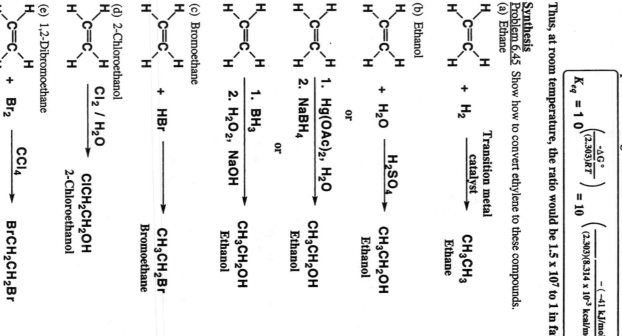

trans-2,2,5,5-Tetramethyl-3-hexene

Synthesis

Problem 6.45   Show how to convert ethylene to these compounds.

(a) Ethane

H₂C=CH₂ + H₂ →[Transition metal catalyst] **CH₃CH₃**
Ethane

(b) Ethanol

H₂C=CH₂ + H₂O →[H₂SO₄] **CH₃CH₂OH**
Ethanol

or

H₂C=CH₂ →[1. Hg(OAc)₂, H₂O] [2. NaBH₄] **CH₃CH₂OH**
Ethanol

or

H₂C=CH₂ →[1. BH₃] [2. H₂O₂, NaOH] **CH₃CH₂OH**
Ethanol

(c) Bromoethane

H₂C=CH₂ + HBr → **CH₃CH₂Br**
Bromoethane

(d) 2-Chloroethanol

H₂C=CH₂ →[Cl₂ / H₂O] **ClCH₂CH₂OH**
2-Chloroethanol

(e) 1,2-Dibromoethane

H₂C=CH₂ + Br₂ →[CCl₄] **BrCH₂CH₂Br**
1,2-Dibromoethane

(f) 1,2-Ethanediol

$$\xrightarrow[\text{H}_2\text{O}_2]{\text{OsO}_4}$$

HOCH$_2$CH$_2$OH
1,2-Ethanediol

(g) Chloroethane

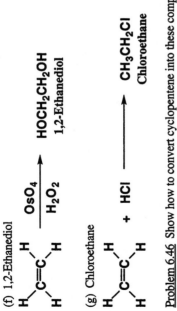

$+$ **HCl** $\longrightarrow$

CH$_3$CH$_2$Cl
Chloroethane

<u>Problem 6.46</u> Show how to convert cyclopentene into these compounds.

(a) *trans*-1,2-Dibromocyclopentane

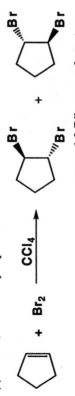

$+$ **Br$_2$** $\xrightarrow{\text{CCl}_4}$

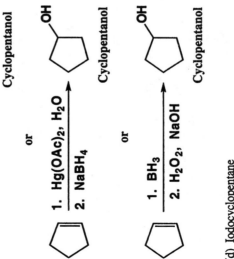

$+$

trans-1,2-Dibromocyclopentane

Note that the product of the reaction is actually a pair of enantiomers.

(b) *cis*-1,2-Cyclopentanediol

$\xrightarrow[\text{H}_2\text{O}_2]{\text{OsO}_4}$

cis-1,2-Cyclopentanediol

(c) Cyclopentanol

$+$ **H$_2$O** $\xrightarrow{\text{H}_2\text{SO}_4}$

Cyclopentanol

or

$\xrightarrow[\text{2. NaBH}_4]{\text{1. Hg(OAc)}_2, \text{ H}_2\text{O}}$

Cyclopentanol

or

$\xrightarrow[\text{2. H}_2\text{O}_2, \text{ NaOH}]{\text{1. BH}_3}$

Cyclopentanol

(d) Iodocyclopentane

$+$ **HI** $\longrightarrow$

Iodocyclopentane

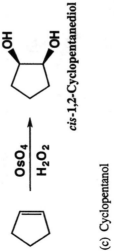

(e) Cyclopentane

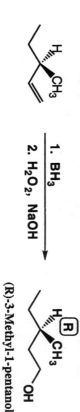

+ H₂

$\xrightarrow{\text{Transition metal}}{\text{catalyst}}$

Cyclopentane

(f) Pentanedial

$\xrightarrow[\text{2. (CH}_3)_2\text{S}]{\text{1. O}_3}$

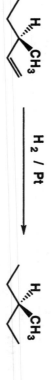

Pentanedial

Reactions that Produce Chiral Compounds

Problem 6.47 State the number and kind of stereoisomers formed when (R)-3-methyl-1-pentene is treated with these reagents.

(R)-3-Methyl-1-pentene

(a) Hg(OAc)₂, H₂O followed by NaBH₄

The alcohol produced in this reaction has a new stereocenter, so the net result is a pair of diastereomers; (2S,3R)-3-methyl-2-pentanol and (2R,3R)-3-methyl-2-pentanol.

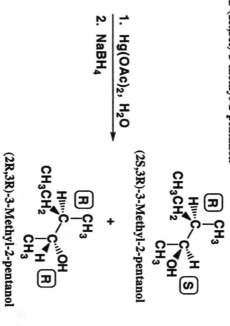

$\xrightarrow[\text{2. NaBH}_4]{\text{1. Hg(OAc)}_2, \text{H}_2\text{O}}$

(2S,3R)-3-Methyl-2-pentanol

+

(2R,3R)-3-Methyl-2-pentanol

(b) H₂/Pt

The alkane produced in this reaction does not have any new stereocenters, and the only product, 3-methylpentane, is not chiral.

$\xrightarrow{\text{H}_2 \text{/ Pt}}$

3-Methyl-1-pentane (achiral)

(c) BH₃ followed by H₂O₂ in NaOH

These reagents give the non-Markovnikov product, so the primary alcohol produced does not have any new stereocenters, and the only product is (R)-3-methyl-1-pentanol.

$\xrightarrow[\text{2. H}_2\text{O}_2, \text{ NaOH}]{\text{1. BH}_3}$

(R)-3-Methyl-1-pentanol

(d) Br$_2$ in CCl$_4$

The dibromide produced in this reaction has a new stereocenter, so the net result is a pair of diastereomers; (2R,3R)-1,2-dibromo-3-methylpentane and (2S,3R)-1,2-dibromo-3-methylpentane.

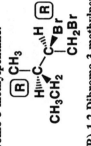

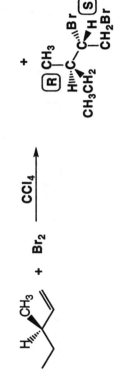

(2R,3R)-1,2-Dibromo-3-methylpentane

+

(2S,3R)-1,2-Dibromo-3-methylpentane

Problem 6.48 Describe the stereochemistry of the bromohydrin formed in each reaction.

(a) *cis*-3-Hexene + Br$_2$/H$_2$O

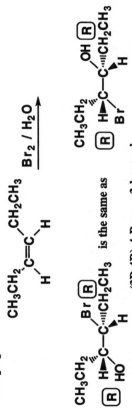

(3R,4R)-4-Bromo-3-hexanol   is the same as   (3R,4R)-4-Bromo-3-hexanol

(3S,4S)-4-Bromo-3-hexanol   is the same as   (3S,4S)-4-Bromo-3-hexanol

Remember that the anti stereochemistry of addition means that the Br and OH groups add to opposite faces of the double bond. There are only two products because of symmetry, the (3R,4R)-4-bromo-3-hexanol and (3S,4S)-4-bromo-3-hexanol

(b) *trans*-3-Hexene + Br$_2$/H$_2$O

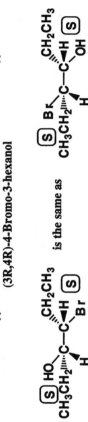

(3R,4S)-4-Bromo-3-hexanol   is the same as   (3R,4S)-4-Bromo-3-hexanol

(3S,4R)-4-Bromo-3-hexanol   is the same as   (3S,4R)-4-Bromo-3-hexanol

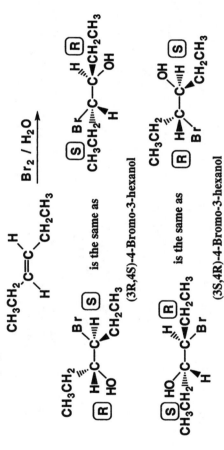

Remember that the anti stereochemistry of addition means that the Br and OH groups add to opposite faces of the double bond. There are only two products because of symmetry, the (3R,4S)-4-bromo-3-hexanol and (3S,4R)-4-bromo-3-hexanol.

Problem 6.49 In each of these reactions, the organic starting material is achiral. The structural formula of the product is given. For each reaction state:
(1) How many stereoisomers are possible for the product.
(2) Which of the possible stereoisomers is/are formed in the reaction shown.
(3) Whether the product is optically active or optically inactive.

(a)

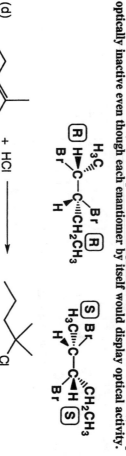

1. Hg(OAc)₂, H₂O

2. NaBH₄

The product molecule has one new stereocenter, and both enantiomers will be formed in equal amounts (racemic mixture). Thus, the product mixture will be optically inactive even though each enantiomer by itself would display optical activity.

(b)

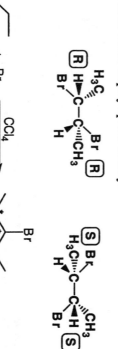

+ Br₂ $\xrightarrow{\text{CCl}_4}$

The product molecule has two new stereocenters for a total of three possible stereoisomers, a meso compound (*cis* addition) and a pair of enantiomers (*trans* addition). However, due to the anti relationship of the added bromine atoms, only the R,R and S,S isomers are formed. These are enantiomers and since they will be produced in equal amounts (racemic mixture), the product mixture will be optically inactive even though each enantiomer by itself would display optical activity.

(c)

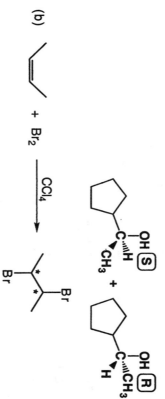

+ Br₂ $\xrightarrow{\text{CCl}_4}$

The product molecule has two new stereocenters for a total of four possible stereoisomers. However, due to the anti relationship of the added bromine atoms, only the R,R and S,S isomers can be formed. These are enantiomers and since they will be produced in equal amounts (racemic mixture), the product mixture will be optically inactive even though each enantiomer by itself would display optical activity.

(d)

+ HCl $\longrightarrow$

The product molecule has no stereocenters so no stereoisomerism is possible. The product mixture will therefore be optically inactive.

(e)

+ Cl₂ in H₂O ⟶

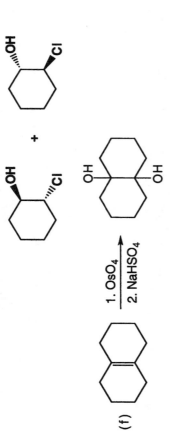

The product molecule has two new stereocenters for a total of four possible stereoisomers. However, due to the anti relationship of the added chlorine and -OH groups, only the *trans* isomers can be formed. These are enantiomers and since they will be produced in equal amounts (racemic mixture), the product mixture will be optically inactive even though each enantiomer by itself would display optical activity.

(f)

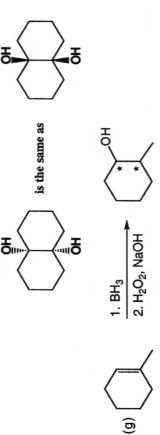

1. OsO₄
2. NaHSO₄

Because of symmetry in the molecule, the product has no stereocenters. However, there is still the possibility of *cis*, *trans* isomers, so two products are possible. Because of the mechanism of the OsO₄ reaction, only the *cis* isomer will be formed. The product is achiral so there is no optical activity.

(g)

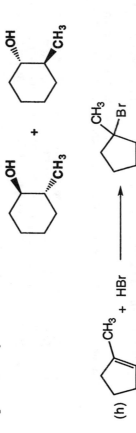

1. BH₃
2. H₂O₂, NaOH

The product molecule has two new stereocenters for a total of four possible stereoisomers. However, due to the syn addition involved with the reaction of borane, the -OH and methyl groups must be *trans* leading to only the two products shown. These are enantiomers and since they will be produced in equal amounts (racemic mixture), the product mixture will be optically inactive even though each enantiomer by itself would display optical activity.

(h)

CH₃ + HBr ⟶

The product has no stereocenters so no stereoisomerism is possible. The product mixture will therefore be optically inactive.

CHAPTER 7

Solutions to the Problems

Problem 7.1 Write the IUPAC name, and where possible, the common name of each compound.

(a)

1-Chloro-2-methylpropane
(Isobutyl chloride)

(b)

(Z)-2-Chloro-2-butene

(c)

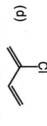

cis-1-tert-Butyl-4-chlorocyclohexane

(d)

2-Chloro-1,3-butadiene

Problem 7.2 Name and draw structural formulas for all monochlorination products formed by treatment of butane with Cl_2. Predict the major product based on the regioselectivity of the reaction of Cl_2 with alkanes.

$$CH_3CH_2CH_2CH_3 \; + \; Cl_2 \; \xrightarrow[\text{or light}]{\text{heat}} \; \text{monochlorobutanes} \; + \; HCl$$

CH₃CHCH₂CH₃
2-Chlorobutane

ClCH₂CH₂CH₂CH₃
1-Chlorobutane

There are 4 secondary hydrogen atoms and 6 primary hydrogen atoms on the molecule. The ratio of reactivity for 2°:1° chlorination is 4:1. Therefore, the predominant product will be the 2-chlorobutane, formed in approximately:

$$\frac{4 \times 4}{(4 \times 4) + (6 \times 1)} \times 100 \; = \; \boxed{73\%}$$

Problem 7.3 Using tables of bond dissociation energies (Appendix 3), calculate $\Delta H°$ for bromination of propane to give 1-bromopropane and hydrogen bromide.

$$CH_3CH_2CH_3 \; + \; Br_2 \; \longrightarrow \; CH_3CH_2CH_2Br \; + \; HBr$$

$\Delta H°$ equals the difference between the bond dissociation energies of bonds made vs. bonds broken in the reaction. One C-H bond [+ 418(100) kJ(kcal)/mol)] and one Br-Br bond [+ 192(46) kJ(kcal)/mol)] were broken while one C-Br [-285(68) kJ(kcal)/mol)] bond and one H-Br bond [-368(88) kJ(kcal)/mol)] were made in this reaction. Thus, for the whole reaction $\Delta H°$ = 418(100) + 192(46) - 285(68) - 368(88) = - 43(10) kJ(kcal)/mol.

Problem 7.4 Write a pair of chain propagation steps for the radical bromination of propane to give 1-bromopropane, and calculate $\Delta H°$ for each propagation step, and for the overall reaction.

Following is a pair of chain propagation steps for this reaction. Of these steps, the first involving hydrogen abstraction, has the higher activation energy.

$$\Delta H^\circ \text{ kJ/mol (kcal/mol)}$$

$$CH_3-CH_2-CH_3 + \cdot Br \longrightarrow CH_3-CH_2-\dot{C}H_2 + H-Br$$
+418   +50   -368
(+100)   (+12)   (-88)

$$CH_3-CH_2-\dot{C}H_2 + Br_2 \longrightarrow CH_3-CH_2-CH_2-Br + \cdot Br$$
+192   -285   -93
(+46)   (-68)   (-22)

$$CH_3-CH_2-CH_3 + Br_2 \longrightarrow CH_3-CH_2-CH_2-Br + H-Br$$
-43
(-10)

Problem 7.5 Given the solution to Example 7.5, predict the structure of the product(s) formed when 3-hexene is treated with NBS?

$$\underset{\overset{|}{H}}{CH_3CHCH=CHCH_2CH_3} + \cdot Br \longrightarrow$$

$$\dot{C}H_3CHCH=CHCH_2CH_3 \longleftrightarrow CH_3CH=CH\dot{C}HCH_2CH_3 + \cdot Br$$

$$\downarrow Br_2 \qquad\qquad\qquad\qquad \downarrow Br_2$$

$$\underset{\overset{|}{Br}}{CH_3\overset{*}{C}HCH=CHCH_2CH_3} \qquad\qquad CH_3CH=CH\overset{*}{C}HCH_2CH_3$$

$$\underset{\overset{|}{Br}}{CH_3CHCH=CHCH_2CH_3} \qquad\qquad CH_3CH=CHCHCH_2CH_3$$

Both of the above products have a stereocenter, so they will be produced as a racemic mixture of enantiomers.

Nomenclature
Problem 7.6 Give IUPAC names for the following compounds. Where stereochemistry is shown, include a designation of configuration in your answer.

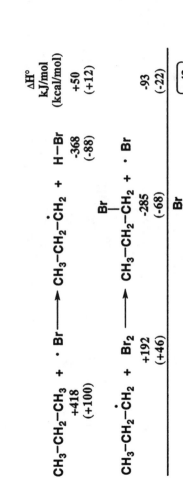

(a)

(Z)-2-Bromo-2-hexene
(trans-2-bromo-2-hexene)

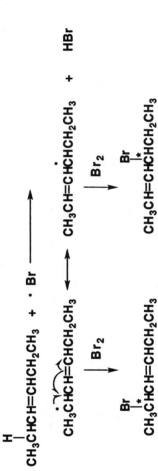

(b)

(R)-3-Bromo-3-methylcyclohexene

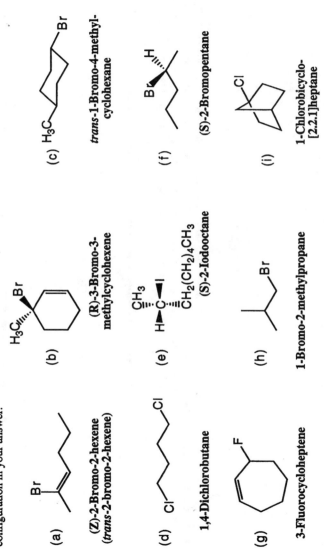

(c)

trans-1-Bromo-4-methyl-cyclohexane

(d)

1,4-Dichlorobutane

(e)

(S)-2-Iodooctane

(f)

(S)-2-Bromopentane

(g)

3-Fluorocycloheptene

(h)

1-Bromo-2-methylpropane

(i)

1-Chlorobicyclo-[2.2.1]heptane

Problem 7.7 Draw structural formulas for the following compounds.

(a) 3-Iodo-1-propene

(b) (R)-2-Chlorobutane

(c) *meso*-2,3-Dibromobutane

CH₂=CHCH₂I

H₃C—C—''''H
CH₃
Cl
CH₃CH₂

H₃C   CH₃
Br''''C—C''''/Br
H   H

(d) *trans*-1-Bromo-3-isopropylcyclohexane

(e) 1-Iodo-2,2-dimethylpropane

(f) Bromocyclobutane

or

CH(CH₃)₂   CH(CH₃)₂

CH₃
H₃C—C—CH₂I
CH₃

Br

Physical Properties

Problem 7.8 Water and methylene chloride are insoluble in each other. When each is added to a test tube, two layers form. Which layer is water and which layer is methylene chloride?

The densities of water and methylene chloride are 1.00 and 1.327 g/mL, respectively. The increased density of the methylene chloride is a consequence of the relatively high atomic weight of the chlorine atoms compared to oxygen, hydrogen, and carbon atoms. Thus, methylene chloride will be the bottom layer.

Problem 7.9 The boiling point of methylcyclohexane (C₇H₁₄, MW 98.2) is 101°C. The boiling point of perfluoromethylcyclohexane (C₇F₁₄, MW 350) is 76°C. Account for the fact that although the molecular weight of perfluoromethylcyclohexane is over 3 times that of methylcyclohexane, its boiling point is lower than that of methylcyclohexane.

This difference is due to the low polarizability of fluorine that is attributed to its small size and the tightness with which its electrons are held.

Problem 7.10 Account for the fact that, among the chlorinated derivatives of methane, chloromethane has the largest dipole moment and tetrachloromethane has the smallest dipole moment.

Name	Molecular Formula	Dipole Moment (debyes: D)
Chloromethane	CH₃Cl	1.87
Dichloromethane	CH₂Cl₂	1.60
Trichloromethane	CHCl₃	1.01
Tetrachloromethane	CCl₄	0

Each C-Cl bond is polar covalent with carbon bearing a partial positive charge and chlorine bearing a partial negative charge. Recall that molecular dipole moments are the vector sum of all the individual bond dipole moments. As shown on the following structures, adjacent C-Cl bond dipoles actually cancel each other to some extent in dichloromethane and trichloromethane, and completely in tetrachloromethane.

Vector sum $\updownarrow$ 1.87 D $\updownarrow$ 1.60 D $\updownarrow$ 1.01 D

Chloromethane   Dichloromethane   Trichloromethane   Tetrachloromethane

Halogenation of Alkanes

Problem 7.11 Name and draw structural formulas for all possible monohalogenation products that might be formed in the following reactions.

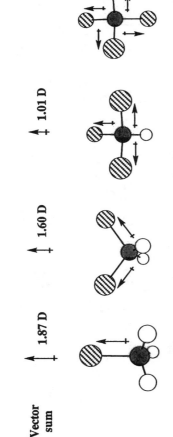

(a) [cyclopentane] + Cl$_2$ $\xrightarrow{\text{light}}$

Chlorocyclopentane

[Cl-cyclopentane]

(b)

CH$_3$
|
CH$_3$CHCHCH$_2$CH$_3$
|
Cl
3-Chloro-2-methyl-
pentane

+ Cl$_2$ $\xrightarrow{\text{light}}$

CH$_2$Cl
|*
CH$_3$CHCH$_2$CH$_2$CH$_3$
1-Chloro-2-methyl-
pentane

CH$_3$
|
CH$_3$CHCH$_2$CHCH$_3$
|
Cl
2-Chloro-4-methyl-
pentane

CH$_3$
|
CH$_3$CCH$_2$CH$_2$CH$_3$
|
Cl
2-Chloro-2-methyl-
pentane

CH$_3$
|
CH$_3$CHCH$_2$CH$_2$CH$_2$Cl
1-Chloro-4-methyl-
pentane

(c)

CH$_3$
|
CH$_3$CHCHCH$_3$
|
CH$_3$

+ Br$_2$ $\xrightarrow{\text{light}}$

CH$_2$Br
|*
CH$_3$CHCHCH$_3$
|
CH$_3$
1-Bromo-2,3-dimethyl-
butane

CH$_3$
|
CH$_3$C—CHCH$_3$
| |
Br CH$_3$
2-Bromo-2,3-dimethyl-
butane

In the above reactions, the products with stereocenters are created as racemic mixtures.

Problem 7.12 Which compounds can be prepared in high yield by regioselective halogenation of an alkane?
(a) 2-Chloropentane   (b) Chlorocyclopentane
(c) 2-Bromo-2-methylheptane   (d) 2-Bromo-3-methylbutane
(e) 2-Bromo-2,4,4-trimethylpentane   (f) Iodoethane

To be made in high yield, the compound must be the only monohalogentation product possible because of symmetry in the starting alkane, or, alternatively, the product must have the halogen on the single most-substituted carbon atom. Thus, (b), (c), and (e) can be prepared in high yield. (f) Cannot be prepared because it would be an endothermic reaction. The other products would be produced along with unsatisfactory amounts of other monohalogenation products. In particular, 3-chloropentane and 2-bromo-2-methylbutane would be major contaminants in preparations of (a) 2-chloropentane and (d) 2-bromo-3-methylbutane, respectively.

Problem 7.13 There are three constitutional isomers of molecular formula C_5H_{12}. When treated with chlorine at 300°C, isomer A gives a mixture of four monochlorination products. Under the same conditions, isomer B gives a mixture of three monochlorination products, and isomer C gives only one monochlorination product. From this information, assign structural formulas to isomers A, B, and C.

Structural formulas for the three alkanes of are:

$$CH_3-CH-CH_2-CH_3 \quad CH_3-CH_2-CH_2-CH_2-CH_3 \quad CH_3-C-CH_3$$

with CH_3 at top of A and CH_3 above and below the central C of C.

A	B	C
2-Methylbutane (Isopentane)	Pentane	2,2-Dimethylpropane (Neopentane)

To arrive at the correct assignments of structural formulas, first write formulas for all monochloroalkanes possible from each structural formula. Then compare these numbers with those observed for A, B, and C. Because isomer B gives three monochlorination products, it must be pentane. By the same reasoning, A must be 2-methylbutane, and C must be 2,2-dimethylpropane.

Problem 7.14 Following is a balanced equation for bromination of propane.

$$CH_3CH_2CH_3 + Br_2 \longrightarrow CH_3\overset{\displaystyle Br}{CH}CH_3 + HBr$$

(a) Using the values for bond dissociation energies given in Appendix A, calculate ΔH° for this reaction.

Formation of 2-bromopropane (isopropyl bromide) by radical bromination of propane is *exothermic* by 59(14) kJ(kcal)/mol.

					ΔH° kJ/mol (kcal/mol)
$CH_3-CH_2-CH_3$	+	Br_2	$\longrightarrow$	$CH_3-\overset{\displaystyle Br}{CH}-CH_3 + H-Br$	
+402		+192		-285 -368	-59
(+96)		(+46)		(-68) (-88)	(-14)

(b) Propose a pair of chain propagation steps and show that they add up to the observed reaction.

Following is a pair of chain propagation steps for this reaction. Of these steps, the first involving hydrogen abstraction has the higher activation energy.

					ΔH° kJ/mol (kcal/mol)
$CH_3-CH_2-CH_3$	+	$\cdot Br$	$\longrightarrow$	$CH_3-\dot{C}H-CH_3 + H-Br$	
+402				-368	+34
(+96)				(-88)	(+8)
$CH_3-\dot{C}H-CH_3$	+	Br_2	$\longrightarrow$	$CH_3-\overset{\displaystyle Br}{CH}-CH_3 + \cdot Br$	
+192				-285	-93
(+46)				(-68)	(-22)
				sum of ΔH° for chain propagation steps:	-59 (-14)

Following is an alternative pair of chain propagation steps. Because of the considerably higher activation energy of the first of these steps, the rate of chain propagation by this mechanism is so low that it is not competitive with the chain mechanism first proposed.

$$\underset{\substack{+402 \\ (+96)}}{CH_3-CH_2-CH_3} + \; \cdot Br \longrightarrow \underset{\substack{-285 \\ (-68)}}{CH_3-\overset{\displaystyle Br}{\overset{|}{CH}}-CH_3} + \underset{\substack{+117 \\ (+28)}}{\cdot H}$$

$$\underset{\substack{+192 \\ (+46)}}{\cdot H \; + \; Br_2} \longrightarrow \underset{\substack{-368 \\ (-88)}}{H-Br} + \underset{\substack{-176 \\ (-42)}}{\cdot \; Br}$$

ΔH° kJ/mol (kcal/mol)

sum of ΔH° for chain propagation steps: | -59 (-14) |

(c) Calculate ΔH° for each chain propagation step.

See answer to part (b)

(d) Which propagation step is rate determining, that is, which crosses the higher potential energy barrier?

See answer to part (b)

Problem 7.15 Write a balanced equation and calculate ΔH° for reaction of CH_4 and I_2 to give CH_3I and HI. Explain why this reaction cannot be used as a method of preparation of iodomethane.

The reaction is not a useful preparation method, because formation of iodomethane (methyl iodide) by radical iodination is *endothermic* by 55(13) kJ(kcal)/mol. It will not occur spontaneously.

$$\underset{\substack{+439 \\ (+105)}}{CH_4} + \underset{\substack{+151 \\ (+36)}}{I_2} \longrightarrow \underset{\substack{-238 \\ (-57)}}{CH_3I} + \underset{\substack{-297 \\ (-71)}}{HI}$$

ΔH° kJ/mol (kcal/mol)

+55 (+13)

Problem 7.16 Following are balanced equations for fluorination of propane to produce a mixture of 1-fluoropropane and 2-fluoropropane.

$$\underset{\text{Propane}}{CH_3CH_2CH_3} + F_2 \longrightarrow \underset{\text{1-Fluoropropane}}{CH_3CH_2CH_2F} + HF$$

$$\underset{\text{Propane}}{CH_3CH_2CH_3} + F_2 \longrightarrow \underset{\text{2-Fluoropropane}}{CH_3\overset{\displaystyle F}{\overset{|}{C}}HCH_3} + HF$$

Assume that each product is formed by a radical chain mechanism.
(a) Calculate ΔH° for each reaction.

Formation of 1-fluoropropane (propyl fluoride) and 2-fluoropropane (isopropyl fluoride) are both exothermic.

$$\underset{\substack{+418 \\ (+100)}}{CH_3-CH_2-CH_3} + \underset{\substack{+159 \\ (+38)}}{F_2} \longrightarrow \underset{\substack{-444 \\ (-106)}}{CH_3-CH_2-CH_2-F} + \underset{\substack{-569 \\ (-136)}}{HF}$$

ΔH° kJ/mol (kcal/mol)

-436 (-104)

$$\underset{\substack{+402 \\ (+96)}}{CH_3-CH_2-CH_3} + \underset{\substack{+159 \\ (+38)}}{F_2} \longrightarrow \underset{\substack{-448 \\ (-107)}}{CH_3-\overset{\displaystyle F}{\overset{|}{CH}}-CH_3} + \underset{\substack{-569 \\ (-136)}}{HF}$$

-456 (-109)

(b) Propose a pair of chain propagation steps for each reaction, and calculate ΔH° for each step.

	ΔH° kJ/mol (kcal/mol)
$CH_3-CH_2-CH_3 + \cdot F \longrightarrow CH_3-CH_2-\dot{C}H_2 + HF$	
+418 (+100)	-569 (-136)	-151 (-36)
$CH_3-CH_2-\dot{C}H_2 + F_2 \longrightarrow CH_3-CH_2-CH_2-F + \cdot F$	
+159 (+38)	-444 (-106)	-285 (-68)

sum of ΔH° for chain propagation steps: **-436 (-104)**

Following is an alternative pair of chain propagation steps.

	ΔH° kJ/mol (kcal/mol)
$CH_3-CH_2-CH_3 + \cdot F \longrightarrow CH_3-\dot{C}H-CH_3 + HF$	
+402 (+96)	-569 (-136)	-167 (-40)
$CH_3-\dot{C}H-CH_3 + F_2 \longrightarrow CH_3-CH-CH_3 + \cdot F$ (with F)	
+159 (+38)	-448 (-107)	-289 (-69)

sum of ΔH° for chain propagation steps: **-456 (-109)**

(c) Reasoning from the Hammond postulate, predict the regioselectivity of radical fluorination relative to that of radical chlorination and bromination.

Because the hydrogen abstraction step in each radical fluorination sequence is highly exothermic, the transition state is reached very early in hydrogen abstraction and the intermediate in this step has very little radical character. Therefore, the relative stabilities of primary versus secondary radicals is of little importance in determination of product. Accordingly predict very low regioselectivity for fluorination of hydrocarbons.

Problem 7.17 As you demonstrated in Problem 7.16, radical fluorination of alkanes is highly exothermic. As per Hammond's postulate, assume that the transition state for radical fluorination is almost identical to starting material. With this assumption, estimate the fraction of each monofluoro product formed in the fluorination of 2-methylbutane.

If it is assumed that the transition state for radical fluorination is almost identical to starting materials, then relative radical stabilities are not important. Thus, all types of carbon atoms (3°, 2°, and 1°) will react with approximately equal rate. The fraction of each monofluoro product will then be determined by the number of each kind of hydrogen present as shown.

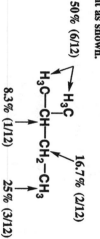

50% (6/12)   16.7% (2/12)   H₃C

8.3% (1/12)   25% (3/12)

$H_3C-CH-CH_2-CH_3$

Problem 7.18 Cyclobutane reacts with bromine to give bromocyclobutane, but bicyclo[1.1.0]butane reacts with bromine to give 1,3-dibromocyclobutane. Account for the differences between the reactions of these two compounds.

   +   Br₂   —heat→   —Br   +   HBr

Cyclobutane   Bromocyclobutane

Bicyclo[1.1.0]butane   1,3-Dibromocyclobutane

The first reaction follows the normal radical chain mechanism. The propagation steps consist of hydrogen atom abstraction followed by reaction with Br₂ to generate the product.

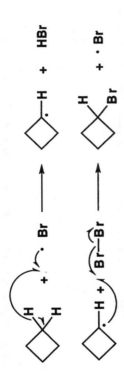

The lower reaction can be explained by a mechanism in which the highly strained bridging bond reacts with the Br radical during the first propagation step. Note that the extreme ring strain of the bicyclic molecule provides the driving force for the first propagation step.

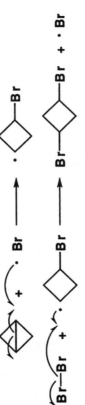

<u>Problem 7.19</u> The first chain propagation step of all radical halogenation reactions we considered in Section 7.5B is abstraction of hydrogen by the halogen atom to give an alkyl radical and HX, as for example

CH₃CH₃ + ·Br ⟶ CH₃CH₂· + HBr

Suppose, instead, that radical halogenation occurs by an alternative pair of chain propagation steps, beginning with this step:

CH₃CH₃ + ·Br ⟶ CH₃CH₂Br + H·

(a) Propose a second chain propagation step. Remember that a characteristic of chain propagation steps is that they add to the observed reaction.

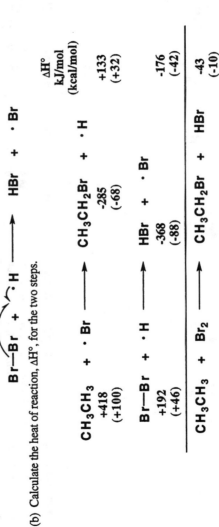

Br—Br + ·H ⟶ HBr + ·Br

(b) Calculate the heat of reaction, ΔH°, for the two steps.

				ΔH° kJ/mol (kcal/mol)
CH₃CH₃ + ·Br ⟶ CH₃CH₂Br + ·H				
+418 (+100)		-285 (-68)		+133 (+32)
Br—Br + ·H ⟶ HBr + ·Br				
+192 (+46)		-368 (-88)		-176 (-42)
CH₃CH₃ + Br₂ ⟶ CH₃CH₂Br + HBr				-43 (-10)

(c) Compare the energetics and relative rates of the set of chain propagation steps in Section 7.5B with the set proposed here.

Here, the first propagation step has a very high activation energy barrier, so it would not compete with the first propagation step proposed in Section 7.5B.

Allylic Halogenation

Problem 7.20 Following is a balanced equation for the allylic bromination of propene.

$$CH_2=CHCH_3 + Br_2 \longrightarrow CH_2=CHCH_2Br + HBr$$

(a) Calculate the heat of reaction, $\Delta H°$, for this conversion.

See the answer to part (b)

(b) Propose a pair of chain propagation steps and show that they add up to the observed stoichiometry.

$$
\begin{array}{c}
 & & & & & & \Delta H° \\
 & & & & & & \text{kJ/mol} \\
 & & & & & & \text{(kcal/mol)}
\end{array}
$$

$$CH_2=CH-CH_3 \; + \; \cdot \, Br \longrightarrow CH_2=CH-\overset{\cdot}{C}H_2 + H-Br$$
$$\begin{array}{cccc} +360 & & & -368 \\ (+86) & & & (-88) \end{array}$$

$$CH_2=CH-\overset{\cdot}{C}H_2 \; + \; Br_2 \longrightarrow CH_2=CH-CH_2Br \; + \; \cdot \, Br$$
$$\begin{array}{cccc} +192 & & & -230 & & -38 \\ (+46) & & & (-55) & & (-9) \end{array}$$

sum of $\Delta H°$ for chain propagation steps: $\boxed{\begin{array}{c} -46 \\ (-11) \end{array}}$

(c) Calculate the $\Delta H°$ for each chain propagation step and show that they add up to the observed $\Delta H°$ for the overall reaction.

See the answer to part (b).

Problem 7.21 Using the table of bond dissociation energy (Appendix 3), estimate the bond dissociation energy of each indicated bond in cyclohexene.

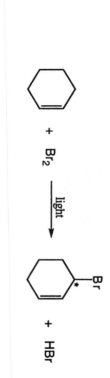

(a)

(b)

(c)

Estimate (a) the bond dissociation energy (BDE) of this secondary site to be **402(96) kJ (kcal)/mol**, (b) BDE of this allylic site to be **360(86) kJ(kcal)/mol**, and (c) BDE of this vinylic site to be **444(106) kJ (kcal)/mol**.

Problem 7.22 Propose a series of chain initiation, propagation, and termination steps for this reaction and estimate its heat of reaction.

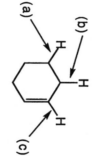

$$+ \; Br_2 \xrightarrow{\text{light}} + \; HBr$$

Initiation:

$$Br_2 \xrightarrow{\text{light}} 2 \; Br\cdot$$

Propagation:

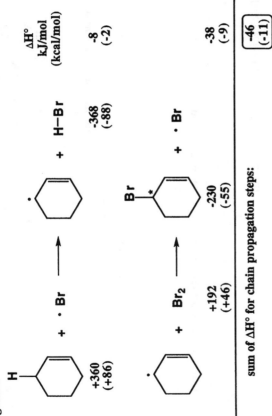

			ΔH° kJ/mol (kcal/mol)

+360
(+86)   -368
(-88)   -8
(-2)

+192
(+46)   -230
(-55)   -38
(-9)

sum of ΔH° for chain propagation steps: **-46
(-11)**

Termination:

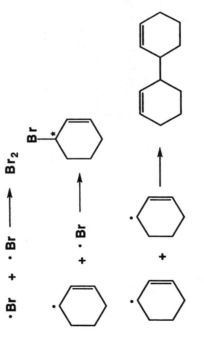

· Br + · Br ⟶ Br₂

Problem 7.23 The major product formed when methylenecyclohexane is treated with NBS in carbon tetrachloride is 1-bromomethylcyclohexene. Account for the formation of this product.

Recall that NBS can be considered a source of Br radicals and Br₂.

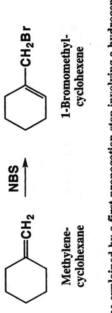

Methylene-
cyclohexane

1-Bromomethyl-
cyclohexene

The above reaction can be explained by a first propagation step involving a hydrogen atom abstraction.

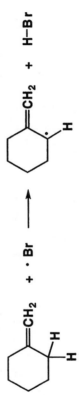

The resulting allylic radical can be represented as the hybrid of two contributing structures. The one on the right is the major contributor because it contains the more stable trisubstituted carbon-carbon double bond.

The second propagation step completes the reaction.

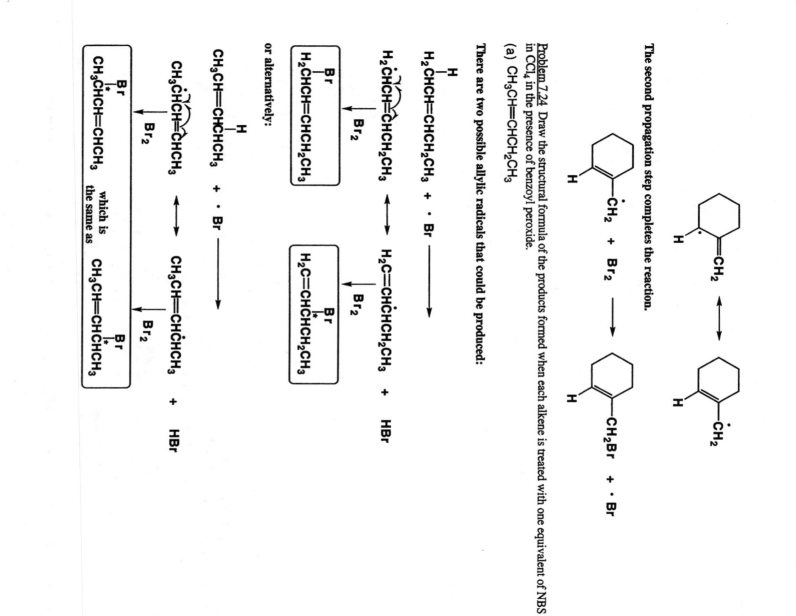

Problem 7.24 Draw the structural formula of the products formed when each alkene is treated with one equivalent of NBS in CCl₄ in the presence of benzoyl peroxide.

(a) CH₃CH=CHCH₂CH₃

There are two possible allylic radicals that could be produced:

H₂ĊHCH=CHCH₂CH₃ + · Br ⟶

H₂ĊHCH=CHCH₂CH₃ ⟷ H₂C=CHĊHCH₂CH₃ + HBr

Br₂

H₂CHCH=CHCH₂CH₃
  |
  Br

H₂C=CHĊHCH₂CH₃
        |*
        Br

or alternatively:

CH₃ĊHCH=CHCH₃ + · Br ⟶

CH₃ĊHCH=CHCH₃ ⟷ CH₃CH=CHĊHCH₃ + HBr

Br₂

CH₃CHCH=CHCH₃
   |
   Br

CH₃CH=CHCHCH₃
           |*
           Br

which is the same as   CH₃CH=CHCHCH₃

(b)

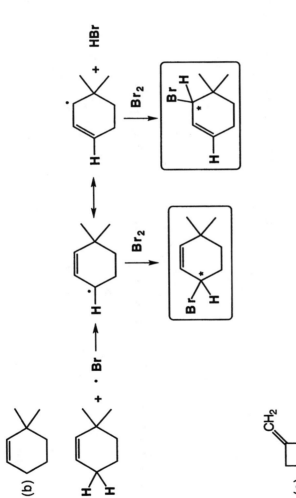

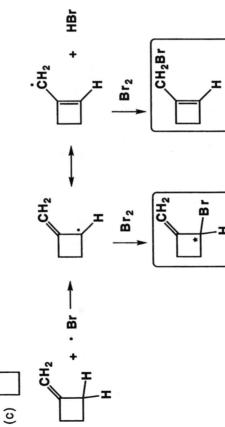

(c)

Note that the products shown above having a stereocenter will be created as a racemic mixture.

Problem 7.25 The activation energy for hydrogen abstraction from ethane by a chlorine atom is 4.2 kJ/mol; that for hydrogen abstraction by a bromine atom is 55 kJ/mol (Section 7.5D). Calculate the ratio of rate constants, k_{Cl}/k_{Br}, for these two reactions.

The difference in activation energies ($\Delta\Delta G^{\ddagger}$) is (4.2 - 55) kJ/mol = -50.8 kJ/mol. Using the equation derived for Problem 6.2 and plugging in the appropriate values at room temperature gives:

$$\Delta\Delta G^{\ddagger} = -2.303 \ RT \log \frac{k_{Cl}}{k_{Br}} = (-5.71 \ kJ/mol)(\log \frac{k_{Cl}}{k_{Br}})$$

$$\log \frac{k_{Cl}}{k_{Br}} = \frac{\Delta\Delta G^{\ddagger}}{-5.71} = \frac{-50.8}{-5.71} = 8.9 \qquad \boxed{\frac{k_{Cl}}{k_{Br}} = 10^{8.9}}$$

Synthesis

Problem 7.26 Calculate the ΔH° for the following reaction step. What can you say regarding the possibility of bromination at a vinylic hydrogen?

					ΔH° kJ/mol (kcal/mol)
$CH_2=CH_2$	+	$\cdot Br$	⟶	$CH_2=CH\cdot$	+	$H-Br$
+444				-368		76
(+106)				(-88)		(18)

This reaction step has such a high ΔH°, 76 kJ/mol, that it will preclude bromination at a vinylic hydrogen.

Problem 7.27 Show reagents and conditions to bring about these conversions.

(a) 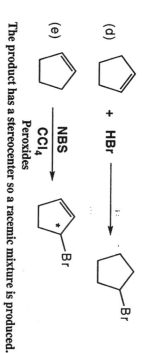 + HCl ⟶

(b) $CH_3CH=CHCH_3$ $\xrightarrow[\text{Peroxides}]{\text{NBS} \atop \text{CCl}_4}$ $CH_3CH=CHCH_2Br$

(c) $CH_3CH=CHCH_3$ + Br_2 $\xrightarrow{\text{CCl}_4}$ $\overset{*}{C}H_3CH-\overset{*}{C}HCH_3$
Br Br

Two new stereocenters are created leading to either a meso compound or a racemic mixture depending on whether the 2-butene starting material is cis or trans.

(d) 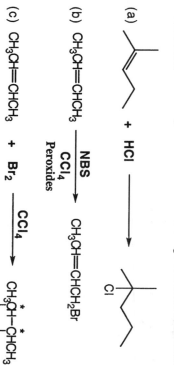 + HBr ⟶

(e)  $\xrightarrow[\text{Peroxides}]{\text{NBS} \atop \text{CCl}_4}$

The product has a stereocenter so a racemic mixture is produced.

Manta ray *Manta birostris*
Bora Bora, French Polynesia

CHAPTER 8
Solutions to the Problems

Problem 8.1 Complete the following nucleophilic substitution reactions. In each, show the electron pair on the nucleophile that forms the new bond to carbon.

(a)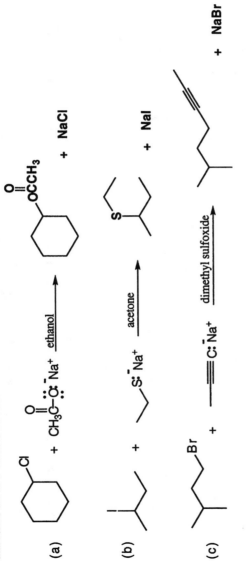

+ NaCl

(b)

+ NaI

(c)

+ NaBr

Problem 8.2 Complete these S_N2 reactions, showing the configuration of the product.

In both cases, the stereochemistry at the site of reaction is due to the nucleophile's backside attack that occurs during an S_N2 reaction.

(a)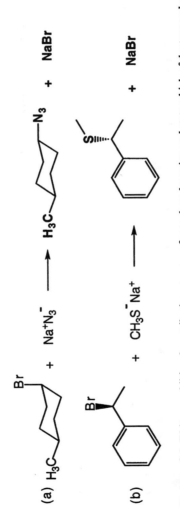

+ NaBr

(b)

+ NaBr

Problem 8.3 Write an additional contributing structure for each carbocation and state which of the two makes the greater contribution to the resonance hybrid.

A more highly substituted carbocation is more stable, so the contributing structure that has the more highly substituted carbocation will make the greater contribution to the resonance hybrid.

(a)

Greater contribution
(2° carbocation)

(b)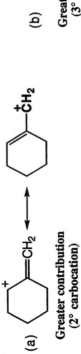

Greater contribution
(3° carbocation)

Problem 8.4 Knowing what you do about the regioselectivity of S_N2 reactions, predict the product of hydrolysis of this compound.

Step 1:

+ H_2O → S_N2; internal nucleophile

[Less-hindered site]

The nucleophilic attack by water on the three-membered ring intermediate will occur on the less-hindered carbon atom.

Step 2:

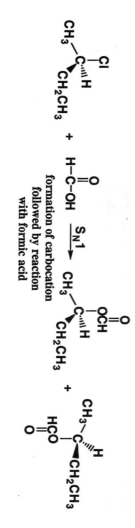

- H^+

$(CH_3)_2NCHCH_2OH$ ≡

Step 3:

A stereocenter is created so a racemic mixture is produced.

Problem 8.5 Write the expected substitution product(s) for each reaction and predict the mechanism by which each product is formed.

(a)

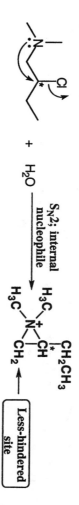

+ NaSH → acetone

The HS⁻ is a good nucleophile and, since the reaction involves a secondary haloalkane with a good leaving group, the reaction mechanism is S_N2 so inversion of configuration is observed.

(b)

+ NaSH → acetone

+ H–C–OH →

R enantiomer

The haloalkane is secondary and chloride is a good leaving group. Formic acid is an excellent ionizing solvent and a poor nucleophile. Therefore, substitution takes place by an S_N1 mechanism and leads to a mixture of enantiomers.

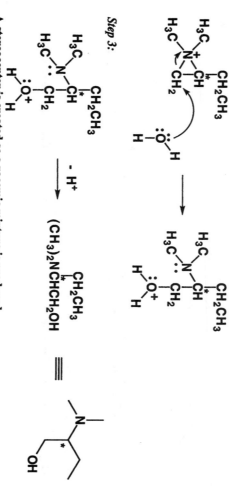

formation of carbocation followed by reaction with formic acid

Problem 8.6 Predict the β-elimination product(s) formed when each chloroalkane is treated with sodium ethoxide in ethanol. If two or more products might be formed, predict which is the major product.

When there is a choice, the more highly substituted alkene will be the major product, as predicted by Zaitsev's rule.

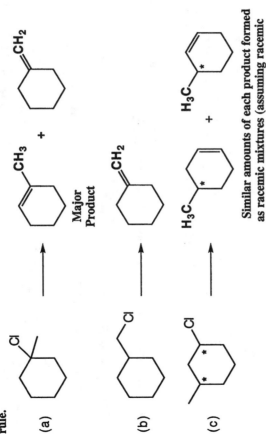

(a)

CH₃   +   CH₂

Major
Product

(b)

(c)

H₃C   +   H₃C

Similar amounts of each product formed as racemic mixtures (assuming racemic starting material).

Problem 8.7 1-Chloro-4-isopropylcyclohexane exists as two stereoisomers: one cis and one trans. Treatment of either isomer with sodium ethoxide in ethanol gives 4-isopropylcyclohexene by an E2 reaction.

$\xrightarrow[\text{CH}_3\text{CH}_2\text{OH}]{\text{CH}_3\text{CH}_2\text{O}^-\text{Na}^+}$

1-Chloro-4-isopropylcyclohexane   4-Isopropylcyclohexene

The cis isomer undergoes E2 reaction several orders of magnitude faster than the trans isomer. How do you account for this experimental observation?

The key to this problem is to think about chair conformations and the orientation of the Cl leaving group. The isopropyl group is the larger substituent on the cyclohexane ring. In the more stable chair conformation of both the cis and trans isomers, it will be in an equatorial position. In the more stable chair conformation of the cis isomer, -Cl is axial and coplanar to -H on adjacent carbons. This chair conformation undergoes β-elimination by an E2 mechanism at either of the two indicatd axial H atoms to give a racemic mixture.

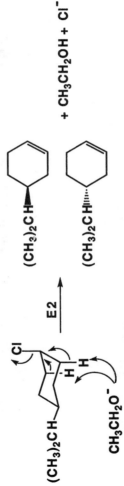

(CH₃)₂CH

CH₃CH₂O⁻

$\xrightarrow{\text{E2}}$

(CH₃)₂CH

(CH₃)₂CH

+ CH₃CH₂OH + Cl⁻

In the more stable chair conformation of the trans isomer, chlorine is equatorial and not coplanar to either -H on an adjacent carbon. Interconversion from this chair to the less stable chair results in the -Cl becoming axial and coplanar to an -H atom. It is this conformation that undergoes E2 elimination to give the cycloalkene. The bottom line is that the trans isomer undergoes E2 reaction more slowly because of the energy required to convert the more stable but E2-unreactive chair, to the less stable but E2-reactive chair. Note that reaction takes place at either of the indicated axial H atoms to generate a racemic mixture.

$(CH_3)_2CH$

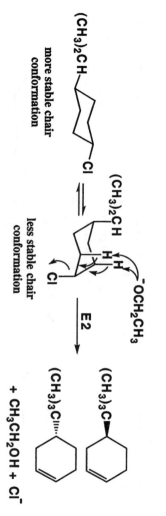

more stable chair
conformation

$(CH_3)_2CH$

less stable chair
conformation

$^-OCH_2CH_3$

E2

$(CH_3)_3C$

$(CH_3)_3C$

$+ CH_3CH_2OH + Cl^-$

Problem 8.8 Predict whether each reaction proceeds predominantly by substitution (S_N1 or S_N2), elimination (E1 or E2), or whether the two compete. Write structural formulas for the major organic product(s).

Reaction (a) will proceed predominantly by elimination (strong base, secondary haloalkane), while reactions (b) and (c) will proceed predominantly by substitution.

(a)

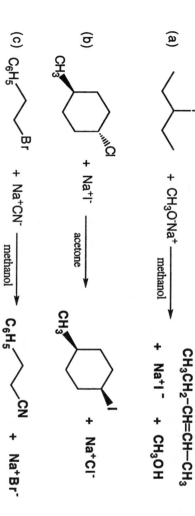

$+ CH_3O^-Na^+$ $\xrightarrow{\text{methanol}}$ $CH_3CH_2-CH=CH-CH_3$ $+ Na^+I^- + CH_3OH$

(b)

$+ Na^+I^-$ $\xrightarrow{\text{acetone}}$

$+ Na^+Cl^-$

(c)

C_6H_5 $+ Na^+CN^-$ $\xrightarrow{\text{methanol}}$ C_6H_5 CN $+ Na^+Br^-$

Nucleophilic Aliphatic Substitution

Problem 8.9 Draw a structural formula for the most stable carbocation of each molecular formula.

For (a) and (b), the most stable cations are the most highly substituted alkyl cations, while for (c) the most stable cation is the most highly substituted resonance stabilized allylic cation. For (d), the most stable cation is resonance stabilized by an adjacent oxygen atom as shown. Spreading the charge over more than one atom has a stabilizing influence on charged species.

(a) $C_4H_9^+$

(b) $C_3H_7^+$

(c) $C_8H_{15}^+$

(d) $C_3H_7O^+$

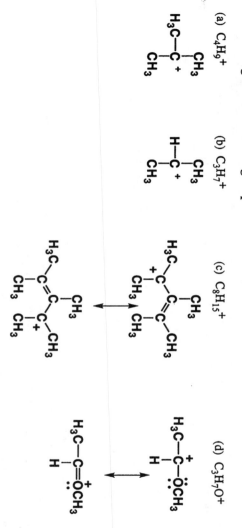

Problem 8.10 Reaction of 1-bromopropane and sodium hydroxide in ethanol follows an S_N2 mechanism. What happens to the rate of this reaction if the
(a) Concentration of NaOH is doubled?

The rate of a bimolecular reaction such as the S_N2 reaction is proportional to the concentrations of both the hydroxide and haloalkane. Thus, if the concentration of hydroxide is doubled, the rate doubles.

(b) Concentrations of both NaOH and 1-bromopropane are doubled?

The rate of a bimolecular reaction, such as the S_N2 reaction, is proportional to the concentrations of both the hydroxide and haloalkane. Thus, if the concentration of both hydroxide and haloalkane are doubled, then the rate quadruples.

(c) Volume of the solution in which the reaction is carried out is doubled?

Doubling the volume lowers the concentration of each reactant by a factor of two, so the rate is slower by a factor of four.

Problem 8.11 From each pair, select the stronger nucleophile.

(a) H_2O or OH^- (b) CH_3COO^- or OH^- (c) CH_3SH or CH_3S^-

$OH^- > H_2O$ $OH^- > CH_3COO^-$ $CH_3S^- > CH_3SH$

(d) Cl^- or I^- (e) Cl^- or I^- (f) CH_3OCH_3 or CH_3SCH_3
in DMSO   in methanol

$Cl^- > I^-$ $I^- > Cl^-$ $CH_3SCH_3 > CH_3OCH_3$

Problem 8.12 Draw the structural formula for the product of each S_N2 reaction. Where configuration of the starting material is given, show the configuration of the product.

(a) $CH_3CH_2CH_2Cl + CH_3CH_2ONa \xrightarrow{\text{ethanol}} CH_3CH_2CH_2OCH_2CH_3 + NaCl$

(b) $(CH_3)_3N: + CH_3I \xrightarrow{\text{acetone}} (CH_3)_4N^+ I^-$

(c) 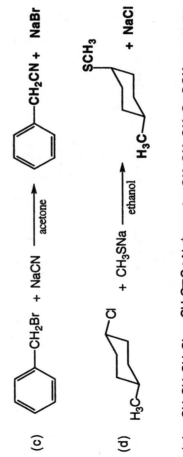 $—CH_2Br + NaCN \xrightarrow{\text{acetone}} —CH_2CN + NaBr$

(d)

(e) $CH_3CH_2CH_2Cl + CH_3C{\equiv}C:^- Na^+ \longrightarrow CH_3CH_2CH_2C{\equiv}CCH_3 + NaCl$

(f)

(g)

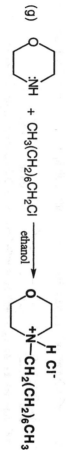

(h)   CH₃CH₂CH₂Br   + NaCN $\xrightarrow{\text{acetone}}$ CH₃CH₂CH₂CN   +   Na⁺Br⁻

Problem 8.13 You were told that each reaction in Problem 8.12 proceeds by an S_N2 mechanism. Suppose you were not told the mechanism. Describe how you could conclude from the structure of the haloalkane, the nucleophile, and the solvent that each reaction is in fact an S_N2 reaction.

(a) A primary halide, good nucleophile/strong base in ethanol, a moderately ionizing solvent all favor S_N2.

(b) Trimethylamine is a moderate nucleophile. A methyl halide in acetone, a weakly ionizing solvent, all work together to favor S_N2.

(c) Cyanide is a good nucleophile. A primary halide in acetone, a weakly ionizing solvent, all work together to favor S_N2.

(d) The alkyl chloride is secondary, so either an S_N1 or S_N2 mechanism is possible. Ethylsulfide ion is a good nucleophile, but weak base. It therefore reacts by an S_N2 pathway.

(e) The sodium salt of the terminal alkyne is a moderate nucleophile, but also a strong base. Because the halide is primary, an S_N2 pathway is favored.

(f) Ammonia is a weak base and moderate nucleophile, and the halide is primary. Therefore S_N2 is favored.

(g) The major factor here favoring an S_N2 pathway is that the leaving group is halide and on a primary carbon.

(h) The cyanide anion is a good nucleophile and bromide is a good leaving group on a primary carbon. Therefore S_N2 is favored.

Problem 8.14 Treatment of 1,3-dichloropropane with potassium cyanide results in formation of propanedinitrile. The rate of this reaction is about 1000 times greater in DMSO than it is in ethanol. Account for this difference in rate.

1,3-Dichloropropane

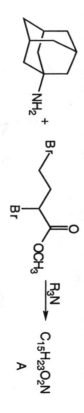

Propanedinitrile

The hydroxyl H atom of ethanol is a hydrogen bond donor, so in ethanol the CN⁻ is strongly solvated via hydrogen bonding. This strong solvation slows down the CN⁻ reaction with the haloalkane. DMSO cannot act as a hydrogen bond donor, so the CN⁻ is not strongly solvated thereby allowing faster reaction with the haloalkane.

Problem 8.15 Treatment of 1-aminoadamantane, C₁₀H₁₇N, with methyl 2,4-dibromobutanoate involves two successive S_N2 reactions and gives compound A. Propose a structural formula for compound A.

There are two successive S_N2 displacement reactions to give the four-membered ring.

The stereochemistry of the product will depend on the stereochemistry of the 2,4-dibromobutanoate starting material

<u>Problem 8.16</u> Select the member of each pair that shows the greater rate of S_N2 reaction with KI in acetone.

The relative rates of S_N2 reactions for pairs of molecules in this problem depend on two factors: (1) bromine is a better leaving group than chlorine, and (2) a primary carbon without β-branching is less hindered and more reactive toward S_N2 substitution than a primary carbon with one, two, or three branches on the β-carbon atom. The molecule that reacts faster is circled.

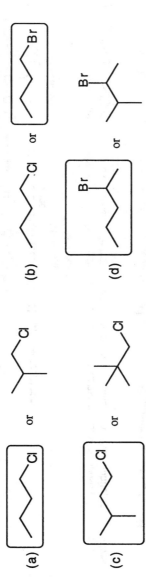

(a)   or   (b)

(c)   or   (d)

<u>Problem 8.17</u> Select the member of each pair that shows the greater rate of S_N2 reaction with KN_3 in acetone.

The compound with the least steric hindrance will reacts with the greater rate. In both pairs, it is the molecule on the left.

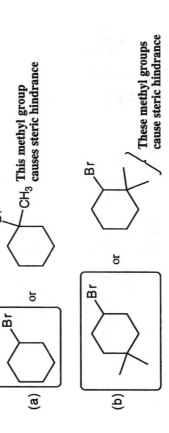

(a)   or

(b)   or

This methyl group causes steric hindrance

These methyl groups cause steric hindrance

<u>Problem 8.18</u> What hybridization best describes the reacting carbon in the S_N2 transition state?

The hybridization state of the reacting carbon is best described as sp^2 in the S_N2 transition state.

<u>Problem 8.19</u> Each carbocation is capable of rearranging to a more stable carbocation. Limiting yourself to a single 1,2-shift, suggest a structure for the rearranged carbocation.

(a)

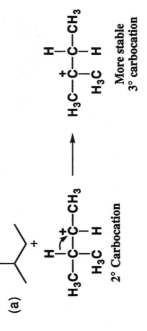

2° Carbocation

More stable
3° carbocation

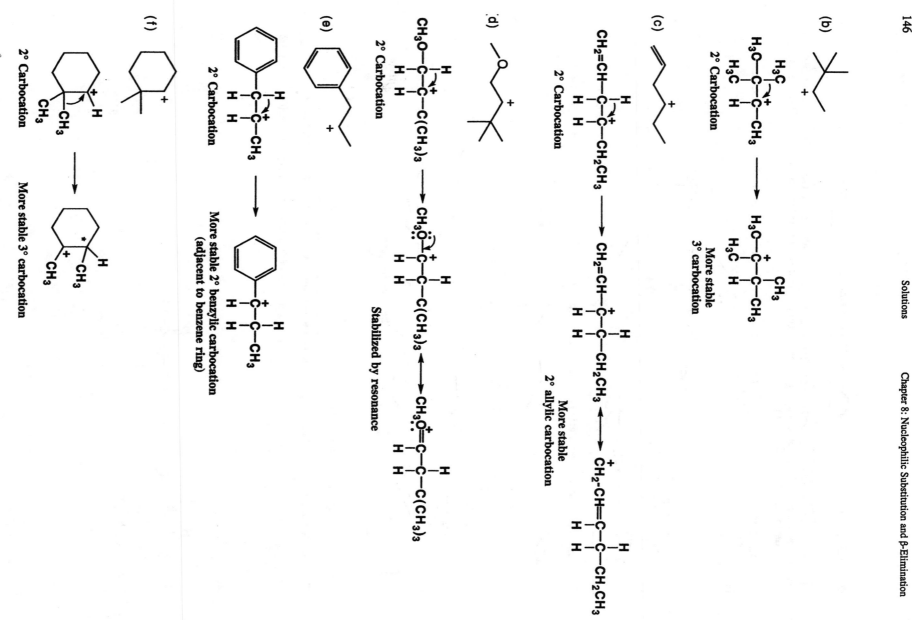

Problem 8.20 Attempts to prepare optically active iodides by nucleophilic displacement on optically active compounds with I⁻ normally produce racemic alkyl iodides. Why are the product alkyl iodides racemic?

Iodide is a good nucleophile as well as a good leaving group. The alkyl iodide that is formed will therefore react with other iodide anions according to an S$_N$2 mechanism. The resulting repeated inversion of configuration leads to full racemization of the product.

Problem 8.21 Draw a structural formula for the product of each S$_N$1 reaction. Where configuration of the starting material is given, show the configuration of the product.

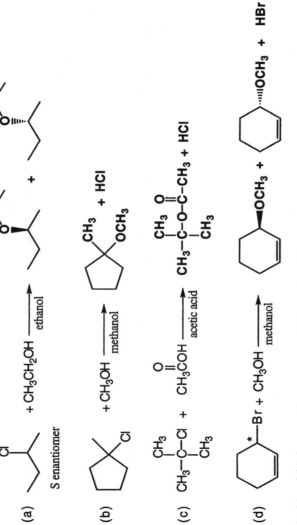

(a)

S enantiomer

+ CH₃CH₂OH →(ethanol)

(b)

+ CH₃OH →(methanol)   CH₃ + HCl, OCH₃

(c)

$CH_3-\overset{CH_3}{\underset{CH_3}{C}}-Cl$ + CH₃COH →(acetic acid) $CH_3-\overset{CH_3}{\underset{CH_3}{C}}-O-\overset{O}{C}-CH_3$ + HCl

(d)

*Br + CH₃OH →(methanol) OCH₃ + ''''OCH₃ + HBr

Problem 8.22 Suppose that you were told that each reaction in Problem 8.21 is a substitution reaction but were not told the mechanism. Describe how you could conclude from the structure of the haloalkane or cycloalkene, the nucleophile, and the solvent that each reaction is in fact an S$_N$1 reaction.

For an S$_N$1 reaction to be favored, a good leaving group and ionizing solvent are needed along with a carbocation intermediate that is relatively stable.

(a) Chlorine is a good leaving group and the resulting secondary carbocation is a relatively stable carbocation intermediate. Ethanol is a moderately ionizing solvent and a poor nucleophile.
(b) Methanol is a moderately ionizing solvent and a poor nucleophile. Chlorine is a good leaving group and the resulting carbocation is tertiary.
(c) Acetic acid is a strongly ionizing solvent and a poor nucleophile. Chlorine is a good leaving group and the resulting carbocation is tertiary.
(d) Methanol is a moderately ionizing solvent and a poor nucleophile. Bromine is a good leaving group, and the resulting carbocation is both secondary and allylic.

Problem 8.23 Alkenyl halides such as vinyl bromide, CH₂=CHBr, undergo neither S$_N$1 nor S$_N$2 reactions. What factors account for this lack of reactivity?

In vinyl bromide, the bromine atom is bonded to an sp² hybridized carbon atom. An S$_N$1 reaction would give a vinyl carbocation, but such a carbocation is high in energy and very difficult to generate. In order to undergo an S$_N$2 reaction, the nucleophile must approach in a direction opposite the C-X bond. This trajectory is not possible for a vinyl halide.

Problem 8.24 Select the member of each pair of compounds that undergoes S$_N$1 solvolysis in aqueous ethanol more rapidly.

Relative rates for each pair of compounds listed in this problem depend on a combination of two factors: (1) bromine is a better leaving group than chlorine and (2) the stability of the resulting carbocation. The molecule that reacts faster is circled.

(a)  or

Formation of a tertiary carbocation is feasible, while formation of a primary carbocation is not.

(b) 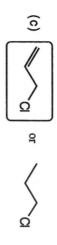 or

Bromine is a better leaving group than chlorine.

(c) 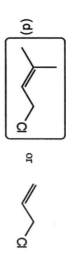 or

Formation of a resonance-stabilized 1° allylic carbocation is feasible, while formation of a primary carbocation is not.

(d) 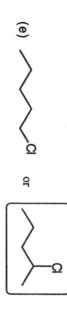 or

The activation energy for formation of the dialkyl allylic carbocation is lower than that for formation of an unsubstituted allylic carbocation.

(e)  or

The activation energy for formation of a secondary carbocation is lower than that for formation of a vinylic carbocation.

(f)  or

The activation energy for formation of an allylic carbocation is lower than that for formation of a vinylic carbocation.

Problem 8.25 Account for the following relative rates of solvolysis under experimental conditions favoring S_N1 reaction.

relative rate of solvolysis (S_N1)	0.2	1	10^9

1-Chloro-1-ethoxymethane (chloromethyl ethyl ether) reacts the fastest by far in a solvolysis reaction because the carbocation produced by loss of the chlorine atom is stabilized by the adjacent ether oxygen atom. Thus, the activation energy for this reaction is significantly lower than for the other two molecules. The most important contributing structures are shown below for the stabilized carbocation.

$$CH_3CH_2\ddot{O}CH_2Cl \longrightarrow CH_3CH_2\ddot{O}\overset{+}{C}H_2 \longleftrightarrow CH_3CH_2\overset{+}{\ddot{O}}=CH_2$$

1-Chloro-2-methoxyethane (2-chloroethyl methyl ether) reacts the slowest because the carbocation produced during the reaction is somewhat destabilized by the ether oxygen atom that is two carbon atoms away. This is because oxygen is more electronegative than carbon, so there is a partial positive charge on the carbon atoms bonded to the oxygen. Thus, the carbocation produced by departure of the chlorine atom is adjacent to this partially positive carbon atom; a destabilizing arrangement.

destabilizing

$$\overset{\delta+}{CH_3}-\overset{\delta-}{\underset{\cdot\cdot}{O}}-\overset{\delta+}{CH_2CH_2Cl} \longrightarrow \overset{\delta+}{CH_3}-\overset{\delta-}{\underset{\cdot\cdot}{O}}-\overset{\delta+}{CH_2CH_2}$$

Please note that in the case of 1-chloro-2-methoxyethane, there is no way to produce contributing structures with any positive charge on the oxygen atom like that shown for 1-chloro-1-ethoxymethane above.

<u>Problem 8.26</u> Not all tertiary halides undergo S_N1 reactions readily. For example, 1-iodobicyclo[2.2.2]octane is very unreactive under S_N1 conditions. What feature of this molecule is responsible for such lack of reactivity?

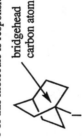

bridgehead
carbon atom

1-Iodobicyclo[2.2.2]octane

In order to form a cation, great angle strain would have to be produced in the molecule. This is because carbocations prefer to be trigonal planar (sp^2 hybridized), and loss of iodine would place a carbocation at the bridgehead position. However, the bicyclic structure of the molecule enforces a tetrahedral geometry at the bridgehead position (109.5° bond angles), thus preventing formation of the carbocation.

<u>Problem 8.27</u> Show how you might synthesize the following compounds from an haloalkane and a nucleophile:

(a)

Treatment of a halocyclohexane with cyanide.

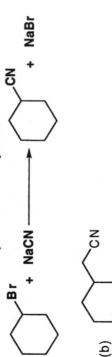

+ NaBr

(b)

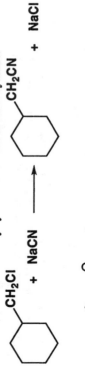

Treatment of chloromethylcyclohexane with sodium cyanide.

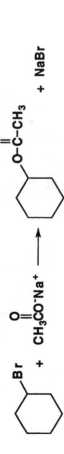

+ NaCl

(c)

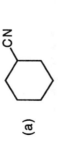

Treatment of a halocyclohexane with the sodium salt of acetic acid.

+ NaBr

(d)

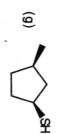

Treatment of a 1-halopentane with sodium hydrosulfide.

$$CH_3(CH_2)_3CH_2Br + HS^-Na^+ \longrightarrow CH_3(CH_2)_3CH_2SH + NaBr$$

(e)

Treatment of a 1-halohexane with the sodium salt of acetylene.

$$CH_3(CH_2)_4CH_2Br + HC\equiv C^-Na^+ \longrightarrow CH_3(CH_2)_5C\equiv CH + NaCl$$

(f)

Treatment of a haloethane with sodium or potassium ethoxide in ethanol.

$$CH_3CH_2O^-Na^+ + CH_3CH_2I \xrightarrow{CH_3CH_2OH} CH_3CH_2OCH_2CH_3 + NaI$$

(g)

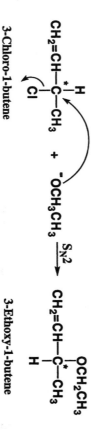

Treatment of the appropriate *trans*-halocyclopentane with sodium hydrosulfide.

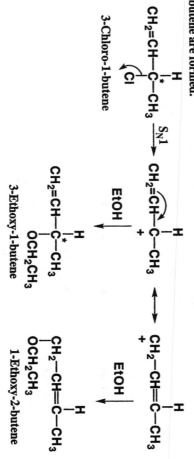

Problem 8.28 3-Chloro-1-butene reacts with sodium ethoxide in ethanol to produce 3-ethoxy-1-butene. The reaction is second order; first order in 3-chloro-1-butene and first order in sodium ethoxide. In the absence of sodium ethoxide, 3-chloro-1-butene reacts with ethanol to produce both 3-ethoxy-1-butene and 1-ethoxy-2-butene. Explain these results.

For the second order reaction with sodium ethoxide, the mechanism is S_N2 so the -OCH$_2$CH$_3$ group ends up only where the -Cl leaving group was attached. The stereochemistry of the product depends on the stereochemistry of the starting material.

$$CH_2=CH-\overset{H}{\underset{Cl}{\overset{|}{\underset{|}{C}}}}-CH_3 \quad + \quad {}^-OCH_3CH_3 \xrightarrow{S_N2} CH_2=CH-\overset{OCH_2CH_3}{\underset{H}{\overset{|}{\underset{|}{\overset{*}{C}}}}}-CH_3 + NaBr$$

3-Chloro-1-butene   3-Ethoxy-1-butene

For the second reaction, the absence of a strong nucleophile allows the S_N1 mechanism to operate. The allylic carbocation intermediate can be attacked at either the 1 or 3 positions, a fact that is readily explained by considering the two predominant contributing structures of this intermediate. Note that both enantiomers of 3-ethoxy-1-butene are formed.

$$CH_2=CH-\overset{H}{\underset{Cl}{\overset{|}{\underset{|}{\overset{*}{C}}}}}-CH_3 \xrightarrow{S_N1} CH_2=CH-\overset{H}{\underset{+}{\overset{|}{C}}}-CH_3 \longleftrightarrow CH_2-CH=\overset{H}{\underset{+}{\overset{|}{C}}}-CH_3$$

3-Chloro-1-butene

$$\xrightarrow{EtOH} CH_2=CH-\overset{H}{\underset{OCH_2CH_3}{\overset{|}{\underset{|}{\overset{*}{C}}}}}-CH_3 \qquad \xrightarrow{EtOH} CH_2-CH=\overset{H}{\underset{OCH_2CH_3}{\overset{|}{\underset{|}{C}}}}-CH_3$$

3-Ethoxy-1-butene   1-Ethoxy-2-butene

Problem 8.29 1-Chloro-2-butene undergoes hydrolysis in warm water to give a mixture of these allylic alcohols. Propose a mechanism for their formation.

$$CH_3CH=CHCH_2Cl \xrightarrow{H_2O} CH_3CH=CHCH_2OH + CH_3CHCH=CH_2$$

1-Chloro-2-butene   2-Buten-1-ol   3-Buten-2-ol

The two products can be explained by an S_N1 mechanism that produces an allylic cation that can react with water at either the 1 or 3 position. Note that 3-buten-2-ol is actually formed as a racemic mixture of enantiomers.

$$CH_3CH=CH-CH_2 \xrightarrow{S_N1} CH_3CH=CH-CH_2^+ \longleftrightarrow CH_3CH-CH=CH_2^+$$

$$\downarrow H_2O \qquad\qquad \downarrow H_2O$$

$$CH_3CH=CHCH_2OH \qquad\qquad CH_3CHCH=CH_2$$
$$\overset{+}{O}H$$

Problem 8.30 The following nucleophilic substitution occurs with rearrangement. Suggest a mechanism for formation of the observed product. If the starting material has the S configuration, what is the configuration of the stereocenter in the product.

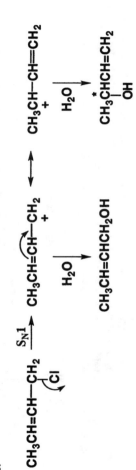

$$\xrightarrow[\;\;H_2O\;\;]{NaOH}$$

An internal nucleophilic reaction leads to the three-membered ring intermediate $C_8H_{18}NCl$ shown, that is then attacked by hydroxide at the less-hindered site to yield the product.

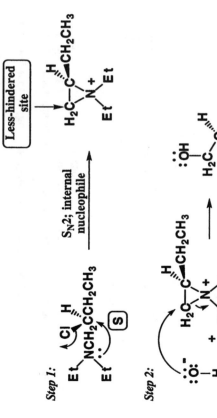

If the starting material has the S configuration, the product will have the R configuration as shown.

Problem 8.31 Propose a mechanisms for the formation of these products in the solvolysis of this alkyl bromide.

The bromide leaves to generate a 2° carbocation, that then rearranges by ring expansion to give the more stable 3° carbocation. Note how this new 3° carbocation with a five-membered ring also has much less ring strain than the strained four-membered ring 2° carbocation. The new 3° carbocation either loses the proton shown to complete an E1 reaction, or reacts with ethanol to complete the S_N1 reaction.

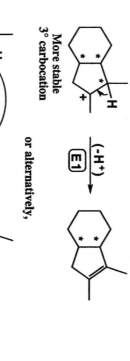

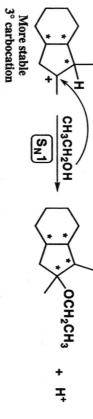

Note that the stereochemistry of the product depends on the stereochemistry of the starting material.

Problem 8.32 Solvolysis of the following bicyclic compound in acetic acid gives a mixture of products, two of which are shown. The leaving group is the anion of a sulfonic acid, ArSO_3H. A sulfonic acid is a strong acid and its anion, ArSO_3^-, is a weak base and a good leaving group. Propose a mechanism for this reaction.

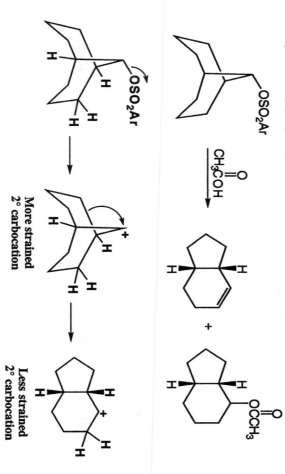

The sulfonate anion leaves to generate a 2° carbocation, that then rearranges to give a new 2° carbocation with less ring strain. The new 2° carbocation either loses the proton shown to complete an E1 reaction, or reacts with acetic acid to complete the S$_N$1 reaction.

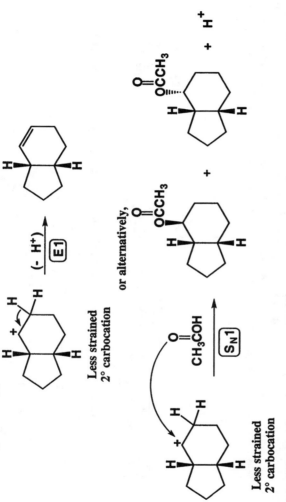

Less strained 2° carbocation

or alternatively,

Problem 8.33 Which compound in each set undergoes more rapid solvolysis when refluxed in ethanol? Show the major product formed from the more reactive compound.

More rapid solvolysis will occur for the molecule that can produce the more stable carbocation or all things being equal, the one with the better leaving group.

(a)
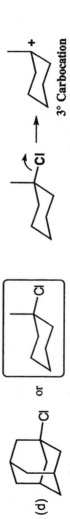

The molecule on the left can make a more stable allylic carbocation.

(b)
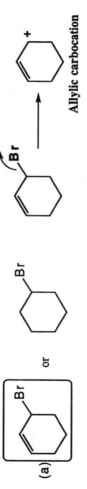

Bromide is a better leaving group than chloride.

(c)

The molecule on the left can make a more stable 3° carbocation.

(d)
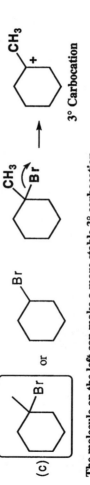

The molecule on the right can make a stable 3° carbocation. In fact, the molecule on the left cannot adopt the favored trigonal planar geometry so it reacts much slower than expected.

Problem 8.34 Account for the relative rates of solvolysis of these compounds in aqueous acetic acid.

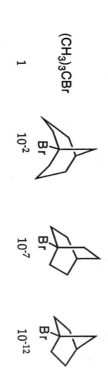

$(CH_3)_3CBr$ 1   Br 10^{-2} Br 10^{-7} Br 10^{-12}

More rapid solvolysis will occur for the molecule that can produce the more stable carbocation.

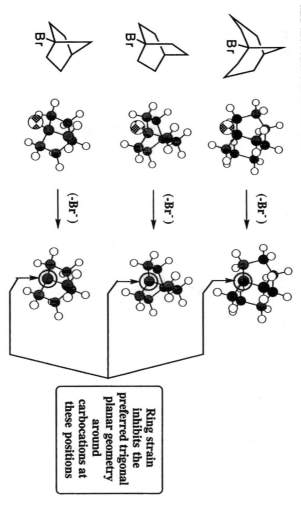

$(CH_3)_3CBr$ (-Br⁻) $(CH_3)_3C^+$ (Side view)

Preferred trigonal planar geometry

The *tert*-butyl carbocation produced upon solvolysis of *tert*-butyl bromide can easily adopt the preferred trigonal planar geometry as shown above. A trigonal planar geometry is preferred because this allows for maximum stabilization (maximum orbital overlap) due to hyperconjugation with the three adjacent methyl groups. As shown below, it is much more difficult for the three bicyclic bromides to produce the preferred trigonal planar geometries due to ring strain. Thus, the activation energies for these reactions are much higher, and the reactions are slower.

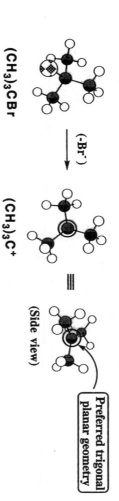

Br   (-Br⁻)

Br   (-Br⁻)

Br   (-Br⁻)

Ring strain inhibits the preferred trigonal planar geometry around carbocations at these positions

Problem 8.35 A comparison of the rates of S$_N$1 solvolysis of the bicyclic compounds (1) and (2) in acetic acid shows that compound (1) reacts 10^{11} times faster than compound (2). Furthermore, solvolysis of (1) occurs with complete retention of configuration: The nucleophile occupies the same position on the one-carbon bridge as did the leaving OSO₂Ar group.

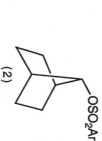

(1)   OSO₂Ar   (2)   OSO₂Ar

(a) Draw structural formulas for the products of solvolysis of each compound.

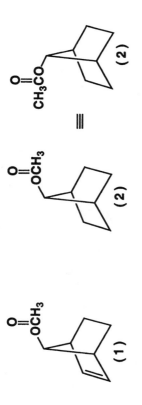

(b) Account for difference in rate of solvolysis of (1) and (2).

The pi bond of (1) is in position to not only assist in the departure of the -OSO₂Ar leaving group, but also stabilize the resulting carbocation intermediate. This stabilization can be visualized by considering the filled pi bonding orbital of the double bond overlapping with the empty 2p orbital of the cation. This interaction has the effect of placing some pi electron density into the empty carbon 2p orbital, thereby stabilizing the positive charge.

(c) Account for complete retention of configuration in the solvolysis of (1).

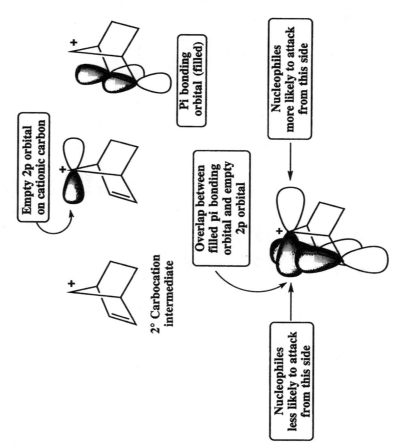

As shown on the figure above, the pi bond will block a nucleophile from attacking the cation on one face. Thus, the nucleophile will approach from the other side, the side that the -OSO₂Ar departed from, leading to predominantly retention of configuration.

β-Eliminations
Problem 8.36 Draw structural formulas for the alkene(s) formed by treatment of each haloalkane or halocycloalkane with sodium ethoxide in ethanol. Assume *anti* and coplanar, and that elimination occurs by an E2 mechanism.

The major and minor products for each E2 reaction are shown. Where there is more than one combination of leaving groups *anti* and coplanar, the major product is the more substituted alkene (the so-called Zaitzev product).

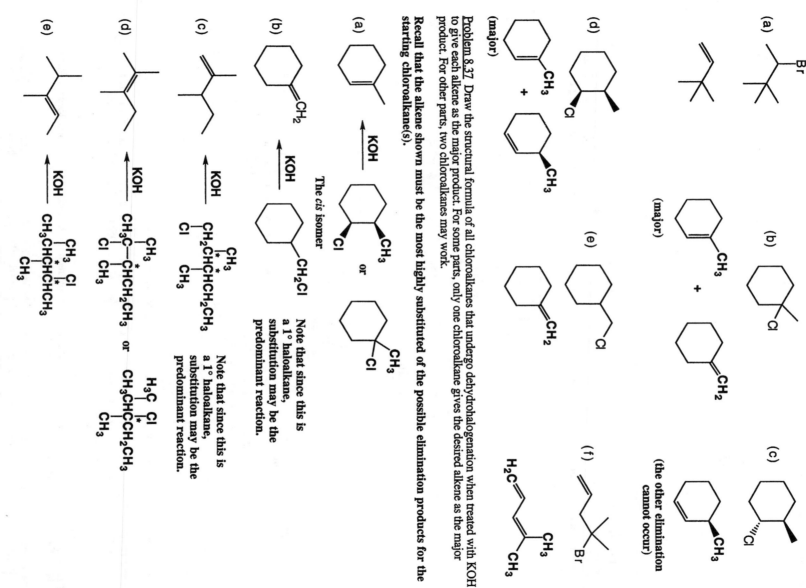

(a)

Br

(b)

(c)
(the other elimination
cannot occur)

(d)

+

CH₃
Cl

(major)

CH₃

(major)

(e)

(f)

H₂C

CH₃

CH₃

Br

Problem 8.37 Draw the structural formula of all chloroalkanes that undergo dehydrohalogenation when treated with KOH to give each alkene as the major product. For some parts, only one chloroalkane gives the desired alkene as the major product. For other parts, two chloroalkanes may work.

Recall that the alkene shown must be the most highly substituted of the possible elimination products for the starting chloroalkane(s).

(a)

KOH

or

CH₃

CH₃
Cl

CH₃
Cl

(b)

CH₂

KOH

CH₂Cl

The cis isomer

Note that since this is
a 1° haloalkane,
substitution may be the
predominant reaction.

(c)

CH₂

KOH

CH₃
CH₂CHCHCH₂CH₃
Cl   CH₃

Note that since this is
a 1° haloalkane,
substitution may be the
predominant reaction.

(d)

KOH

CH₃
CH₃C—CHCH₂CH₃
Cl   CH₃

or

H₃C Cl
CH₃CHCCH₂CH₃
CH₃

(e)

KOH

CH₃ Cl
CH₃CHCHCHCH₃
CH₃

Note that for (c), (d), and (e) any of the possible steroisomers would work.

Problem 8.38 Following are diastereomers (A) and (B) of 3-bromo-3,4-dimethylhexane. On treatment with sodium ethoxide in ethanol, each gives 3,4-dimethyl-3-hexene as the major product. One of these diastereomers gives the (E)-alkene, and the other gives the (Z)-alkene. Which diastereomer gives which alkene? Account for the stereoselectivity of each β-elimination.

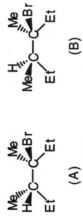

(A)   (B)

Rotate the given conformation of each stereoisomer into a conformation in which Br and H are anti and coplanar and then made to undergo E2 elimination. You will find that diastereomer (A) gives the (E)-isomer and diastereomer (B) gives the (Z)-isomer.

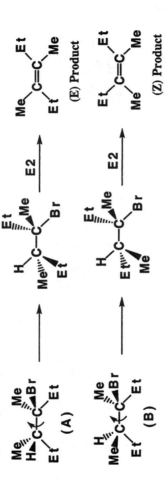

(A)

(B)

Problem 8.39 Treatment of the following stereoisomer of 1-bromo-1,2-diphenylpropane with sodium ethoxide in ethanol gives a single stereoisomer of 1,2-diphenylpropene. Predict whether the product has the E configuration or the Z configuration.

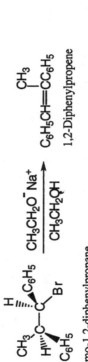

1-Bromo-1,2-diphenylpropane

1,2-Diphenylpropene

Problem 8.40 Elimination of HBr from 2-bromonorbornane gives only 2-norbornene and no 1-norbornene. How do you account for the regiospecificity of this dehydrohalogenation? In answering this question, you may find it helpful to examine molecular models of both 1-norbornene and 2-norbornene on the CD and analyze the strain in each.

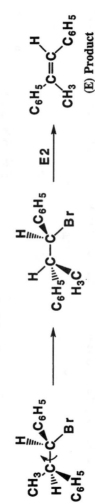

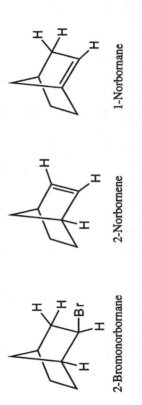

2-Bromonorbornane   2-Norbornene   1-Norbornane

1-Norbornene is not formed for at least two reasons. First, it is not possible to achieve the preferred *anti* and coplanar geometry of the hydrogen and Br atom that would lead to 1-norbornene. Second, the alkene in 1-norbornene has considerably more angle strain compared with 2-norbornene.

Problem 8.41 Which compound reacts faster when refluxed with potassium *tert*-butoxide in *tert*-butyl alcohol, *cis*-1-bromo-3-isopropylcyclohexane or *trans*-1-bromo-3-isopropylcyclohexane? Draw the structure of the expected product from the faster reacting compound.

The isopropyl group is the larger substituent on the cyclohexane ring. In the more stable chair conformation of both the *cis* and *trans* isomers, it will be in an equatorial position. In the more stable chair conformation of the *trans* isomer, -Cl is axial and coplanar to -H on adjacent carbons. This chair conformation readily undergoes β-elimination by an E2 mechanism so this is the fastest reaction.

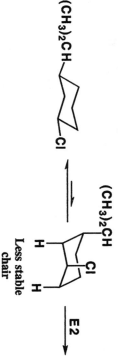

More stable chair

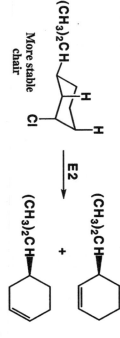

In the more stable chair conformation of the *cis* isomer, chlorine is equatorial and therefore not coplanar to either -H on adjacent carbon atoms. Interconversion from this chair to the less stable chair is required for the -Cl to become axial and coplanar to an adjacent -H atom. It is this conformation that undergoes E2 elimination to give the cycloalkene. The bottom line is that the *cis* isomer undergoes E2 reaction more slowly because of the energy required to convert the more stable chair, but E2-unreactive chair, to the less stable, but E2-reactive, chair. Note this reaction will lead to the same products as above.

Less stable chair

Note that in the above, only one of the possible enantiomers was shown as starting material for simplicity.

Substitution Versus Elimination

Problem 8.42 Consider the following statements in reference to SN1, SN2, E1, and E2 reactions of haloalkanes. To which mechanism(s), if any, does each statement apply?

(a) Involves a carbocation intermediate SN1, E1

(b) Is first order in haloalkane and first order in nucleophile SN2

(c) Involves inversion of configuration at the site of substitution SN2

(d) Involves retention of configuration at the site of substitution None. SN1, however, may proceed with predominantly racemization, but some retention.

(e) Substitution at a stereocenter gives predominantly a racemic product SN1

(f) Is first order in haloalkane and zero order in base E1

(g) Is first order in haloalkane and first order in base E2

(h) Is greatly accelerated in protic solvents of increasing polarity SN1, E1

(i) Rearrangements are common SN1, E1

(j) Order of reactivity is 3° > 2° > 1° SN1, E2, E1

(k) Order of reactivity is methyl > 1° > 2° > 3° SN2

Problem 8.43 Arrange these haloalkanes in order of increasing ratio of E2 to S$_N$2 products observed on reaction of each with sodium ethoxide in ethanol.

(a) CH$_3$CH$_2$Br   (b) CH$_3$CHCH$_2$Br   (c) CH$_3$CCH$_2$CH$_3$ (d) CH$_3$CHCH$_2$CH$_2$Br

with CH$_3$ above (b), CH$_3$ above (c) with Cl below, CH$_3$ above (d)

More E2 will occur the more hindered the site of reaction and/or the more substituted (stable) the product alkene. Thus, listed in order from least to most E2 product:

CH$_3$CH$_2$Br < CH$_3$CHCH$_2$CH$_2$Br < CH$_3$CHCH$_2$Br < CH$_3$CCH$_2$CH$_3$

with CH$_3$ above third, CH$_3$ above and Cl below fourth

Problem 8.44 Draw a structural formula for the major organic product of each reaction and specify the most likely mechanism for formation of the product you have drawn.

The substitution and elimination products for the reactions are given in bold. In each case, the different parameters discussed in the chapter are considered including the type of haloalkane (primary, secondary, tertiary, etc.) and the relative strength of the nucleophile/base.

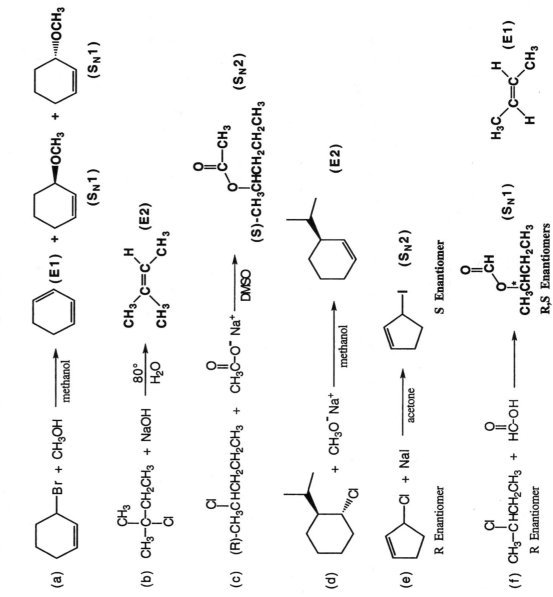

(g) $CH_3CH_2O^-$ Na + $CH_2=CHCH_2Cl$ $\xrightarrow{\text{ethanol}}$ $CH_3CH_2OCH_2CH=CH_2$ (S_N2)

Problem 8.45 When *cis*-4-chlorocyclohexanol is treated with sodium hydroxide in ethanol, it gives only the substitution product *trans*-1,4-cyclohexanediol (1). Under the same reaction conditions, *trans*-4-chlorocyclohexanol gives 3-cyclohexenol (2) and the bicyclic ether (3).

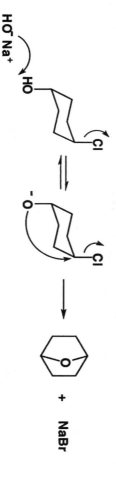

cis-4-Chloro-
cyclohexanol

(1)

trans-4-Chloro-
cyclohexanol

(2)

(3)

(a) Propose a mechanism for formation of product (1), and account for its configuration.

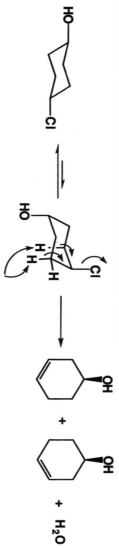

Inversion of configuration is observed because of an S_N2 mechanism.

(b) Propose a mechanism for formation of product (2).

The reaction takes place by an E2 mechanism. The molecule must adopt the chair conformation that places both the HO- and Cl-groups in the axial position in order for the reaction to occur. Note that either of the two axial H atoms indicated can be removed to create a racemic mixture.

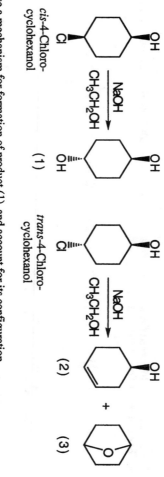

(c) Account for the fact that the bicyclic ether product (3) is formed from the trans isomer but not from the cis isomer.

The bicyclic ether product (3) is formed from an intramolecular backside attack of the deprotonated axial hydroxyl group upon an axial chlorine atom. Only the *trans* isomer can adopt the diaxial orientation necessary for this process.

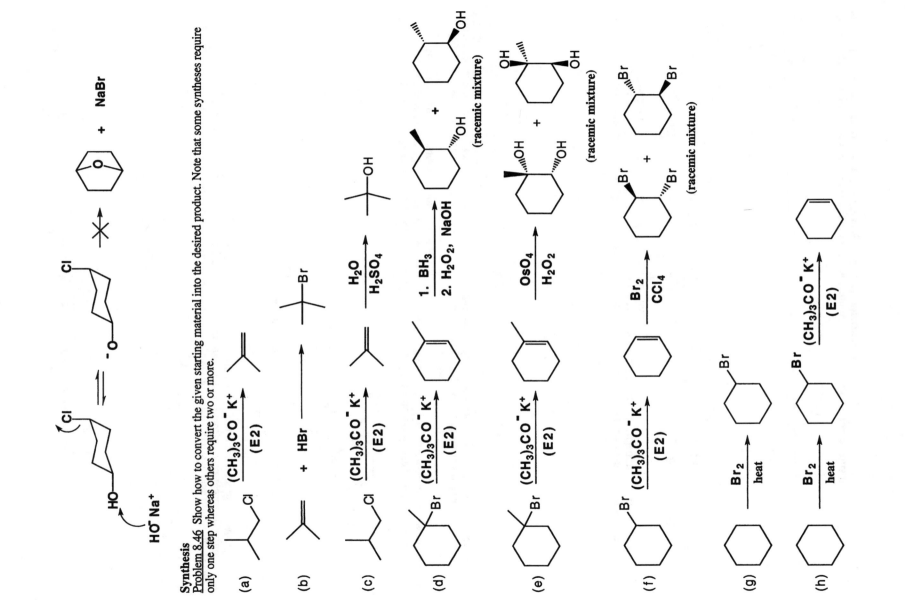

Synthesis

<u>Problem 8.46</u> Show how to convert the given starting material into the desired product. Note that some syntheses require only one step whereas others require two or more.

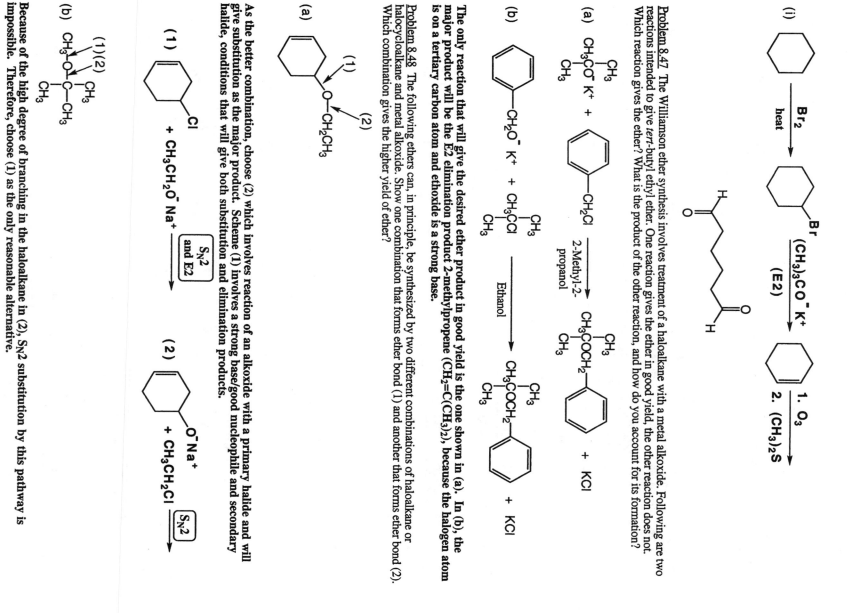

(i)

Br₂ / heat

(CH₃)₃CO⁻ K⁺ (E2)

1. O₃
2. (CH₃)₂S

Problem 8.47 The Williamson ether synthesis involves treatment of a haloalkane with a metal alkoxide. Following are two reactions intended to give *tert*-butyl ethyl ether. One reaction gives the ether in good yield, the other reaction does not. Which reaction gives the ether? What is the product of the other reaction, and how do you account for its formation?

(a) CH₃CO⁻ K⁺ + ⟶ CH₂Cl ⟶ 2-Methyl-2-propanol ⟶ + KCl

(b) CH₂O⁻ K⁺ + CH₃CCl ⟶ Ethanol ⟶ + KCl

The only reaction that will give the desired ether product in good yield is the one shown in (a). In (b), the major product will be the E2 elimination product 2-methylpropene (CH₂=C(CH₃)₂), because the halogen atom is on a tertiary carbon atom and ethoxide is a strong base.

Problem 8.48 The following ethers can, in principle, be synthesized by two different combinations of haloalkane or halocycloalkane and metal alkoxide. Show one combination that forms ether bond (1) and another that forms ether bond (2). Which combination gives the higher yield of ether?

(a)

(1)
(2)

As the better combination, choose (2) which involves reaction of an alkoxide with a primary halide and will give substitution as the major product. Scheme (1) involves a strong base/good nucleophile and secondary halide, conditions that will give both substitution and elimination products.

(1) + CH₃CH₂O⁻ Na⁺ [Sₙ2 and E2]

(2) + CH₃CH₂Cl [Sₙ2]

(b)

(1)(2)

Because of the high degree of branching in the haloalkane in (2), Sₙ2 substitution by this pathway is impossible. Therefore, choose (1) as the only reasonable alternative.

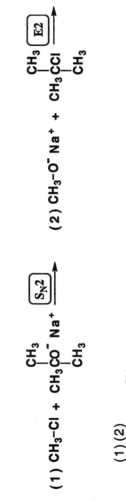

(1) CH_3-Cl + CH_3CO^- Na^+ $\boxed{S_N2}$ →

CH₃
|
CH₃CO—CH₂CCH₃
| |
CH₃   CH₃

CH₃
|
(2) CH_3-O^- Na^+ + CH_3CCl $\boxed{E2}$ →
|
CH₃

(c) $H_2C=CHCH_2-O-CH_2CCH_3$

Because of the high degree of branching on the β-carbon in the haloalkane in (2), S_N2 substitution by this pathway is prevented. Therefore, choose (1) as the only reasonable alternative.

CH₃
|
(1) $H_2C=CHCH_2Cl$ + $CH_3CCH_2O^-$ Na^+ $\boxed{S_N2}$ →
|
CH₃

CH₃
|
(2) $H_2C=CHCH_2O^-$ Na^+ + CH_3CCH_2Cl $\boxed{No\ reaction}$ →
|
CH₃

Problem 8.49 Propose a mechanism for this reaction.

$ClCH_2CH_2OH$ $\xrightarrow{Na_2CO_3,\ H_2O}$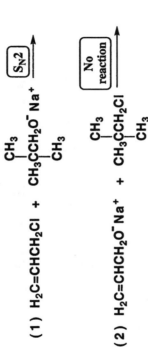

The mechanism of this reaction involves an initial deprotonation of the hydroxyl group, followed by an intramolecular S_N2 reaction to give the epoxide.

Step 1

$ClCH_2CH_2OH$ + Na_2CO_3 ⇌ $ClCH_2CH_2O^-$ Na^+ + $NaHCO_3$

Step 2

Cl͡CH₂CH₂O⁻ Na⁺ $\xrightarrow{S_N2}$

O
/ \
H₂C—CH₂ + NaCl

Problem 8.50 Each of these compounds can be synthesized by an S_N2 reaction. Suggest a combination of haloalkane and nucleophile that will give each product.

In the following reactions, the Br atom could be replaced with Cl or I, except for (f) and (k) in which Cl is required.

(a) CH_3OCH_3 ⟶ CH_3O^- Na^+ + CH_3Br

(b) CH_3SH ⟶ HS^- Na^+ + CH_3Br

(c) $CH_3CH_2CH_2PH_2$ ⟶ PH_3 + $CH_3CH_2CH_2Br$

(d) CH_3CH_2CN ⟶ Na^+ CN^- + CH_3CH_2Br

(e) $CH_3SCH_2C(CH_3)_3$ ⟶ $(CH_3)_3CCH_2S^-$ Na^+ + CH_3Br

Solutions

(f) $(CH_3)_3NH^+$ Cl^- $\longrightarrow$ $(CH_3)_2NH$ + CH_3Cl

(g) $C_6H_5\overset{O}{\overset{\|}{C}}OCH_2C_6H_5$ $\longrightarrow$ $C_6H_5\overset{O}{\overset{\|}{C}}O^-Na^+$ + $C_6H_5CH_2Br$

(h) (R)-$CH_3\overset{N_3}{\underset{|}{C}}HCH_2CH_2CH_3$ $\longrightarrow$ $Na^+N_3^-$ + (S)-$CH_3\overset{Br}{\underset{|}{C}}HCH_2CH_2CH_3$

(i) $CH_2=CHCH_2OCH(CH_3)_2$ $\longrightarrow$ $(CH_3)_2CHO^-$ Na^+ + $CH_2=CHCH_2Br$

(j) $CH_2=CHCH_2OCH_2CH=CH_2$ $\longrightarrow$ $CH_2=CHCH_2O^-$ Na^+ + $CH_2=CHCH_2Br$

(k) 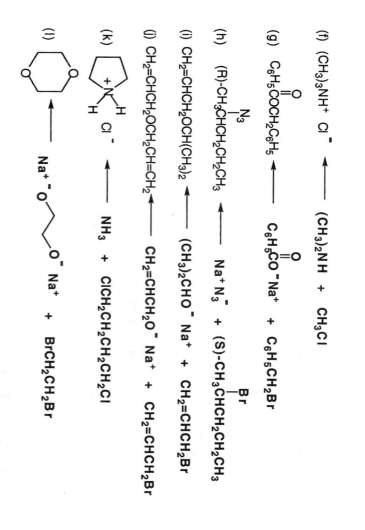 $\longrightarrow$ NH_3 + $ClCH_2CH_2CH_2CH_2Cl$

(l) 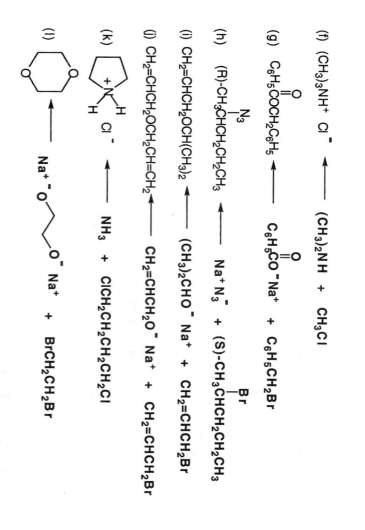 $\longrightarrow$ Na^+ ^-O O^- Na^+ + $BrCH_2CH_2Br$

Green sea turtle *Chelonia mydas*
Maui, Hawaii

CHAPTER 9
Solutions to the Problems

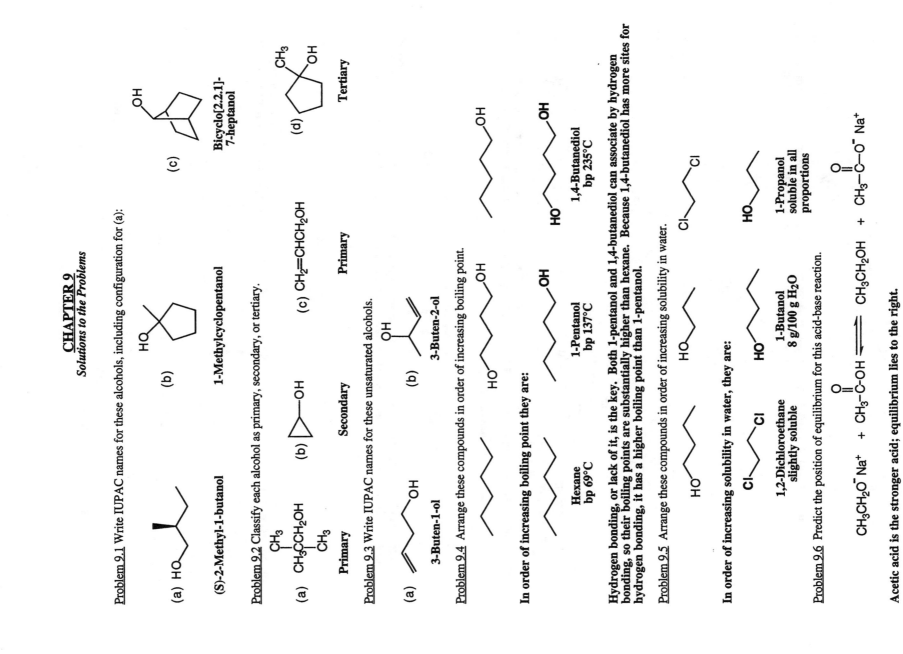

Problem 9.1 Write IUPAC names for these alcohols, including configuration for (a):

(a) HO⟍⟋

(b)

(c)

(S)-2-Methyl-1-butanol   1-Methylcyclopentanol   Bicyclo[2.2.1]-7-heptanol

(d)

Tertiary

Problem 9.2 Classify each alcohol as primary, secondary, or tertiary.

(a) CH₃CCH₂OH   (b) ▷—OH   (c) CH₂=CHCH₂OH

Primary   Secondary   Primary

Problem 9.3 Write IUPAC names for these unsaturated alcohols.

(a)   (b)

3-Buten-1-ol   3-Buten-2-ol

Problem 9.4 Arrange these compounds in order of increasing boiling point.

In order of increasing boiling point they are:

Hexane   1-Pentanol   1,4-Butanediol
bp 69°C   bp 137°C   bp 235°C

Hydrogen bonding, or lack of it, is the key. Both 1-pentanol and 1,4-butanediol can associate by hydrogen bonding, so their boiling points are substantially higher than hexane. Because 1,4-butanediol has more sites for hydrogen bonding, it has a higher boiling point than 1-pentanol.

Problem 9.5 Arrange these compounds in order of increasing solubility in water.

In order of increasing solubility in water, they are:

Cl⟍⟋Cl   1-Butanol   1-Propanol
                 8 g/100 g H₂O   soluble in all
1,2-Dichloroethane   proportions
slightly soluble

Problem 9.6 Predict the position of equilibrium for this acid-base reaction.

$$CH_3CH_2O^- Na^+ \; + \; CH_3-\overset{O}{\underset{\|}{C}}-OH \; \rightleftharpoons \; CH_3CH_2OH \; + \; CH_3-\overset{O}{\underset{\|}{C}}-O^- \, Na^+$$

Acetic acid is the stronger acid; equilibrium lies to the right.

$$CH_3CH_2O^- \ Na^+ \ + \ CH_3-\overset{\displaystyle O}{\overset{\|}{C}}-OH \ \rightleftharpoons \ CH_3CH_2OH \ + \ CH_3-\overset{\displaystyle O}{\overset{\|}{C}}-O^- \ Na^+$$

(stronger base) pK_a 4.76 (stronger acid) pK_a 15.9 (weaker acid)   (weaker base)

Problem 9.7 Show how you might convert (R)-2-butanol to (S)-2-butanethiol via a tosylate.

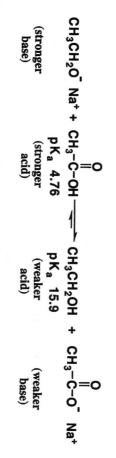

(R)-2-Pentanol   ?   (S)-2-Pentanethiol

In the first step, the -OH group is converted to the good leaving group -OTs by treatment with tosyl chloride (Ts-Cl) in pyridine.

Step 1:

(R)-2-Pentanol   +   Ts-Cl $\xrightarrow{\text{Pyridine}}$

Next the good nucleophile HS⁻ is used to carry out an S$_N$2 reaction and displace the -OTs group. Note that the inversion of stereochemistry seen with S$_N$2 reactions insures that the desired (S) isomer of 2-pentanethiol is produced.

Step 2:

$\xrightarrow[\text{(inversion)}]{S_N2}$ +   HS⁻ Na⁺

(S)-2-Pentanethiol

Problem 9.8 Draw structural formulas for the alkenes formed by acid-catalyzed dehydration of these alcohols. Where isomeric alkenes are possible, predict which is the major product.

(a)

$\xrightarrow[\text{dehydration}]{\text{acid-catalyzed}}$

(b)

Major product

Problem 9.9 Propose a mechanism to account for this reaction.

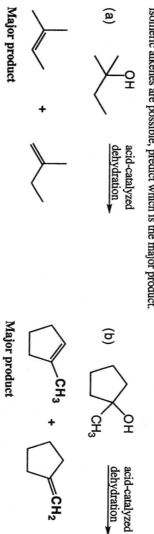

$\xrightarrow[\text{warm}]{H_2SO_4}$ + H_2O

Major product

This reaction can best be explained as an acid catalyzed E1 reaction with a carbocation rearrangement. The first steps involve protonation of the -OH group, followed by loss of H_2O to give a 2° carbocation.

Step 1:

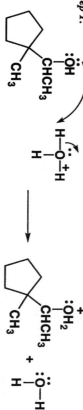

Step 2:

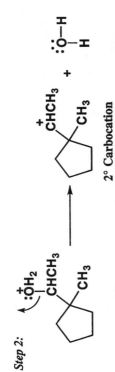

2° Carbocation

This 2° carbocation then rearranges to a more stable 3° carbocation by expanding to a six-membered ring. The 3° carbocation completes the E1 reaction by losing a proton to generate the product alkene.

Step 3:

2° Carbocation

More stable
3° carbocation

Step 4:

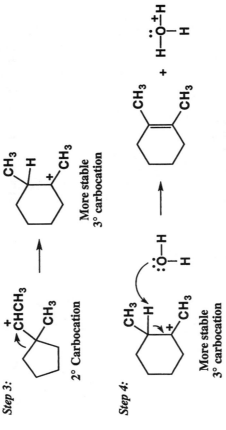

More stable
3° carbocation

Problem 9.10 Propose a mechanism to account for the following transformation:

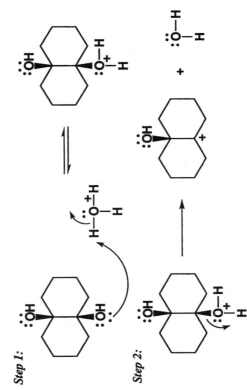

$$\xrightarrow[-H_2O]{H_2SO_4}$$

Protonation of either hydroxyl group followed by departure of water gives a tertiary carbocation in the first two steps.

Step 1:

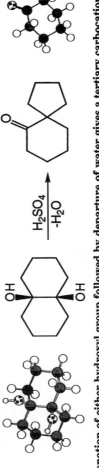

Step 2:

Migration of a pair of electrons from an adjacent bond and loss of a proton to give a ketone gives the observed product.

Step 3:

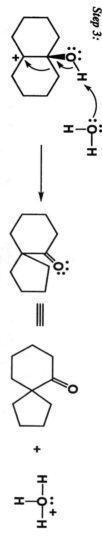

Problem 9.11 Draw the product of treatment of each alcohol in Example 9.11 with chromic acid.

Primary alcohols are oxidized by chromic acid (H_2CrO_4) to give carboxylic acids, while secondary alcohols give ketones.

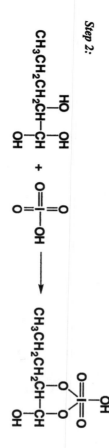

(a)

1-Hexanol   OH + H_2CrO_4 ⟶ Hexanoic acid

(b)

2-Hexanol   + H_2CrO_4 ⟶ 2-Hexanone

(c)

Cyclohexanol   OH + H_2CrO_4 ⟶ Cyclohexanone

Problem 9.12 α-Hydroxyketones and α-hydroxyaldehydes are also oxidized by treatment with periodic acid. It is the not the α-hydroxyketone or aldehyde, however, that undergoes reaction with periodic acid, but rather the compound formed by addition of a molecule of water to the carbonyl group of the α-hydroxyketone or aldehyde. Write a mechanism for the oxidation of this α-hydroxyaldehyde with HIO_4.

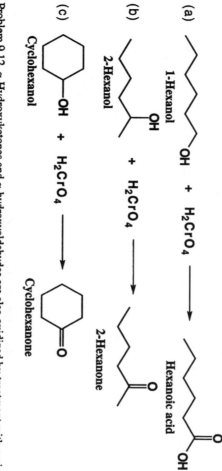

$$CH_3CH_2CH_2CH-CH \xrightarrow[\text{2. } HIO_4]{\text{1. } H_2O}$$

The aldehyde group reacts with water to give the following molecule, a geminal diol.

Step 1:

$$CH_3CH_2CH_2CH-CH + H_2O \rightleftharpoons CH_3CH_2CH_2CH-CH$$

This geminal diol then reacts with HIO_4 according to the usual mechanism to give butanal plus formic acid.

Step 2:

$$CH_3CH_2CH_2CH-CH + O=I-OH \longrightarrow CH_3CH_2CH_2CH-CH$$

Step 3:

$$CH_3CH_2CH_2CH-CH \longrightarrow CH_3CH_2CH_2CH + HCOH + HIO_3$$

Butanal   Formic acid

Problem 9.13 Write IUPAC names for these thiols (including stereochemistry if relevant)

(a)

3-Methyl-1-butanethiol

(b)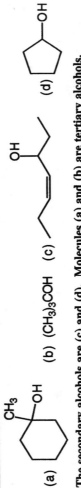

(Z)-4-Hexene-2-thiol

Structure and Nomenclature

Problem 9.14 Which are secondary alcohols?

(a)    (b) $(CH_3)_3COH$ (c) [structure]   (d) [cyclopentanol]

The secondary alcohols are (c) and (d). Molecules (a) and (b) are tertiary alcohols.

Problem 9.15 Name these compounds.

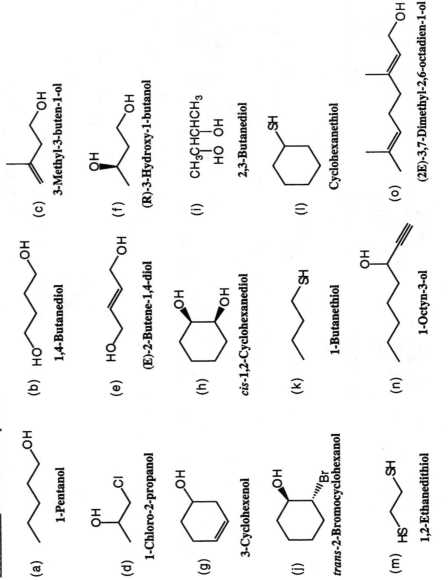

(a) 1-Pentanol

(b) 1,4-Butanediol

(c) 3-Methyl-3-buten-1-ol

(d) 1-Chloro-2-propanol

(e) (E)-2-Butene-1,4-diol

(f) (R)-3-Hydroxy-1-butanol

(g) 3-Cyclohexenol

(h) cis-1,2-Cyclohexanediol

(i) 2,3-Butanediol

(j) trans-2-Bromocyclohexanol

(k) 1-Butanethiol

(l) Cyclohexanethiol

(m) 1,2-Ethanedithiol

(n) 1-Octyn-3-ol

(o) (2E)-3,7-Dimethyl-2,6-octadien-1-ol

Problem 9.16 Write structural formulas for the following:
(a) Isopropyl alcohol   (b) Propylene glycol   (c) 5-Methyl-2-hexanol

(d) 2-Methyl-2-propyl-1,3-propanediol   (e) 1-Chloro-2-hexanol   (f) *cis*-3-Isobutylcyclohexanol

(g) 2,2-Dimethyl-1-propanol   (h) 2-Mercaptoethanol   (i) Allyl alcohol

(j) *trans*-2-Vinylcyclohexanol   (k) (Z)-5-Methyl-2-hexen-1-ol   (l) 2-Propyn-1-ol

(m) 3-Chloro-1,2-propanediol   (n) *cis*-3-Pentene-1-ol   (o) Bicyclo[2.2.1]-heptan-7-ol

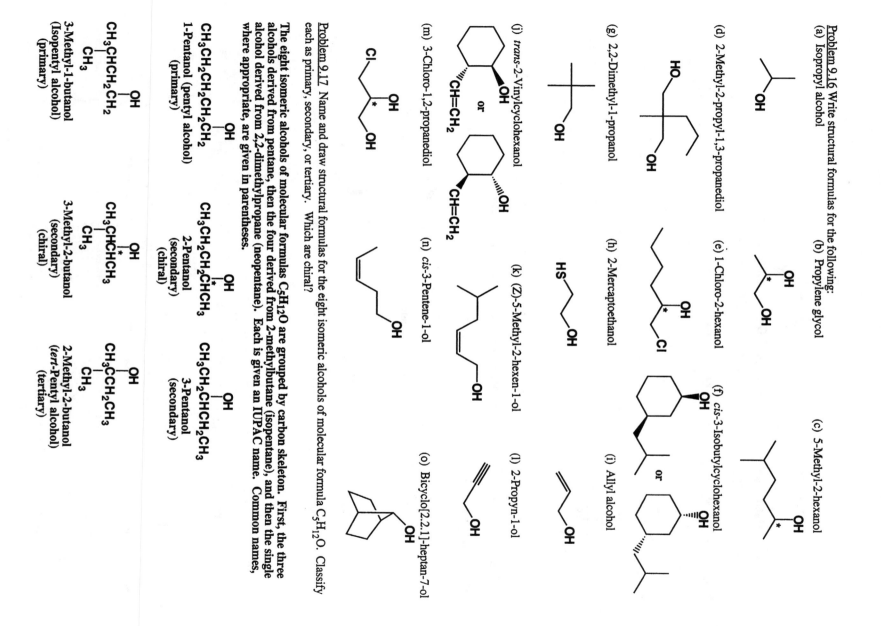

Problem 9.17 Name and draw structural formulas for the eight isomeric alcohols of molecular formula $C_5H_{12}O$. Classify each as primary, secondary, or tertiary. Which are chiral?

The eight isomeric alcohols of molecular formulas $C_5H_{12}O$ are grouped by carbon skeleton. First, the three alcohols derived from pentane, then the four derived from 2-methylbutane (isopentane), and then the single alcohol derived from 2,2-dimethylpropane (neopentane). Each is given an IUPAC name. Common names, where appropriate, are given in parentheses.

CH₃CH₂CH₂CH₂CH₂
OH
1-Pentanol (pentyl alcohol)
(primary)

CH₃CH₂CH₂CHCH₃
OH
2-Pentanol
(secondary)
(chiral)

CH₃CH₂CHCH₂CH₃
OH
3-Pentanol
(secondary)

CH₃CHCH₂CH₂
OH CH₃
3-Methyl-1-butanol
(isopentyl alcohol)
(primary)

CH₃CHCHCH₃
OH CH₃
3-Methyl-2-butanol
(secondary)
(chiral)

CH₃CCH₂CH₃
OH CH₃
2-Methyl-2-butanol
(*tert*-Pentyl alcohol)
(tertiary)

CH₃CHCH₂CH₃
OH
2-Methyl-2-butanol
(*tert*-Pentyl alcohol)
(tertiary)

$CH_3CH_2CH_2CH_3$

OH
|
$CH_2\overset{*}{C}HCH_2CH_3$
|
CH_3
2-Methyl-1-butanol
(primary)
(chiral)

CH_3
|
CH_3CCH_2OH
|
CH_3
2,2-Dimethyl-1-propanol
(Neopentyl alcohol)
(primary)

Physical Properties of Alcohols and Thiols

Problem 9.18 Arrange these compounds in order of increasing boiling point. (Values in °C are -42, 78, 117, and 198.)

(a) $CH_3CH_2CH_2CH_2OH$ (b) CH_3CH_2OH (c) $HOCH_2CH_2OH$ (d) $CH_3CH_2CH_3$

In order of increasing boiling point they are:

$CH_3CH_2CH_3$	CH_3CH_2OH	$CH_3CH_2CH_2CH_2OH$	$HOCH_2CH_2OH$
Propane	Ethanol	1-Butanol	Ethylene glycol
bp -42°C	bp 78°C	bp 117°C	bp 198°C

The keys for this problem are hydrogen bonding and size. Propane cannot make any hydrogen bonds, so it has the lowest boiling point by far. Ethanol and 1-butanol can each make hydrogen bonds through their single -OH group. However, 1-butanol is larger, so it will have a higher boiling point than ethanol. Ethylene glycol has two -OH groups per molecule with which to make hydrogen bonds, so it will have the highest boiling point of this set.

Problem 9.19 Arrange these compounds in order of increasing boiling point. (Values in °C are -42, -24, 78, and 118.)

(a) CH_3CH_2OH (b) CH_3OCH_3 (c) $CH_3CH_2CH_3$ (d) CH_3COOH

In order of increasing boiling point they are:

$CH_3CH_2CH_3$	CH_3OCH_3	CH_3CH_2OH	CH_3CO_2H
Propane	Dimethyl ether	Ethanol	Acetic acid
bp -42°C	bp -24°C	bp 78°C	bp 118°C

The keys for this problem are hydrogen bonding, polarity, and size. We know from the last problem that propane and ethanol have boiling points of -42°C and 78°C, respectively. Dimethyl ether is polar, but cannot make hydrogen bonds. Therefore, it makes sense that dimethyl ether has a boiling point (-24°C) that is higher than propane, but lower than ethanol. Acetic acid can make strong hydrogen bonds and has a higher molecular weight than ethanol, so it makes sense that acetic acid has the highest boiling point of this set.

Problem 9.20 Compounds that contain an N-H group associate by hydrogen bonding.

(a) Do you expect this association to be stronger or weaker than that of compounds containing an O-H group?

Weaker. The O-H bond is more polar, because the difference in electronegativity between N and H is less than the difference between O and H. Thus, the degree of intermolecular interaction between compounds containing an N-H group is less than that between compounds containing an -OH group.

Bond	Difference in electronegativity
N-H	3.0 - 2.1 = 0.9
O-H	3.5 - 2.1 = 1.4

(b) Based on your answer to part (a), which would you predict to have the higher boiling point, 1-butanol or 1-butanamine?

$CH_3CH_2CH_2CH_2OH$	$CH_3CH_2CH_2CH_2NH_2$
1-Butanol	1-Butanamine
bp 117°C	bp 78°C

The stronger the hydrogen bonds, the higher the boiling point because hydrogen bonds in the liquid state must be broken upon boiling. Therefore, 1-butanol, with the stronger hydrogen bonds, will have the higher boiling point.

Problem 9.21 Which compounds can participate in hydrogen bonding with water?

The molecules in bold can participate in hydrogen bonding with water. The hydrogen bond donor and acceptor sites are labeled.

(a) $CH_3CH_2CH_2OCH_2CH_2OH$

(b) $(CH_3CH_2)_2NH$

(c) $CH_3CH=CHCH_3$

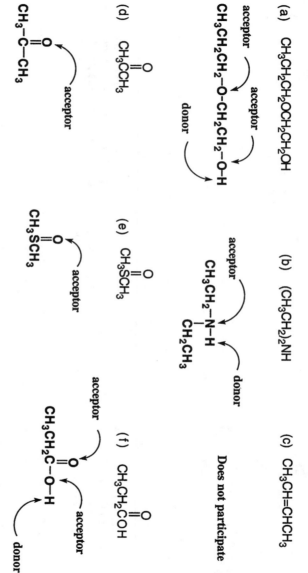

acceptor   acceptor

$CH_3CH_2CH_2-O-CH_2CH_2-O-H$

donor

acceptor

CH_3CH_2-N-H

CH_2CH_3

donor

Does not participate

(d) CH_3CCH_3 (with O double bond)

acceptor

CH_3-C-CH_3 (with O)

(e) CH_3SCH_3 (with O double bond)

acceptor

CH_3SCH_3 (with O)

(f) CH_3CH_2COH (with O double bond)

acceptor

CH_3CH_2C-O-H (with O)

acceptor

donor

Problem 9.22 From each pair of compounds, select the one that is more soluble in water.

(a) CH_2Cl_2 or CH_3OH

Methanol, CH_3OH, is soluble in all proportions in water. Dichloromethane, CH_2Cl_2, is insoluble. The highly polar -OH group of methanol is capable of participating both as a hydrogen bond donor and hydrogen bond acceptor with water and, therefore, interacts strongly with water by intermolecular association. No such interaction is possible with dichloromethane.

(b) CH_3CCH_3 or CH_3CCH_3 (one with CH_2)

Propanone (acetone), CH_3COCH_3, is soluble in water in all proportions. 2-Methylpropene (isobutylene) is insoluble in water. Acetone has a large dipole moment and can function as a hydrogen bond acceptor from water.

(c) CH_3CH_2Cl or   NaCl

NaCl is the more soluble. Chloroethane is insoluble in water. Following is a review of some of the general water solubility rules developed in General Chemistry. For these rules, soluble is defined as dissolving greater than 0.10 mol/L. Slightly soluble is dissolving between 0.01 mol/L and 0.10 mol/L.
1. Sodium, potassium, and ammonium salts of halogens or nitrates are soluble.
2. Silver, lead, and mercury(I) salts of halogens are insoluble.
Thus, applying Rule 1, NaCl is soluble in water. Chloroethane (ethyl chloride) is a nonpolar organic compound and insoluble in water.

(d) $CH_3CH_2CH_2SH$ or $CH_3CH_2CH_2OH$

The alcohol is more soluble in water. Sulfur is less electronegative than oxygen, so an S-H bond is less polarized than an O-H bond. Hydrogen bonding is therefore weaker with thiols than alcohols, so the alcohol is more able to interact with water molecules through hydrogen bonding.

(e) $CH_3CH_2CHCH_2CH_3$ (with OH)   or $CH_3CH_2CCH_2CH_3$ (with O)

The alcohol is more able to interact with water through hydrogen bonding, and is more soluble in water. The hydroxyl group has both a hydrogen bond donor and acceptor (the oxygen and hydrogen atoms of the -OH group, respectively), while the carbonyl group has only a hydrogen bond acceptor (the oxygen atom).

Problem 9.23 Arrange the compounds in each set in order of decreasing solubility in water.
(a) Ethanol; butane; diethyl ether

CH₃CH₂OH   CH₃CH₂OCH₂CH₃   CH₃CH₂CH₂CH₃

soluble in all   8 g/100 mL water   insoluble in water
proportions

In general, the more strongly a molecule can take part in hydrogen bonding with water, the greater the molecule is able to interact with the water molecules and dissolve. Only ethanol can act both as a donor and acceptor of hydrogen bonds with water. Diethyl ether can act as an acceptor of hydrogen bonds. Butane can act as neither a donor nor an acceptor of hydrogen bonds.

(b) 1-Hexanol; 1,2-hexanediol; hexane

CH₂CHCH₂CH₂CH₂CH₃   CH₃CH₂CH₂CH₂CH₂CH₂OH   CH₃CH₂CH₂CH₂CH₂CH₃
OH OH

1,2-Hexanediol molecules can take part in more hydrogen bonds with water than the 1-hexanol, because the diol has two -OH groups. Hexane has no polar bonds and thus cannot take part in any dipole-dipole interactions with water molecules.

Problem 9.24 Each compound given in this problem is a common organic solvent. From each pair of compounds, select the solvent with the greater solubility in water.

Solubility in water increases with increasing hydrogen bonding ability and decreases with increasing surface area of hydrophobic groups such as alkyl groups.

(a) CH₂Cl₂ or CH₃CH₂OH   (b) CH₃CH₂OCH₂CH₃ or CH₃CH₂OH

CH₃CH₂OH   CH₃CH₂OH

(c) CH₃CCH₃ or CH₃CH₂OCH₂CH₃   (d) CH₃CH₂OCH₂CH₃ or CH₃(CH₂)₃CH₃
      ‖O

CH₃CCH₃   CH₃CH₂OCH₂CH₃
  ‖O

Problem 9.25 The decalinols A and B can be equilibrated using aluminum isopropoxide in 2-propanol (isopropyl alcohol) containing a small amount of acetone. Assume a value of ΔG° (equatorial - axial) for cyclohexanol is 4.0 kJ(0.95 kcal)/mol, calculate the percent of each isomer in the equilibrium mixture at 25°C.

$$\text{Al[OCH(CH}_3)_2]_3$$
$$\xrightleftharpoons[\text{acetone}]{}$$

A   B

At equilibrium, the relative amounts of A and B are given by the equation:

$$\Delta G° = -RT\ln K_{eq} \qquad \text{where} \quad K_{eq} = \frac{[B]}{[A]}$$

$$\log K_{eq} = \frac{-\Delta G°}{(2.303)RT}$$

Taking the antilog of both sides gives:

$$K_{eq} = 10^{\left(\frac{-\Delta G^\circ}{(2.303)RT}\right)}$$

Plugging in the values for ΔG°, R, and 298 K gives the final answer:

$$K_{eq} = 10^{\left(\frac{-4,000 \text{ J/mol}}{(2.303)(8.314 \text{ J} \cdot \text{K}^{-1} \cdot \text{mol}^{-1})(298 \text{ K})}\right)} = 10^{-0.70} = 0.20$$

Thus, the molecule with an equatorial hydroxyl group (A) will predominate in a 5 to 1 ratio.

Acidity of Alcohols

Problem 9.26 Complete the following acid-base reactions. In addition, show all valence electrons on the interacting atoms and show by the use of curved arrows the flow of electrons in each reaction.

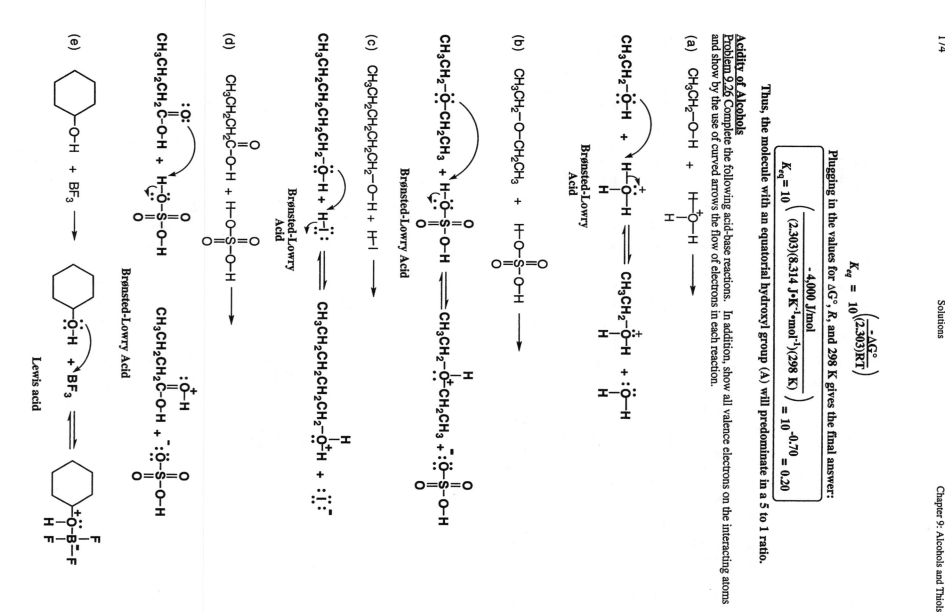

(a) CH₃CH₂—O—H + H—Ö—H ⟶

(b) CH₃CH₂—O—CH₂CH₃ + H—O—S—O—H ⟶

Brønsted-Lowry Acid

(c) CH₃CH₂CH₂CH₂CH₂—O—H + H—I ⟶

Brønsted-Lowry Acid

(d) CH₃CH₂CH₂C—O—H + H—O—S—O—H ⟶

Brønsted-Lowry Acid

(e) [cyclohexanol] —O—H + BF₃ ⟶

Lewis acid

(f)   CH₃CH=CHĊHCH₃ + H–O–H ⟶

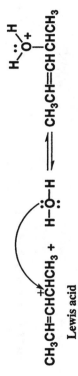

CH₃CH=CHĊHCH₃ + H–Ö–H ⇌ CH₃CH=CHCHCH₃

Lewis acid

__Problem 9.27__ Select the stronger acid from each pair and explain your reasoning. For each stronger acid, write a structural formula for its conjugate base.

(a)   H₂O   or $\boxed{H_2CO_3}$ (b)   CH₃OH   or $\boxed{CH_3COOH}$

(c) $\boxed{CH_3CH_2OH}$ or   CH₃C≡CH   (d)   CH₃CH₂OH   or $\boxed{CH_3CH_2SH}$

Relative acidity rankings can be predicted by examining the relative stability of the anions produced upon deprotonation of the different acidic groups. *The more stable anion will be derived from deprotonation of the stronger acid.*
For (a), the carbonyl group of carbonic acid allows for resonance stabilization of the deprotonated bicarbonate anion relative to the hydroxide anion.
For (b), the carbonyl group of acetic acid allows for resonance stabilization of the deprotonated carboxylate anion relative to the methoxide anion.
For (c), oxygen is a more electronegative element (farther to the right on the Periodic Table) than carbon, adding greater stability to the deprotonated alkoxide anion relative to the deprotonated alkyne.
For (d), sulfur is below oxygen in the Periodic Table and is a larger atom. Larger atoms are more able to carry negative charge, so the deprotonated thiolate anion is more stable than the alkoxide anion.
Under each acid is given its pKₐ. Recall that acid strength increases with decreasing values of pKₐ.

	weaker acid	stronger acid	conjugate base of stronger acid
(a)	H₂O pKₐ 15.7	H₂CO₃ pKₐ 6.36	HCO₃⁻
(b)	CH₃OH pKₐ 15.5	CH₃COOH pKₐ 4.76	CH₃COO⁻
(c)	CH₃C≡CH pKₐ 25	CH₃CH₂OH pKₐ 15.9	CH₃CH₂O⁻
(d)	CH₃CH₂OH pKₐ 15.9	CH₃CH₂SH pKₐ 8.5	CH₃CH₂S⁻

__Problem 9.28__ From each pair, select the stronger base. For each stronger base, write the structural formula of its conjugate acid.

Recall that the stronger base will have the weaker conjugate acid.

(a)   OH⁻   or   CH₃O⁻   (each in H₂O)   HO⁻ ⟶ HOH
                                                 stronger   weaker
                                                  base   acid

This is a close one, but HOH is a weaker acid than CH₃OH because HOH has a less-electronegative H atom attached to the OH group.

(b) $CH_3CH_2O^-$ or $CH_3C\equiv C^-$ $CH_3C\equiv C^-$ $\longrightarrow$ $CH_3C\equiv CH$
                                                                                                                                                    stronger   weaker
                                                                                                                                                    base   acid

Carbon is less electronegative (farther to the left on the Periodic Table) than oxygen, so the alkyne is the weaker acid.

(c) $CH_3CH_2S^-$ or $CH_3CH_2O^-$ $CH_3CH_2O^-$ $\longrightarrow$ CH_3CH_2OH
                                                                                                                                                    stronger   weaker
                                                                                                                                                    base   acid

Oxygen is above sulfur in the Periodic Table, so the alcohol is the weaker acid.

(d) $CH_3CH_2O^-$ or NH_2^- NH_2^- $\longrightarrow$ NH_3
                                                                                                                                                    stronger   weaker
                                                                                                                                                    base   acid

Nitrogen is less electronegative (farther to the left on the Periodic Table) than oxygen, so ammonia is the weaker acid.

Problem 9.29 In each equilibrium, label the stronger acid, the stronger base, the weaker acid, and the weaker base. Also estimate the position of each equilibrium.

(a) $CH_3CH_2O^-$ + $CH_3C\equiv CH$ $\rightleftharpoons$ CH_3CH_2OH + $CH_3C\equiv C^-$

$CH_3CH_2O^-$ + $CH_3C\equiv CH$ $\rightleftharpoons$ CH_3CH_2OH + $CH_3C\equiv C^-$ $K_{eq} = 10^{-9.1}$
weaker base   weaker acid   pKa 25
                            pKa 15.9   stronger acid   stronger base

Oxygen is a more electronegative element (farther to the right on the Periodic Table) than carbon, so the alcohol is the stronger acid.

(b) $CH_3CH_2O^-$ +   HCl $\rightleftharpoons$ CH_3CH_2OH + Cl^-

$CH_3CH_2O^-$ + HCl $\longrightarrow$ CH_3CH_2OH + Cl^- $K_{eq} = 10^{22.5}$
stronger   stronger   pKa 15.9
base   acid   weaker
pKa -7   acid   weaker
                                                                          base

Chlorine is more electronegative (farther to the right on the Periodic Table) than oxygen, so HCl is the stronger acid.

(c) $CH_3C(=O)OH$ + $CH_3CH_2O^-$ $\rightleftharpoons$ $CH_3C(=O)O^-$ + CH_3CH_2OH

$CH_3C(=O)OH$ + $CH_3CH_2O^-$ $\longrightarrow$ $CH_3C(=O)O^-$ + CH_3CH_2OH $K_{eq} = 10^{11.1}$
pKa 4.76   stronger   pKa 15.9
stronger   base   weaker   weaker
acid   base   acid

The carbonyl group of acetic acid inductively weakens the O-H bond and stabilizes the deprotonated anion through resonance.

Reactions of Alcohols

Problem 9.30 Write equations for the reaction of 1-butanol with each reagent. Where you predict no reaction, write NR.

(a) Na metal

2 $CH_3CH_2CH_2CH_2OH$ +   2 Na $\longrightarrow$ 2 $CH_3CH_2CH_2CH_2O^-$ Na^+ + H_2

(b) HBr, heat

$$CH_3CH_2CH_2CH_2OH + HBr \xrightarrow{heat} CH_3CH_2CH_2CH_2Br + H_2O$$

(c) HI heat

$$CH_3CH_2CH_2CH_2OH + HI \xrightarrow{heat} CH_3CH_2CH_2CH_2I + H_2O$$

(d) PBr$_3$

$$3\ CH_3CH_2CH_2CH_2OH + PBr_3 \longrightarrow 3\ CH_3CH_2CH_2CH_2Br + P(OH)_3$$

(e) SOCl$_2$, pyridine

$$CH_3CH_2CH_2CH_2OH + SOCl_2 \xrightarrow{pyridine} CH_3CH_2CH_2CH_2Cl + SO_2 + HCl$$

(f) K$_2$Cr$_2$O$_7$, H$_2$SO$_4$, H$_2$O, heat

$$CH_3CH_2CH_2CH_2OH + K_2Cr_2O_7 \xrightarrow[\substack{H_2O \\ heat}]{H_2SO_4} CH_3CH_2CH_2COH \ (=O) + Cr^{3+}$$

(g) HIO$_4$
NR

(h) PCC

$$CH_3CH_2CH_2CH_2OH + PCC \longrightarrow CH_3CH_2CH_2CH\ (=O) + Cr^{3+}$$

(i) CH$_3$SO$_2$Cl, pyridine

$$CH_3CH_2CH_2CH_2OH + ClSCH_3\ (O=S=O) \xrightarrow{pyridine} CH_3CH_2CH_2CH_2OSCH_3\ (O=S=O) + HCl$$

Problem 9.31 Write equations for the reaction of 2-butanol with each reagent listed in Problem 9.30. Where you predict no reaction, write NR.

(a) Na metal

$$2\ CH_3CH_2\overset{*}{C}HCH_3\ (OH) + 2\ Na \longrightarrow 2\ CH_3CH_2\overset{*}{C}HCH_3\ (O^- Na^+) + H_2$$

(b) HBr, heat

$$CH_3CH_2\overset{*}{C}HCH_3\ (OH) + HBr \xrightarrow{heat} CH_3CH_2\overset{*}{C}HCH_3\ (Br) + H_2O$$

(c) HI, heat

$$CH_3CH_2\overset{*}{C}HCH_3\ (OH) + HI \xrightarrow{heat} CH_3CH_2\overset{*}{C}HCH_3\ (I) + H_2O$$

(d) PBr$_3$

$$3\ CH_3CH_2\overset{*}{C}HCH_3\ (OH) + PBr_3 \longrightarrow 3\ CH_3CH_2\overset{*}{C}HCH_3\ (Br) + P(OH)_3$$

(e) SOCl$_2$, pyridine

$$CH_3CH_2\overset{*}{C}HCH_3\ (OH) + SOCl_2 \xrightarrow{pyridine} CH_3CH_2\overset{*}{C}HCH_3\ (Cl) + SO_2 + HCl$$

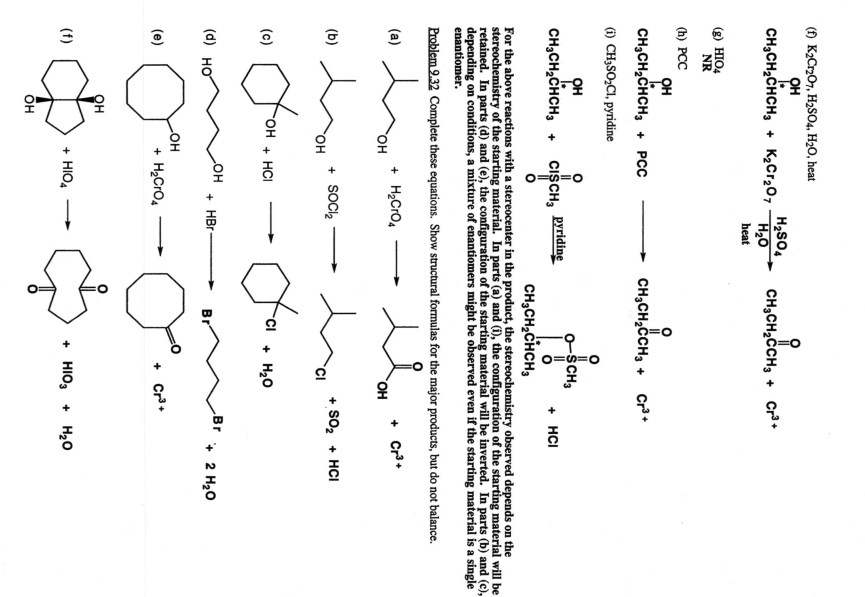

(f) $K_2Cr_2O_7$, H_2SO_4, H_2O, heat

$$CH_3CH_2\overset{OH}{\underset{*}{C}H}CH_3 + K_2Cr_2O_7 \xrightarrow[\text{heat}]{\underset{H_2O}{H_2SO_4}} CH_3CH_2\overset{O}{C}CH_3 + Cr^{3+}$$

(g) HIO_4
NR

(h) PCC

(i) CH_3SO_2Cl, pyridine

$$CH_3CH_2\overset{OH}{\underset{*}{C}H}CH_3 + PCC \longrightarrow CH_3CH_2\overset{O}{C}CH_3 + Cr^{3+}$$

$$CH_3CH_2\overset{OH}{\underset{*}{C}H}CH_3 + Cl\overset{O}{\underset{O}{S}}CH_3 \xrightarrow{pyridine} CH_3CH_2\overset{*}{C}HCH_3 \text{—}O\text{—}\overset{O}{\underset{O}{S}}CH_3 + HCl$$

For the above reactions with a stereocenter in the product, the stereochemistry observed depends on the stereochemistry of the starting material. In parts (a) and (i), the configuration of the starting material will be retained. In parts (d) and (e), the configuration of the starting material will be inverted. In parts (b) and (c), depending on conditions, a mixture of enantiomers might be observed even if the starting material is a single enantiomer.

Problem 9.32 Complete these equations. Show structural formulas for the major products, but do not balance.

(a)   + H_2CrO_4 ⟶

(b)   + $SOCl_2$ ⟶

(c)   + HCl ⟶

(d)   + HBr ⟶

(e)   + H_2CrO_4 ⟶

(f)   + HIO_4 ⟶

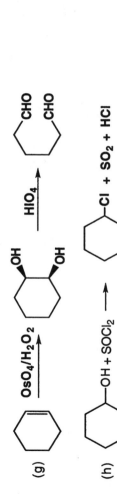

(g)

(h)

Problem 9.33 When (R)-2-butanol is left standing in aqueous acid, it slowly loses its optical activity. Account for this observation.

In aqueous acid, (R)-2-butanol is in equilibrium with a small amount of the protonated form, that subsequently loses H_2O to create an achiral cation. The achiral cation reacts with another water molecule to give back 2-butanol. The key to this question is that the achiral cation can react with H_2O on either face of the ion, to give either (R)-2-butanol or (S)-2-butanol. Thus, a pure sample of (R)-2-butanol will gradually turn into a racemic mixture of (R)-2-butanol and (S)-2-butanol, thereby losing optical activity.

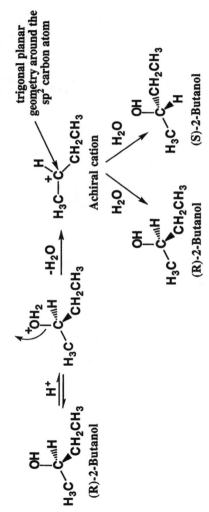

Problem 9.34 Two diastereomeric sets of enantiomers, A/B and C/D, exist for 3-bromo-2-butanol. When enantiomer A or B is treated with HBr, only racemic 2,3-dibromobutane is formed; no meso isomer is formed. When enantiomer C or D is treated with HBr, only meso 2,3-dibromobutane is formed; no racemic 2,3-dibromobutane is formed. Account for these observations. (*Hint*: Consider neighboring group participation (Section 8.5) and the type of intermediate that could produce this stereoselectivity).

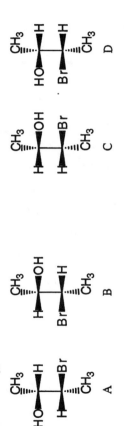

The -OH group of the starting materials will be protonated as usual, and H_2O will depart to produce a carbocation. The key to this problem is realizing that a bromine atom adjacent to a carbocation will form a three-membered ring bromonium ion intermediate that then dictates an *anti* geometry for the incoming bromide nucleophile as shown.

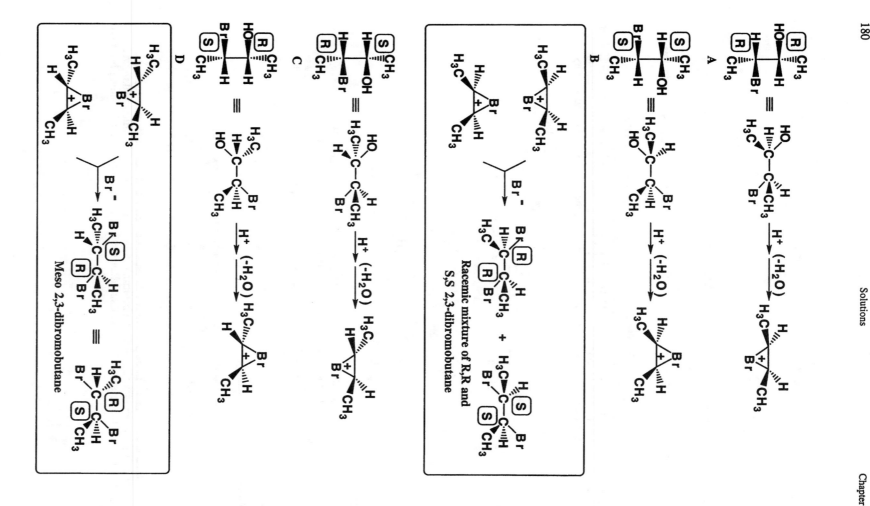

Problem 9.35 Acid-catalyzed dehydration of 3-methyl-2-butanol gives three alkenes: 2-methyl-2-butene, 3-methyl-1-butene, and 2-methyl-1-butene. Propose a mechanism to account for the formation of each product.

The 3-methyl-2-butanol is protonated on the -OH group, followed by loss of water to produce a 2° carbocation.

Step 1:

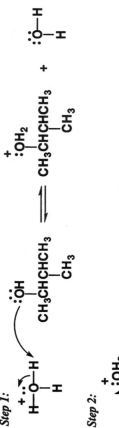

Step 2:

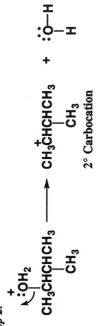

This 2° carbocation can lose either of two protons to give 2-methyl-2-butene or 3-methyl-1-butene.

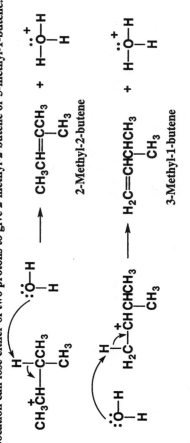

Alternatively, the 2° carbocation can rearrange to give a more stable 3° carbocation.

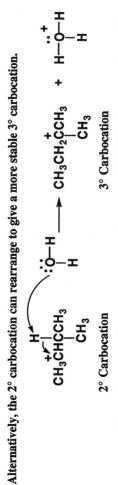

This 3° carbocation can lose either of 2 protons to give 2-methyl-2-butene or 2-methyl-1-butene.

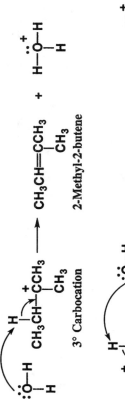

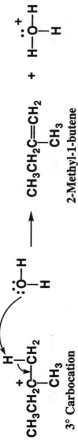

Problem 9.36 Show how you might bring about the following conversions. For any conversion involving more than one step, show each intermediate compound formed.

(a)

H₃PO₄
heat

+ H₂O

The most common laboratory methods for dehydration of an alcohol to an alkene involve heating the alcohol with either 85% phosphoric acid or concentrated sulfuric acid.

(b)

We have not seen any reaction that can switch an -OH group from one carbon atom to an adjacent carbon atom. We have seen, however, reactions by which we can (1) bring about dehydration of an alcohol to an alkene and then (2) hydrate the alkene to an alcohol in the following way.

H₃PO₄
heat

+ H₂O

H₂SO₄

(c)

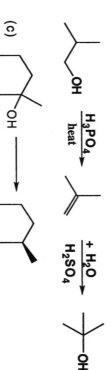

Acid-catalyzed dehydration of this tertiary alcohol to 1-methylcyclohexene followed by hydroboration/oxidation to form *trans*-2-methylcyclohexanol.

H₃PO₄
heat

1. BH₃
2. H₂O₂, NaOH

(racemic mixture)

(d)

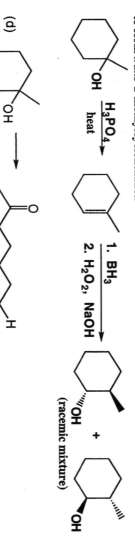

Acid-catalyzed dehydration of the tertiary alcohol as in part (c) followed by oxidative cleavage of the carbon-carbon double bond using osmium tetroxide in the presence of periodic acid, or ozonolysis followed by work-up in the presence of dimethyl sulfide.

H₃PO₄
heat

1) OsO₄/H₂O₂
2) HIO₄

(e)

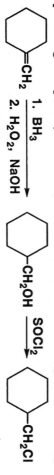

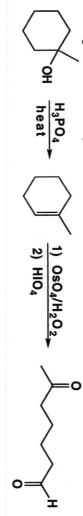

Hydroboration of the alkene followed by oxidation of the resulting organoborane with alkaline hydrogen peroxide gives a primary alcohol. Reaction of this alcohol with SOCl₂ gives the desired primary chloride.

1. BH₃
2. H₂O₂, NaOH

SOCl₂

(f)

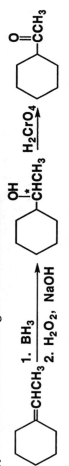

Hydroboration of the alkene followed by oxidation in alkaline hydrogen peroxide gives a secondary alcohol. Oxidation of this alcohol with chromic acid in aqueous sulfuric acid, by chromic acid in pyridine, or by pyridinium chlorochromate gives the desired ketone.

(g) $CH_3(CH_2)_6CH_2OH \longrightarrow CH_3(CH_2)_6CH$

This conversion can be accomplished through oxidation with PCC.

$CH_3(CH_2)_6CH_2OH \xrightarrow{PCC} CH_3(CH_2)_6CH$

(h)

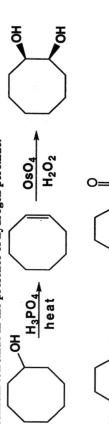

Acid-catalyzed dehydration of the secondary alcohol to an alkene followed by oxidation to the *cis*-glycol using osmium tetroxide in the presence of hydrogen peroxide.

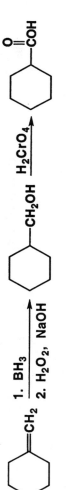

(i)

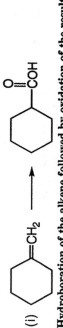

Hydroboration of the alkene followed by oxidation of the resulting organoborane with alkaline hydrogen peroxide gives a primary alcohol. Oxidation of this alcohol with H_2CrO_4 gives the desired carboxylic acid.

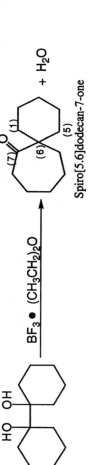

Pinacol Rearrangement

Problem 9.37 Propose a mechanism for the following pinacol rearrangement catalyzed by boron trifluoride etherate:

$\xrightarrow[BF_3 \bullet (CH_3CH_2)_2O]{}$

$+ H_2O$

Spiro[5.6]dodecan-7-one

Step 1:

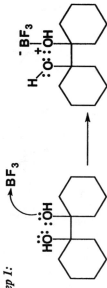

Step 2:

Step 3:

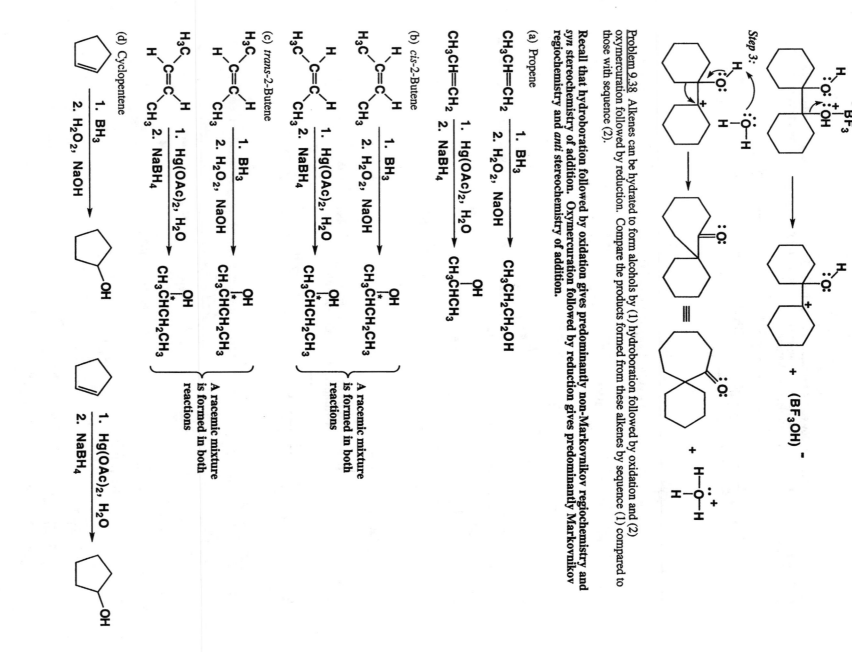

Problem 9.38 Alkenes can be hydrated to form alcohols by (1) hydroboration followed by oxidation and (2) oxymercuration followed by reduction. Compare the products formed from these alkenes by sequence (1) compared to those with sequence (2).

Recall that hydroboration followed by oxidation gives predominantly non-Markovnikov regiochemistry and *syn* stereochemistry of addition. Oxymercuration followed by reduction gives predominantly Markovnikov regiochemistry and *anti* stereochemistry of addition.

(a) Propene

$CH_3CH=CH_2$ 1. BH_3 / 2. H_2O_2, NaOH → $CH_3CH_2CH_2OH$

$CH_3CH=CH_2$ 1. $Hg(OAc)_2$, H_2O / 2. $NaBH_4$ → CH_3CHCH_3 (OH)

(b) *cis*-2-Butene

1. BH_3 / 2. H_2O_2, NaOH → $CH_3CHCH_2CH_3$ (OH*) — A racemic mixture is formed in both reactions

1. $Hg(OAc)_2$, H_2O / 2. $NaBH_4$ → $CH_3CHCH_2CH_3$ (OH*)

(c) *trans*-2-Butene

1. BH_3 / 2. H_2O_2, NaOH → $CH_3CHCH_2CH_3$ (OH*) — A racemic mixture is formed in both reactions

1. $Hg(OAc)_2$, H_2O / 2. $NaBH_4$ → $CH_3CHCH_2CH_3$ (OH*)

(d) Cyclopentene

1. BH_3 / 2. H_2O_2, NaOH → (cyclopentanol)

1. $Hg(OAc)_2$, H_2O / 2. $NaBH_4$ → (cyclopentanol)

(e) 1-Methylcyclohexene

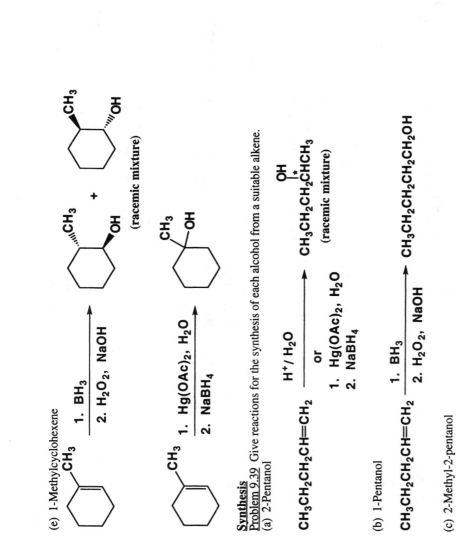

(racemic mixture)

Synthesis
Problem 9.39 Give reactions for the synthesis of each alcohol from a suitable alkene.

(a) 2-Pentanol

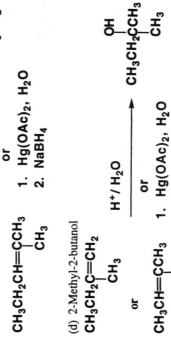

(b) 1-Pentanol

$$CH_3CH_2CH_2CH=CH_2 \xrightarrow[\text{2. } H_2O_2, \text{ NaOH}]{\text{1. } BH_3} CH_3CH_2CH_2CH_2CH_2OH$$

(c) 2-Methyl-2-pentanol

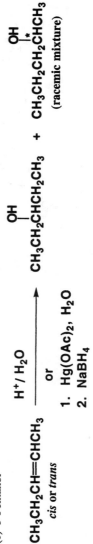

(d) 2-Methyl-2-butanol

(e) 3-Pentanol

(f) 3-Ethyl-3-pentanol

$(CH_3CH_2)_2C=CHCH_3$ $\xrightarrow{\text{H}^+/\text{H}_2\text{O}}$

or

1. $Hg(OAc)_2$, H_2O
2. $NaBH_4$

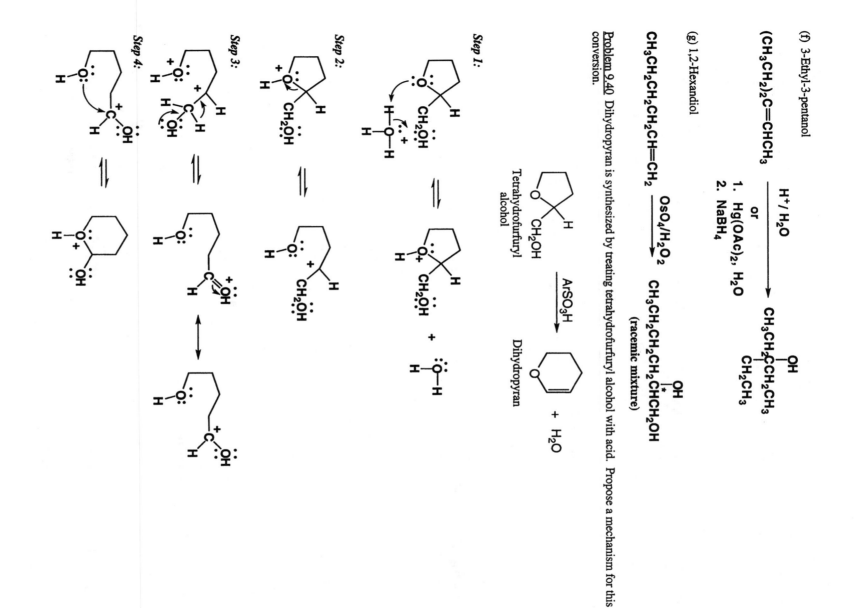

(g) 1,2-Hexandiol

$CH_3CH_2CH_2CH_2CH=CH_2$ $\xrightarrow{\text{OsO}_4/\text{H}_2\text{O}_2}$ $CH_3CH_2CH_2CH_2CHCH_2OH$
(racemic mixture)

<u>Problem 9.40</u> Dihydropyran is synthesized by treating tetrahydrofurfuryl alcohol with acid. Propose a mechanism for this conversion.

Tetrahydrofurfuryl alcohol $\xrightarrow{\text{ArSO}_3\text{H}}$ Dihydropyran $+$ H_2O

Step 1:

Step 2:

Step 3:

Step 4:

Step 5:

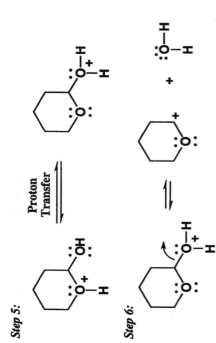

Proton
Transfer

Step 6:

Step 7:

<u>Problem 9.41</u>   (a)   How many stereoisomers are possible for 4-methyl-1,2-cyclohexanediol?

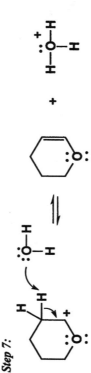

4-Methyl-1,2-cyclohexanediol

There are 2³ or eight stereoisomers, since there are three stereocenters in this molecule and no planes of symmetry.

(b)   Which of the possible stereoisomers are formed by oxidation of 4-methylcyclohexene with osmium tetroxide?

There are two *cis-trans* isomers possible for the diol formed by oxidation of 4-methylcyclohexene. Because of the stereoselectivity of oxidation by osmium tetroxide under these conditions, the two -OH groups are *cis* to each other, and both are either *cis* or *trans* to the methyl group.

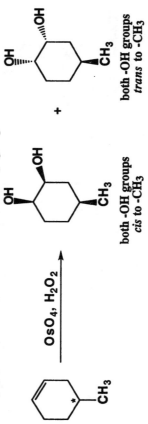

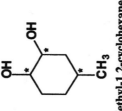

both -OH groups
cis to -CH₃

both -OH groups
trans to -CH₃

(c)   Is the product formed in part (b) optically active or optically inactive?

The products will be optically active if the starting material is either pure 4R or 4S enantiomer, but the problem doesn't specify configuration so assume 4R,4S (racemic) and therefore an optically inactive product.

Problem 9.42 Show how to convert propene to each of these compounds. Use any inorganic reagents as necessary.

(a) Propane

$$CH_3CH=CH_2 + H_2 \xrightarrow[\text{catalyst}]{\text{Transition metal}} CH_3CH_2CH_3$$

(b) 1,2-Propanediol

$$CH_3CH=CH_2 \xrightarrow[H_2O_2]{OsO_4} \underset{\substack{| \\ OH}}{CH_3\overset{*}{C}HCH_2OH}$$

(racemic mixture)

(c) 1-Propanol

$$CH_3CH=CH_2 \xrightarrow[\text{2. } H_2O_2, \text{ NaOH}]{\text{1. } BH_3} CH_3CH_2CH_2OH$$

(d) 2-Propanol

$$CH_3CH=CH_2 \xrightarrow[\text{heat}]{H_2O, H_2SO_4} \underset{\substack{| \\ OH}}{CH_3CHCH_3}$$

(e) Propanal

$$CH_3CH=CH_2 \xrightarrow[\text{2. } H_2O_2, \text{ NaOH}]{\text{1. } BH_3} CH_3CH_2CH_2OH \xrightarrow{PCC} \underset{\substack{\| \\ O}}{CH_3CH_2CH}$$

(f) Propanone

$$CH_3CH=CH_2 \xrightarrow[\text{heat}]{H_2O, H_2SO_4} \underset{\substack{| \\ OH}}{CH_3CHCH_3} \xrightarrow[H_2O, \text{ heat}]{\substack{K_2Cr_2O_7 \\ H_2SO_4}} \underset{\substack{\| \\ O}}{CH_3CCH_3}$$

(g) Propanoic acid

$$CH_3CH=CH_2 \xrightarrow[\text{2. } H_2O_2, \text{ NaOH}]{\text{1. } BH_3} CH_3CH_2CH_2OH \xrightarrow[H_2O, \text{ heat}]{\substack{K_2Cr_2O_7 \\ H_2SO_4}} \underset{\substack{\| \\ O}}{CH_3CH_2COH}$$

(h) 1-Bromo-2-propanol

$$CH_3CH=CH_2 \xrightarrow{Br_2/H_2O} \underset{\substack{| \\ OH}}{CH_3\overset{*}{C}HCH_2Br}$$

(racemic mixture)

(i) 3-Chloropropene

$$CH_3CH=CH_2 \xrightarrow[\text{heat}]{Cl_2} ClCH_2CH=CH_2$$

(Allylic halogenation)

(j) 1,2,3-trichloropropane

$$CH_3CH=CH_2 \xrightarrow[\text{heat}]{Cl_2} ClCH_2CH=CH_2 \xrightarrow{Cl_2} ClCH_2CHClCH_2Cl$$

(Allylic halogenation)   (Normal halogenation)

(k) 1-Chloropropane

$$CH_3CH=CH_2 \xrightarrow[\text{2. } H_2O_2, \text{ NaOH}]{\text{1. } BH_3} CH_3CH_2CH_2OH \xrightarrow[\text{Pyridine}]{SOCl_2} CH_3CH_2CH_2Cl$$

(l) 2-Chloropropane

$$CH_3CH=CH_2 \xrightarrow[H_2O, H_2SO_4]{} \overset{\overset{\displaystyle OH}{|}}{CH_3CHCH_3} \xrightarrow[\text{Pyridine}]{SOCl_2} \overset{\overset{\displaystyle Cl}{|}}{CH_3CHCH_3}$$

or

$$CH_3CH=CH_2 \xrightarrow{HCl} \overset{\overset{\displaystyle Cl}{|}}{CH_3CHCH_3}$$

(m) 2-Propen-1-ol

$$CH_3CH=CH_2 \xrightarrow[\text{(Allylic}]{\text{NBS}} \xrightarrow[\text{halogenation)}]{\text{Peroxides}} BrCH_2CH=CH_2 \xrightarrow[S_N2]{\text{NaOH}} HOCH_2CH=CH_2$$

(n) Propenal

$$CH_3CH=CH_2 \xrightarrow[\text{(Allylic}]{\text{NBS}} \xrightarrow[\text{halogenation)}]{\text{Peroxides}} BrCH_2CH=CH_2 \xrightarrow[S_N2]{\text{NaOH}} HOCH_2CH=CH_2 \xrightarrow{\text{PCC}} \overset{\overset{\displaystyle O}{\parallel}}{H}CCH=CH_2$$

Problem 9.43 Show how to bring about this conversion in good yield.

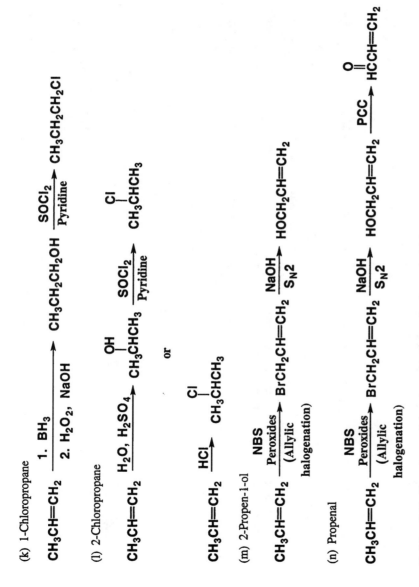

The alcohol is first converted to the alkene by treatment with acid, then the resulting alkene is treated with O_3. Alternatively, (not shown) the alkene could be treated with OsO_4 / H_2O_2 followed by HIO_4.

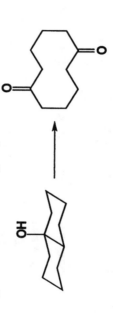

Problem 9.44 The tosylate of a primary alcohol normally undergoes an S_N2 reaction with hydroxide ion to give a primary alcohol. Reaction of this tosylate, however, gives a compound of molecular formula $C_7H_{12}O$. Propose a structural formula for this compound and a mechanism for its formation.

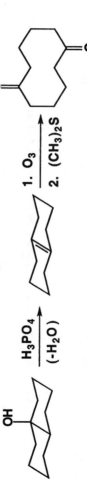

In this reaction, the HO⁻ is acting as a base, not a nucleophile. This makes sense because proton transfer reactions are generally so fast. The alkoxide produced by deprotonation of the alcohol carries out an intramolecular S_N2 attack to give the bicyclic product, $C_7H_{12}O$.

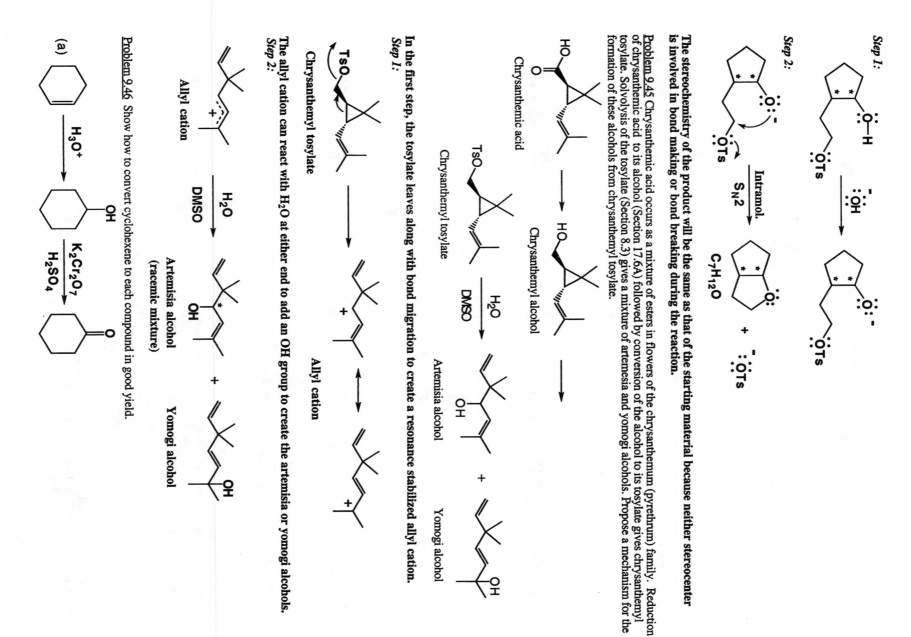

Step 1:

Step 2:

The stereochemistry of the product will be the same as that of the starting material because neither stereocenter is involved in bond making or bond breaking during the reaction.

Problem 9.45 Chrysanthemic acid occurs as a mixture of esters in flowers of the chrysanthemum (pyrethrum) family. Reduction of chrysanthemic acid to its alcohol (Section 17.6A) followed by conversion of the alcohol to its tosylate gives chrysanthemyl tosylate. Solvolysis of the tosylate (Section 8.3) gives a mixture of artemisia and yomogi alcohols. Propose a mechanism for the formation of these alcohols from chrysanthemyl tosylate.

Chrysanthemic acid

Chrysanthemyl alcohol

Chrysanthemyl tosylate

Artemisia alcohol   Yomogi alcohol

In the first step, the tosylate leaves along with bond migration to create a resonance stabilized allyl cation.

Step 1:

Chrysanthemyl tosylate

Allyl cation

The allyl cation can react with H₂O at either end to add an OH group to create the artemisia or yomogi alcohols.

Step 2:

Allyl cation

Artemisia alcohol   Yomogi alcohol
(racemic mixture)

Problem 9.46 Show how to convert cyclohexene to each compound in good yield.

(a)

(b)

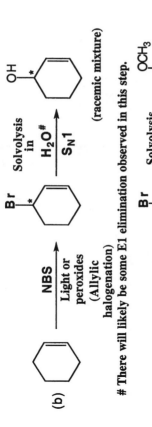

NBS
Light or
peroxides
(Allylic
halogenation)

Solvolysis
in
$H_2O^{\#}$
S_N1

(racemic mixture)

There will likely be some E1 elimination observed in this step.

(c)

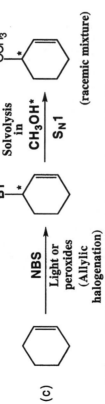

NBS
Light or
peroxides
(Allylic
halogenation)

Solvolysis
in
CH_3OH^*
S_N1

(racemic mixture)

There will likely be some E1 elimination observed in this step.

(d)

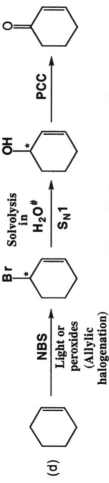

NBS
Light or
peroxides
(Allylic
halogenation)

Solvolysis
in
$H_2O^{\#}$
S_N1

PCC

There will likely be some E1 elimination observed in this step.

(e)

1. O_3
2. $(CH_3)_2S$

$\overset{O}{\underset{\parallel}{HC}}(CH_2)_4\overset{O}{\underset{\parallel}{CH}}$

or alternatively,

OsO_4
H_2O_2

HIO_4

$\overset{O}{\underset{\parallel}{HC}}(CH_2)_4\overset{O}{\underset{\parallel}{CH}}$

Molecular Modeling

Problem 9.47 Oxymercuration of an alkene followed by reduction with sodium borohydride is regioselective. Oxymercuration of this bicycloalkene followed by reduction gives a single alcohol in better than 95% yield. Propose a structural formula for the alcohol formed. Examine the model of the alkene on the CD and see if you can determine which face of the double bond is more accessible to the oxymercuration reagent.

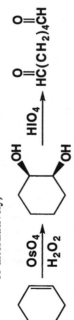

Following is the structural formula of the bicyclic alcohol formed by oxymercuration/reduction.

First, given the known regioselectivity of oxymercuration, predict that -OH adds to the more substituted carbon atom of the double bond. To account for the stereoselectivity, predict that the bridged mercurinium ion intermediate will form on the less hindered side. That is, the same side as the one-carbon bridge (from the top

of the molecule as it is drawn in the problem) rather than on the same side as the two-carbon bridge (from the bottom side as the molecule is drawn in the problem). Approach of the water molecule must be from the backside of the mercurinium ion, that is, the side with the two carbon bridge, leading to the observed product.

Side view:

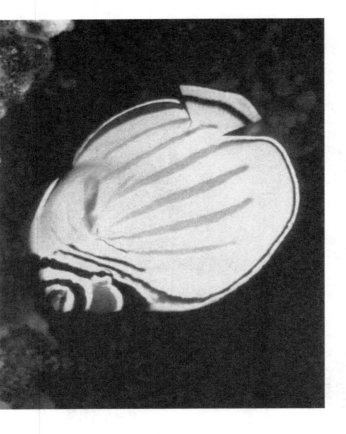

Less hindered face

More hindered face

Bridging mercurinium ion forms on this face

$(OAc)Hg^+ - CH_2$

Water approaches from this direction

$\xrightarrow[\text{2. } NaBH_4]{H_2O}$

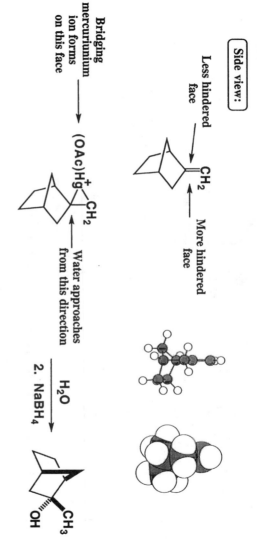

CH₃
OH

Ornate butterflyfish *Chaetodon ornatissimus*
Lanai, Hawaii

Chapter 10: Alkynes

Solutions

193

CHAPTER 10
Solutions to the Problems

Problem 10.1 Write the IUPAC name for each compound.

(a)

1-Octyne

(b)

3,3-Dimethyl-1-butyne

(c)

2-Methyl-3-butyn-2-ol

Problem 10.2 Write the common name for each compound.

(a)

Diisopropylacetylene

(b)

Cyclohexylacetylene

(c)

Butylacetylene

Problem 10.3 Draw a structural formula for an alkene and dichloroalkane of the given molecular formula that yields the indicated alkyne by each reaction sequence.

The key to this problem is to realize that wherever the triple bond is in the carbon chain, that is where the double bond was.

(a) C_6H_{12} $\xrightarrow{Cl_2}$ $C_6H_{12}Cl_2$ $\xrightarrow{2NaNH_2}$

C_6H_{12}

$CH_3CH_2CH=CHCH_2CH_3$ $\xrightarrow{Cl_2}$ $CH_3CH_2\overset{*}{C}H-\overset{*}{C}HCH_2CH_3$

(cis or trans)

$\begin{matrix} | & | \\ Cl & Cl \end{matrix}$

$C_6H_{12}Cl_2$

(b) C_7H_{14} $\xrightarrow{Cl_2}$ $C_7H_{14}Cl_2$ $\xrightarrow{2NaNH_2}$

CH_3
$|$
$CH_3-CH=CH-C-CH_3$
$|$
CH_3

(cis or trans)

C_7H_{14}

$\xrightarrow{Cl_2}$

CH_3
$|$
$CH_3-\overset{*}{C}H-\overset{*}{C}H-CH-C-CH_3$
$\begin{matrix} | & | & | \\ Cl & Cl & CH_3 \end{matrix}$

$C_7H_{14}Cl_2$

$\xrightarrow{2NaNH_2}$

Problem 10.4 Draw the structural formula for a hydrocarbon of the given molecular formula that undergoes hydroboration-oxidation to give the indicated product:

(a) 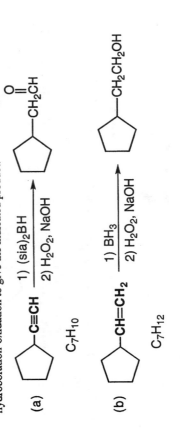 $C≡CH$ $\xrightarrow[\text{2) H}_2O_2,\text{ NaOH}]{\text{1) (sia)}_2BH}$

C_7H_{10}

(b) $-CH=CH_2$ $\xrightarrow[\text{2) H}_2O_2,\text{ NaOH}]{\text{1) BH}_3}$

C_7H_{12}

Problem 10.5 Hydration of 2-pentyne gives a mixture of two ketones, each of molecular formula $C_5H_{10}O$. Propose structural formulas for these two ketones and for the enol from which each is derived.

The two ketones are 2-pentanone and 3-pentanone.

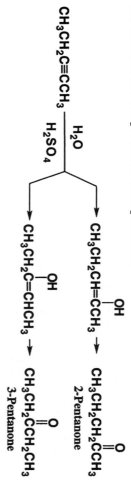

$$CH_3CH_2C{\equiv}CCH_3 \xrightarrow[H_2SO_4]{H_2O}$$

$$CH_3CH_2CH{=}C(OH)CH_3 \longrightarrow CH_3CH_2CH_2\overset{O}{C}CH_3 \quad \text{2-Pentanone}$$

$$CH_3CH_2C(OH){=}CHCH_3 \longrightarrow CH_3CH_2\overset{O}{C}CH_2CH_3 \quad \text{3-Pentanone}$$

Problem 10.6 Show how the synthetic scheme in Example 10.6 might be modified to give
(a) 1-Heptanol

$$HC{\equiv}CH \xrightarrow[\text{2. } CH_3(CH_2)_3CH_2Br]{\text{1. } NaNH_2} HC{\equiv}C(CH_2)_4CH_3 \xrightarrow[\text{2. } H_2O_2,\ NaOH]{\text{1. } BH_3} HOCH_2(CH_2)_5CH_3 \quad \text{1-Heptanol}$$

A reduction step can be included using the Lindlar catalyst to give an alkene, that can then be converted to the primary alcohol on the less-substituted carbon of the double bond using hydroboration.

$$HC{\equiv}CH \xrightarrow[\text{2. } CH_3(CH_2)_3CH_2Br]{\text{1. } NaNH_2} HC{\equiv}C(CH_2)_4CH_3 \xrightarrow[\text{(Lindlar catalyst)}]{H_2 \;\; Pd/CaCO_3}$$

(b) 2-Heptanol

$$HC{\equiv}CH \xrightarrow[\text{2. } CH_3(CH_2)_3CH_2Br]{\text{1. } NaNH_2} HC{\equiv}C(CH_2)_4CH_3 \xrightarrow[\text{2. } NaBH_4]{\text{1. } Hg(OAc)_2,\ H_2O}$$

Alkene structure:

$$\begin{array}{c} H \\ \backslash \\ C{=}C \\ / \qquad \backslash \\ H \qquad (CH_2)_4CH_3 \\ H \end{array} \xrightarrow[\text{(Lindlar catalyst)}]{H_2 \;\; Pd/CaCO_3}$$

$$CH_3\overset{*}{C}H(OH)(CH_2)_4CH_3 \quad \text{2-Heptanol}$$

This is the same as part (a), but the last step uses oxymercuration/reduction to place the -OH group on the more substituted carbon of the double bond.

Structure and Nomenclature
Problem 10.7 Write IUPAC names for the following compounds:

(a) $CH_3C{\equiv}C{-}\underset{\underset{CH_3}{|}}{\overset{\overset{CH_3}{|}}{C}}{-}CH_3$

4,4-Dimethyl-2-pentyne
(*tert*-Butylmethylacetylene)

(b) $HC{\equiv}CCH_2Br$

3-Bromopropyne

(c) [cyclopentane ring]$-C{\equiv}CH$

Ethynylcyclopentane
(Cyclopentylacetylene)

(d) $HC{\equiv}CCH_2CH_2CH_2C{\equiv}CH$

1,6-Heptadiyne

(e) $CH_3(CH_2)_5C{\equiv}CCH_2OH$

2-Nonyn-1-ol

(f) $CH_3(CH_2)_6C{\equiv}CH$

1-Nonyne

(g) $CH_3C{\equiv}CCH_2OH$

2-Butyn-1-ol

(h) $CH_3(CH_2)_7C{\equiv}C(CH_2)_7COOH$

9-Octadecynoic acid

Problem 10.8 Draw structural formulas for each compound:

(a) 3-Hexyne

CH₃CH₂C≡CCH₂CH₃

(b) Vinylacetylene

CH₂=CH−C≡CH

(c) 3-Chloro-1-butyne

Cl
|*
HC≡CCHCH₃

(d) 5-Isopropyl-3-octyne

CH(CH₃)₂
|*
CH₃CH₂C≡CCHCH₂CH₂CH₃

(e) 3-Pentyn-2-ol

*
CH₃C≡CCHCH₃
|
OH

(f) 2-Butyne-1,4-diol

HOCH₂C≡CCH₂OH

(g) 2,2,5-Trimethyl-3-pentyne

CH₃ CH₃
| |
CH₃CHC≡CCCH₃
       |
       CH₃

(h) *tert*-Butylmethylacetylene

CH₃
|
CH₃C≡CCCH₃
|
CH₃

(i) Cyclodecyne

Problem 10.9 Predict all bond angles about each circled atom.

(a) CH₃−C≡Ⓒ−CH₃

180°

CH₃−C≡Ⓒ−CH₃

(b) CH₂=Ⓒ−C≡CH

120°

CH₂=Ⓒ−C≡CH

(c) CH₂=Ⓒ−CH−CH₃

180°

CH₂=Ⓒ−CH−CH₃

(d) Ⓒ H₂=CH−CH=CH₂

120°

Ⓒ H₂=CH−CH=CH₂

Problem 10.10 State the orbital hybridization of each circled atom.

(a) CH₃−C≡Ⓒ−CH₃

sp

CH₃−C≡Ⓒ−CH₃

(b) CH₂=Ⓒ−C≡CH

sp²

CH₂=Ⓒ−C≡CH

(c) CH₂=Ⓒ−CH−CH₃

sp

CH₂=Ⓒ−CH−CH₃

(d)

O=C=O

sp

O=C=O

Problem 10.11 Describe each circled carbon-carbon bond in terms of the overlap of atomic orbitals.

(a) CH₃—C≡C—CH₃

σ sp-sp³

CH₃—C≡C—CH₃

(b) CH₂=CH—C≡CH

σ sp²-sp

CH₂=CH—C≡CH

(c) CH₂=C=CH—CH₃

σ sp²-sp²

CH₂=C=CH—CH₃

π 2p-2p

CH₂=C=CH—CH₃

(d) CH₂=CH—CH=CH₂

σ sp²-sp²

CH₂=CH—CH=CH₂

Problem 10.12   Enanthotoxin (The Merck Index, 12th Ed., #3608) is an extremely poisonous organic compound found in hemlock water dropwort, which is reputed to be the most poisonous plant in England. It is believed that no British plant has been responsible for more fatal accidents. The most poisonous part of the plant is the roots, which resemble small white carrots, giving the plant the name "five finger death." Also poisonous are its leaves, which look like parsley. Enanthotoxin is thought to interfere with the Na⁺ current in nerve cells, which leads to convulsions and death.

Can show *cis-trans* isomerism

How many stereoisomers are possible for enanthotoxin?

There is one tetrahedral stereocenter (marked with an *) and three double bonds that can show *cis-trans* isomerism, so there are 2⁴ or 16 possible stereoisomers.

Preparation of Alkynes

Problem 10.13 Show how to prepare each alkyne from the given starting material. In part (c), "D" indicates deuterium. Deuterium-containing reagents such as BD3, D₂O, CH₃CH₂OD, and CH3COOD are available commercially.

(a) CH₃CH₂CH₂CH=CH₂ ⟶ CH₃CH₂CH₂C≡CH

First treat 1-pentene with either bromine (Br₂) or chlorine (Cl₂) to form a 1,2-dihalopentane. Then carry out a double dehydrohalogenation with three moles sodium amide (NaNH₂) followed by an aqueous workup to form 1-pentyne.

$$CH_3(CH_2)_2CH=CH_2 \xrightarrow{Br_2} CH_3(CH_2)_2\overset{\overset{Br}{|}}{C}H-\overset{\overset{Br}{|}}{C}H_2 \xrightarrow[\text{2. H}_2O]{\text{1. 3NaNH}_2} CH_3CH_2CH_2C \equiv CH$$

(b) CH₃(CH₂)₅CHCH₃ ⟶ CH₃(CH₂)₄C≡CCH₃

        |
        Cl

First convert 2-chlorooctane to 2-octene by dehydrohalogenation with sodium amide. This is an example of a β-elimination reaction. Then follow the procedure in (a) to convert an alkene to an alkyne, namely addition of Br₂ or Cl₂ followed by a double dehydrohalogenation.

$$CH_3(CH_2)_5\overset{*}{C}HCH_3 \xrightarrow{\text{NaNH}_2} CH_3(CH_2)_4CH=CHCH_3 \xrightarrow[\text{2) 2NaNH}_2]{\text{1) Br}_2} CH_3(CH_2)_4C \equiv CCH_3$$

        |
        Cl

(c) $CH_3CH_2CH_2C\equiv CH \longrightarrow CH_3CH_2CH_2C\equiv CD$

First form the acetylide anion with sodium amide and then react the anion with a deuterium donor such as D_2O or CH_3CH_2OD.

$$CH_3CH_2CH_2C\equiv CH \xrightarrow{NaNH_2} CH_3CH_2CH_2C\equiv C^-\ Na^+ \xrightarrow{D_2O} CH_3CH_2CH_2C\equiv CD$$

<u>Problem 10.14</u> If a catalyst could be found that would establish an equilibrium between 1,2-butadiene and 2-butyne, what would be the ratio of the more stable isomer to the less stable isomer at 25°C?

$$CH_2=C=CHCH_3 \rightleftharpoons CH_3C\equiv CCH_3 \qquad \Delta G° = -16.7\ kJ\ (-4.0\ kcal)/mol$$

At equilibrium the relative amounts of each form are given by the equation:

$$\Delta G° = -RT\ln K_{eq} \qquad \text{where} \quad K_{eq} = \frac{[2\text{-butyne}]}{[1,2\text{-butadiene}]}$$

Converting to base 10 and rearranging gives

$$\log K_{eq} = \frac{-\Delta G°}{(2.303)RT}$$

Taking the antilog of both sides gives:

$$K_{eq} = 10^{\left(\frac{-\Delta G°}{(2.303)RT}\right)}$$

Plugging in the values for $\Delta G°$, R, and 298 K gives the final answer:

$$\boxed{K_{eq} = 10^{\left(\frac{-(-16,700\ J/mol)}{(2.303)(8.314\ J\cdot K^{-1}\cdot mol^{-1})(298\ K)}\right)} = 10^{2.93} = 8.6 \times 10^2}$$

Because $\Delta G°$ is negative, it means the alkyne is favored, and if a catalyst could be found for the interconversion, the ratio of alkyne to allene would be 860 to 1.

Reactions of Alkynes
<u>Problem 10.15</u> Complete each acid-base reaction and predict whether the position of equilibrium lies toward the left or the right.

Recall that equilibrium favors formation of the weaker acid, weaker base pair.

(a) $CH_3C\equiv CH + (CH_3)_3CO^- K^+ \xrightarrow{(CH_3)_3COH} CH_3C\equiv C^-\ K^+ + (CH_3)_3COH$

pK$_a$ 25 $\qquad\qquad\qquad\qquad\qquad\qquad\qquad\qquad\qquad\quad$ pK$_a$ 18
(weaker $\qquad\qquad$ (weaker $\qquad\qquad\qquad\qquad$ (stronger $\qquad\qquad$ (stronger
acid) $\qquad\qquad\quad$ base) $\qquad\qquad\qquad\qquad\qquad$ base) $\qquad\qquad\qquad$ acid)

Oxygen is more electronegative (farther to the right on the Periodic Table) than carbon so the alcohol is the stronger acid and equilibrium lies toward the left.

(b) $CH_2=CH_2 + Na^+NH_2^- \xrightarrow{NH_3\ (l)} CH_2=CH^-\ Na^+ + NH_3$

pK$_a$ 44 $\qquad\qquad\qquad\qquad\qquad\qquad\qquad\qquad\qquad\quad$ pK$_a$ 38
(weaker $\qquad\qquad$ (weaker $\qquad\qquad\qquad\qquad$ (stronger $\qquad\qquad$ (stronger
acid) $\qquad\qquad\quad$ base) $\qquad\qquad\qquad\qquad\qquad$ base) $\qquad\qquad\qquad$ acid)

Nitrogen is more electronegative (farther to the right on the Periodic Table) than carbon so ammonia is the stronger acid and equilibrium lies toward the left.

(c) $CH_3C\equiv CCH_2OH + Na^+NH_2^- \xrightarrow{NH_3\ (l)} CH_3C\equiv CCH_2O^-\ Na^+ + NH_3$

pK$_a$ ~16 $\qquad\qquad\qquad\qquad\qquad\qquad\qquad\qquad\qquad\quad$ pK$_a$ 38
(stronger $\qquad\qquad$ (stronger $\qquad\qquad\qquad\qquad$ (weaker $\qquad\qquad$ (weaker
acid) $\qquad\qquad\quad$ base) $\qquad\qquad\qquad\qquad\qquad$ base) $\qquad\qquad\qquad$ acid)

Oxygen is more electronegative (farther to the right on the Periodic Table) than nitrogen so the alcohol is the stronger acid and equilibrium lies toward the right.

Problem 10.16 Draw structural formulas for the major product(s) formed by reaction of 3-hexyne with each of these reagents. Where you predict no reaction, write NR.

(a) H₂(excess) / Pt

(b) H₂ / Lindlar catalyst

(c) Na in NH₃ (liquid)

CH₃CH₂CH₂CH₂CH₂CH₃

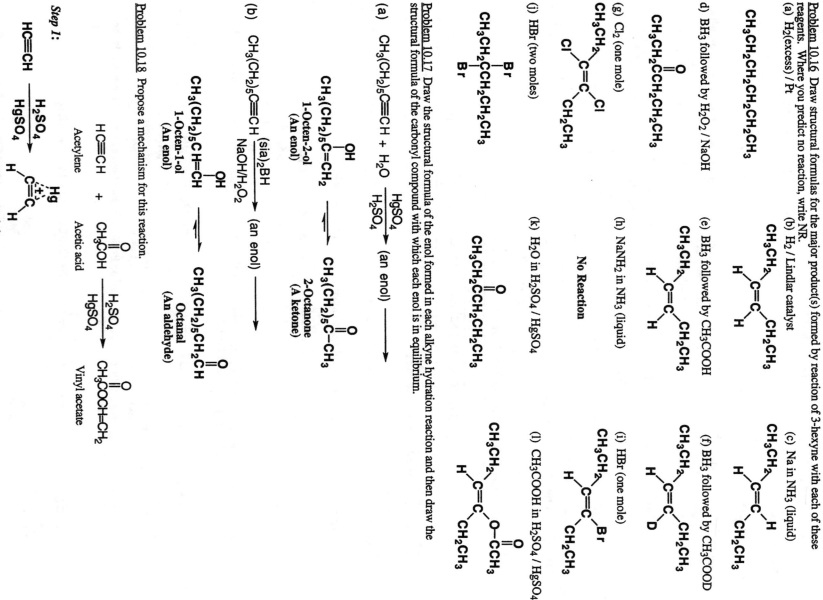

(d) BH₃ followed by H₂O₂ / NaOH

(e) BH₃ followed by CH₃COOH

(f) BH₃ followed by CH₃COOD

(g) Cl₂ (one mole)

(h) NaNH₂ in NH₃ (liquid)

No Reaction

(i) HBr (one mole)

(j) HBr (two moles)

(k) H₂O in H₂SO₄ / HgSO₄

(l) CH₃COOH in H₂SO₄ / HgSO₄

Problem 10.17 Draw the structural formula of the enol formed in each alkyne hydration reaction and then draw the structural formula of the carbonyl compound with which each enol is in equilibrium.

(a) CH₃(CH₂)₅C≡CH + H₂O →(HgSO₄ / H₂SO₄) (an enol) →

1-Octen-2-ol (An enol)

CH₃(CH₂)₅C(OH)=CH₂

2-Octanone (A ketone)

CH₃(CH₂)₅C—CH₃

(b) CH₃(CH₂)₅C≡CH →((sia)₂BH ; NaOH/H₂O₂) (an enol) →

1-Octen-1-ol (An enol)

CH₃(CH₂)₅CH=CH(OH)

Octanal (An aldehyde)

CH₃(CH₂)₅CH₂CH

Problem 10.18 Propose a mechanism for this reaction.

HC≡CH + CH₃COOH →(H₂SO₄ / HgSO₄) CH₃COOCH=CH₂

Acetylene   Acetic acid   Vinyl acetate

Step 1:

HC≡CH →(H₂SO₄ / HgSO₄) Bridged mercurinium ion intermediate

Step 2:

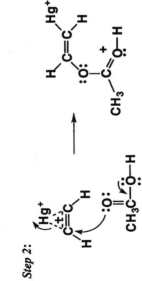

Step 3:

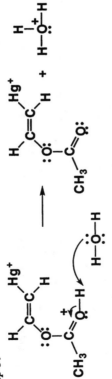

Step 4:

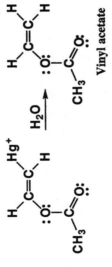

Vinyl acetate

Syntheses

Problem 10.19 Show how to convert 9-octadecynoic acid to the following:

$$CH_3(CH_2)_7C \equiv C(CH_2)_7COOH$$

9-Octadecynoic acid

(a) (E)-9-Octadecenoic acid (eliadic acid)

Chemical reduction of the alkyne with two moles sodium in liquid ammonia converts the alkyne to an E alkene. Note that the carboxyl group is unaffected by these conditions.

$$CH_3(CH_2)_7C \equiv C(CH_2)_7COOH \xrightarrow[\text{then HCl}]{\substack{2Na \\ NH_3(l)}}$$

9-Octadecynoic acid

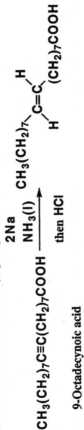

(b) (Z)-9-Octadecenoic acid (oleic acid)

Reduction of the alkyne with one mole of hydrogen with Lindlar's catalyst gives a Z alkene. The carboxyl group is unaffected by these conditions.

$$CH_3(CH_2)_7C \equiv C(CH_2)_7COOH \xrightarrow[\text{Pd/CaCO}_3]{H_2}$$

9-Octadecynoic acid

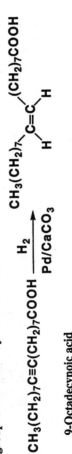

(c) 9,10-Dihydroxyoctadecanoic acid

Either the E alkene or the Z alkene can be converted to the glycol by oxidation with OsO_4/H_2O_2.

$$CH_3(CH_2)_7CH=CH(CH_2)_7COOH \xrightarrow[H_2O_2]{OsO_4} CH_3(CH_2)_7\overset{*}{C}H-\overset{*}{C}H(CH_2)_7COOH$$
$$\underset{HO\quad\;\;OH}{}$$

9-Octadecenoic acid
(cis or trans)

(d) Octadecanoic acid

Reduction of either the E alkene or the Z alkene with one mole H_2 in the presence of a Ni, Pd, or Pt catalyst, or reduction of the alkyne with two moles H_2 gives the desired product.

$$CH_3(CH_2)_7CH=CH(CH_2)_7COOH \xrightarrow{H_2 \atop Ni} CH_3(CH_2)_7CH_2-CH_2(CH_2)_7COOH$$

9-Octadecenoic acid
(cis or trans)

$$CH_3(CH_2)_7C\equiv C(CH_2)_7COOH \xrightarrow{2H_2 \atop Ni} CH_3(CH_2)_7CH_2-CH_2(CH_2)_7COOH$$

9-Octadecynoic acid

The stereochemistry of the products depends on whether cis or trans 9-octadecenoic acid is used as starting material.

Problem 10.20 For small-scale and consumer welding applications, many hardware stores sell cylinders of MAAP gas, which is a mixture of propyne (methylacetylene) and 1,2-propadiene (allene), with other hydrocarbons. How would you prepare the methylacetylene/allene mixture from propene in the laboratory?

This gas mixture could be prepared from a double dehydrohalogenation of 1,2-dibromopropane. As described in section 10.6, 1,2-propadiene (allene) is a side product of the reaction, being derived from β-elimination of the intermediate 2-bromopropene.

$$CH_3-CH=CH_2 \xrightarrow{Br_2} CH_3-\overset{*}{C}HBr-CH_2Br \xrightarrow[2.\ H_2O]{1.\ 3NaNH_2} CH_3-C\equiv CH + CH_2=C=CH_2$$

Propene   1,2-Dibromopropane   Propyne   1,2-Propadiene
                                                                               (Allene)

Problem 10.21 Show reagents and experimental conditions you might use to convert propyne into each product. Some of these syntheses can be done in one step. Others require two or more steps.

(a)
$$CH_3-C\equiv CH \xrightarrow{Br_2} \underset{Br\ Br}{CH_3C=CH} \xrightarrow{Br_2} \underset{Br\ Br}{\overset{Br\ Br}{CH_3-C-CH}}$$

Addition of two moles Br_2 to propyne gives 1,1,2,2-tetrabromopropane.

(b)
$$CH_3-C\equiv CH \xrightarrow{HBr} \underset{Br}{CH_3C=CH_2} \xrightarrow{HBr} \underset{Br}{\overset{Br}{CH_3-C-CH_3}}$$

Addition of two moles HBr occurs by electrophilic addition and gives first 2-bromopropene and then 2,2-dibromopropane.

(c)
$$CH_3C\equiv CH + H_2O \xrightarrow[HgSO_4]{H_2SO_4} \left[\underset{CH_3C=CH_2}{\overset{OH}{}} \right] \longrightarrow \overset{O}{\underset{CH_3C-CH_3}{}}$$

Acid-catalyzed hydration of the alkyne gives an enol which is in equilibrium, by keto-enol tautomerism, with the isomeric ketone, in this case propanone (acetone).

(d) CH₃CH₂—C—H (with =O above C)

$$CH_3C{\equiv}CH \xrightarrow[\text{2) NaOH/H}_2\text{O}_2]{\text{1) (sia)}_2\text{B H}} \left[CH_3CH{=}CH \;(\text{with OH}) \right] \longrightarrow CH_3CH_2CH \;(\text{with =O})$$

Hydroboration with (sia)₂BH or other hindered derivative of borane followed by oxidation with alkaline hydrogen peroxide gives an enol which is in equilibrium, by keto-enol tautomerism, with the isomeric aldehyde, in this case propanal.

<u>Problem 10.22</u> Show reagents and experimental conditions you might use to convert each starting material into the desired product. Some of these syntheses can be done in one step. Others require two or more steps.

(a)

$$\xrightarrow[\text{NH}_3(l)]{\text{2Na}}$$

Chemical reduction of the alkyne with sodium in liquid ammonia gives (E)-2-hexene.

(b)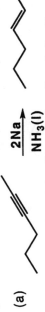

$$\xrightarrow[\text{2) CH}_3\text{CO}_2\text{H}]{\text{1) BH}_3}$$

Hydroboration of the internal alkyne followed by reaction of the organoborane with acetic acid gives (Z)-2-hexene.

$$+ \;H_2 \xrightarrow[\substack{\text{Lindlar} \\ \text{catalyst}}]{\text{Pd/CaCO}_3}$$

Alternatively, catalytic reduction with hydrogen in the presence of Lindlar catalyst gives the (Z)-alkene.

(c)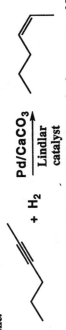

$$\xrightarrow[\substack{\text{Transition metal} \\ \text{catalyst}}]{\text{H}_2}$$

This reduction can be carried out with H₂ in the presence of a transition metal catalyst such as Pd.

(d)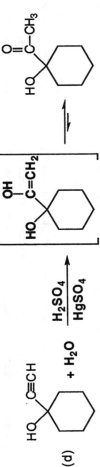

$$+ \;H_2O \xrightarrow[\text{HgSO}_4]{\text{H}_2\text{SO}_4}$$

Acid-catalyzed hydration of the carbon-carbon triple bond followed by keto-enol tautomerism of the resulting enol gives the desired ketone. Note that the tertiary alcohol in the starting material is unaffected by these conditions.

<u>Problem 10.23</u> Show how to convert 1-butyne to each of these compounds.

(a) **CH₃CH₂C≡CH + NaNH₂** ⟶ **CH₃CH₂C≡C:⁻ Na⁺ + NH₃**

The anion can be formed using sodium amide, NaNH₂.

(b) **CH₃CH₂C≡CH** $\xrightarrow{\text{NaNH}_2}$ **CH₃CH₂C≡C:⁻ Na⁺** $\xrightarrow{\text{D}_2\text{O}}$ **CH₃CH₂C≡CD**

Formation of the terminal acetylide anion followed by reaction with a deuterium donor such as D₂O gives 1-deutero-1-butyne.

(c) **CH₃CH₂C≡CH** $\xrightarrow[\text{2. CH}_3\text{CO}_2\text{D}]{\text{1. (sia)}_2\text{B H}}$

Hydroboration with this disubstituted derivative of diborane followed by reaction of the organoborane with deuteroacetic acid gives the desired 1-deutero-1-butene.

(d)   CH₃CH₂C≡CH $\xrightarrow[\text{2. CH}_3\text{CO}_2\text{H}]{\text{1. (sia)}_2\text{B D}}$

Hydroboration of the terminal alkyne with deuteroborane adds deuterium to the more substituted carbon of the alkyne. Reaction of the deuterated organoborane with acetic acid gives the 2-deutero-1-butene.

Problem 10.24 Rimantadine is effective in preventing infections caused by the influenza A virus and in treating established illnesses (The Merck Index, 12th ed., #8390). It is thought to exert its antiviral effect by blocking a late stage in the assembly of the virus. Rimantadine is synthesized from adamantane by the following sequence. We discuss the chemistry of Step 5 in Chapter 21

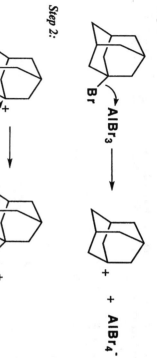

Adamantane   1-Bromoadamantane

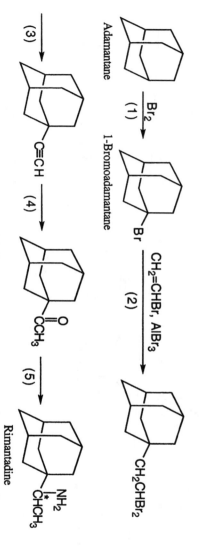

Rimantadine
(an antiviral agent)

(a) Describe experimental conditions to bring about Step 1. By what type of mechanism does this reaction occur? Account for the regioselectivity of bromination in Step 1.

This reaction occurs via a radical bromination (Br₂, heat or hν), with the bromine ending up on the tertiary carbon atom.

(b) Propose a mechanism for Step 2. *Hint:* As we shall see in Section 21.1A, reaction of a bromoalkane such as 1-bromoadamantane with aluminum bromide (a Lewis acid, Section 4.3) results in the formation of a carbocation and AlBr₄⁻. Assume that adamantyl cation is formed in Step 2, and proceed from there to describe a mechanism.

The bromoethene pi electrons attack the admantyl cation, to create a new cation that captures a bromide from AlBr₄⁻ to yield dibromoethyl-adamantane.

Step 1:

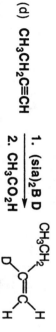

Step 2:

Step 3:

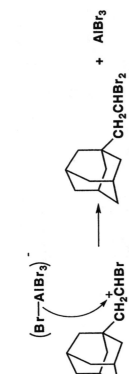

(c) Account for the regioselectivity of carbon-carbon bond formation in Step 2.

The carbon atom without the halogen ends up attached to the adamantane group, because this allows formation of the more stable intermediate with the cation adjacent to the halogen atom. The formation of this cation has a lower activation energy because of resonance stabilization of the cation provided by the halogen.

(d) Describe experimental conditions to bring about Step 3.

This reaction occurs via double dehydrohalogenation using a strong base such as NaNH₂.

(e) Describe experimental conditions to bring about Step 4.

This transformation occurs via an acid-catalyzed hydration of the alkyne using H₂O, H₂SO₄, and HgSO₄. The initially formed enol equilibrates to the more stable keto form.

Problem 10.25 Show reagents and experimental conditions required to bring about the following transformations.

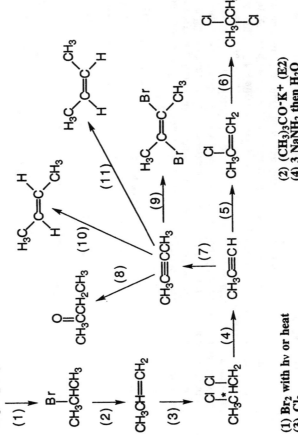

(1) Br₂ with hν or heat
(2) (CH₃)₃CO⁻K⁺ (E2)
(3) Cl₂
(4) 3 NaNH₂ then H₂O
(5) HCl
(6) HCl
(7) NaNH₂ then CH₃I
(8) HgSO₄, H₂SO₄, H₂O
(9) Br₂
(10) Li or Na in liquid NH₃
(11) H₂, Pd/CaCO₃ (Lindlar catalyst)

Problem 10.26 Show reagents to bring about each conversion.

$$CH_3(CH_2)_5C{\equiv}CH \xrightarrow{\ (a)\ } CH_3(CH_2)_5C{\equiv}C^-\ Na^+ \xrightarrow{\ (b)\ } CH_3(CH_2)_5C{\equiv}CCH_2CH_3$$

(d) $CH_3(CH_2)_5CH_2CH \overset{O}{\|}$

(e) $CH_3(CH_2)_5CCH_3 \overset{O}{\|}$

(c)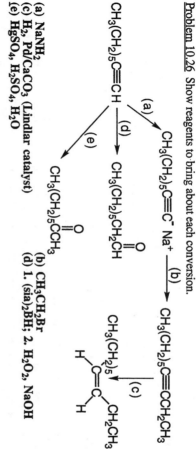

(a) NaNH₂
(b) CH₃CH₂Br
(c) H₂, Pd/CaCO₃ (Lindlar catalyst)
(d) 1. (sia)₂BH; 2. H₂O₂, NaOH
(e) HgSO₄, H₂SO₄, H₂O

Problem 10.27 Which of these alkynes can be prepared in good yield by monoalkylation or dialkylation of acetylene? For each that cannot, explain why not.
(a) 3-Methyl-1-butyne

$HC{\equiv}C:^-\ Na^+$ +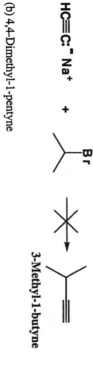

3-Methyl-1-butyne

(b) 4,4-Dimethyl-1-pentyne

$HC{\equiv}C:^-\ Na^+$ +

4,4-Dimethyl-1-pentyne

(c) 2-Octyne

$HC{\equiv}C:^-\ Na^+$

1. CH₃Br
2. NaNH₂
3. CH₂(CH₂)₃CH₂Br

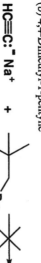

2-Octyne

The acetylide anion is a strong base as well as a good nucleophile, so good yields can only be obtained when acetylide anions react with methyl or primary alkyl halides. Secondary or tertiary alkyl halides give large amounts of elimination via the E2 mechanism. The preparation of (a) could therefore not be carried out in high yield, because preparation of 3-methyl-1-butyne would require reaction of an acetylide anion with a 2-halopropane, a secondary alkyl halide. 4,4-Dimethyl-1-pentyne(b) can also not be made, but for a different reason. It would have to be made from the acetylide anion reacting with 2,2-dimethyl-1-halopropane, also known as the neopentyl alkyl halide. This alkyl halide is too sterically hindered to undergo nucleophilic substitution. The molecule in (c) could be made in high yield, since sequential reactions with a methyl and a primary alkyl halide, halomethane and 1-halopentane, respectively, are required.

Problem 10.28 Propose a synthesis for (Z)-9-tricosene (muscalure, The Merck Index, 12th Ed., # 6388), the sex pheromone for the common house fly (*Musca domestica*) starting with acetylene and haloalkanes as sources of carbon atoms.

$$CH_3(CH_2)_7\underset{H}{\overset{}{C}}{=}\underset{H}{\overset{}{C}}(CH_2)_{12}CH_3$$

Muscalure

Alkylation of acetylene with the two alkyl halides followed by reduction using H₂ and the Lindlar catalyst gives muscalure.

$$HC{\equiv}CH \xrightarrow[\text{2. } CH_3(CH_2)_6CH_2Br]{\text{1. NaNH}_2} CH_3(CH_2)_7C{\equiv}CH \xrightarrow[\text{2. } CH_3(CH_2)_{11}CH_2Br]{\text{1. NaNH}_2}$$

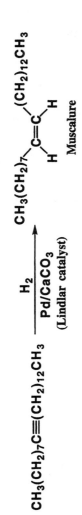

Problem 10.29 Propose a synthesis of each compound starting from acetylene and any necessary organic and inorganic reagents.

(a) 4-octyne

$$HC\equiv CH \xrightarrow[\text{2. CH}_3\text{CH}_2\text{CH}_2\text{Br}]{\text{1. NaNH}_2} HC\equiv CCH_2CH_2CH_3 \xrightarrow[\text{2. CH}_3\text{CH}_2\text{CH}_2\text{Br}]{\text{1. NaNH}_2}$$

$$CH_3CH_2CH_2C\equiv CCH_2CH_2CH_3$$

(b) 4-Octanone

$$CH_3CH_2CH_2C\equiv CCH_2CH_2CH_3 \xrightarrow[\text{HgSO}_4]{\text{H}_2\text{O, H}_2\text{SO}_4} CH_3CH_2CH_2CH_2\overset{\displaystyle O}{\overset{\|}{C}}CH_2CH_2CH_3$$

(From (a))

(c) *cis*-4-Octene

$$CH_3CH_2CH_2C\equiv CCH_2CH_2CH_3 \xrightarrow[\text{Pd/CaCO}_3\ \text{(Lindlar catalyst)}]{\text{H}_2}$$

(From (a))

(d) *trans*-4-Octene

$$CH_3CH_2CH_2C\equiv CCH_2CH_2CH_3 \xrightarrow[\text{NH}_3\ (l)]{\text{Li or Na}}$$

(From (a))

(e) 4-Octanol

(From (c))

or

(From (d))

$$\xrightarrow[\text{2. NaBH}_4]{\text{1. Hg(OAc)}_2,\ \text{H}_2\text{O}} CH_3CH_2CH_2CH_2CH_2\underset{*}{\overset{OH}{C}}HCH_2CH_2CH_3$$

(racemic mixture)

(f) meso-4,5-Octanediol

(From (c))

$$\xrightarrow[\text{H}_2\text{O}_2]{\text{OsO}_4} CH_3CH_2CH_2\overset{HO}{\underset{H}{C}}\overset{OH}{\underset{H}{C}}CH_2CH_2CH_3$$

Note that only the *cis* isomer gives the meso product, the *trans* isomer gives a pair of R,R and S,S enantiomers.

Problem 10.30 Show how to prepare each compound from ethylene.

(a) 1,2-Dichloroethane

$$H_2C=CH_2 \xrightarrow[CCl_4]{Cl_2} ClCH_2CH_2Cl$$

(b) Chloroethylene (vinyl chloride)

From (a)

$$ClCH_2CH_2Cl \xrightarrow[\text{2. } H_2O]{\text{1. } 3NaNH_2} HC\equiv CH \xrightarrow{\text{1 mole } HCl} \overset{Cl}{\underset{H}{C}}=\overset{H}{\underset{H}{C}}$$

(c) 1,1-Dichloroethane

From (a)

$$ClCH_2CH_2Cl \xrightarrow[\text{2. } H_2O]{\text{1. } 3NaNH_2} HC\equiv CH \xrightarrow{\text{2 moles } HCl} Cl_2CHCH_3$$

Problem 10.31 Show how to bring about this conversion.

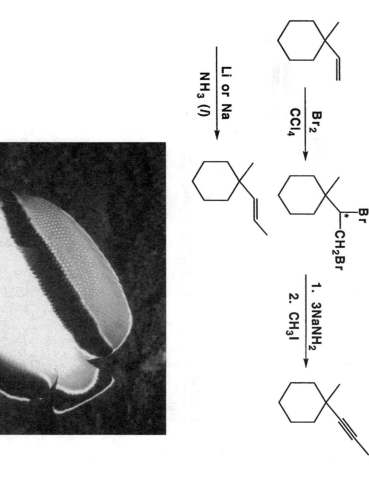

The alkene is first converted to the dibromide, which is converted to the alkyne. The alkyne is reacted with base and the resulting acetylide anion treated with methyl iodide to give the alkyne with the proper number of carbon atoms. Metal reduction is used to give the desired *trans* final product.

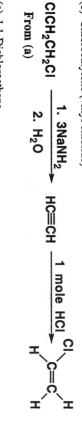

$$\xrightarrow[CCl_4]{Br_2}$$

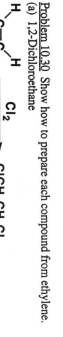

$$\xrightarrow[\text{2. } CH_3I]{\text{1. } 3NaNH_2}$$

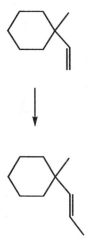

$$\xrightarrow[NH_3 \ (l)]{Li \text{ or } Na}$$

Bandit angelfish *Holacanthus arcuatus*, Lanai, Hawaii

CHAPTER 11
Solutions to the Problems

Problem 11.1 Write IUPAC and common names for these ethers.

(a) CH$_3$CHCH$_2$OCH$_2$CH$_3$
     |
     CH$_3$

1-Ethoxy-2-methylpropane
(Ethyl isobutyl ether)

(b) CH$_3$OCH$_2$CH$_2$OCH$_3$

1,2-Dimethoxyethane
(Ethylene glycol dimethyl ether)

(c)

cis-1,2-Diethoxycyclohexane

Problem 11.2 Arrange these compounds in order of increasing boiling point.

Boiling point increases with an increasing number of hydrogen bonding groups. In order of increasing boiling point, they are:

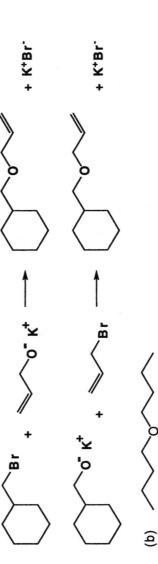

84°C   125°C   198°C

Problem 11.3 Show how you might use the Williamson ether synthesis to prepare these ethers:

(a)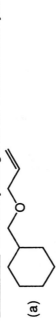

There are two combinations of reagents that could produce the desired ether in high yield because both involve primary haloalkanes and are thus unlikely to eliminate via an E2 pathway

Br + ...O⁻K⁺ ⟶ ... + K⁺Br⁻

...O⁻K⁺ + Br ⟶ ... + K⁺Br⁻

(b)

Br + ...O⁻K⁺ ⟶ ... + K⁺Br⁻

Treatment of 1-bromobutane with sodium butoxide gives dibutyl ether.

Problem 11.4 Show how ethyl hexyl ether might be prepared by a Williamson ether synthesis.

Using the Williamson ether synthesis, ethyl hexyl ether could be synthesized by either of the two following routes:

CH$_3$CH$_2$Br + CH$_3$(CH$_2$)$_4$CH$_2$O⁻Na⁺

or   ⟶ CH$_3$CH$_2$OCH$_2$(CH$_2$)$_4$CH$_3$

CH$_3$(CH$_2$)$_4$CH$_2$Br + CH$_3$CH$_2$O⁻Na⁺

Problem 11.5 Account for the fact that treatment of *tert*-butyl methyl ether with a limited amount of concentrated HI gives methanol and *tert*-butyl iodide rather than methyl iodide and *tert*-butyl alcohol.

The first step in the reaction involves protonation of the ether oxygen to give an oxonium ion intermediate. Cleavage of the oxonium ion intermediate on one side gives methanol and the *tert*-butyl cation. Cleavage on the other side gives a methyl cation and *tert*-butyl alcohol. Because of the greater stability of the tertiary carbocation, cleavage to give methanol and *tert*-butyl cation is favored. Reaction of the tertiary cation with the iodide completes the reaction.

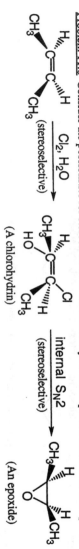

cleavage to form a
tertiary carbocation

$CH_3-\overset{\cdot\cdot}{\underset{H}{O}}: + \overset{+}{C}-CH_3$ more stable

cleavage to form a
primary carbocation

$H-\overset{\cdot\cdot}{\underset{\cdot\cdot}{O}}:-\underset{CH_3}{\overset{CH_3}{C}}-CH_3 + {}^+CH_3$ much less stable

$\overset{CH_3}{\underset{CH_3}{C}}{}^+ \xrightarrow{I^-} I-\underset{CH_3}{\overset{CH_3}{C}}-CH_3$

There is an alternative pathway for formation of product, namely reaction of the oxonium ion with iodide ion by an S_N2 pathway on the less-hindered methyl carbon to give iodomethane and the 2-methyl-2-propanol (*tert*-butyl alcohol). The fact is that the S_N1 pathway by way of a tertiary carbocation has a lower activation energy (a faster rate) than reaction of the oxonium ion with iodide ion by an S_N2 pathway so the S_N1 reaction predominates.

Problem 11.6 Draw structural formulas for the major products of these reactions:

(a)  $H_3C-\underset{CH_3}{\overset{CH_3}{C}}-O-CH_3 \; + \; HBr \;$ (excess) $\; \longrightarrow \; H_3C-\underset{CH_3}{\overset{CH_3}{C}}-Br \; + \; CH_3Br \; + \; H_2O$

(b) 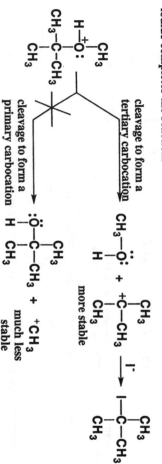 $+ \; HBr$ (excess) $\; \longrightarrow \; BrCH_2CH_2CH_2CH_2CH_2Br \; + \; H_2O$

Problem 11.7 Why is the use of the *tert*-butyl protecting group limited to protection of primary alcohols?

The *tert*-butyl ether is prepared from the primary alcohol and isobutylene in the presence of an acid catalyst. If a secondary or tertiary alcohol were used in this reaction, dehydration of the alcohol to give an alkene would be a major side reaction, preventing a high yield.

Problem 11.8 Consider the possibilities for stereoisomerism in the halohydrin and epoxide formed in Example 11.8.

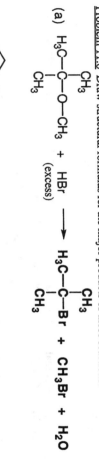

(a) How many stereoisomers are possible for the chlorohydrin? Which of the possible chlorohydrins are formed by reaction of *trans*-2-butene with Cl_2/H_2O?

This chlorohydrin has two stereocenters and, by the 2^n rule, there are four possible stereoisomers - two pairs of enantiomers.

$CH_3-\overset{*}{C}H-\overset{*}{C}H-CH_3$
$Cl OH$

However, given the anti stereoselectivity of halohydrin formation, only one pair of these enantiomers is formed from *trans*-2-butene.

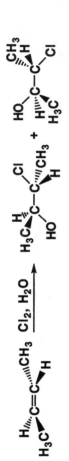

(b) How many stereoisomers are possible in this reaction sequence?

There are three stereoisomers possible for this epoxide; a pair of enantiomers and a meso compound. However, because of the stereoselectivity (requirement for anti, planar geometry) of the β–elimination reaction, it is the pair of enantiomers that will be produced in the reaction sequence that began with *trans*-2-butene.

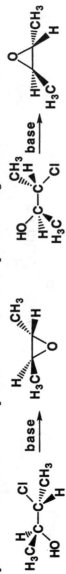

<u>Problem 11.9</u> Draw the expected products of Sharpless epoxidation of each allylic alcohol using (+)-diethyl tartrate as the chiral catalyst.

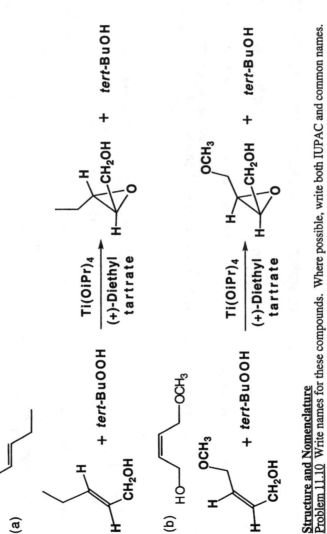

<u>Structure and Nomenclature</u>
<u>Problem 11.10</u> Write names for these compounds. Where possible, write both IUPAC and common names.

Common names are listed in parentheses.

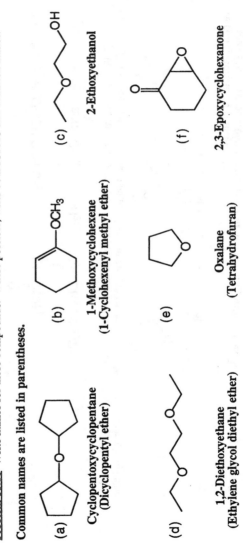

(a)

Cyclopentoxycyclopentane
(Dicyclopentyl ether)

(b)

1-Methoxycyclohexene
(1-Cyclohexenyl methyl ether)

(c)

2-Ethoxyethanol

(d)

1,2-Diethoxyethane
(Ethylene glycol diethyl ether)

(e)

Oxalane
(Tetrahydrofuran)

(f)

2,3-Epoxycyclohexanone

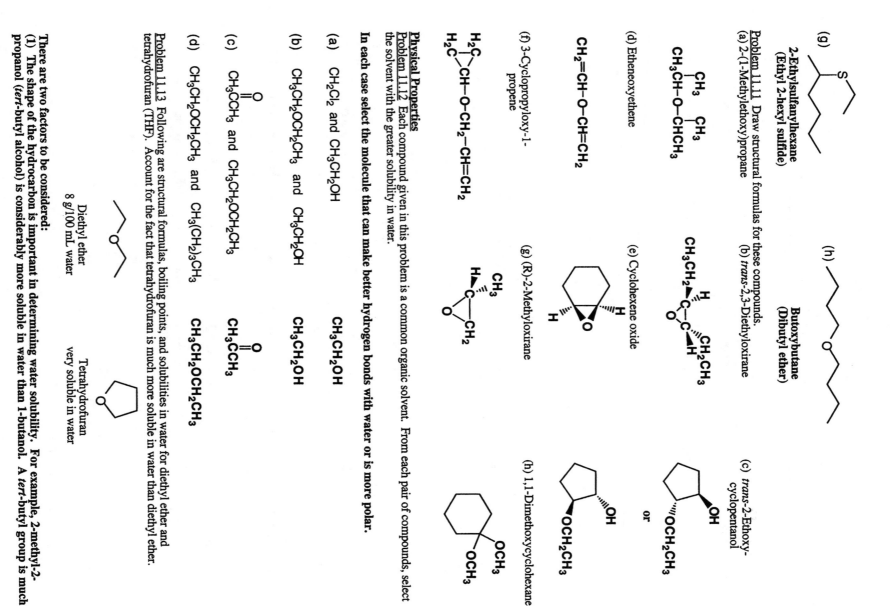

(g)

**2-Ethylsulfanylhexane
(Ethyl 2-hexyl sulfide)**

(h) **Butoxybutane
(Dibutyl ether)**

Problem 11.11 Draw structural formulas for these compounds.

(a) 2-(1-Methylethoxy)propane

(b) *trans*-2,3-Diethyloxirane

CH_3 CH_3
CH_3CH—O—$CHCH_3$

CH_3CH_2—C—C—CH_2CH_3 (with H's, O)

(c) *trans*-2-Ethoxy-cyclopentanol

OH / OCH_2CH_3

or

OCH_2CH_3

(d) Etheneoxyethene

CH_2=CH—O—CH=CH_2

(e) Cyclohexene oxide

(f) 3-Cyclopropyloxy-1-propene

H_2C
      CH—O—CH_2—CH=CH_2
H_2C

(g) (R)-2-Methyloxirane

CH_3 / C / O / CH_2 (with H's)

(h) 1,1-Dimethoxycyclohexane

OCH_3 / OCH_3

Physical Properties

Problem 11.12 Each compound given in this problem is a common organic solvent. From each pair of compounds, select the solvent with the greater solubility in water.

(a) CH_2Cl_2 and CH_3CH_2OH CH_3CH_2OH

(b) $CH_3CH_2OCH_2CH_3$ and CH_3CH_2OH CH_3CH_2OH

(c) CH_3CCH_3 (=O) and $CH_3CH_2OCH_2CH_3$ CH_3CCH_3 (=O)

(d) $CH_3CH_2OCH_2CH_3$ and $CH_3(CH_2)_3CH_3$ $CH_3CH_2OCH_2CH_3$

In each case select the molecule that can make better hydrogen bonds with water or is more polar.

Problem 11.13 Following are structural formulas, boiling points, and solubilities in water for diethyl ether and tetrahydrofuran (THF). Account for the fact that tetrahydrofuran is much more soluble in water than diethyl ether.

Diethyl ether
8 g/100 mL water

Tetrahydrofuran
very soluble in water

There are two factors to be considered:

(1) The shape of the hydrocarbon is important in determining water solubility. For example, 2-methyl-2-propanol (*tert*-butyl alcohol) is considerably more soluble in water than 1-butanol. A *tert*-butyl group is much

more compact than a butyl group and, consequently, there is less disruption of water hydrogen bonding when *tert*-butyl alcohol is dissolved in water than when 1-butanol is dissolved in water. Similarly, the hydrocarbon portion of tetrahydrofuran (THF) is more compact than that of diethyl ether which increases the solubility of THF in water compared to diethyl ether.

(2) The oxygen atom of THF is more accessible for hydrogen bonding and solvation by water than the oxygen atom of diethyl ether. This greater accessibility arises because the hydrocarbon chains bonded to oxygen in THF are "tied back" whereas those on the oxygen atom of diethyl ether have more degrees of freedom and consequently present more steric hindrance to solvation.

Problem 11.14 Because of the Lewis base properties of ether oxygen atoms, crown ethers are excellent complexing agents for Na$^+$, K$^+$, and NH$_4^+$. What kind of molecule might serve as a complexing agent for Cl$^-$ or Br$^-$?

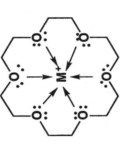

18-crown-6 with a generic metal M$^+$

To build an analogous system for complexation of Cl$^-$ it is necessary to first realize that the chloride ion is a Lewis base. Instead of a cavity with electron pair donors (as in crown ethers) we need a system with electron pair acceptors (or Lewis acid sites), such as the hydrogen atoms of protonated amines.

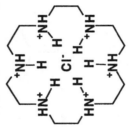

The size of the cavity will determine which anions will be complexed preferentially.

Preparation of Ethers
Problem 11.15 Write equations to show a combination of reactants to prepare each ether. Which ethers can be prepared in good yield by a Williamson ether synthesis? If there are any that cannot be prepared by the Williamson method, explain why not.

In each case, choose reagents that utilize a methyl halide (best), primary alkyl halide (2nd best), or a secondary alkyl halide (3rd best).

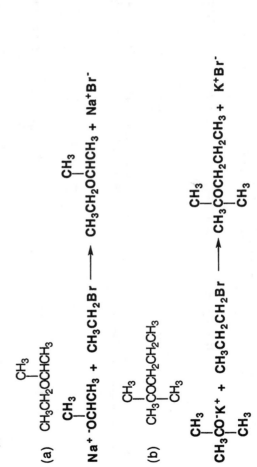

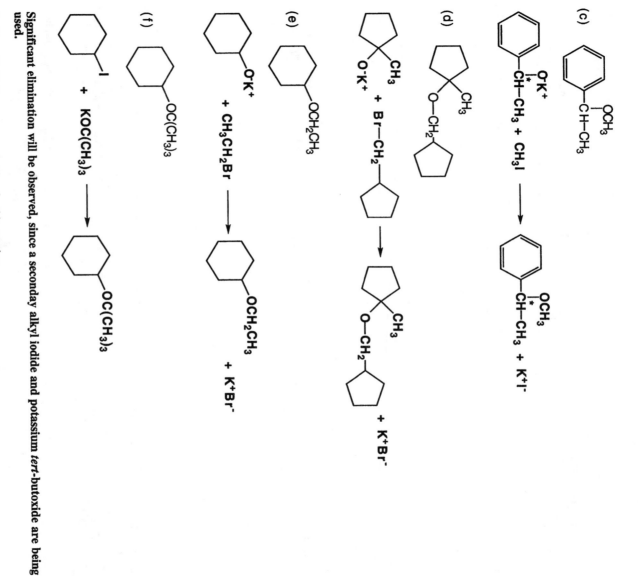

(c)

(d)

(e)

(f)

Significant elimination will be observed, since a secondary alkyl iodide and potassium *tert*-butoxide are being used.

Problem 11.16 Propose a mechanism for this reaction.

The following mechanism is in three steps: (1) protonation of the alkene to form a resonance-stabilized carbocation, (2) reaction of the carbocation with methanol to give an oxonium ion, and (3) loss of a proton to give the ether.

Step 1:

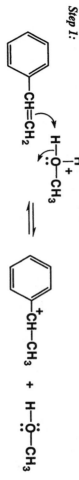

Step 2:

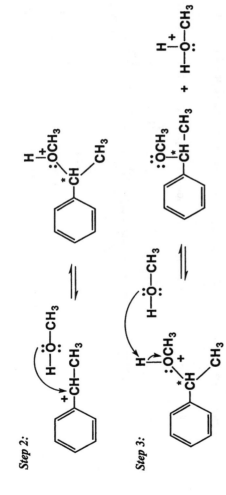

Step 3:

Reactions of Ethers
Problem 11.17　Draw structural formulas for the products formed when each compound is refluxed in concentrated HI.

Since an excess of HI is used, you can assume that the alcohol products initially produced will be converted to alkyl iodides under these conditions.

(a) 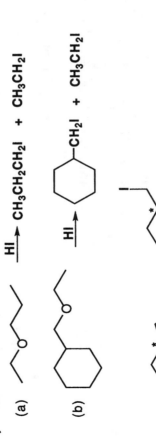 $\xrightarrow{HI}$ CH₃CH₂CH₂I + CH₃CH₂I

(b) $\xrightarrow{HI}$ [cyclohexyl]–CH₂I + CH₃CH₂I

(c) $\xrightarrow{HI}$

(d) $\xrightarrow{HI}$ 2

Problem 11.18　Following is an equation for the reaction of diisopropyl ether and oxygen to form a hydroperoxide.

Diisopropyl ether + O₂ ⟶ (A hydroperoxide)

Formation of an ether hydroperoxide is a radical chain reaction.
(a) Write a pair of chain propagation steps that accounts for the formation of this ether hydroperoxide. Assume that initiation is by a radical, R•.

Initiation:

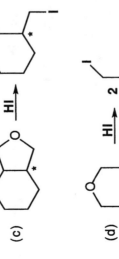

CH₃CH–O–CHCH₃ + R• ⟶ CH₃CH–O–ĊCH₃ + H–R
　CH₃　　CH₃　　　　　　CH₃　　CH₃

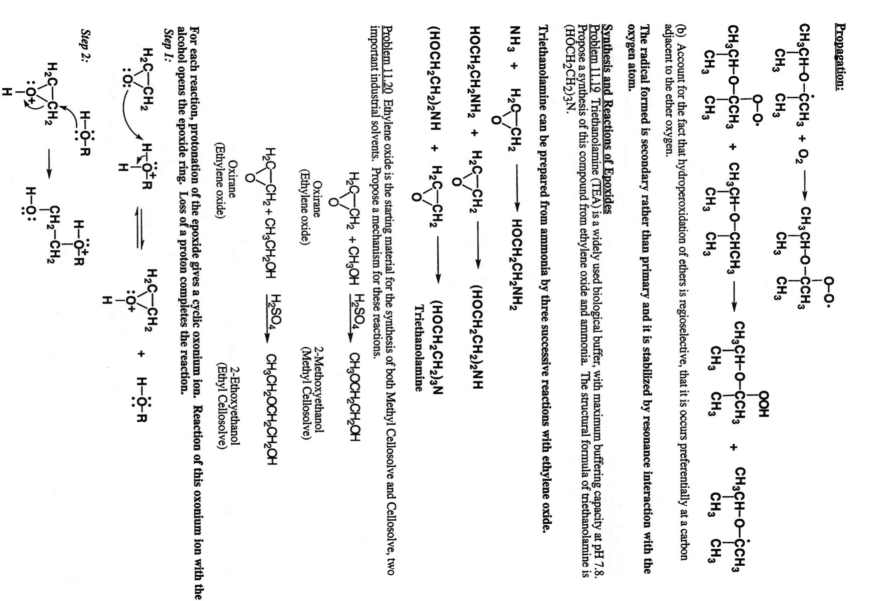

Step 3:

Problem 11.21 Ethylene oxide is the starting material for the synthesis of 1,4-dioxane. Propose a mechanism for each step in this synthesis.

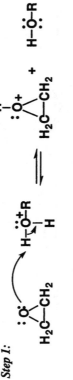

1,4-dioxane

The mechanism for this reaction involves an acid-catalyzed attack of the diol on the epoxide followed by protonation of one of the terminal hydroxyl groups, and displacement of water by the other alcohol.

Step 1:

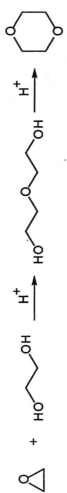

Step 2:

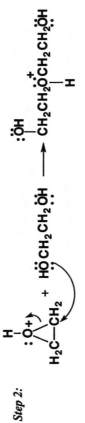

Step 3:

Step 4:

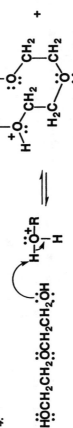

Step 5:

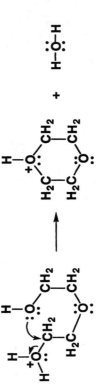

Step 6:

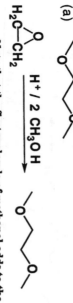

Problem 11.22 Propose a synthesis for each ether starting with ethylene oxide and any readily available alcohols.

(a)

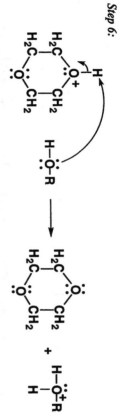

$$H_2C\text{—}CH_2 \xrightarrow{\ H^+ \ / \ 2 \ CH_3OH\ }$$

Note that the first molecule of methanol adds to the ethylene oxide, then the second reacts to replace the OH group via an acid-catalyzed substitution mechanism.

(b)

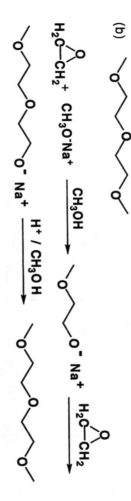

$$H_2C\text{—}CH_2^+ \ CH_3O^-Na^+ \xrightarrow{\ CH_3OH\ }$$

Problem 11.23 Propose a synthesis for 18-crown-6. If a base is used in your synthesis, does it make a difference if it is lithium hydroxide or potassium hydroxide?

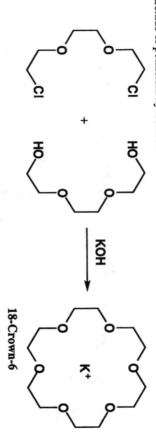

18-Crown-6

Because 18-crown-6 binds K⁺ the best, KOH should be used so that the crown ether can form around the K⁺ ion. This type of strategy is called the template approach and is used in a variety of similar situations.

The above two pieces can be synthesized as follows:

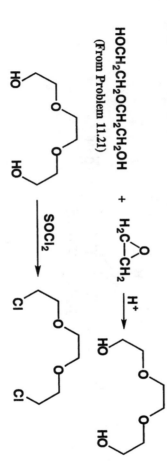

$$HOCH_2CH_2OCH_2CH_2OH$$
(From Problem 11.21)

Problem 11.24 Predict the structural formula of the major product of the reaction of 2,2,3-trimethyloxirane with each set of reagents.

Nucleophiles in the absence of acid catalysis react via an S$_N$2 mechanism, so the nucleophile ends up on the less hindered, that is less substituted, carbon atom of the epoxide. The acid-catalyzed reaction of epoxides occurs in an S$_N$1-like fashion, so substitution occurs at the site of the more stable cation, namely the more substituted carbon atom of the epoxide. In both cases, the stereochemistry of addition is *trans* to the epoxide oxygen atom. Note that because 2,2,3-trimethyloxirane is actually a pair of enantiomers, the products of each reaction are a pair of enantiomers as well.

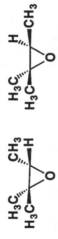

2,2,3-Trimethyloxirane (a pair of enantiomers)

(a) CH$_3$OH/CH$_3$O$^-$ Na$^+$

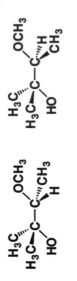

(b) CH$_3$OH/H$^+$

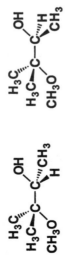

Problem 11.25 The following equation shows the reaction of *trans*-2,3-diphenyloxirane with hydrogen chloride in benzene to form 2-chloro-1,2-diphenylethanol.

trans-2,3-Diphenyloxirane   2-Chloro-1,2-diphenylethanol

(a) How many stereoisomers are possible for 2-chloro-1,2-diphenylethanol?

There are two stereocenters in the molecule so there are 2^2 or 4 possible stereoisomers.

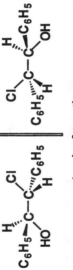

A pair of enantiomers
(from addition of HCl to *trans*-2,3-diphenyloxirane)

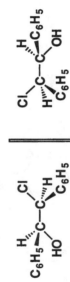

A pair of enantiomers
(from addition of HCl to *cis*-2,3-diphenyloxirane)

(b) Given that opening of the epoxide ring in this reaction is stereoselective, predict which of the possible stereoisomers of 2-chloro-1,2-diphenylethanol is/are formed in the reaction.

Only the first (upper) pair of enantiomers is formed.

Problem 11.26 Propose a mechanism to account for this rearrangement.

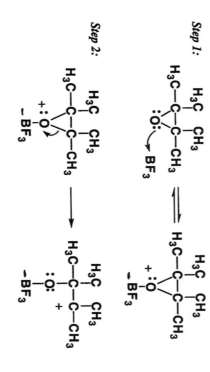

Reaction between boron trifluoride, a Lewis acid, and the epoxide oxygen, a Lewis base, forms an oxonium ion. The ring then opens followed by rearrangement of the resulting secondary carbocation and subsequent loss of boron trifluoride to give the observed ketone.

Step 1:

Step 2:

Step 3:

Step 4:

Problem 11.27 Following is the structural formula for an epoxide derived from 9-methyldecalin. Acid-catalyzed hydrolysis of this epoxide gives a trans diol. Of the two possible trans diols, only one is formed. How do you account for this stereospecificity?

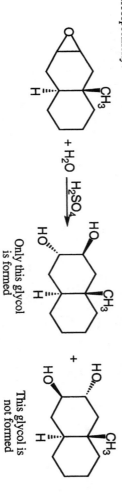

+ H_2O $\xrightarrow{H_2SO_4}$

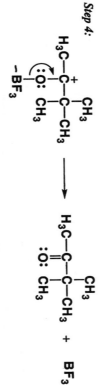

Only this glycol is formed

This glycol is not formed

The key to this problem is that opening of the epoxide is stereospecific; the incoming nucleophile and the leaving protonated epoxide oxygen must be anti coplanar. In a cyclohexane ring, anti coplanar corresponds to *trans* and diaxial. An accurate model of the glycol formed will show that in it, the two -OH groups are diaxial and, therefore, *trans* and coplanar. In the alternative glycol, the -OH groups are also *trans*, but because they are diequatorial, they are not coplanar.

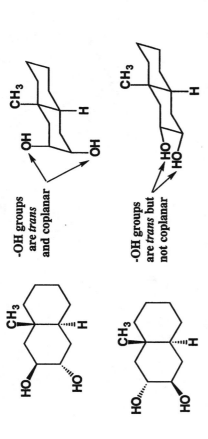

-OH groups
are *trans*
and coplanar

-OH groups
are *trans* but
not coplanar

Problem 11.28 Following are two reaction sequences for converting 1,2-diphenylethylene into 2,3-diphenyloxirane.

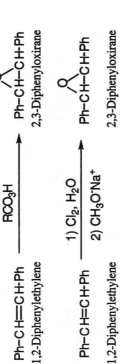

Ph–CH=CH-Ph $\xrightarrow{\text{RCO}_3\text{H}}$ Ph–CH–CH-Ph

1,2-Diphenylethylene   2,3-Diphenyloxirane

Ph–CH=CH-Ph $\xrightarrow[\text{2) CH}_3\text{O}^-\text{Na}^+]{\text{1) Cl}_2,\ \text{H}_2\text{O}}$ Ph–CH–CH-Ph

1,2-Diphenylethylene   2,3-Diphenyloxirane

Suppose that the starting alkene is *trans*-1,2-diphenylethylene.
(a) What is the configuration of the oxirane formed in each sequence?

In each case, the oxirane has the *trans* configuration.

(b) Does the oxirane formed in either sequence rotate the plane of polarized light? Explain.

Neither product will rotate the plane of polarized light. The *trans* isomer can exist as a pair of enantiomers. In each reaction, the product formed is a racemic mixture and will not be optically active.

Problem 11.29 The following enantiomer of a chiral epoxide is an intermediate in the synthesis of the insect pheromone frontalin. Show how can this enantiomer be prepared from an allylic alcohol precursor, using the Sharpless epoxidation.

It will probably be helpful to construct models of the molecules involved. The key to this problem is to use the following diagram of how the Sharpless epoxidation operates, then identify which face of the alkene must react to give the indicated epoxide stereoisomer.

Comparing the above diagram to the desired product it is clear that (-)-diethyl tartrate is the correct choice for this epoxidation reaction.

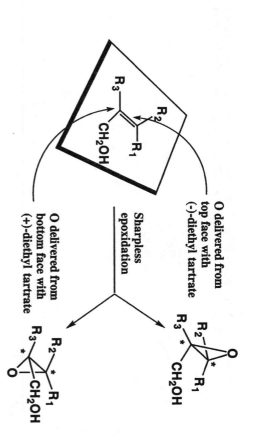

O delivered from
top face with
(-)-diethyl tartrate

Sharpless
epoxidation

O delivered from
bottom face with
(+)-diethyl tartrate

Problem 11.30 Human white cells produce an enzyme called myeloperoxidase. This enzyme catalyzes the reaction between hydrogen peroxide and chloride ion to produce hypochlorous acid, HOCl, which reacts as if it is Cl+OH-. When attacked by white cells, cholesterol gives a chlorohydrin as the major product.

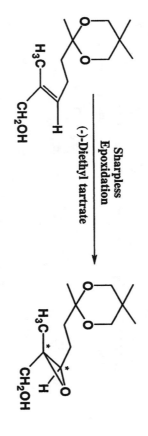

Sharpless
Epoxidation

(-)-Diethyl tartrate

Cholesterol

White blood
cells

(a) Propose a mechanism for this reaction. Account for both the regioselectivity and the stereoselectivity.

The mechanism involves reaction of the double bond of cholesterol with HOCl to produce a chloronium ion intermediate, that is attacked by HO- to produce the chlorohydrin.

Step 1:

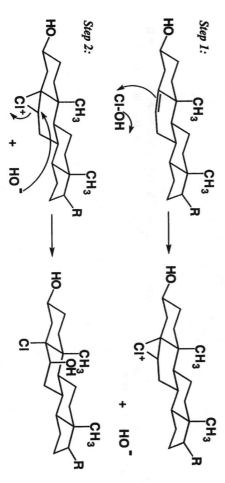

Cl-OH

Step 2:

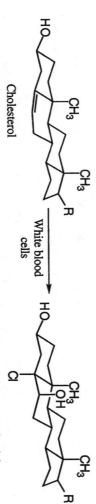

Cl(+)

+ HO-

The *trans* stereochemistry of the chlorohydrin product is the result of the backside attack that is required by the chloronium ion intermediate. The fact that the chlorine atom ends up on the bottom face of the molecule is because this is the less-hindered face of the double bond. The axial methyl group adjacent to the double bond provides a steric barrier to reaction on the top face.

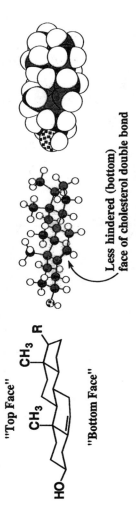

"Top Face"

"Bottom Face"

Less hindered (bottom) face of cholesterol double bond

The HO· nucleophile attacks the secondary carbon atom of the chloronium ion, as opposed to the tertiary carbon atom, again because of steric hindrance caused by the axial methyl group preventing access to the tertiary carbon atom. Note that attack by HO· must be from the top face of the molecule, anti to the chlorine atom.

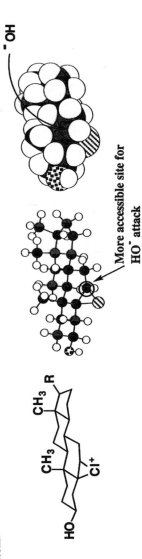

More accessible site for HO· attack

(b) On standing or (much more rapidly) on treatment with base, the chlorohydrin is converted to an epoxide. Show the structure of the epoxide and a mechanism for its formation.

The OH group of the chlorohydrin is deprotonated by the base, allowing for epoxide formation as chloride anion departs. The backside attack of the oxygen atom dictates that the epoxide product will have the oxygen atom on what we are calling the top face of the molecule.

Step 1:

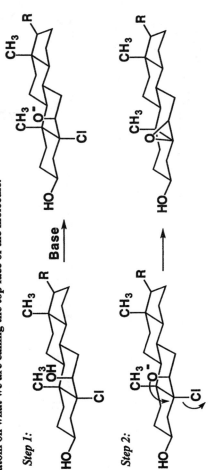

Base

Step 2:

Problem 11.31 Propose a mechanism for the following acid-catalyzed rearrangement.

H₂SO₄, THF

The mechanism of this rearrangement involves initial protonation of the epoxide oxygen atom, making the epoxide electrophilic enough to react with the weakly nucleophilic pi bond of the adjacent triple bond. The triple bond reacts at the more reactive, more highly substituted tertiary side of the protonated epoxide to give the six-membered ring intermediate.

Step 1:

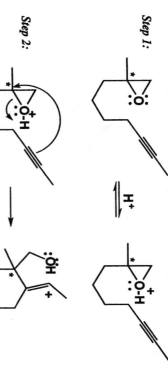

Attack of the alcohol upon the carbocation of this intermediate followed by loss of a proton completes the reaction.

Step 2:

Step 3:

Step 4:

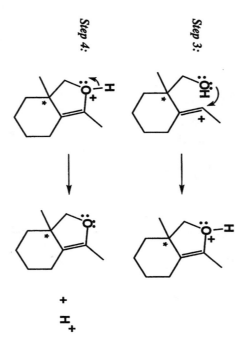

Synthesis

Problem 11.32 Show reagents and experimental conditions to synthesize the following compounds from 1-propanol. Any derivative of 1-propanol prepared in an earlier part of this problem may then be used for a later synthesis.

(a) Propanal

$$CH_3CH_2CH_2OH + PCC \longrightarrow CH_3CH_2CH{\overset{\displaystyle O}{\parallel}}$$

(b) Propanoic acid

$$CH_3CH_2CH_2OH \xrightarrow[\substack{H_2O, \text{ heat}}]{\substack{K_2Cr_2O_7 \\ H_2SO_4}} CH_3CH_2C{\overset{\displaystyle O}{\parallel}}OH$$

(c) Propene

$$CH_3CH_2CH_2OH \xrightarrow[\text{heat}]{H_3PO_4} CH_3CH{=}CH_2$$

(d) 2-Propanol

$$CH_3CH{=}CH_2 \xrightarrow[\substack{H_2SO_4}]{H_2O} CH_3CHCH_3{\overset{\displaystyle OH}{\vert}}$$

From (c)

(e) 2-Bromopropane

$$CH_3CHCH_3 \quad \text{or} \quad H_2C=CHCH_3 + HBr \xrightarrow{\text{heat}} CH_3CHCH_3$$
   |OH |Br

From (d)

(f) 1-Chloropropane

$$CH_3CH_2CH_2OH \xrightarrow[\text{Pyridine}]{SOCl_2} CH_3CH_2CH_2Cl + SO_2 + HCl$$

(g) 1,2-Dibromopropane

$$CH_3CH=CH_2 \xrightarrow[CCl_4]{Br_2} CH_3CHCH_2Br$$
                                |Br

From (c)   (racemic mixture)

(h) Propyne

$$CH_3CHCH_2Br \xrightarrow[\text{2. } H_2O]{\text{1. } 3NaNH_2} H_3CC\equiv CH$$
   |Br

From (g)

(i) 2-Propanone

$$CH_3CHCH_3 \xrightarrow[H_2O,\ \text{heat}]{\substack{K_2Cr_2O_7 \\ H_2SO_4}} CH_3CCH_3$$
   |OH ||O

From (d)

(j) 1-Chloro-2-propanol

$$CH_3CH=CH_2 \xrightarrow{Cl_2\ /\ H_2O} CH_3CHCH_2Cl$$
                                |OH
                           (racemic mixture)

From (c)

(k) Methyloxirane

$$CH_3CHCH_2Cl \xrightarrow{NaOH} H_2C\text{—}CHCH_3$$
   |OH   \O/

From (j)   (racemic mixture assuming racemic starting material)

or

$$CH_3CH=CH_2 \xrightarrow{RCO_3H} H_2C\text{—}CHCH_3$$
                                \O/
                          (racemic mixture)

From (c)

(l) Dipropyl ether

$$CH_3CH_2CH_2Cl + CH_3CH_2CH_2O^- Na^+ \longrightarrow CH_3CH_2CH_2OCH_2CH_2CH_3$$

From (f)   From 1-propanol and Na

(m) Isopropyl propyl ether

$$CH_3CH_2CH_2Cl + CH_3CHCH_3 \longrightarrow CH_3CHCH_3$$
                          |O^- Na^+ |OCH_2CH_2CH_3

From (f)   From (d) and Na

(n) 1-Mercapto-2-propanol

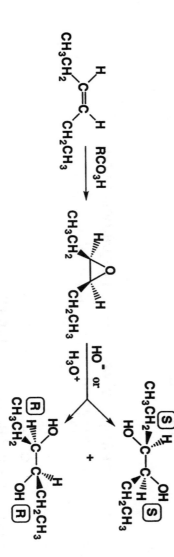

H₂C—CHCH₃ →(HS⁻ Na⁺ / H₂O) $\underset{*}{CH_3CHCH_2SH}$ with OH
From (k)

(racemic mixture)

(o) 1-Amino-2-propanol

H₂C—CHCH₃ →(NH₃) $\underset{*}{CH_3CHCH_2NH_2}$ with OH
From (k)

(racemic mixture)

(p) 1,2-Propanediol

CH₃CH=CH₂ →(OsO₄ / H₂O₂) $\underset{*}{CH_3CH-CH_2}$ with HO OH

From (c)

(racemic mixture)

or

H₂C—CHCH₃ →(HO⁻ or / H₃O⁺) CH_3CH-CH_2 with HO OH
From (k)

(racemic mixture)

Note that the products in g, j, k, n, o, and p will be racemic mixtures.

The syn geometry of addition observed with OsO₄ leads to meso 3,4-hexanediol.

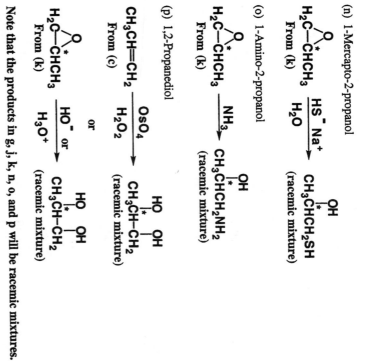

Problem 11.33 Starting with *cis*-3-hexene, show how to prepare the following:
(a) Meso 3,4-hexanediol

(b) Racemic 3,4-hexanediol

The anti geometry produced via epoxidation followed by hydrolysis gives a racemic mixture of enantiomers.

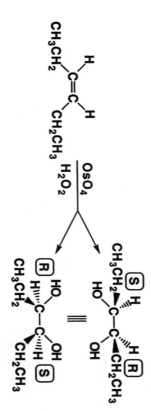

Problem 11.34 Show reagents and experimental conditions to convert cycloheptene to the following. Any compound made in an earlier part of this problem may be used as an intermediate in any following conversion.

(a)

Oxidation of cycloheptene with a peroxyacid such as trifluoroperacetic acid.

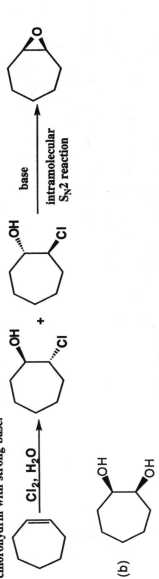

As an alternative, addition of HOCl to cycloheptene to form a chlorohydrin followed by treatment of the chlorohydrin with strong base.

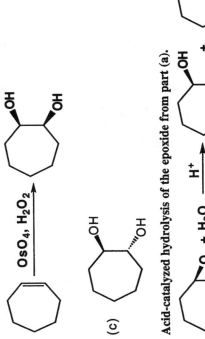

Oxidation of cycloheptene with osmium tetroxide in the presence of hydrogen peroxide.

(b)

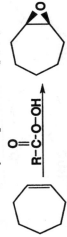

(c)

Acid-catalyzed hydrolysis of the epoxide from part (a).

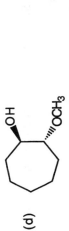

(d)

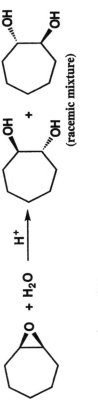

Treatment of the epoxide from part (a) with methanol in the presence of an acid catalyst, or alternatively, treatment with methanol in the presence of sodium methoxide.

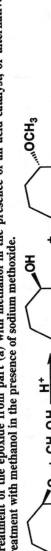

(e)

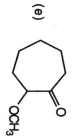

Oxidation of the secondary alcohol of part (d) using chromium trioxide in pyridine or pyridinium chlorochromate or H₂CrO₄.

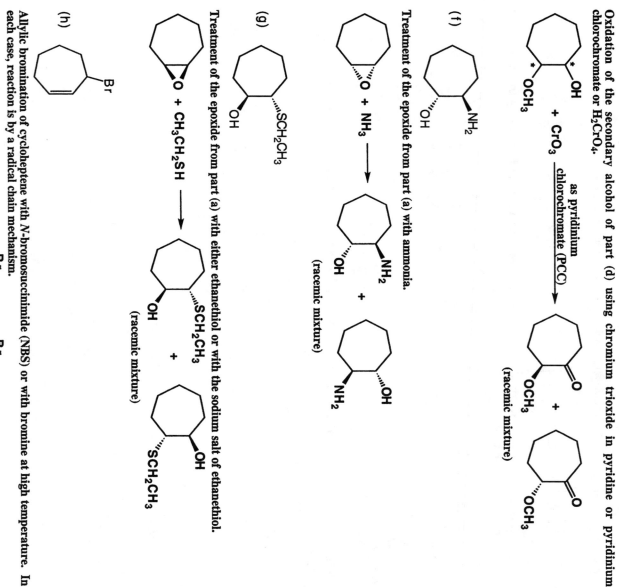

(f)

Treatment of the epoxide from part (a) with ammonia.

(g)

Treatment of the epoxide from part (a) with either ethanethiol or with the sodium salt of ethanethiol.

(h)

Allylic bromination of cycloheptene with N-bromosuccinimide (NBS) or with bromine at high temperature. In each case, reaction is by a radical chain mechanism.

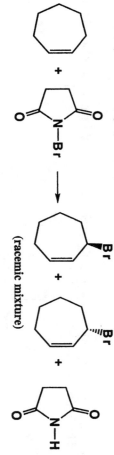

(i)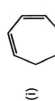

Dehydrohalogenation of product (h) with a strong base such as potassium *tert*-butoxide.

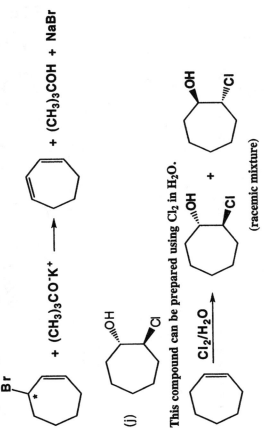

Br
+ (CH₃)₃CO⁻K⁺ ⟶   + (CH₃)₃COH + NaBr

(j)

This compound can be prepared using Cl₂ in H₂O.

Cl₂/H₂O

(racemic mixture)

(k)

NaSH

As a first step, hydration of cycloheptene to cycloheptanol. Hydration can be accomplished by (1) acid-catalyzed hydration, (2) oxymercuration followed by reduction with sodium borohydride, or (3) hydroboration followed by oxidation of the organoborane intermediate with alkaline hydrogen peroxide. Treatment of cycloheptanol with thionyl chloride and then sodium hydrosulfide gives the product.

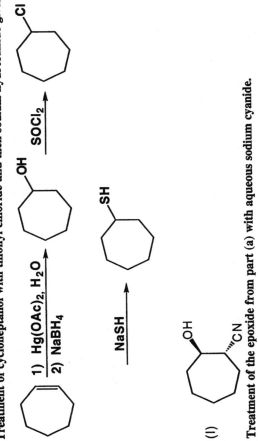

1) Hg(OAc)₂, H₂O
2) NaBH₄

SOCl₂

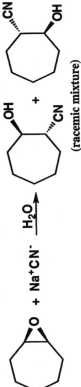

(l)

Treatment of the epoxide from part (a) with aqueous sodium cyanide.

O + Na⁺CN⁻ ⟶ H₂O

(racemic mixture)

(m)

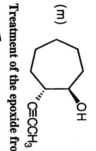

Treatment of the epoxide from part (a) with the sodium salt of propyne followed by treatment with H₂O.

$$\xrightarrow[\text{2. H}_2\text{O}]{1.\ \text{CH}_3\text{C}\equiv\text{C}^-\ \text{Na}^+}$$

(racemic mixture)

(n) $HC(CH_2)_5CH$

Oxidation of the cis glycol from part (b) with periodic acid

+ HIO₄ ⟶ $HC(CH_2)_5CH$ + HIO₃

Alternatively, oxidation of cycloheptene with ozone followed by treatment of the ozonide with dimethyl sulfide.

$$\xrightarrow[\text{2) CH}_3\text{SCH}_3]{\text{1) O}_3}\ HC(CH_2)_5CH$$

Problem 11.35 Show reagents to bring about each reaction:

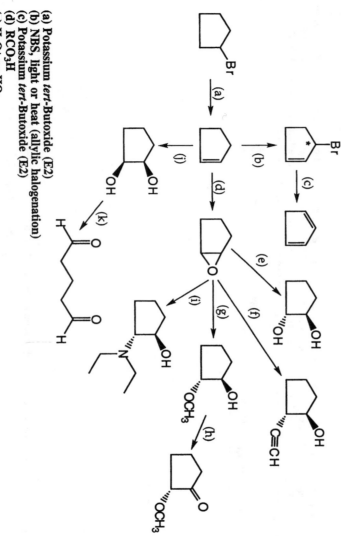

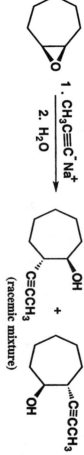

(a) Potassium *tert*-Butoxide (E2)
(b) NBS, light or heat (allylic halogenation)
(c) Potassium *tert*-Butoxide (E2)
(d) RCO₃H
(e) H₃O⁺ or HO⁻
(f) 1. HC≡C⁻Na⁺ 2. H₂O
(g) CH₃O⁻Na⁺ / CH₃OH
(h) PCC or H₂CrO₄
(i) (CH₃CH₂)₂NH
(j) OsO₄ / H₂O₂
(k) HIO₄

Note that the stereoisomers shown after reactions (e), (f), (g), (h), and (i) represent only one enantiomer of the racemic mixture that will actually be formed in the reactions. Reaction (b) will also give a racemic mixture, as will reaction (h), if the the starting material is racemic.

<u>Problem 11.36</u> Propose a synthesis of the following alcohol from styrene and 1-chloro-3-methyl-2-butene.

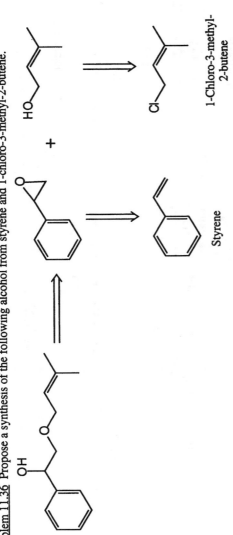

The 1-chloro-3-methyl-2-butene is converted to the alcohol in base then the alkoxide by reaction with Na metal. The styrene is converted to the epoxide, then reacted with the alkoxide followed by an aqueous workup to give the final product. Notice that the last step requires reaction of the nucleophile at the less-hindered carbon of the epoxide, so an acid-catalyzed step could not be used. The final product is actually a racemic mixture.

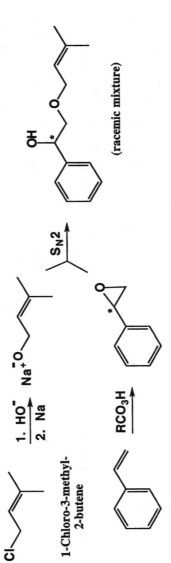

<u>Problem 11.37</u> Starting with acetylene and ethylene oxide as the only sources of carbon atoms, show how to prepare these compounds.
(a) 3-Butyn-1-ol

$$HC{\equiv}CH \xrightarrow[NH_3\ (l)]{NaNH_2} HC{\equiv}C^-\ Na^+ \xrightarrow[\substack{1.\ H_2C-CH_2\\ 2.\ H_2O}]{} HC{\equiv}CCH_2CH_2OH$$

(b) 3-Hexyn-1,6-diol

Because the acidity of the -OH group will interfere with formation of the acetylide anion in base, it must be protected by a group such as the *tert*-butyl group shown below. Note how the *tert*-butyl ether is created in the presence of an anhydrous acid catalyst, usually sufluric acid, then removed in aqueous HCl at the end of the reaction sequence.

$Na^+ \ {}^-C{\equiv}CCH_2CH_2O$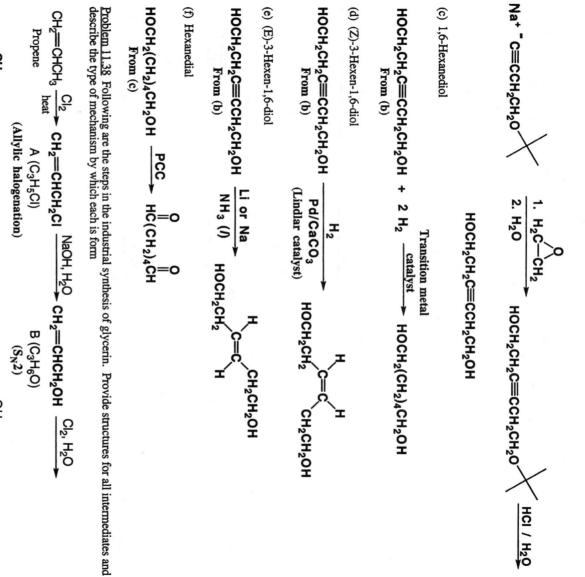

$\xrightarrow[\text{2. } H_2O]{\text{1. } H_2C-CH_2}$ $HOCH_2CH_2C{\equiv}CCH_2CH_2O$ $\xrightarrow{HCl \ / \ H_2O}$

$HOCH_2CH_2C{\equiv}CCH_2CH_2OH$

(c) 1,6-Hexanediol

(d) (Z)-3-Hexen-1,6-diol

$HOCH_2CH_2C{\equiv}CCH_2CH_2OH \ + \ 2 \ H_2$ $\xrightarrow[\text{catalyst}]{\text{Transition metal}}$ $HOCH_2(CH_2)_4CH_2OH$

From (b)

$HOCH_2CH_2C{\equiv}CCH_2CH_2OH$ $\xrightarrow{H_2}$
From (b) (Lindlar catalyst) Pd/CaCO$_3$

(e) (E)-3-Hexen-1,6-diol

$HOCH_2C{\equiv}CCH_2CH_2OH$ $\xrightarrow[NH_3 \ (l)]{Li \ or \ Na}$
From (b)

(f) Hexanedial

$HOCH_2(CH_2)_4CH_2OH$ $\xrightarrow{PCC}$ $HC(CH_2)_4CH$
From (c)

Problem 11.38 Following are the steps in the industrial synthesis of glycerin. Provide structures for all intermediates and describe the type of mechanism by which each is form

$CH_2{=}CHCH_3$ $\xrightarrow[\text{heat}]{Cl_2}$ $CH_2{=}CHCH_2Cl$ $\xrightarrow{NaOH, \ H_2O}$ $CH_2{=}CHCH_2OH$ $\xrightarrow{Cl_2, \ H_2O}$
Propene A (C_3H_5Cl) B (C_3H_6O)
(Allylic halogenation) (S_N2)

$\underset{\text{(Hydrohalogenation)}}{\underset{\text{C ($C_3H_7ClO_2$)}}{ClCH_2CHCH_2OH}}$ $\xrightarrow[\text{heat}]{Ca(OH)_2}$ $H_2C{-}CHCH_2OH$ $\xrightarrow{H_2O, \ HCl}$ $HOCH_2CHCH_2OH$
D ($C_3H_6O_2$) 1,2,3-Propanetriol
(Internal displacement (glycerol, glycerin)
by an alkoxide) (Acid-catalyzed
epoxide hydrolysis)

Notice that because of symmetry the product glycerin is not chiral, even though the last two synthetic intermediates are chiral and racemic.

Problem 11.39 Gossyplure, the sex pheromone of the pink bollworm, is the acetic ester of 7,11-hexadecadien-1-ol (The Merck Index, 12th ed., #4548). The active pheromone has the Z configuration at the C7-C8 double bond and is a mixture of E, Z isomers at the C11-C12 double bond. Shown here is the Z,E isomer.

(7Z,11E)-7,11-hexadecadienyl acetate

Following is a retrosynthetic analysis for (7Z,11E)-7,11-hexadecadien-1-ol, which then led to a successful synthesis of gossyplure.

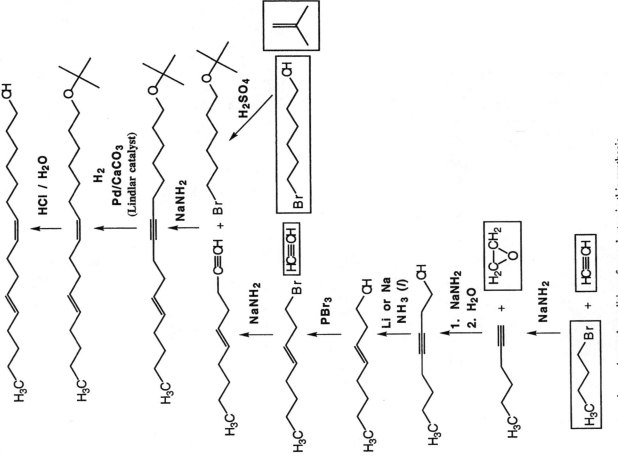

(a) Suggest reagents and experimental conditions for each step in this synthesis.

Reasonable reagents are shown on the above scheme.

(b) Why is it necessary to protect the -OH group of 6-bromohexanol?

A hydroxyl group is considerably more acidic than an alkyne. Thus, left unprotected, the hydroxyl group would be deprotonated by NaNH₂, creating a nucleophilic alkoxide anion that would displace bromine to give an ether.

(c) How might you modify this synthesis to prepare the 7Z,11Z isomer of gossyplure?

Lindlar's catalyst would be used in place of the Li or Na in NH₃ for the reduction of the 3-octyn-2-ol.

Problem 11.40 Epichlorhydrin is a valuable synthetic intermediate because each of its three carbons contains a reactive group. We shall see many examples of its use in the problems in the interchapter section "Problems from Medicinal Chemistry". Epichlorohydrin is synthesized industrially in the following series of three reactions.

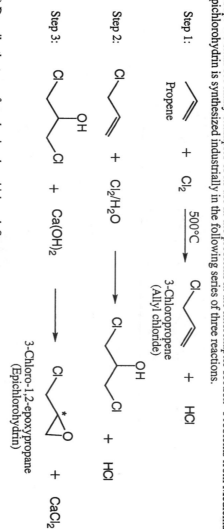

Step 1:

Propene + Cl₂ →(500°C) Cl⟍⟍ + HCl
3-Chloropropene (Allyl chloride)

Step 2:

Cl⟍⟍ + Cl₂/H₂O → [structure with Cl, OH, Cl] + HCl

Step 3:

[Cl–CH₂–CH(OH)–CH₂Cl] + Ca(OH)₂ → 3-Chloro-1,2-epoxypropane (Epichlorohydrin) + CaCl₂

(a) Describe the type of mechanism by which each Step occurs.

Step 1: Propene is reacting by a radical chain reaction involving an allylic radical intermediate. The heat breaks the chlorine bond, producing enough chlorine radicals (Cl·) to initiate the radical chain process.
Step 2: This is an electrophilic addition to an alkene. The initially formed chloronium ion is attacked by water and the product alcohol is formed following loss of a proton.
Step 3: The calcium hydroxide acts as a base to deprotonate the alcohol. The alkoxide then takes part in an internal Sₙ2 reaction that displaces the chlorine atom from the backside to give the product epoxide.

(b) Account for the regiospecificity of Steps 1 and 2.

Step 1: This is a radical process due to the very high temperature used. Initial hydrogen atom abstraction will occur to generate the most stable radical intermediate. The most stable possible radical intermediate is the resonance stabilized allyl radical intermediate that is formed by hydrogen abstraction from the methyl group of propene, leading to the observed 3-chloropropene product.
Step 2: This reaction is not a radical addition because high temperature is not used. The reaction involves reaction of the alkene with chlorine to generate a chloronium ion intermediate. Water attacks this intermediate at the more highly substituted carbon atom in accord with Markovnikov's rule, placing the chlorine atom on the terminal carbon atom and the hydroxyl group on the central carbon atom.

Pyramid butterflyfish *Hemitaurichthys polyepis*, Lanai, Hawaii

CHAPTER 12
Solutions to the Problems

Problem 12.1   Calculate the energy of red light (680 nm) in kilojoules per mole. Which form of radiation carries more energy, infrared radiation of wavelength 2.50 μm or red light of wavelength 680 nm?

Combining the two equations given in the text gives:

$$E = h\nu = h\left(\frac{c}{\lambda}\right)$$

Plugging in the appropriate values gives the desired answer:

$$E = \frac{(3.99 \times 10^{-13} \text{ kJ-sec-mol}^{-1})(3.00 \times 10^{8} \text{ m-sec}^{-1})}{680 \times 10^{-9} \text{ m}} \quad \boxed{= 176 \text{ kJ-mol}^{-1}}$$

Notice how the units canceled to give the final answer in kJ-mol^{-1}. The longer the wavelength, the lower the energy, thus red light (680 nm) carries more energy than infrared radiation of 2.50 μm.

Problem 12.2   Which is higher in energy?
(a) Infrared radiation of 1715 cm^{-1} or 2800 cm^{-1}?

The higher the wavenumber, the higher the energy. As a result, 2800 cm^{-1} is higher energy than 1705 cm^{-1}.

(b) Radiofrequency radiation 300 MHz or 60 Hz?

Energy is directly proportional to frequency, so 300 MHz is higher energy than 60 Hz.

Problem 12.3   Without doing the calculation, which member of each pair do you expect to occur at the higher wavenumber?
(a) C=O or C=C stretching?

We must assume that C=O and C=C have similar force constants, since no actual values were provided. Using this assumption, the C=C bond vibrational absorption is predicted to occur at a higher frequency. The atomic weight of C is lower than O, and the frequency of vibration is predicted to be higher for bonds with atoms of lower atomic weight.

(b) C=O or C-O stretching?

Double bonds have higher force constants than single bonds, so the C=O bond will have a stretching frequency that occurs at a higher frequency than C-O.

(c) C≡C and C=O stretching?

Triple bonds have higher force constants than double bonds, so the C≡C bond will have a stretching frequency that occurs at a higher frequency than C=O.

(d) C-H or C-Cl stretching?

Assuming that C-H and C-Cl have similar force constants, then the C-H will have an absorbance at a higher frequency because the atomic weight of H is much smaller than Cl.

Problem 12.4   A compound shows strong, very broad IR absorption in the region 3300-3600 cm^{-1} and strong absorption at 1715 cm^{-1}. What functional groups account for these absorptions?

The very broad IR absorption centered at 3300-3600 cm^{-1} corresponds to O-H stretching and the strong absorption at 1715 cm^{-1} corresponds to a carbonyl C=O stretch. The functional group responsible for these absorptions is the -COOH group.

Problem 12.5   Propanoic acid and methyl ethanoate are constitutional isomers. Show how to distinguish between them by IR spectroscopy.

CH₃CH₂COH (with O double bond)
Propanoic acid

CH₃COCH₃ (with O double bond)
Methyl ethanoate
(Methyl acetate)

The key difference between these constitutional isomers is the OH group of propanoic acid that is not present in methyl ethanoate. The -OH group will show a strong, broad O-H stretch in the IR spectrum of propanoic acid

between 2500-3300 cm⁻¹ that will be absent in the spectrum of methyl ethanoate. Also, the strong C=O stretching absorption will be between 1700-1725 cm⁻¹ for propanoic acid, but 1735-1800 cm⁻¹ for methyl ethanoate. Note that both spectra will have C-H stretching absorptions near 3000 cm⁻¹ and C-O stretching absorptions near 1100 cm⁻¹.

Infrared Spectra

Problem 12.6 Following are infrared spectra of methylenecyclopentane and 2,3-dimethyl-2-butene. Assign each compound its correct spectrum.

Methylenecyclopentane   2,3-Dimethyl-2-butene

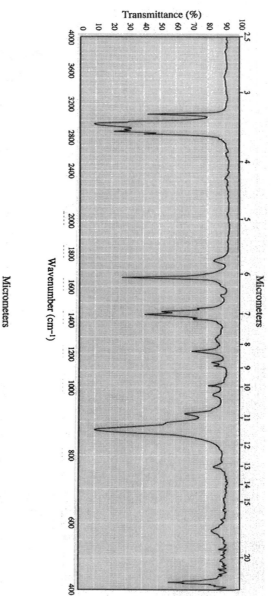

Transmittance (%)

Wavenumber (cm⁻¹)

Micrometers

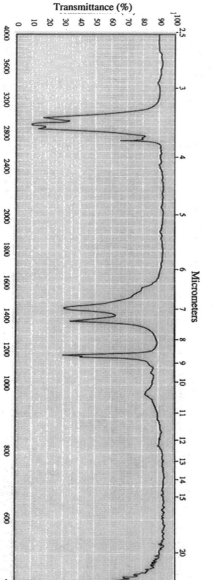

Transmittance (%)

Wavenumber (cm⁻¹)

Micrometers

Both molecules have several C-H bonds and thus both spectra have C-H stretches and C-H bending vibrations at 2900 cm⁻¹ and 1450 cm⁻¹, respectively. The alkene in methylenecyclopentane is unsymmetrical and therefore has a permanent dipole, so the C=C stretching will have a prominent band at 1654 cm⁻¹ as seen in the upper spectrum. On the other hand, 2,3-dimethyl-2-butene has a symmetrically substituted carbon-carbon double bond with no permanent dipole, so no C=C stretching should be prominent. In addition, the four methyl groups of 2,3-dimethyl-2-butene should give a prominent CH₃ bending band at 1375 cm⁻¹ and 1450 cm⁻¹ as is seen in the lower spectrum.

Thus, the first (upper) spectrum corresponds to methylenecyclopentane and the second (lower) spectrum corresponds to 2,3-dimethyl-2-butene.

Problem 12.7 Following are infrared spectra of nonane and 1-hexanol. Assign each compound its correct spectrum.

Nonane

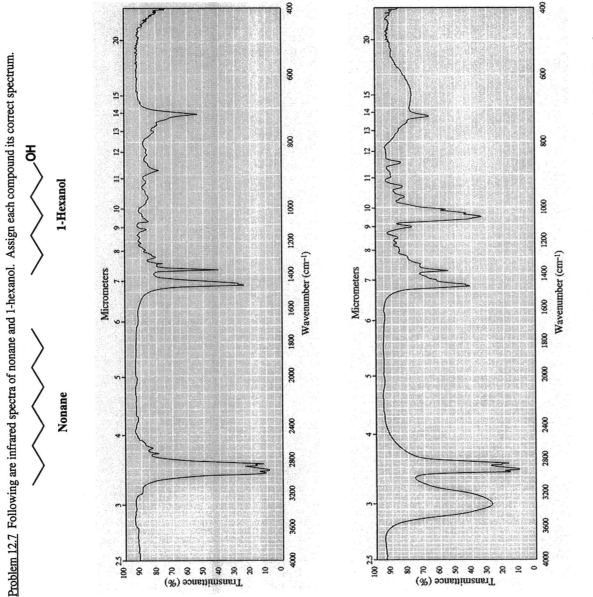

OH

1-Hexanol

Both compounds have C-H bonds, so both spectra have C-H stretches and bends at 2900 cm⁻¹ and 1450 cm⁻¹, respectively. Furthermore, both compounds have methyl groups so both spectra have methyl bending vibrations at 1375 cm⁻¹ and 1450 cm⁻¹. On the other hand, the 1-hexanol has an OH group, that will give rise to an O-H and C-O stretching vibrations at 3340 cm⁻¹ and 1050 cm⁻¹, respectively.

The 3340 cm⁻¹ and 1050 cm⁻¹ features are in the second (lower) spectrum, so this second spectrum must correspond to the 1-hexanol and the first (upper) spectrum must correspond to nonane.

Problem 12.8 Following are infrared spectra of 2-methyl-1-butanol and *tert*-butyl methyl ether. Assign each compound its correct spectra.

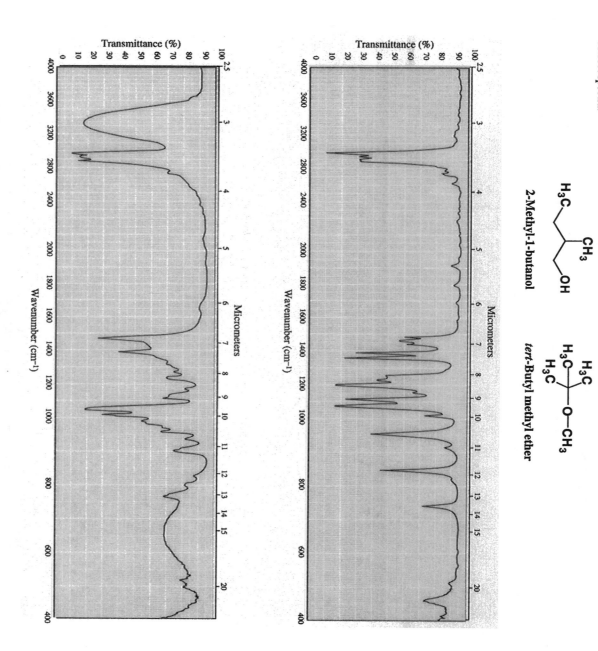

2-Methyl-1-butanol

tert-Butyl methyl ether

The molecules are extremely similar except for the -OH group present in the 2-methyl-1-butanol. The characteristic O-H stretch is present at 3300 - 3600 cm⁻¹ in the lower spectrum.

The broad peak at 3300 - 3600 cm⁻¹ verifies that the second (lower) spectrum corresponds to 2-methyl-1-butanol. Therefore, the first (upper) spectrum corresponds to *tert*-butyl methyl ether.

Problem 12.9 The IR C≡C stretching absorption in symmetrical alkynes is usually absent. Why is this so?

For a bond vibration to have a corresponding absorption of high intensity in the IR, that bond must have a dipole moment, and absorption of energy must result in a large change in that dipole moment. For symmetrical alkynes, there is no dipole moment associated with the C≡C bond, and thus no large change in dipole moment associated with bond stretching. Thus, the IR band is usually absent.

Problem 12.10 Explain the fact that the C–O stretch in ethers and esters occurs at 1000-1100 cm^{-1} when the C is sp^3 hybridized, but at 1250 cm^{-1} when it is sp^2 hybridized.

When the C atom is sp^2 hybridized, it has more S character compared to an sp^3 hybridized atom. As can be seen in the table of bond dissociation energies (Appendix 3), all things being equal, sp2 hybridized atoms will make stronger bonds than sp3 hybridized atoms. Stronger bonds absorb infrared radiation at higher energy, thus higher frequency.

Problem 12.11 A compound has strong infrared absorption bands at the following frequencies. Suggest likely functional groups that may be present.

(a) 1735, 1250, and 1100 cm^{-1}.   (b) 1745 cm^{-1} but not 1000-1250 cm^{-1} (c) 1710 and 2500-3400 (broad) cm^{-1} .
    An ester   **A ketone**   **A carboxyl group**

(d) A single band at about 3300 cm^{-1} ,   (e) 3600 and 1050 cm^{-1}.   (f) 1100 but not 3300-3650 cm^{-1} .
    A secondary amine   **An alcohol**   **An ether**

Problem 12.12 Show how IR spectroscopy can be used to distinguish between the compounds in each set.

The key to using IR spectroscopy for distinguishing between molecules is to look for characteristic peaks that arise from functional groups unique to one of the structures. Characteristic peaks are listed under each molecule. Note that all of the following will have C-H stretching peaks at 2850-3000 cm^{-1} and bending peaks near 1450 cm^{-1}, so these have been ignored. Also, the methyl group bending peaks at 1375 cm^{-1} and 1450 cm^{-1} are generally ignored, unless they will be useful as in (d) and (f).

(a) ＯＨ   and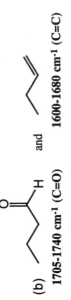

1050 cm^{-1} 1°(C-O)   1000-1100 cm^{-1} (C-O)
3200-3500 cm^{-1} (O-H)

The alcohol will be identifiable by the broad O-H peak at 3200-3500 cm^{-1}.

(b) 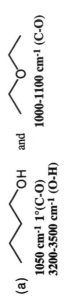   and

1705-1740 cm^{-1} (C=O)   1600-1680 cm^{-1} (C=C)

The ketone will be identifiable by the C=O peak at 1705-1740 cm^{-1}, while the alkene will have a characteristic peak at 1600-1680 cm^{-1}.

(c) 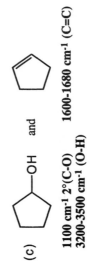 ＯＨ   and

1100 cm^{-1} 2°(C-O)   1600-1680 cm^{-1} (C=C)
3200-3500 cm^{-1} (O-H)

The alcohol will be identifiable by the broad O-H peak at 3200-3500 cm^{-1} and C-O peak at 1100 cm^{-1}, while the alkene will have a characteristic peak at 1600-1680 cm^{-1}.

(d) 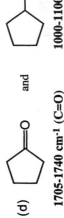   and   ＯＣＨ$_3$

1705-1740 cm^{-1} (C=O)   1000-1100 cm^{-1} (C-O)
                                     1375, 1450 cm^{-1} (-CH$_3$)

The ketone will be identifiable by the C=O peak at 1705-1740 cm⁻¹, while the ether will have a characteristic C-O peak at 1000-1100 cm⁻¹ and methyl bending peaks at 1375 cm⁻¹ and 1450 cm⁻¹.

(e)

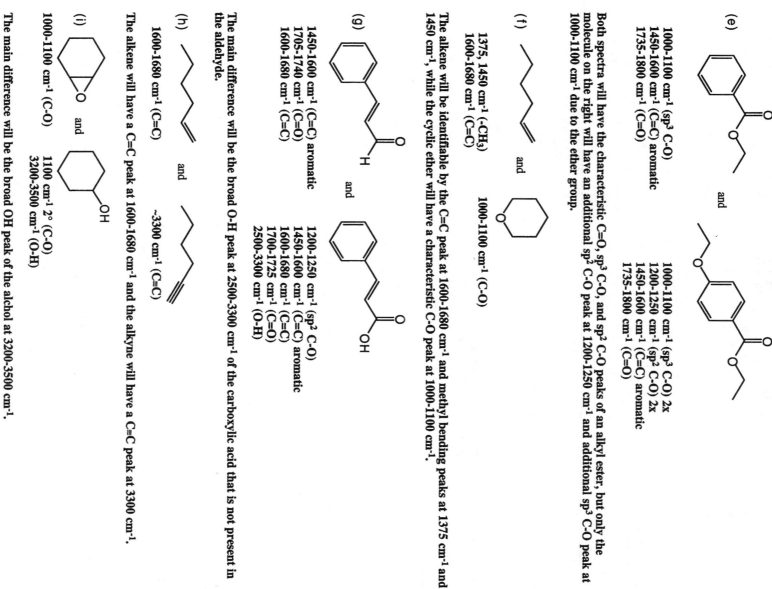

1000-1100 cm⁻¹ (sp³ C-O)
1450-1600 cm⁻¹ (C=C) aromatic
1735-1800 cm⁻¹ (C=O)

1000-1100 cm⁻¹ (sp³ C-O) 2x
1200-1250 cm⁻¹ (sp² C-O) 2x
1450-1600 cm⁻¹ (C=C) aromatic
1735-1800 cm⁻¹ (C=O)

Both spectra will have the characteristic C=O, sp³ C-O, and sp² C-O peaks of an alkyl ester, but only the molecule on the right will have an additional sp² C-O peak at 1200-1250 cm⁻¹ and additional sp³ C-O peak at 1000-1100 cm⁻¹ due to the ether group.

(f)   and

1375, 1450 cm⁻¹ (-CH₃)   1000-1100 cm⁻¹ (C-O)
1600-1680 cm⁻¹ (C=C)

The alkene will be identifiable by the C=C peak at 1600-1680 cm⁻¹ and methyl bending peaks at 1375 cm⁻¹ and 1450 cm⁻¹, while the cyclic ether will have a characteristic C-O peak at 1000-1100 cm⁻¹.

(g)   and

1450-1600 cm⁻¹ (C=C) aromatic
1705-1740 cm⁻¹ (C=O)
1600-1680 cm⁻¹ (C=C)

1200-1250 cm⁻¹ (sp² C-O)
1450-1600 cm⁻¹ (C=C) aromatic
1600-1680 cm⁻¹ (C=C)
1700-1725 cm⁻¹ (C=O)
2500-3300 cm⁻¹ (O-H)

The main difference will be the broad O-H peak at 2500-3300 cm⁻¹ of the carboxylic acid that is not present in the aldehyde.

(h)   and

1600-1680 cm⁻¹ (C=C)   ~3300 cm⁻¹ (C≡C)

The alkene will have a C=C peak at 1600-1680 cm⁻¹ and the alkyne will have a C≡C peak at 3300 cm⁻¹.

(i)   and

1000-1100 cm⁻¹ (C-O)   1100 cm⁻¹ 2° (C-O)
                             3200-3500 cm⁻¹ (O-H)

The main difference will be the broad OH peak of the alchol at 3200-3500 cm⁻¹.

(j)

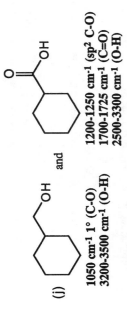

1050 cm^{-1} 1° (C-O)	1200-1250 cm^{-1} (sp^2 C-O)
3200-3500 cm^{-1} (O-H)	1700-1725 cm^{-1} (C=O)
	2500-3300 cm^{-1} (O-H)

and

The main difference will be the carboxylic acid C=O peak at 1700-1725 cm^{-1}.

**Bluestripe butterflyfish *Chaetodon fremblii*
Kauai, Hawaii**

CHAPTER 13
Solutions to the Problems

Problem 13.1 Calculate the ratio of nuclei in the higher spin state to those in the lower spin state, N_h/N_l, for ^{13}C at 25°C in an applied field strength of 7.05 T. The difference in energy between the higher and lower nuclear spin states in this applied field is approximately 0.030 J (0.00715 cal)/mol.

The important equation relates the change in energy of two spin states to their equilibrium concentrations:

$$\Delta G° = -2.303RT\log\frac{N_h}{N_l}$$

Rearranging this expression in terms of N_h/N_l gives:

$$\log\frac{N_h}{N_l} = \frac{-\Delta G°}{2.303RT}$$

Substituting in the appropriate values for R ($8.314 \text{ J·deg}^{-1}\text{·mol}^{-1}$), T (298 K) and $\Delta G°$ (0.030 J·mol⁻¹) gives:

$$\log\frac{N_h}{N_l} = \frac{-0.030 \text{ J·mol}^{-1}}{2.303(8.314 \text{ J·deg}^{-1}\text{·mol}^{-1})(298 \text{ deg})} = -5.258 \times 10^{-6}$$

$$\boxed{\frac{N_h}{N_l} = 0.9999879 = \frac{1.0000000}{1.0000121}}$$

Problem 13.2 State the number of sets of equivalent hydrogens in each compound and the number of hydrogens in each set.

Numbers have been added to the carbon atoms of the structures to aid in referring to specific hydrogens. Use the "test atom" approach if you have trouble understanding the answers.

(a) 3-Methylpentane

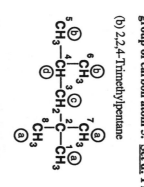

There are four sets of equivalent hydrogens. Set a: 6 hydrogens from the methyl groups of carbon atoms 1 and 5. Set b: 4 hydrogens from the -CH₂- groups of carbon atoms 2 and 4. Set c: 3 hydrogens from the methyl group of carbon atom 5. Set d: 1 hydrogen from the -CH- group of carbon atom 3.

(b) 2,2,4-Trimethylpentane

There are four sets of equivalent hydrogens. Set a: 9 hydrogens from the methyl groups of carbon atoms 1, 7, and 8. Set b: 6 hydrogens from the methyl groups of carbon atoms 5 and 6. Set c: 2 hydrogens from the -CH₂- group of carbon atom 3. Set d: 1 hydrogen from the -CH- group of carbon atom 4.

Problem 13.3 Each compound gives only one signal in its ^{1}H-NMR spectrum. Propose a structural formula for each compound.

In order for these molecules to give a single absorption peak, each of the hydrogen nuclei must be in an identical environment. This will only occur in symmetrical molecules.

(a) C_3H_6O (b) C_5H_{10} (c) C_5H_{12}

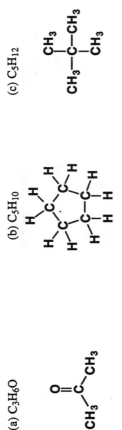

(d) $C_4H_6Cl_4$

CH_3-CCl_2-CCl_2-CH_3

Problem 13.4 The line of integration of the two signals in the ^{1}H-NMR spectrum of a ketone of molecular formula $C_7H_{14}O$ rises 62 and 10 chart divisions, respectively. Calculate the number of hydrogens giving rise to each signal, and propose a structural formula for this ketone.

The ratio of signals is approximately 6:1, which corresponds to a 12:2 ratio of hydrogens. Thus, the larger signal represents 12 hydrogens and the smaller signal represents 2 hydrogens. A structure consistent with this assignment is 2,4-dimethyl-3-pentanone as shown below:

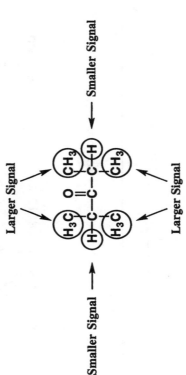

Problem 13.5 Following are two constitutional isomers of molecular formula $C_4H_8O_2$.

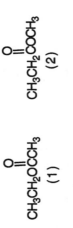

(a) Predict the number of signals in the ^{1}H-NMR spectrum of each isomer.

Each isomer will have three signals.

(b) Predict the ratio of areas of the signals in each spectrum.

In each spectrum, the ratio of areas of the three signals will be 3:2:3.

(c) Show how to distinguish between these isomers on the basis of chemical shift.

The -CH_3 singlet signals will be diagnostic. Isomer (1) is the only one with a -CH_3 attached to the carbonyl carbon atom, while isomer (2) is the only one with a -CH_3 attached to an ester oxygen atom. Therefore, the spectrum of isomer (1) will be the one with a -CH_3 singlet at δ 2.1-2.3, and the spectrum of isomer (2) will be the one with a -CH_3 singlet at δ 3.7-3.9.

Problem 13.6 Following are pairs of constitutional isomers. Predict the number of signals in the ^1H-NMR spectrum of each isomer and the splitting pattern of each signal.

(a) CH₃OCH₂CCH₃   and   CH₃CH₂COCH₃
            ‖   ‖
            O   O

The molecule on the left will have three signals that are all singlets, and the molecule on the right will have three signals with splitting patterns as indicated.

singlet singlet singlet   triplet quartet singlet
       a   b   c   a   b   c
    CH₃OCH₂CCH₃   CH₃CH₂COCH₃
              ‖   ‖
              O   O

(b) CH₃CCH₃   and   ClCH₂CH₂CH₂Cl
        |
     Cl—C—Cl
        |
        Cl

The molecule on the left will have one signal and the molecule on the right will have two signals with splitting patterns as indicated.

        singlet   triplet quintet triplet
          a   a   b   a
       Cl   ClCH₂CH₂CH₂Cl
       |
    CH₃CCH₃
       |
       Cl

Problem 13.7 Following is the spectrum of 2-butanol. Explain why the CH₂ protons appear as a complex multiplet instead of a simple quintet.

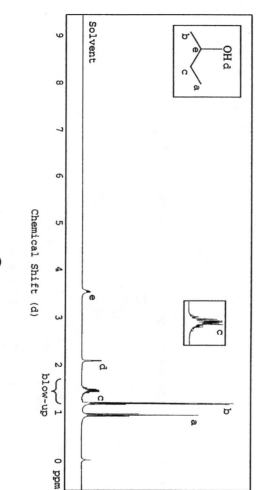

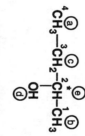

Carbon atom 2 is a stereocenter, and the two hydrogens labeled as group c on carbon atom 3 are diastereotopoic. Thus, these two protons are in different environments and give rise to different chemical shifts, greatly complicating the corresponding signals in the observed spectra.

Problem 13.8 Explain how to distinguish between the members of each pair of constitutional isomers based on the number of signals in the proton-decoupled ^{13}C-NMR spectrum of each member.

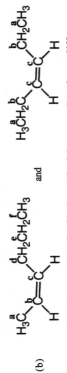

(a)

and

These molecules can be distinguished because they have different numbers of nonequivalent carbon nuclei and thus will have different numbers of ^{13}C-NMR signals. Different signals are indicated by different letters on the above structures. The molecule on the left has higher symmetry and will have 5 signals corresponding to the carbon atoms labeled as a - e, while the molecule on the right has less symmetry and will have 7 signals corresponding to the carbon atoms labeled as a - g.

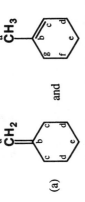

(b)

and

These molecules can also be distinguished because they have different numbers of nonequivalent carbon nuclei and thus will have different numbers of ^{13}C-NMR signals. Different signals are indicated by different letters on the above structures. The molecule on the right has higher symmetry and will only have 3 signals corresponding to the carbon atoms labeled as a - c, while the molecule on the left has less symmetry and will have 6 signals corresponding to the carbon atoms labeled as a - f.

Problem 13.9 Assign all signals in the ^{13}C-NMR spectrum of 4-methyl-2-pentanone.

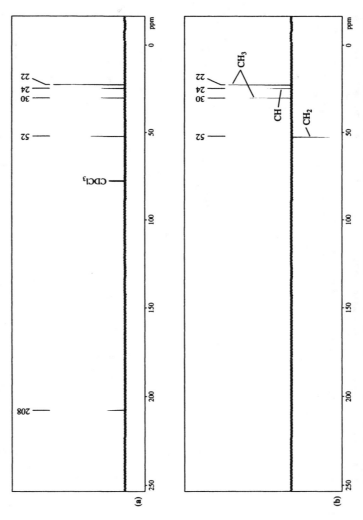

(a)

(b)

The appropriate peaks have been assigned using the DEPT spectra to distinguish the CH$_3$, CH$_2$ and CH groups. The different CH$_3$ groups were distinguished by the fact that the CH$_3$ group adjacent to the carbonyl group is less shielded (higher chemical shift) than the other two methyl groups, which are equivalent. The carbonyl group carbon atom has a partial positive charge, thus explaining the relative lack of shielding observed for the adjacent carbon nuclei.

$$\underset{30}{CH_3}\overset{O}{\underset{208}{C}}\underset{52}{CH_2}\underset{24}{CH}(\underset{22}{CH_3})_2$$

4-Methyl-2-pentanone

Problem 13.10 Calculate the index of hydrogen deficiency of cyclohexene, C_6H_{10}, and account for this deficiency by reference to its structural formula.

The molecular formula for cyclohexene is C_6H_{10}. The molecular formula for the reference compound with 6 carbon atoms is C_6H_{14}. Thus the index of hydrogen deficiency is $(14-10)/2$ or 2. This makes sense since cyclohexene has one ring and one pi bond.

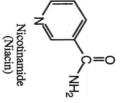

Cyclohexene

Problem 13.11 The index of hydrogen deficiency of niacin is 5. Account for this index of hydrogen deficiency by reference to the structural formula of niacin.

Nicotinamide
(Niacin)

The index of hydrogen deficiency of niacin is 5 because there are four pi bonds and one ring in the structure.

Index of Hydrogen Deficiency

Problem 13.12 Complete the following table:

Class of Compound	General Molecular Formula	Index of Hydrogen Deficiency	Reason for Hydrogen Deficiency
alkane	C_nH_{2n+2}	0	(reference hydrocarbon)
alkene	C_nH_{2n}	1	one pi bond
alkyne	C_nH_{2n-2}	2	two pi bonds
alkadiene	C_nH_{2n-2}	2	two pi bonds
cycloalkane	C_nH_{2n}	1	one ring
cycloalkene	C_nH_{2n-2}	2	one ring and one pi bond
bicycloalkane	C_nH_{2n-2}	2	two rings

Problem 13.13 Calculate the index of hydrogen deficiency of these compounds:

(a) Aspirin, $C_9H_8O_4$

$(20-8)/2 = 6$

(b) Ascorbic acid (vitamin C), $C_6H_8O_6$

$(14-8)/2 = 3$

(c) Pyridine, C_5H_5N

$(13-5)/2 = 4$
(nitrogen correction)

(d) Urea, CH_4N_2O

$(6-4)/2 = 1$
(nitrogen correction)

(e) Cholesterol, $C_{27}H_{46}O$

$(56-46)/2 = 5$

(f) Dopamine, $C_8H_{11}NO_2$

$(19-11)/2 = 4$
(nitrogen correction)

Interpretation of ^{1}H-NMR and ^{13}C-NMR Spectra

Problem 13.14 Complete the following table. Which nucleus requires the least energy to flip its spin at this applied field? Which nucleus requires the most energy?

Nucleus	Applied field (T)	Radiofrequency (MHz)	Energy (J/mol)
^{1}H	7.05	300	0.120
^{13}C	7.05	75.5	0.302
^{19}F	7.05	282	0.113

Based on the entries in the table, the ^{13}C requires the least energy to flip its spin and the ^{1}H requires the most.

Problem 13.15 The natural abundance of ^{13}C is only 1.1%. Furthermore, its sensitivity in NMR spectroscopy (a measure of the energy difference between a spin aligned with or against an external magnetic field) is only 1.6% that of ^1H. What are the relative signal intensities expected for the ^1H-NMR and ^{13}C-NMR spectra of the same sample of $Si(CH_3)_4$?

A given ^{13}C signal is $(0.011)(0.016) = 0.000176$ as strong as a given ^1H signal. There are three times as many H atoms as C atoms in $Si(CH_3)_4$, so overall the ratio of H to C signals is $1 : (0.000176/3) = 1 : 0.000059$. (Note that this is the same as 17,000 to 1)

Problem 13.16 Following are structural formulas for three constitutional isomers of molecular formula $C_7H_{16}O$ and three sets of ^{13}C-NMR spectral data. Assign each constitutional isomer its correct spectral data.

	Spectrum 1	Spectrum 2	Spectrum 3
(a) $CH_3CH_2CH_2CH_2CH_2CH_2CH_2OH$	74.66	70.97	62.93
	30.54	43.74	32.79
OH	7.73	29.21	31.86
(b) $CH_3CH_2CCH_2CH_2CH_3$		26.60	29.14
CH_3		23.27	25.75
		14.09	22.63
OH			14.08
(c) $CH_3CH_2CCH_2CH_3$			
CH_2CH_3			

These constitutional isomers are most readily distinguished by the number of sets of nonequivalent carbon atoms and thus different ^{13}C signals. Using the following analysis, it can be seen that compound (a) has 7 sets of nonequivalent carbon atoms corresponding to Spectrum 3, compound (b) has 6 sets of nonequivalent carbon atoms corresponding to Spectrum 2, and compound (c) has 3 sets of nonequivalent carbon atoms corresponding to Spectrum 1.

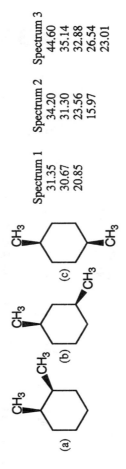

$$\underset{g}{CH_3}\underset{f}{CH_2}\underset{e}{CH_2}\underset{d}{CH_2}\underset{c}{CH_2}\underset{b}{CH_2}\underset{a}{CH_2}OH$$

$$\underset{c}{CH_3}\underset{b}{CH_2}\underset{a}{C}\underset{b}{CH_2}\underset{c}{CH_2}\underset{b}{CH_3}$$
OH
CH_2CH_3

$$\underset{e}{CH_3}\underset{a}{C}\underset{b}{CH_2}\underset{c}{CH_2}\underset{d}{CH_2}\underset{f}{CH_2}CH_3$$
OH
CH_3
e

Problem 13.17 Following are structural formulas for the *cis* isomers of 1,2-, 1,3-, and 1,4-dimethylcyclohexane and three sets of ^{13}C-NMR spectral data. Assign each constitutional isomer its correct spectral data.

(a) [structure: cyclohexane with CH3, CH3]

(b) [structure: cyclohexane with CH3, CH3]

(c) [structure: cyclohexane with CH3, CH3]

	Spectrum 1	Spectrum 2	Spectrum 3
	31.35	34.20	44.60
	30.67	31.30	35.14
	20.85	23.56	32.88
		15.97	26.54
			23.01

These constitutional isomers are most readily distinguished by the number of sets of nonequivalent carbon atoms and thus different ^{13}C signals. Using the following analysis, it can be seen that compound (a) has 4 sets of nonequivalent carbon atoms corresponding to Spectrum 2, compound (b) has 5 sets of nonequivalent carbon atoms corresponding to Spectrum 3, and compound (c) has 3 sets of nonequivalent carbon atoms corresponding to Spectrum 1. The different sets of equivalent carbon atoms are indicated by the letters.

Problem 13.18 Following are structural formulas, dipole moments, and ^{1}H-NMR chemical shifts for acetonitrile, fluoromethane, and chloromethane.

$CH_3-C{\equiv}N$	CH_3-F	CH_3-Cl
Acetonitrile	Fluoromethane	Chloromethane
3.92 D	1.85 D	1.87 D
δ 1.97	δ 4.26	δ 3.05

(a) How do you account for the fact that the dipole moments of fluoromethane and chloromethane are almost identical even though fluorine is considerably more electronegative than chlorine?

Recall that dipole moment is proportional to the partial charge times the distance of that charge separation. Fluorine is a much smaller atom than chlorine, so it makes shorter bonds leading to relatively short charge separation distances. Thus, the differences in bond lengths almost exactly offsets the differences in electronegativities between fluorine and chlorine and the dipole moments come out almost the same.

(b) How do you account for the fact that the dipole moment of acetonitrile is considerably greater than that of either fluoromethane or chloromethane?

Again, the key is distance. The acetonitrile has partial charge distributed over more atoms and thus a larger distance than fluoromethane or chloromethane.

(c) How do you account for the fact that the chemical shift of the methyl hydrogens in acetonitrile is considerably less than that for either fluoromethane or chloromethane? (*Hint*: Consider the magnetic induction in the pi bonds of acetonitrile.)

A magnetic field is induced in the pi system of the nitrile that is against the applied field, thus decreasing the chemical shift.

Problem 13.19 Following are three compounds of molecular formula $C_4H_8O_2$, and three ^{1}H-NMR spectra. Assign each compound its correct spectrum and assign all signals to their corresponding hydrogens.

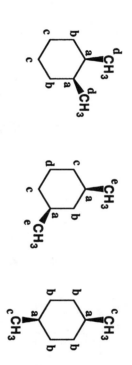

$$CH_3COCH_2CH_3$$
(1)

$$HOCH_2CH_2CH_3$$
(with C=O)
(2)

$$CH_3OCCH_2CH_3$$
(with C=O)
(3)

For the spectral interpretations in the rest of this chapter the chemical shift (δ) is shown on the structure adjacent to the appropriate hydrogen atom.

Spectrum A corresponds to compound 2: ^{1}H-NMR δ 8.1 (1H, singlet, H-C(O)-), 4.2 (2H, triplet, -O-CH$_2$-), 1.7 (2H, multiplet, -CH$_2$-), 1.0 (3H, triplet, -CH$_3$).

$$\overset{8.1}{H}C-O-\overset{4.2}{C}H_2\overset{1.7}{C}H_2\overset{1.0}{C}H_3$$
(with C=O)
(2)
Compound A

Spectrum B corresponds to compound 3: ^{1}H-NMR δ 3.7 (3H, singlet, CH$_3$-C(O)-), 2.3 (2H, quartet, -C(O)-CH$_2$-), 1.2 (3H, triplet, -CH$_3$).

$$CH_3^{3,7}-O-\overset{\overset{O}{\|}}{C}\overset{2,3}{C}H_2\overset{1,2}{C}H_3$$

(3)

Compound B

Spectrum C corresponds to compound 1: ^{1}H-NMR δ 4.1 (2H, quartet, -O-CH$_2$-), 2.0 (3H, singlet, CH$_3$-C-), 1.2 (3H, triplet, -CH$_3$).

$$\overset{2,0}{C}H_3\overset{\overset{O}{\|}}{C}-O-\overset{4,1}{C}H_2\overset{1,2}{C}H_3$$

(1)

Compound C

Compound A:

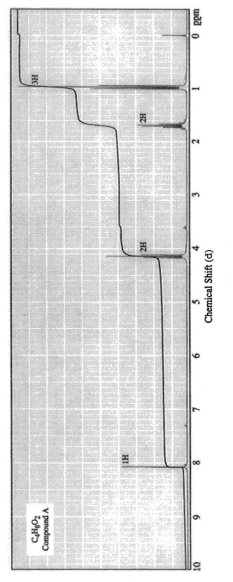

C$_4$H$_8$O$_2$
Compound A

Compound B:

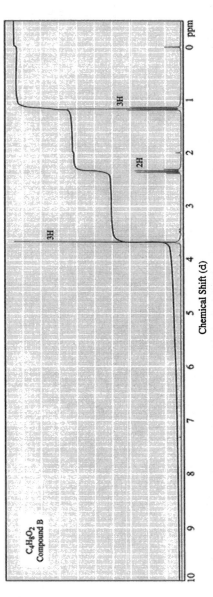

C$_4$H$_8$O$_2$
Compound B

Compound C:

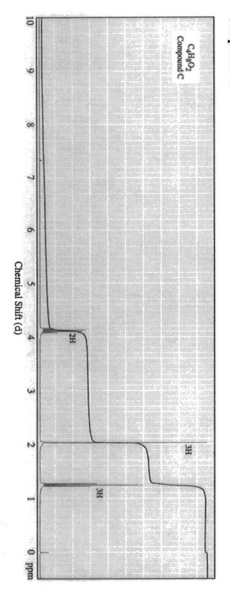

C₄H₈O₂
Compound C

Chemical Shift (δ)

Problem 13.20 Following are ¹H-NMR spectra for compounds D, E, and F, each of molecular formula C₆H₁₂. Each readily decolorizes a solution of Br₂ in CCl₄. Propose structural formulas for compounds D, E, and F, and account for the observed patterns of signal splitting.

Each of the compounds has an index of hydrogen deficiency of 1, in the form of a double bond as evidenced by the reaction with Br₂. The rest of the detailed structures can be deduced from the spectra.

Compound D:

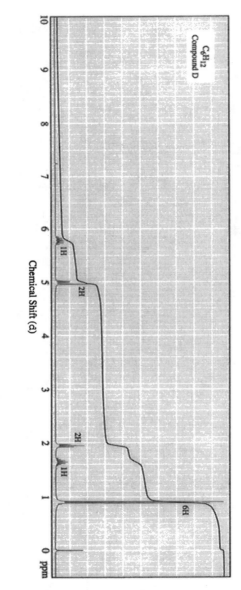

C₆H₁₂
Compound D

Chemical Shift (δ)

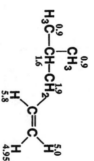

H₃C—CH—CH₂
  |
  CH₃

¹H-NMR δ 5.8 (1H, multiplet; this is more complex than expected because the adjacent vinylic hydrogens are not equivalent, -CH=), 4.95-5.0 (2H, multiplet, =CH₂; this is asymmetric because these two vinylic hydrogens are not equivalent and the hydrogen *trans* to the hydrogen on the other vinylogous carbon has the larger signal splitting so it is the signal at 5.0), 1.9 (2H, multiplet, doublet of doublets, -CH₂-), 1.6 (1H, multiplet, -CH-), 0.9 (6H, one doublet, -CH₃).

Compound E:

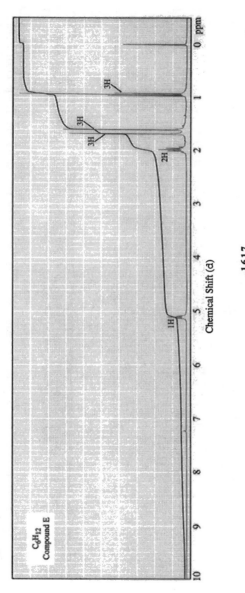 

¹H-NMR δ 5.1 (1H, triplet, -CH=), 2.0 (2H, multiplet, -CH₂-), 1.6 and 1.7 (6H, two singlets, =C(CH₃)₂), 0.9 (3H, triplet, -CH₃)

$$\overset{0.9}{CH_3}-\overset{2.0}{CH_2}-\overset{5.1}{CH}=C-\overset{1.6,1.7}{CH_3}$$
$$\underset{CH_3}{\overset{1.6,1.7}{|}}$$

Compound F:

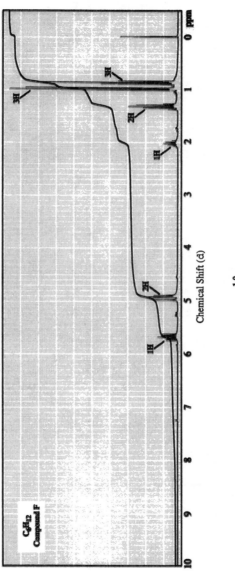 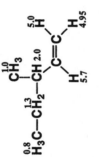

$$\overset{0.8}{H_3C}-\overset{1.3}{CH_2}-\overset{2.0}{CH}\overset{1.0}{\underset{CH_3}{|}}=C\overset{5.0}{\underset{5.7}{\overset{H}{<}}}\overset{H}{\underset{4.95}{}}$$

¹H-NMR δ 5.7 (1H, multiplet; this is more complex than expected because the adjacent vinylic hydrogens are not equivalent, -CH=), 4.9-5.0 (2H, multiplet, =CH₂; this is asymmetric because these two vinylic hydrogens are not equivalent and the hydrogen *trans* to the hydrogen on the other vinylogous carbon has the larger signal splitting so it is the signal at 5.0), 2.0 (1H, multiplet, -CH-), 1.3 (2H, multiplet, -CH₂-), 1.0 (3H, doublet, -CH-CH₃), 0.8 (3H, triplet, -CH₂-CH₃)

Problem 13.21 Following are 1H-NMR spectra for compounds G, H, and I, each of molecular formula $C_5H_{12}O$. Each is a liquid at room temperature, slightly soluble in water, and reacts with sodium metal with the evolution of a gas. (a) Propose structural formulas of compounds G, H, and I.

The index of hydrogen deficiency is 0 for these molecules, so there are no rings or double bonds. The fact that the compounds are slightly soluble in water and react with sodium metal indicates that each molecule has an -OH group. The chemical shifts associated with each set of hydrogens are indicated on the structures.

Compound G:

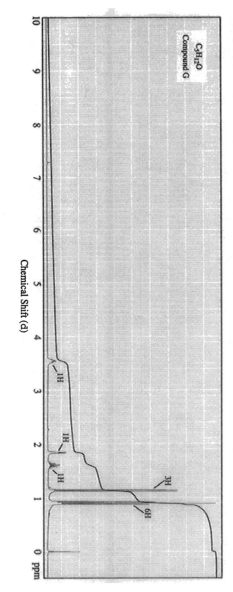

Compound H:

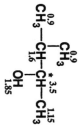

$$\begin{array}{c} \overset{0.9}{CH_3} \\ | \\ CH_3 - \overset{*\ 3.5}{CH} - CH - CH_3 \\ {}_{0.9}\quad {}_{1.6}\quad | \quad {}_{1.15} \\ OH \\ {}_{1.85} \end{array}$$

^{1}H-NMR δ 3.5 (1H, multiplet, -CH-OH-), 1.85 (1H, doublet, -OH), 1.6 (1H, multiplet, -CH-(CH$_3$)$_2$), 1.15 (3H, doublet, -C(OH)-CH$_3$), 0.9 (6H, overlapping doublets; this is more complex than expected because it is adjacent to a stereocenter -CH-(CH$_3$)$_2$).

Compound H:

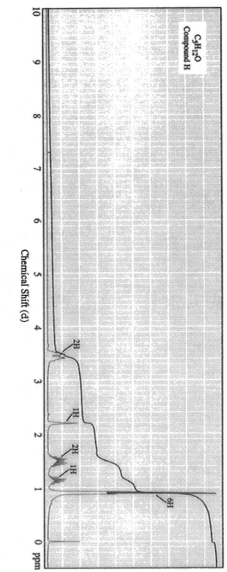

$$\begin{array}{c} \overset{0.8-0.9}{CH_3} \\ {}_{0.8-0.9}\;{}_{1.4-1.6}\;|\; {}_{3.4-3.5} \\ CH_3CH_2CHCH_2OH \\ {}_{2.2} \end{array}$$

1H-NMR δ 3.4-3.5 (2H, multiplet; this is more complex than expected because it is adjacent to a stereocenter, -CH₂-OH), 2.2 (1H, broad triplet, -OH), 1.4-1.6 (2H, multiplet; this is more complex than expected because it is adjacent to a stereocenter, CH₃-CH₂-), 1.1 (1H, multiplet, -CH-), 0.8-0.9 (6H, broad multiplet, both -CH₃ groups).

Compound I:

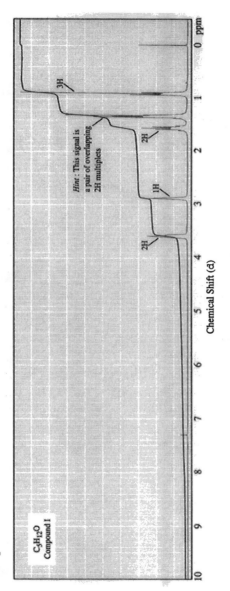

$$CH_3CH_2CH_2CH_2CH_2OH$$
0.9   1.4   1.4   1.55   3.6   2.9

1H-NMR δ 3.6 (2H, broad multiplet, -CH₂-OH), 2.9 (1H, broad peak, -OH), 1.55 (2H, multiplet, -CH₂-OH), 1.4 (4H, multiplet, CH₃-CH₂-CH₂- and CH₃-CH₂-CH₂-), 0.9 (3H, triplet, -CH₃)

(b) Explain why there are four lines between 0.86 and 0.90 for Compound G.

Carbon atom 2 (with the -OH group) is a stereocenter. That makes the two methyl groups diastereotopic, so they have different chemical shifts. The four lines are actually two doublets.

(c) Explain why the 2-H multiplets at 1.5 and 3.5 ppm for compound H are so complex.

Carbon atom 2 is a stereocenter. The stereocenter makes the adjacent -CH₂ protons diastereotopic, so the signals are complex.

Problem 13.22 Propose a structural formula for compound J, molecular formula C₃H₆O, consistent with the following ¹H-NMR spectrum:

Compound J:

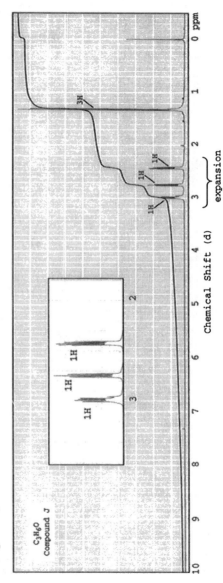

From the molecular formula, there is an index of hydrogen deficiency of 1 indicating that there is one ring or pi bond.

¹H-NMR δ 3.0 (1H, multiplet, H$_c$), 2.4 and 2.75 (2H, multiplets; these hydrogens are not equivalent because the carbon-carbon bonds cannot rotate freely), 1.3 (3H, doublet, -CH₃)

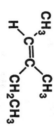

Problem 13.23 Compound K, molecular formula C₆H₁₄O, readily undergoes acid-catalyzed dehydration when warmed with phosphoric acid to give compound L, molecular formula C₆H₁₂, as the major organic product. The ¹H-NMR spectrum of compound K shows signals at δ 0.90 (t, 6H), 1.12 (s, 3H), 1.38 (s, 1H), and 1.48 (q, 4H). The ¹³C-NMR spectrum of compound K shows signals at 72.98, 33.72, 25.85, and 8.16. Deduce the structural formulas of compounds K and L.

From the molecular formula, there is a hydrogen deficiency index of 0, so there are no rings or pi bonds in compound K. From the ¹³C-NMR peak at 72.98, there is a carbon bonded to an -OH group. The rest of the structure can be deduced from the ¹H-NMR spectrum. The chemical shifts associated with each set of hydrogens are indicated on the structure.

$$\underset{CH_3}{\overset{0.90}{CH_3}}\underset{CH_3}{\overset{1.48}{CH_2}}\overset{\underset{|}{\overset{1.38}{OH}}}{C}\underset{CH_3}{\overset{1.48}{CH_2}}\overset{0.90}{CH_3}$$

Dehydration of compound K gives the following alkene as compound L:

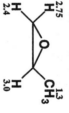

¹H-NMR δ 1.48 (4H, quartet, -CH₂), 1.38 (1H, singlet, -OH), 1.12 (3H, singlet, -C(OH)-CH₃), 0.90 (6H, triplet, both -CH₂-CH₃ groups)

Problem 13.24 Compound M, molecular formula C₅H₁₀O, readily decolorizes Br₂ in CCl₄, and is converted by H₂/Ni into compound N, molecular formula C₅H₁₂O. Following is the ¹H-NMR spectrum of compound M. The ¹³C-NMR spectrum of compound M shows signals at 146.12, 110.75, 71.05, and 29.38. Deduce the structural formulas of compounds M and N.

Compound M:

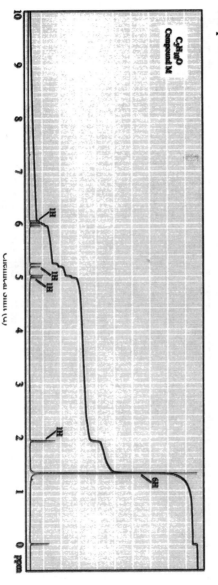

From the reaction with Br_2 and H_2/Ni it is clear the compound M has a carbon-carbon double bond. These conclusions are supported by the ^{13}C-NMR signals corresponding to the sp^2 carbons at δ 146.12 and 110.75. There is also a carbon bonded to an -OH group judging from the signal at δ 71.05. The rest of the structure is deduced from the ^1H-NMR spectrum.

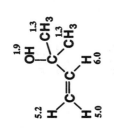

^1H-NMR δ 6.0 (1H, doublet of doublets), 5.0 and 5.2, (2H, doublet of doublets; the vinylic hydrogens are not equivalent and the hydrogen *trans* to the hydrogen on the other vinylogous carbon has the larger signal splitting so it is the signal at 5.2), 1.9 (1H, singlet, -OH), 1.3 (6H, singlet, C-$(CH3)2$)

Upon hydrogenation, compound M is reduced to the alcohol shown below as compound N:

$$CH_3\text{-}CH_2\text{-}\overset{\overset{\displaystyle OH}{|}}{\underset{\underset{\displaystyle CH_3}{|}}{C}}\text{-}CH_3$$

Problem 13.25   Following is the ^1H-NMR spectrum of compound O, molecular formula C_7H_{12}. Compound O reacts with bromine in carbon tetrachloride to give a compound of molecular formula $C_7H_{12}Br_2$. The ^{13}C-NMR spectrum of compound L shows signals at 150.12, 106.43, 35.44, 28.36, and 26.36. Deduce the structural formula of compound L.

Compound O:

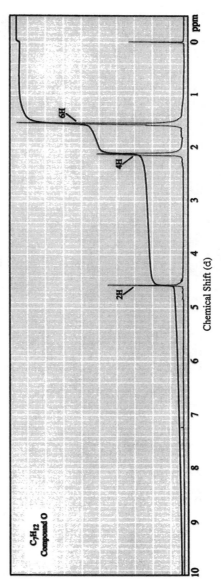

Chemical Shift (d)

The molecular formula indicates that there is an index of hydrogen deficiency of 2, so there are two rings and/or pi bonds. The ^{13}C-NMR indicates there is only one double bond because there are only two resonances corresponding to sp^2 carbon atoms (150.12 and 106.43). Therefore, compound O must have one ring and one pi bond.

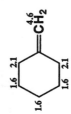

^1H-NMR δ 4.6 (2H, singlet, =CH_2), 2.1 (4H, broad peak, two -CH_2- groups), 1.6 (6H, broad peak, three -CH_2- groups)

Problem 13.26 Treatment of compound P with BH_3 followed by H_2O_2/NaOH gives compound Q. Following are ^1H-NMR spectra for compounds P and Q along with ^{13}C-NMR spectral data. From this information, deduce structural formulas for compounds P and Q

$$C_7H_{12} \xrightarrow[\text{2) } H_2O_2, \text{ NaOH}]{\text{1) } BH_3} C_7H_{14}O$$
$$(P) \qquad\qquad\qquad (Q)$$

	^{13}C-NMR	
	(P)	(Q)
	132.38	72.71
	32.12	37.59
	29.14	28.13
	27.45	22.68

Compound P:

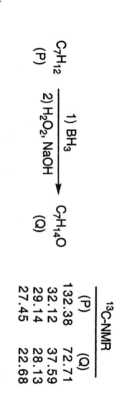

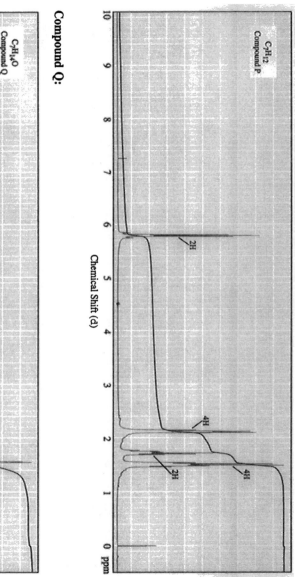

Compound Q:

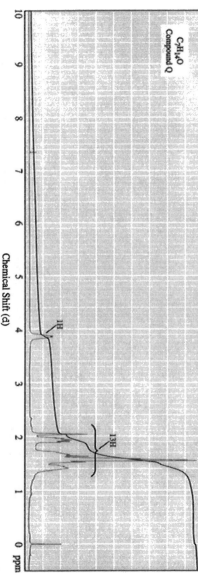

The molecular formula for P indicates that it has an index of hydrogen deficiency of 2 so that is has two rings and/or pi bonds. The ^{13}C-NMR spectral data shows that there is an sp^2 carbon atom (132.38). Since there must be two sp^2 carbon atoms to make a pi bond, then the molecule must be symmetric so that both sp^2 carbon atoms are equivalent. This also explains why there are so few other ^{13}C-NMR signals. Since there is presumably only one pi bond, then there must be one ring in the molecule. The rest of the structure can be deduced from the ^1H-NMR spectrum.

¹H-NMR δ 5.8 (2H, triplet, both =CH-), 2.1 (4H, multiplet, two -CH₂- groups), 1.7 (2H, quintet, the unique -CH₂- group), 1.5 (4H, multiplet, two -CH₂- groups).

Given the structural formula for P, it is clear that compound Q would be the hydroboration/oxidation product, namely the alcohol shown below. This structure is consistent with the ¹³C-NMR spectral data provided as well as the ¹H-NMR spectrum.

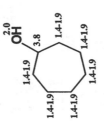

¹H-NMR δ 3.8 (1H, broad peak, -C(OH)H-), 2.0 (1H, sharp singlet, -OH), 1.4-1.9 (12H, broad multiplets, all the remaining hydrogens on the ring. The peaks are so broad and the patterns so complex because this ring does not have a double bond to hold it rigid, so it has a great deal of flexibility)

<u>Problem 13.27</u> The ¹H-NMR of Compound R, $C_6H_{14}O$, consists of two signals: δ 1.1 (doublet) and δ 3.6 (septet) in the ratio 6:1. Propose a structural formula for compound R consistent with this information.

There is a hydrogen deficiency index of 0, so there are no rings or pi bonds in compound R. The simplicity of the ¹H-NMR spectrum indicates a highly level of symmetry in the molecule, with each methyl group being bonded to a carbon with a single hydrogen atom. The only structure consistent with all of this information is the following ether. The chemical shifts associated with each set of hydrogens are indicated on the structure.

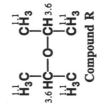

Compound R

<u>Problem 13.28</u> Following are eight structural formulas along with the ¹³C-NMR and DEPT spectral data. Given this data, assign each carbon in each compound its correct ¹³C chemical shift.

(a)

$$\overset{13.40}{C}H_3\overset{43.22}{C}H_2\overset{26.46}{C}H_2\overset{\overset{\displaystyle Br}{|}}{C}H\overset{21.00}{C}H_3$$
         51.55

¹³C	DEPT
51.55	CH
43.22	CH₂
26.46	CH₂
21.00	CH₃
13.40	CH₃

(b)

$$\overset{22.47}{C}H_3 \quad \overset{108.33}{C}H_2$$
$$\overset{30.56}{C}H_3\overset{}{C}H_2\overset{147.70}{C}=\overset{}{C}H_2$$
         12.23

¹³C	DEPT
147.70	–
108.33	CH₂
30.56	CH₂
22.47	CH₃
12.23	CH₃

(c)

$$\overset{115.26}{C}H_2=\overset{137.81}{C}H\overset{43.35}{C}H\overset{28.12}{C}H_2\overset{\overset{\displaystyle 22.26}{\overset{\displaystyle |}{C}H_3}}{C}H\overset{22.26}{C}H_3$$

¹³C	DEPT
137.81	CH
115.26	CH₂
43.35	CH₂
28.12	CH
22.26	CH₃

(d)

28.72
CH₃
|
CH₃-C-CH₂Br
| 49.02
CH₃
33.15
28.72

13C	DEPT
49.02	CH₂
33.15	-
28.72	CH₃

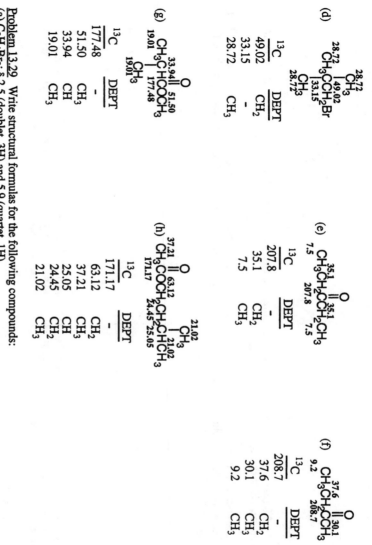

(e)

35.1   O   35.1
CH₃CH₂CCH₂CH₃
7.5 207.8 7.5

13C	DEPT
207.8	-
35.1	CH₂
7.5	CH₃

(f)

37.6   O   30.1
CH₃CH₂CCH₃
9.2   208.7

13C	DEPT
208.7	-
37.6	CH₂
30.1	CH₃
9.2	CH₃

(g)

33.94 O 51.50
CH₃CHCOCH₃
19.01
CH₃
19.01

13C	DEPT
177.48	-
51.50	CH₃
33.94	CH
19.01	CH₃

(h)

37.21   O   21.02
CH₃COCH₂CH₂CHCH₃
171.17   21.02
24.45 25.05   CH₃

13C	DEPT
171.17	-
63.12	CH₂
37.21	CH₃
25.05	CH
24.45	CH₂
21.02	CH₃

Problem 13.29 Write structural formulas for the following compounds:

(a) C₂H₄Br₂: δ 2.5 (doublet, 3H) and 5.9 (quartet, 1H)

2.5 5.9
CH₃-CHBr₂

(b) C₄H₈Cl₂: δ 1.60 (doublet, 3H), 2.15 (multiplet, 2H), 3.72 (triplet, 2H), and 4.27 (multiplet, 1H)

1.6   4.27   2.15 3.72
CH₃-CHCl-CH₂-CH₂Cl

(c) C₅H₈Br₄: δ 3.6 (singlet, 8H)

3.6
CH₂Br
| 3.6
CH₂Br-C-CH₂Br
| 3.6
CH₂Br

(d) C₄H₈O: δ 1.0 (triplet, 3H), 2.1 (singlet, 3H), and 2.4 (quartet, 2H)

1.0 2.4 O 2.1
CH₃CH₂-C-CH₃

(e) C₄H₈O₂: δ 1.2 (triplet, 3H), 2.1 (singlet, 3H) and 4.1 (quartet, 2H); contains an ester group

2.1 O
CH₃-C-O-CH₂-CH₃
4.1 1.2

(f) C₄H₈O₂: δ 1.2 (triplet, 3H), 2.3 (quartet, 2H) and 3.6 (singlet, 3H); contains an ester group

1.2 2.3 O
CH₃-CH₂-C-O-CH₃
3.6

(g) C_4H_9Br: δ 1.1 (doublet, 6H), 1.9 (multiplet, 1H), and 3.4 (doublet, 2H)

$$\overset{1.1}{CH_3} \\ \overset{1.1}{CH_3}-\overset{1.9}{CH}-\overset{3.4}{CH_2Br}$$

(h) $C_6H_{12}O_2$: δ 1.5 (singlet, 9H) and 2.0 (singlet, 3H)

$$\overset{O}{\underset{\parallel}{C}} \\ \overset{2.0}{CH_3}-C-O-\overset{1.5}{\underset{CH_3}{C}}-\overset{1.5}{CH_3} \\ \quad\quad\quad\quad\quad \overset{1.5}{CH_3}$$

(i) $C_7H_{14}O$: δ 0.9 (triplet, 6H), 1.6 (sextet, 4H), and 2.4 (triplet, 4H)

$$\overset{0.9}{CH_3}-\overset{1.6}{CH_2}-\overset{2.4}{CH_2}-\overset{O}{\underset{\parallel}{C}}-\overset{2.4}{CH_2}-\overset{1.6}{CH_2}-\overset{0.9}{CH_3}$$

(j) $C_5H_{10}O_2$: δ 1.2 (doublet, 6H), 2.0 (singlet, 3H) and 5.0 (septet, 1H)

$$\overset{O}{\underset{\parallel}{C}} \overset{5.0}{\underset{H}{|}} \\ \overset{2.0}{CH_3}-C-O-\overset{5.0}{C}-\overset{1.2}{CH_3} \\ \quad\quad\quad\quad\quad \overset{1.2}{CH_3}$$

(k) $C_5H_{11}Br$: δ 1.1 (singlet, 9H) and 3.2 (singlet, 2H)

$$\overset{1.1}{CH_3} \\ \overset{1.1}{CH_3}-\overset{}{\underset{\underset{\overset{1.1}{CH_3}}{|}}{C}}-\overset{3.2}{CH_2Br}$$

(l) $C_7H_{15}Cl$: δ 1.1 (singlet, 9H) and 1.6 (singlet, 6H)

$$\overset{1.1}{CH_3}\;\overset{1.6}{CH_3} \\ \overset{1.1}{CH_3}-\overset{}{\underset{\underset{\overset{1.1}{CH_3}}{|}}{C}}-\overset{}{\underset{\underset{\overset{1.6}{CH_3}}{|}}{C}}-Cl$$

Problem 13.30 The percent s character of carbon participating in a C-H bond can be established by measuring the ^{13}C-^1H coupling constant and using the relationship

$$\text{percent s character} = 0.2\ J(^{13}C\text{-}^1H)$$

The ^{13}C-^1H coupling constant observed for methane, for example, is 125 Hz, which gives 25% s character, the value expected for an sp^3 hybridized carbon atom.
(a) Calculate the expected ^{13}C-^1H coupling constant in ethylene and acetylene.

For ethylene and acetylene the carbon atoms are sp^2 and sp hybridized and thus 33% and 50% s character, respectively. Using the above equation gives coupling constants of 165 Hz and 250 Hz, respectively.

(b) In cyclopropane, the ^{13}C-^1H coupling constant is 160 Hz. What is the hybridization of carbon in cyclopropane?

The carbon atoms in cyclopropane are (0.2)(160) = 32% s character. This corresponds roughly to an sp^2 hybridized carbon atom.

Problem 13.31 Ascaridole is a natural product that has been used to treat intestinal worms. Explain why the two methyls on the isopropyl group in ascaridole appear as 4 lines of equal intensity, with two sets of two each separated by 7 Hz.

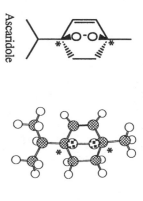

Ascaridole

Ascaridole is a chiral molecule, having two stereocenters as shown. As a result, the NMR spectrum is complex and the two methyl groups of the isopropyl groups are split. The best way to think about it is that the two methyl groups are themselves diastereotopic, thus giving rise to split signals.

Problem 13.32 The ^{13}C-NMR spectrum of 3-methyl-2-butanol shows signals at d 17.88 (CH$_3$), 18.16 (CH$_3$), 20.01 (CH$_3$), 35.04 (carbon-3), and 72.75 (carbon-2). Account for the fact that each methyl group in this molecule gives a different signal.

$$\begin{array}{c}
17.88 \\
\overset{5}{CH_3} 72.75 \\
\overset{18.16}{\underset{}{CH_3}} \underset{35.04}{\overset{4}{\underset{}{|}}} \overset{3}{CH} \overset{2}{\underset{*}{CH}} \overset{1}{\underset{20.01}{CH_3}}
\end{array}$$

Because of the stereocenter at carbon-2, the two methyl groups of carbon-4 and carbon-5 are diastereotopic and thus have two different chemical shifts. In addition, carbon-1 is not equivalent to 4 or 5, so it also has a unique chemical shift.

Humphead wrasse *Cheilinus undulatus*
Rangiroa, French Polynesia

CHAPTER 14
Solutions to the Problems

Problem 14.1 Calculate the nominal mass of each ion. Unless otherwise indicated, use the mass of the most abundant isotope of each element.

(a) $[CH_3Br]^{+\bullet}$ (b) $[CH_3\,^{81}Br]^{+\bullet}$ (c) $[^{13}CH_3Br]^{+\bullet}$

94   96   95

Problem 14.2 Propose a structural formula for the cation of m/z 41 observed in the mass spectrum of methylcyclopentane.

The cation of m/z 41 must have a molecular formula of C_3H_5. This corresponds to the stable allyl cation:

$$CH_2=CH-CH_2^+$$

Problem 14.3 The low-resolution mass spectrum of 2-pentanol shows 15 peaks. Account for the formation of the peaks at m/z 73, 70, 55, 45, 43, and 41.

2-Pentanol has a molecular formula of $C_5H_{12}O$ and thus a nominal mass of 88.

$$CH_3-CH_2-CH_2-\overset{\overset{\displaystyle OH}{|}}{C}H-CH_3$$

2-Pentanol

Loss of a methyl radical would leave a fragment of mass 73:

$$CH_3-CH_2-CH_2-\overset{\overset{\displaystyle +OH}{\|}}{C}H$$

Loss of water results in a fragment of mass 70:

$$[CH_3-CH_2-CH_2-CH=CH_2]^{+\bullet}\quad [CH_3-CH_2-CH=CH-CH_3]^{+\bullet}$$

The above alkene radical cations could then lose a methyl radical to give the following fragments of mass 55:

$$[CH_2-CH_2-CH=CH_2]^+\quad [CH_2-CH=CH-CH_3]^+$$

An alkyl radical could break off from the alcohol to give a fragment of mass 45:

$$\overset{\overset{\displaystyle +OH}{\|}}{C}H-CH_3$$

The original alcohol cation radical could have fragmented in such as way as to generate the following cation with mass 43:

$$CH_3-CH_2-CH_2^+$$

The allyl cation has a mass of 41:

$$CH_2=CH-CH_2^+$$

Problem 14.4 Draw acceptable Lewis structures for the molecular ion (radical cation) formed from the following molecules when each is bombarded by high-energy electrons in a mass spectrometer.

(a)

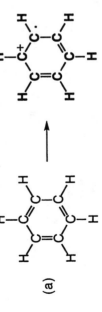

(b)

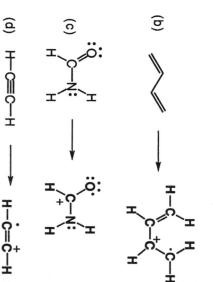

(c)

$H-C\equiv C-H \longrightarrow H-C\equiv \overset{+}{C}-H$

(d)

Problem 14.5 The molecular ion for compounds containing only C, H, and O always has an even mass-to-charge value. Why is this so? What can you say about mass-to-charge ratio of ions that arise from fragmentation of one bond in the molecular ion? From fragmentation of two bonds in the molecular ion?

Stable organic molecules composed of C, H, and O with filled valences will always have an even number of hydrogen atoms because carbon and oxygen make an even number of bonds (4 and 2, respectively). Try making some molecular formulas for yourself to verify this. This fact, combined with the even-numbered atomic weights of carbon and oxygen, guarantee an even mass-to-charge ratio of molecular ions. If one bond is broken in the molecular ion, then a cation and a radical are formed. The cation is the only species observed in the mass spectrum, and this will have an odd mass-to-charge ratio. If the molecular ion fragments so that two bonds are broken, then a new radical cation is formed, and the new radical cation will have an even mass-to-charge ratio.

Problem 14.6 For which compounds containing a heteroatom (an atom other than carbon or hydrogen) does the molecular ion have an even-numbered mass and for which does it have an odd-numbered mass?

(a) A chloroalkane of molecular formula $C_nH_{2n+1}Cl$

The molecular ion has an even-numbered mass. The C and H atoms will add up to an odd-numbered mass, since there will be an odd number of hydrogen atoms. The chlorine provides an additional odd-numbered mass as either of the two most abundant isotopes (35 and 37, respectively).

(b) A bromoalkane of molecular formula $C_nH_{2n+1}Br$

The molecular ion has an even-numbered mass. Again, the C and H atoms add up to an odd-numbered mass, and bromine provides an additional odd-numbered mass as either of the two most abundant isotopes (79 and 81, respectively).

(c) An alcohol of molecular formula $C_nH_{2n+1}OH$

The molecular ion has an even-numbered mass. There is an even number of hydrogen atoms, so the C and H atoms add up to an even-numbered mass. The most abundant isotope of oxygen (16) is also even, so the entire molecular ion has an even-numbered mass.

(d) A primary amine of molecular formula $C_nH_{2n+1}NH_2$.

The molecular ion has an odd-numbered mass. These compounds have an odd number of hydrogens, so the C and H atoms add up to an odd-numbered mass. The nitrogen contributes an even-numbered mass (14) so the molecular ion will have an odd-numbered mass.

(e) A thiol of molecular formula $C_nH_{2n+1}SH$.

The molecular ion has an even-numbered mass. These compounds have an even number of hydrogens, so the C and H atoms add up to an even-numbered mass. The sulfur contributes an even-numbered mass (32 or 34) so the entire molecular ion has an even-numbered mass.

Problem 14.7 The so-called nitrogen rule states that if a compound has an odd number of nitrogen atoms, the value of m/z for its molecular ion will be an odd number. Why is this so?

Nitrogen atoms have an even-numbered mass, but one lone pair of electrons in the neutral state. Thus, they make an odd number of bonds, namely three, to other atoms. As a result, compounds with with an odd number of nitrogens in the neutral state will have an odd-numbered m/z ratio for the molecular ion.

Problem 14.8 Both $C_6H_{10}O$ and C_7H_{14} have the same nominal mass, namely 98. Show how these compounds can be distinguished by the m/z ratio of their molecular ions in high resolution mass spectrometry.

The values in Table 14.1 can be used to calculate precise masses for each molecular ion.

$C_6H_{10}O$: 6(12.0000) + 10(1.00783) + 15.9949 = 98.0732

C_7H_{14}: 7(12.0000) + 14(1.00783) = 98.1096

A high resolution mass spectrum will distinguish these compounds based on this small difference in the expected m/z ratio of the molecular ions.

Problem 14.9 Show how the compounds of molecular formula C_6H_9N and C_5H_5NO can be distinguished by the m/z ratio of their molecular ions in high resolution mass spectrometry.

The values in Table 14.1 can be used to calculate precise masses for each molecular ion.

C_6H_9N: 6(12.0000) + 9(1.00783) + 14.0031 = 95.0736

C_5H_5NO: 5(12.0000) + 5(1.00783) + 14.0031 + 15.9949 = 95.0372

A high resolution mass spectrum will distinguish these compounds based on this small difference in the expected m/z ratio of the molecular ions.

Problem 14.10 What rule would you expect for the m/z values of fragment ions resulting from the cleavage of one bond in a compound with an odd number of nitrogen atoms?

As stated in Problem 14.7, molecules having an odd number of nitrogen atoms have an odd-numbered m/z ratio for the molecular ion. Cleavage of one bond of a molecular ion would generate a cation and a radical, only the cation of which would be observed in the mass spectrum. If the cation fragment retained an odd number of nitrogen atoms, then the fragment would have an even m/z ratio. If the cation fragment contained an even number of nitrogen atoms (or zero), then the fragment would have an odd m/z ratio.

Problem 14.11 Determine the probability of the following in a natural sample of ethane.

(a) One carbon in an ethane molecule is ^{13}C?

The relative abundance of ^{13}C listed in Table 14.1 is 1.11 atoms of ^{13}C for every 100 atoms of ^{12}C. This corresponds to a probability that any one atom is a ^{13}C atom of 1.11/(100 + 1.11) = 0.0110. There are two ways to have this situation since either of the carbon atoms could be the ^{13}C so the final answer is 2(0.0110) = 0.0220. Converting to percent by multiplying by 100 means that there is a 2.20% chance that one of the carbon atoms in ethane is ^{13}C.

(b) Both carbons in an ethane molecule are ^{13}C?

The probability that both of the two carbon atoms in ethane is ^{13}C is $(0.0110)^2 = 1.21 \times 10^{-4}$. Converting to percent by multiplying by 100 means that there is a 0.012% chance that both of the carbon atoms in ethane is ^{13}C.

(c) Two hydrogens in an ethane molecule are replaced by deuterium atoms?

The relative abundance of ^2H listed in Table 14.1 is 0.016 atoms of ^2H for every 100 atoms of ^1H. This corresponds to a probability of 0.016/(100 + 0.016) = 1.6 x 10^{-4} that any given H atom is a ^2H atom. The probability for having two ^2H (1.6 x 10^{-4})^2 = 2.56 x 10^{-8}. Converting to percent by multiplying by 100 means that there is a 2.56 x 10^{-6}% chance that two of the hydrogen atoms in ethane are ^2H.

Problem 14.12 The molecular ions of both $C_5H_{10}S$ and $C_6H_{14}O$ appear at m/z 102 in low-resolution mass spectrometry. Show how determination of the correct molecular formula can be made from the appearance and relative intensity of the (M + 2) peak of each compound.

As shown in Table 14.1, ^{16}O occurs in greater than 99.7% abundance, so no (M + 2) peak will be observable for the case of $C_6H_{14}O$. On the other hand, sulfur has one isotope, ^{32}S, that is 95% abundant and another isotope 2 amu higher, namely ^{34}S, that has an abundance of 4.2%.

Thus, if the low-resolution mass spectrum has an (M + 2) peak that is 4.2% the height of the molecular ion peak, the compound must be $C_5H_{10}S$.

Problem 14.13 In Section 14.3, we saw several examples of fragmentation of molecular ions to give resonance-stabilized cations. Make a list of these resonance-stabilized cations, and write important contributing structures of each. Estimate the relative importance of the contributing structures in each set.

The major contributor of the propargyl cation has the positive charge on the less-electronegative sp² hybridized carbon atom.

H₂C=CH—$\overset{+}{C}$H₂ ⟷ H₂$\overset{+}{C}$—CH=CH₂   HC≡C—$\overset{+}{C}$H₂ ⟷ H$\overset{+}{C}$=C=CH₂
Equal Contributor   Equal Contributor   Major Contributor   Minor Contributor

      Allyl cation   Propargyl cation

The more stable contributor of the oxonium ion is predicted to be the one that has the positive charge on oxygen due to having filled valence shells on both carbon and oxygen.

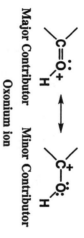

Major Contributor   Minor Contributor

Oxonium ion

Problem 14.14 Carboxylic acids often give a strong fragment ion at m/z (M - 17). What is the likely structure of this cation? How might it be formed? Show by drawing contributing structures that this cation is stabilized by resonance.

Loss of 17 results from losing the -OH group of the carboxylic acid. This produces an acylium ion:

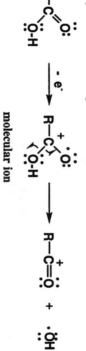

These ions are probably derived from fragmentation of the molecular ion as follows:

$$\text{R—C} \overset{\displaystyle \ddot{O}\!:}{\underset{\displaystyle \ddot{O}\text{-H}}{}} \quad \xrightarrow{-\,e^-} \quad \text{R—}\overset{+}{\text{C}} \overset{\displaystyle \ddot{O}\!:}{\underset{\displaystyle \ddot{O}\text{-H}}{}} \quad \longrightarrow \quad \text{R—}\overset{+}{\text{C}}{=}\ddot{O}\!: \; + \; \cdot\ddot{O}\text{H}$$

molecular ion

Problem 14.15 For primary amines with no branching on the carbon bearing the nitrogen, the base peak occurs at m/z 30. What cation does this peak represent ? How is it formed? Show by drawing contributing structures that this cation is stabilized by resonance.

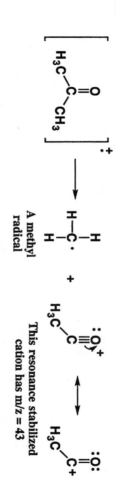

A radical

This resonance stabilized cation has m/z = 30

Problem 14.16 The base peak in the mass spectrum of propanone (acetone) occurs at m/z 43. What cation does this peak represent?

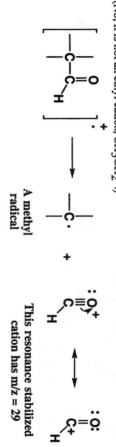

A methyl radical

This resonance stabilized cation has m/z = 43

Problem 14.17 A characteristic peak in the mass spectrum of most aldehydes occurs at m/z 29. What cation does this peak represent? (No, it is not an ethyl cation, CH3CH2⁺)

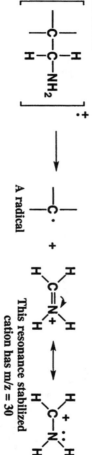

A methyl radical

This resonance stabilized cation has m/z = 29

Problem 14.18 Predict the relative intensities of the M and M + 2 peaks for

The following are based on the values in Table 14.1.

(a) CH_3CH_2Cl (b) CH_3CH_2Br

100 : 32.5   100 : 98

(c) $BrCH_2CH_2Br$

We will assume that the M + 2 peak is due to the ^{81}Br isotope. To get a mass of M + 2 there must be only one ^{81}Br present. Using the formula given in the text: %(M + 2) = (98)(2) = 196. Thus 100 : 196 are the expected relative intensities.

(d) CH_3CH_2SH

100 : 4.40

Problem 14.19 The mass spectrum of Compound A shows the molecular ion at m/z 85, an M + 1 peak at m/z 86 of approximately 6% abundance relative to M, and an M + 2 peak at m/z 87 of less than 0.1% abundance relative to M.
(a) Propose a molecular formula for Compound A.

The odd mass indicates an odd number of nitrogen atoms. Assuming only C, H, and N atoms, the 6% abundance of an M + 1 peak is primarily due to ^{13}C. Therefore guess 5 carbon atoms and 1 nitrogen atom giving a molecular formula of $C_5H_{11}N$. Just as a check, we can estimate the relative abundance of the M + 1 peak as %(M + 1) = (1.11)(5) + (0.016)(11) + (0.38)(1) = 6.1%.

(b) Draw at least 10 possible structural formulas for this molecular formula.

This molecular formula contains either one pi bond or one ring. The following structures are representative examples out of the large number of possible stable molecules that are consistent with this formula.

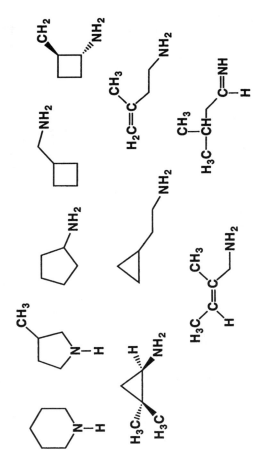

Problem 14.20 The mass spectrum of Compound B, a colorless liquid, shows these peaks in its mass spectrum. Determine the molecular formula of Compound B and propose a structural formula for it.

m/z	Relative Abundance
43	100 (base)
78	23.6 (M)
79	1.00
80	7.55
81	0.25

The ratio of the M to M + 2 peaks is 23.6/7.55 = 3.13 to 1. This is approximately the same ratio as seen with ^{35}Cl to ^{37}Cl, so there must be a chlorine atom in the molecule. The base peak at 43 corresponds to either a propyl cation or isopropyl cation. Therefore, the molecular formula for compound E must be C_3H_7Cl. The two structures consistent with this formula are:

$$CH_3CH_2CH_2Cl \qquad CH_3\overset{\overset{\displaystyle Cl}{|}}{C}HCH_3$$

Problem 14.21 Write molecular formulas for the five possible molecular ions of m/z 88 containing the elements C, H, N, and O.

$C_3H_4O_3 = 36 + 4 + 48 = 88$ $C_3H_8N_2O = 36 + 8 + 28 + 16 = 88$ $C_4H_8O_2 = 48 + 8 + 32 = 88$
$C_4H_{12}N_2 = 48 + 12 + 28 = 88$ $C_5H_{12}O = 60 + 12 + 16 = 88$

Problem 14.22 Write molecular formulas for the five possible molecular ions of m/z 100 containing only the elements C, H, N, and O.

Because of the nitrogen rule, there must be an even number of nitrogen atoms or no nitrogen atoms in the molecular formulas.

$C_4H_8N_2O = 48 + 8 + 28 + 16 = 100$ $C_5H_8O_2 = 60 + 8 + 32 = 100$ $C_5H_{12}N_2 = 60 + 12 + 28 = 100$
$C_6H_{12}O = 72 + 12 + 16 = 100$ $C_7H_{16} = 84 + 16 = 100$

Problem 14.23 The molecular ion in the mass spectrum of 2-methyl-1-pentene appears at m/z 84. Propose structural formulas for the prominent peaks at m/z 69, 55, 41, and 29.

$$\underset{\text{2-Methyl-1-pentene}}{CH_2{=}\overset{\overset{\displaystyle CH_3}{|}}{C}{-}CH_2{-}CH_2{-}CH_3}$$

The peak at m/z 69 results from loss of a methyl radical:

$$CH_2{=}\overset{+}{C}{-}CH_2{-}CH_2{-}CH_3$$

The peak at m/z 55 results from loss of an ethyl group to generate an allylic cation:

$$CH_2{=}\overset{\overset{\displaystyle CH_3}{|}}{\overset{+}{C}}{-}CH_2 \qquad \text{or} \qquad CH_2{=}\overset{\overset{\displaystyle CH_3}{|}}{C}{-}CH_2{-}CH_2{-}\overset{+}{CH_2}$$

The peak at m/z 41 corresponds to the allyl cation:

$$CH_2{=}CH{-}\overset{+}{CH_2}$$

The peak at m/z 29 corresponds to the ethyl cation:

$$CH_3{-}\overset{+}{CH_2}$$

Problem 14.24 Following is the mass spectrum of 1,2-dichloroethane.
(a) Account for the appearance of an (M + 2) peak with approximately two-thirds the intensity of the molecular ion peak.

The relative abundance of ^{37}Cl is 32.5 (Table 14.1) so it makes sense that the M + 2 peak observed in 1,2-dichloroethane is 2(32.5) = 65%.

(b) Predict the intensity of the (M + 4) peak.

The probability of having a ^{37}Cl at both Cl atoms is $(32.5/100)^2 = 0.105$, which corresponds to 10.5% of the molecular ion peak.

(c) Propose structural formulas for the cations of m/z 64, 63, 62, 51, 49, 27, and 26.

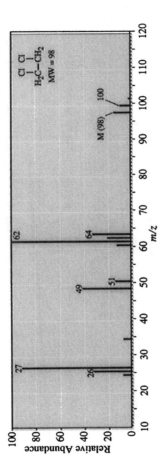

CH₂Cl-CH₂Cl
1,2-Dichloroethane

The peaks at m/z 64, 63, and 62 correspond to the following structures:

$$\left[^{37}Cl-CH=CH_2 \right]^{+\cdot} \quad \left[Cl-{}^{13}CH=CH_2 \right]^{+\cdot} \quad or \quad \left[Cl-CH={}^{13}CH_2 \right]^{+\cdot}$$

$m/z = 64$ $\qquad\qquad$ $m/z = 63$

$$\left[^{35}Cl-CH=CH_2 \right]^{+\cdot}$$

$m/z = 62$

The peaks at m/z 51 and 49 correspond to the following structures:

$$^{37}Cl-CH_2^+ \qquad\qquad ^{35}Cl-CH_2^+$$

$m/z = 51$ $\qquad\qquad$ $m/z = 49$

The peaks at m/z 27 and 26 correspond to the following structures:

$$^+CH=CH_2 \qquad\qquad \left[HC\equiv CH \right]^{+\cdot}$$

$m/z = 27$ $\qquad\qquad$ $m/z = 26$

<u>Problem 14.25</u> Following is the mass spectrum of 1-bromobutane.
(a) Account for the appearance of the (M + 2) peak of approximately 95% of the intensity of the molecular ion peak.

As detailed in Table 14.1, the most abundant isotope of bromine has a mass of 79 amu. Bromine has another isotope with a mass of 81 amu that has a natural abundance of 49.31%. The (M+ 2) peak results from the presence of a ⁸¹Br atom in the 1-bromobutane molecule.

(b) Propose structural formulas for the cations of m/z 57, 41, and 29.

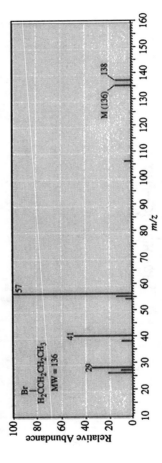

The peak at m/z 57 results from loss of a bromine atom:

$$\overset{+}{CH_2}\text{-}CH_2\text{-}CH_2\text{-}CH_3$$

The peak at m/z 41 corresponds to the allyl cation:

$$CH_2\text{=}CH\text{-}\overset{+}{CH_2}$$

The peak at m/z 29 corresponds to the ethyl cation:

$$CH_3\text{-}\overset{+}{CH_2}$$

<u>Problem 14.26</u> Following is the mass spectrum of bromocyclopentane. The molecular ion m/z 148 is of such low intensity that it does not appear in this spectrum. Assign structural formulas for the cations of m/z 69 and 41.

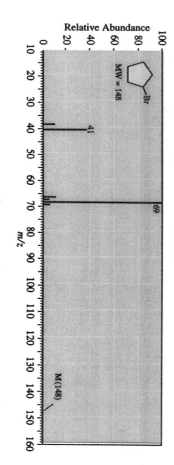

The peak at m/z 69 corresponds to the loss of the bromine atom to give the cyclopentyl cation:

The peak at m/z 41 corresponds to the allyl cation:

$$CH_2\text{=}CH\text{-}\overset{+}{CH_2}$$

<u>Problem 14.27</u> Following is the mass spectrum of 3-methyl-2-butanol. The molecular ion m/z 88 does not appear in this spectrum. Propose structural formulas for the cations of m/z 45, 43, and 41.

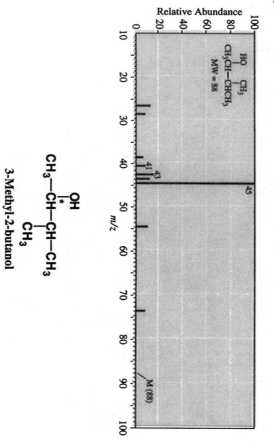

3-Methyl-2-butanol

The peak at m/z 45 corresponds to the following resonance stabilized cation:

$$CH_3-CH=\overset{+}{\underset{H}{O}}H$$

The peak at m/z 43 corresponds to the isopropyl cation:

$$CH_3-\overset{+}{\underset{CH_3}{CH}}$$

The peak at m/z 41 corresponds to the allyl cation:

$$CH_2=CH-\overset{+}{C}H_2$$

<u>Problem 14.28</u> The following is the mass spectrum of a compound C, C_3H_8O. Compound C is infinitely soluble in water, undergoes reaction with sodium metal with the evolution of a gas, and undergoes reaction with thionyl chloride to give a water-insoluble chloroalkane. Propose a structural formula for compound C, and write equations for each of its reactions.

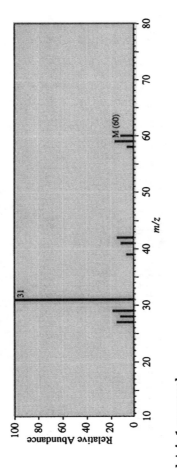

Compound A is 1-propanol.

$$CH_3-CH_2-CH_2OH$$
Compound C

In the mass spectrum, the peak at m/z 31 could only have come from 1-propanol due to the following fragmentation.

$$CH_3-CH_2-\overset{OH}{\underset{|}{CH_2}} \longrightarrow CH_3CH_2\cdot + \overset{OH}{\underset{|}{^+CH_2}}$$
$$m/z\ 31$$

The reaction with sodium liberates molecular hydrogen:

$$2\ CH_3-CH_2-CH_2OH\ +\ 2Na\ \longrightarrow\ 2\ CH_3-CH_2-CH_2-O^-\ Na^+\ +\ H_2$$

The reaction with thionyl chloride produces 1-chloropropane, SO_2, and H-Cl.

$$CH_3-CH_2-CH_2OH\ +\ SOCl_2\ \longrightarrow\ CH_3-CH_2-CH_2Cl\ +\ SO_2\ +\ H-Cl$$

<u>Problem 14.29</u> Following are mass spectra for the constitutional isomers 2-pentanol and 2-methyl-2-butanol. Assign each isomer its correct spectrum.

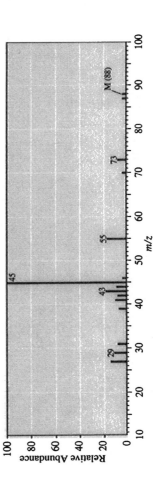

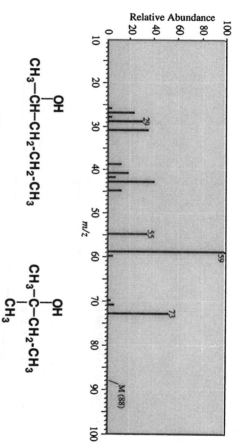

The main difference in the above spectra is the base peak at 45 in the first spectrum that is only about 10% of the base peak in the second spectrum. This peak corresponds to the following resonance stabilized cation:

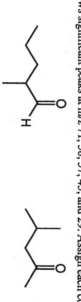

$$CH_3-\overset{+}{\underset{||}{O}H} \longleftrightarrow CH_3-CH\overset{..}{\underset{+}{O}H}$$

This species can be readily produced by loss of the propyl group from 2-pentanol. It is unlikely that 2-methyl-2-butanol could produce such a cation, thus the upper mass spectrum is that of 2-pentanol and the lower mass spectrum is that of 2-methyl-2-butanol. Also, the base peak in the second spectrum has a m/z of 59 that corresponds to the following fragmentation.

$$\underset{\underset{CH_3}{|}}{\overset{\overset{OH}{|}}{CH_3-C-CH_2-CH_3}} \longrightarrow \underset{\underset{CH_3}{|}}{\overset{\overset{+OH}{||}}{CH_3-C}} + \cdot CH_2CH_3$$

Problem 14.30 2-Methylpentanal and 4-methyl-2-pentanone are constitutional isomers of molecular formula $C_6H_{12}O$. Each shows a molecular ion peak in its mass spectrum at m/z 100. Spectrum A shows significant peaks at m/z 85, 58, 57, 43, and 42. Spectrum B shows significant peaks at m/z 71, 58, 57, 43, and 29. Assign each compound its correct spectrum.

2-Methylpentanal   4-Methyl-2-pentanone

Some expected α-cleavage and McClafferty rearrangement fragments can be predicted as follows.

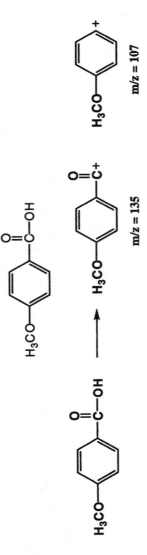

m/z = 85
(α-Cleavage)

m/z = 58
(McClafferty)

m/z = 43
(α-Cleavage)

The α-cleavage fragments at m/z = 85 (Spectrum A) and 29 (Spectrum B) are particularly diagnostic and serve to verify that Spectrum A corresponds to 4-methyl-2-pentanone, and Spectrum B corresponds to 2-methylpentanal.

Problem 14.31 Account for the presence of peaks at m/z 135 and 107 in the mass spectrum of 4-methoxybenzoic acid (*p*-anisic acid).

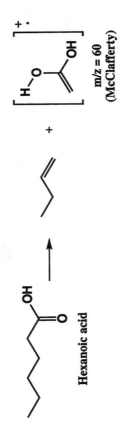

m/z = 45
(α-Cleavage)

m/z = 135

m/z = 107

Problem 14.32 Account for the presence of the following peaks in the mass spectrum of hexanoic acid.
(a) m/z 60

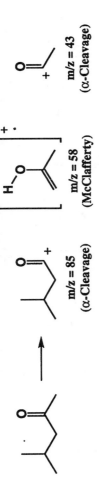

Hexanoic acid

m/z = 60
(McClafferty)

(b) a series of peaks differing by 14 amu at m/z 45, 59, 73, and 87

The key to identifying this series of fragments is that the difference between them, namely 14 amu, corresponds to CH₂ groups. The series starts with the m/z 45 fragment that results from α-cleavage.

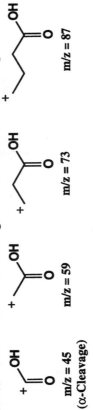

m/z = 45
(α-Cleavage)

m/z = 59

m/z = 73

m/z = 87

(c) a series of peaks differing by 14 amu at m/z 29, 43, 57, and 71

The key to identifying this series of fragments is once again that the difference between them, namely 14 amu corresponds to CH₂ groups. The series starts with the m/z 29 ethyl cation.

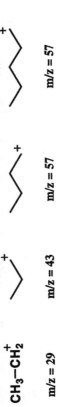

CH₃—CH₂⁺

m/z = 29

m/z = 43

m/z = 57

m/z = 57

Problem 14.33 All methyl esters of long-chain aliphatic acids (for example, methyl tetradecanoate, $C_{13}H_{27}COOCH_3$) show significant fragment ions at m/z 74, 59, and 31. What are the structures of these ions? How are they formed?

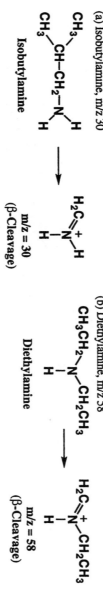

$$R-C\overset{O}{\underset{OCH_3}{\Vert}} \longrightarrow$$

m/z = 74
(McClafferty)

m/z = 59
(α-Cleavage)

m/z = 31
(α-Cleavage)

Problem 14.34 Propylbenzene, $C_6H_5CH_2CH_2CH_3$, and isopropyl benzene, $C_6H_5CH(CH_3)_2$, are constitutional isomers of molecular formula C_9H_{12}. One of these compounds shows prominent peaks in its mass spectrum at m/z 120 and 105. The other shows prominent peaks at m/z 120 and 91. Which compound has which spectrum?

The peak at m/z 120 is the parent ion in both cases. The peak at m/z 105 corresponds to the methyltropylium ion that can be most easily formed from isopropyl benzene following the loss of a single methyl group. The m/z 91 peak is the tropylium ion that can most easily formed from propylbenzene following loss of an ethyl group.

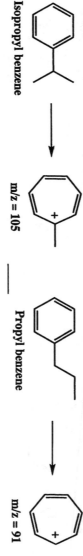

Isopropyl benzene   m/z = 105

Propyl benzene   m/z = 91

Problem 14.35 Account for the formation of the base peaks in these mass spectra.
(a) Isobutylamine, m/z 30   (b) Diethylamine, m/z 58

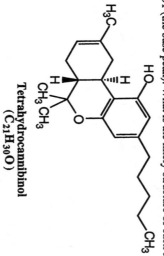

Isobutylamine

$$H_2C\overset{+}{=}\underset{H}{N}-H$$

m/z = 30
(β-Cleavage)

$CH_3CH_2-\underset{H}{N}-CH_2CH_3$

Diethylamine

$$H_2C\overset{+}{=}\underset{H}{N}-CH_2CH_3$$

m/z = 58
(β-Cleavage)

Problem 14.36 Because of the sensitivity of mass spectrometry, it is often used to detect the presence of drugs in blood, urine, or other biological fluids. Tetrahydrocannibinol (nominal mass 314), a component of marijuana, exhibits two strong fragment ions at m/z 246 and 231 (the base peak). What is the likely structure of each ion?

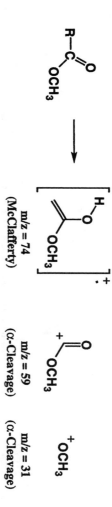

Tetrahydrocannibinol
($C_{21}H_{30}O$)

The peak at m/z 246 is probably the result of a reverse Diels-Alder reaction as described in section 14.4B. In this case, the cyclohexene ring in the top left of the structure is split from the rest of the molecule to leave the radical cation shown below. The peak at m/z 231 results from loss of a methyl group from the m/z 246 radical cation to give a tertiary cation as shown.

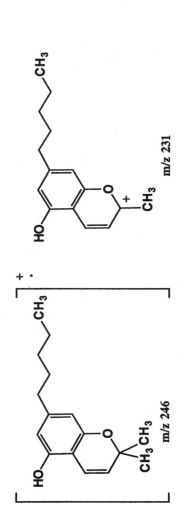

Problem 14.37 Electrospray mass spectrometry is a recently developed technique for looking at large molecules with a mass spectrometer. In this technique, molecular ions, each associated with one or more H+ ions, are prepared under mild conditions in the mass spectrometer. As an example, a protein (P) with a molecular weight of 11,812 gives clusters of the type $(P + 8H)^{8+}$, $(P + 7H)^{7+}$, and $(P + 6H)^{6+}$. At what mass-to-charge values do these three clusters appear in the mass spectrum?

The key to this question is to notice that these ions have multiple charges. Because ions are recorded in a mass spectrum according to their mass divided by their total charge, the mass-to-charge values for these ions are calculated by dividing their total mass (11,812 + the number of protons) by their total charge (8, 7, and 6, respectively). The final answers are thus $(11,812 + 8)/8 = \underline{1477.5 \text{ m/z}}$, $(11,812 + 7)/7 = \underline{1688.4 \text{ m/z}}$, and $(11,812 + 6)/6 = \underline{1969.7 \text{ m/z}}$.

Moorish idol *Zanclus cornutus*
Kona coast, Hawaii

CHAPTER 15
Solutions to the Problems

Problem 15.1 Explain how these Grignard reagents will react with molecules of their own kind to "self-destruct".

Grignard reagents are strong bases. In both cases the Grignard reagent will "self-destruct" due to an acid-base reaction.

(a) HOCH₂CH₂CH₂MgBr

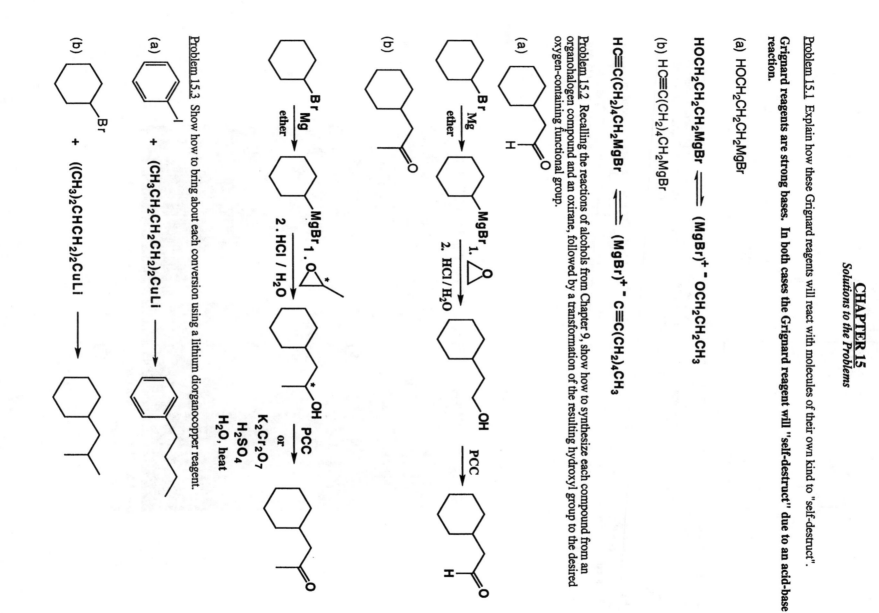

(b) HC≡C(CH₂)₄CH₂MgBr

HOCH₂CH₂CH₂MgBr ⇌ (MgBr)⁺ ⁻ OCH₂CH₂CH₃

(b) HC≡C(CH₂)₄CH₂MgBr

HC≡C(CH₂)₄CH₂MgBr ⇌ (MgBr)⁺ ⁻ C≡C(CH₂)₄CH₃

Problem 15.2 Recalling the reactions of alcohols from Chapter 9, show how to synthesize each compound from an organohalogen compound and an oxirane, followed by a transformation of the resulting hydroxyl group to the desired oxygen-containing functional group.

(a)

(b)

Problem 15.3 Show how to bring about each conversion using a lithium diorganocopper reagent.

(a)

(b)

Problem 15.4 Show how to prepare each Gilman reagent in Example 15.4 from an appropriate alkyl halide and each epoxide from an appropriate alkene.

$$CH_3CH_2Br \xrightarrow{2Li} CH_3CH_2Li + LiBr$$

$$2\ CH_3CH_2Li \xrightarrow{CuI} [(CH_3CH_2)_2Cu]Li + LiI$$

or alternatively,

$$C_6H_5CH_2Br \xrightarrow{2Li} C_6H_5CH_2Li + LiBr$$

$$2\ C_6H_5CH_2Li \xrightarrow{CuI} [(C_6H_5CH_2)_2Cu]Li + LiI$$

Problem 15.5 Show how you might prepare each compound by a Heck reaction using methyl 2-propenoate as the starting alkene.

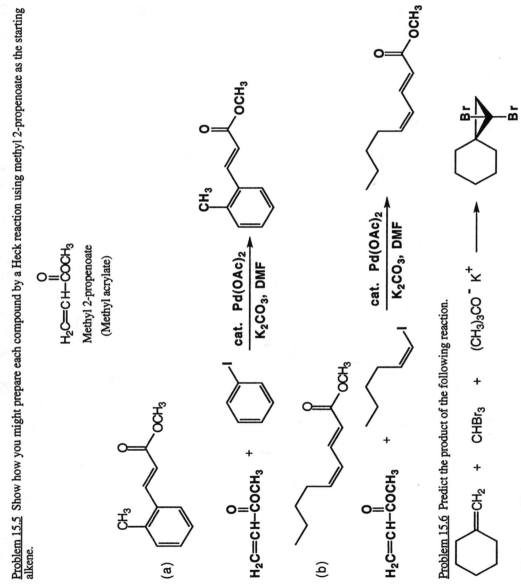

(a)

(b)

Problem 15.6 Predict the product of the following reaction.

Problem 15.7 Show how the following compound could be prepared from any compound containing ten carbons or fewer.

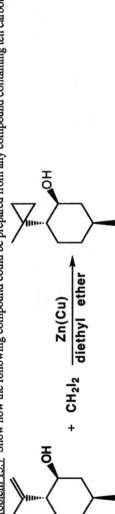

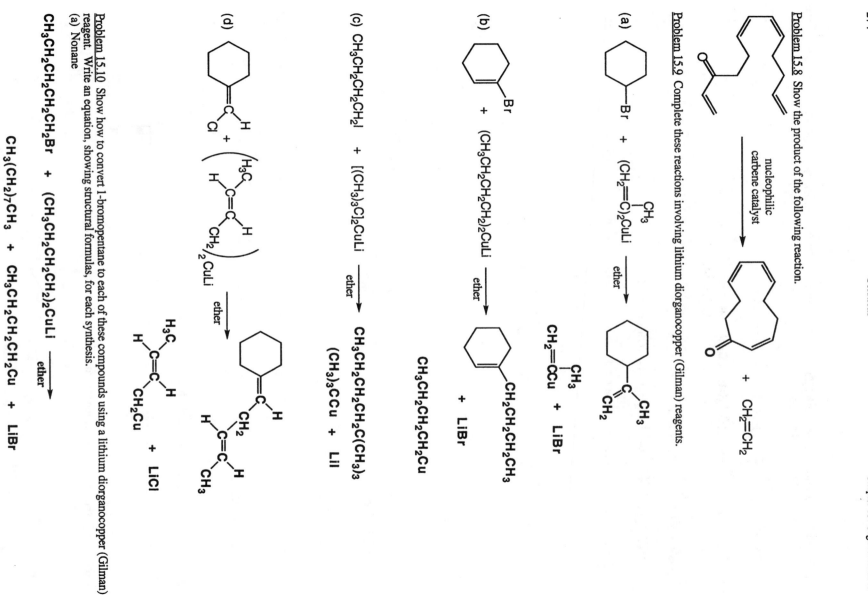

Problem 15.8 Show the product of the following reaction.

Problem 15.9 Complete these reactions involving lithium diorganocopper (Gilman) reagents.

Problem 15.10 Show how to convert 1-bromopentane to each of these compounds using a lithium diorganocopper (Gilman) reagent. Write an equation, showing structural formulas, for each synthesis.
(a) Nonane

$$CH_3CH_2CH_2CH_2CH_2Br + (CH_3CH_2CH_2CH_2CH_2)_2CuLi \xrightarrow{ether}$$

$$CH_3(CH_2)_7CH_3 + CH_3CH_2CH_2CH_2CH_2Cu + LiBr$$
Nonane

(b) 3-Methyloctane

$$CH_3CH_2CH_2CH_2CH_2Br \ + \ \left(\underset{CH_3CH_2\overset{*}{C}H}{\overset{CH_3}{|}} \right)_2 CuLi \ \xrightarrow{\text{ether}}$$

$$\underset{\text{3-Methyloctane}}{CH_3CH_2\overset{\overset{\overset{CH_3}{|}}{*}}{C}HCH_2CH_2CH_2CH_2CH_3} \ + \ CH_3CH_2\overset{\overset{\overset{CH_3}{|}}{*}}{C}HCu \ + \ LiBr$$

Note how the 3-methyloctane has one stereocenter, so a racemic mixture will be produced assuming the Gilman reagent starting material is racemic.

(c) 2,2-Dimethylheptane

$$CH_3CH_2CH_2CH_2CH_2Br \ + \ [(CH_3)_3C]_2CuLi \ \xrightarrow{\text{ether}}$$

$$\underset{\text{2,2-Dimethylheptane}}{\overset{\overset{CH_3}{|}}{\underset{\overset{|}{CH_3}}{CH_3\overset{}{C}CH_2CH_2CH_2CH_2CH_3}}} \ + \ (CH_3)_3CCu \ + \ LiBr$$

(d) 1-Heptene

$$CH_3CH_2CH_2CH_2CH_2Br \ + \ (CH_2{=}CH)_2CuLi \ \xrightarrow{\text{ether}}$$

$$\underset{\text{1-Heptene}}{CH_2{=}CHCH_2CH_2CH_2CH_2CH_3} \ + \ CH_2{=}CHCu \ + \ LiBr$$

(e) 1-Octene

$$CH_3CH_2CH_2CH_2CH_2Br \ + \ (CH_2{=}CHCH_2)_2CuLi \ \xrightarrow{\text{ether}}$$

$$\underset{\text{1-Octene}}{CH_2{=}CHCH_2CH_2CH_2CH_2CH_2CH_3} \ + \ CH_2{=}CHCH_2Cu \ + \ LiBr$$

Problem 15.11 In Problem 15.10, you used a series of lithium diorganocopper (Gilman) reagents. Show how to prepare each Gilman reagent from an appropriate alkyl or vinylic halide.

(a) $2 \ CH_3CH_2CH_2CH_2X \ \xrightarrow{2 \ Li} \ \xrightarrow[\text{ether}]{CuI} \ (CH_3CH_2CH_2CH_2)_2CuLi \ + \ LiX$

(b) $2 \ CH_3CH_2\overset{\overset{CH_3}{|}}{\underset{\overset{|}{X}}{\overset{*}{C}H}} \ \xrightarrow{2 \ Li} \ \xrightarrow[\text{ether}]{CuI} \ \left(CH_3CH_2\overset{\overset{CH_3}{|}}{\overset{*}{C}H} \right)_2 CuLi \ + \ LiX$

(c) $2 \ (CH_3)_3CX \ \xrightarrow{2 \ Li} \ \xrightarrow[\text{ether}]{CuI} \ [(CH_3)_3C]_2CuLi \ + \ LiX$

(d) $2 \ \underset{H}{\overset{H}{{>}}}C{=}C\underset{X}{\overset{H}{{<}}} \ \xrightarrow{2 \ Li} \ \xrightarrow[\text{ether}]{CuI} \ (CH_2{=}CH)_2CuLi \ + \ LiX$

(e) $2 \ \underset{H}{\overset{H}{{>}}}C{=}C\underset{CH_2X}{\overset{H}{{<}}} \ \xrightarrow{2 \ Li} \ \xrightarrow[\text{ether}]{CuI} \ (CH_2{=}CHCH_2)_2CuLi \ + \ LiX$

Problem 15.12 Show how to prepare each compound from the given starting compound through the use of a lithium diorganocopper (Gilman) reagent.

(a) 4-Methylcyclopentene from 4-bromocyclopentene

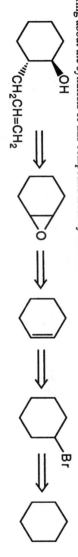

4-Bromocyclo-
pentene

+ (CH₃)₂CuLi $\xrightarrow{\text{ether}}$

CH₃

4-Methylcyclo-
pentene

+ CH₃Cu

+ LiBr

(b) (Z)-2-Undecene from (Z)-1-bromopropene

$$\begin{array}{c} H \quad\quad H \\ C=C \\ H_3C \quad\quad Br \end{array}$$

(Z)-1-Bromo-
propene

+ [CH₃(CH₂)₆CH₂]₂CuLi $\xrightarrow{\text{ether}}$

$$\begin{array}{c} H \quad\quad H \\ C=C \\ H_3C \quad\quad CH_2(CH_2)_6CH_3 \end{array}$$

(Z)-2-Undecene

CH₃(CH₂)₆CH₂Cu

+

LiBr

(c) 1-Butylcyclohexene from 1-iodocyclohexene

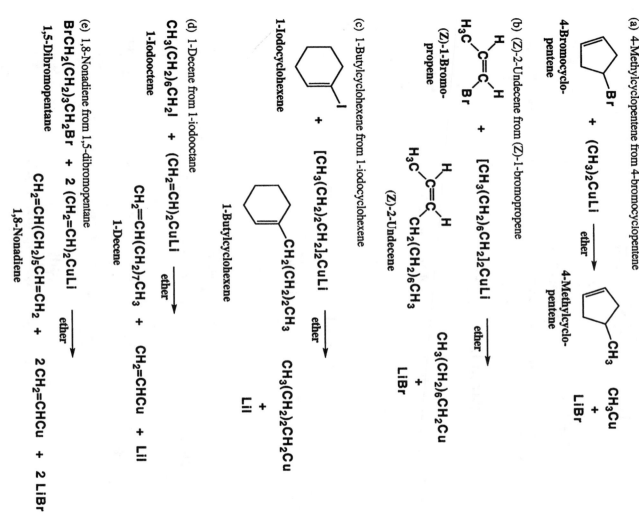

1-Iodocyclohexene

+ [CH₃(CH₂)₂CH₂]₂CuLi $\xrightarrow{\text{ether}}$

CH₂(CH₂)₂CH₃

1-Butylcyclohexene

CH₃(CH₂)₂CH₂Cu

+

LiI

(d) 1-Decene from 1-iodooctane

CH₃(CH₂)₆CH₂I + (CH₂=CH)₂CuLi $\xrightarrow{\text{ether}}$

1-Iodooctane

CH₂=CH(CH₂)₇CH₃ + CH₂=CHCu + LiI

1-Decene

(e) 1,8-Nonadiene from 1,5-dibromopentane

BrCH₂(CH₂)₃CH₂Br + 2 (CH₂=CH)₂CuLi $\xrightarrow{\text{ether}}$

1,5-Dibromopentane

CH₂=CH(CH₂)₅CH=CH₂ + 2 CH₂=CHCu + 2 LiBr

1,8-Nonadiene

Problem 15.13 The following is a retrosynthetic scheme for the preparation of *trans*-2-allylcyclohexanol. Show reagents to bring about the synthesis of this compound from cyclohexane.

OH
CH₂CH=CH₂

⟹ O ⟹ ⟹ Br ⟹

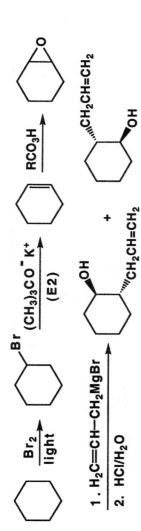

Note that in this reaction scheme two enantiomers are formed as a racemic mixture. Also, a Gilman reagent could have been used to react with the epoxide in the last step.

Problem 15.14   Complete these equations.

(a) $CH_3CH_2CH_2C\equiv CH$ + CH_3CH_2MgBr $\xrightarrow{\text{diethyl ether}}$ $CH_3CH_2CH_2C\equiv C^- \, MgBr^+$ + CH_3CH_3

(b)

(c)   + $CHBr_3$ + $(CH_3)_3CO^- K^+$ ⟶

(d)

(e)

Note that in (b) and (d) racemic mixtures are produced.

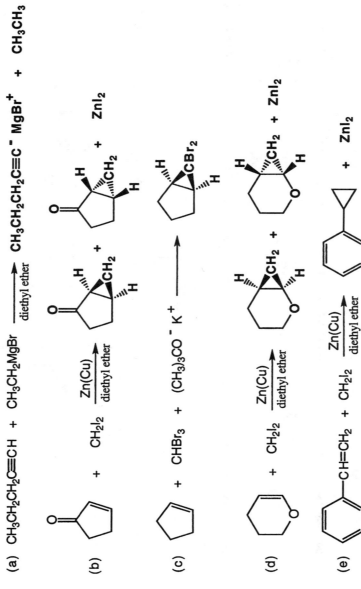

Problem 15.15   Reaction of following cycloalkene with the Simmons-Smith reagent is stereospecific and gives only the isomer shown. Suggest a reason for this stereospecificity.

Recall that the reaction mechanism of the Simmons-Smith reaction is formation of the following Zn species.

As shown in the following three-dimensional models, the -OH group is suspended above one face of the double bond.

The Zn atom of the reagent acts as a Lewis acid and forms a weak complex with the Lewis base oxygen atom of the -OH group. This positions the reagent above the double bond and directs attack of the carbenoid from the top face, leading to the observed product.

Zn atom coordinates to the O atom

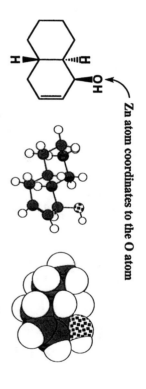

Problem 15.16 Show how the following compound can be prepared in good yield.

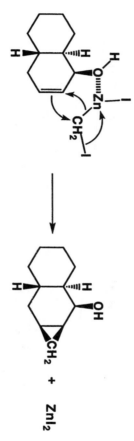

The product will be obtained in high yield from reacting the corresponding alkene with CHCl₃ and hindered base.

$$+ \ CHCl_3 \ + (CH_3)_3CO^- \ K^+ \ \longrightarrow$$

Note how the predominant product is the one shown, in which the dichlorocarbene attacks the less hindered face of the double bond.

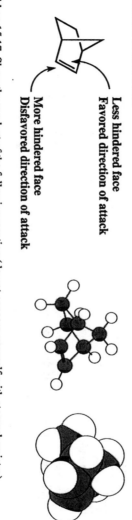

Less hindered face
Favored direction of attack

More hindered face
Disfavored direction of attack

Problem 15.17 Show the product of the following reaction (do not concern yourself with stereochemistry).

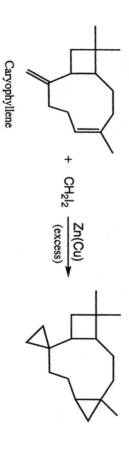

Caryophyllene

$$+ \ CH_2I_2 \ \xrightarrow[\text{(excess)}]{Zn(Cu)}$$

Problem 15.18 As has been demonstrated in the text, when the starting alkene has CH$_2$ as its terminal group, the Heck reaction is highly stereoselective for formation of the E isomer as illustrated in this example. Here, the benzene ring is abbreviated C$_6$H$_5$. Show how the mechanism proposed in the text allows you to account for this stereoselectivity.

$$CH_2=CH-COCH_3 + C_6H_5Br \xrightarrow[\text{(CH}_3\text{CH}_2)_3\text{N}]{\text{Pd(OAc)}_2, 2\text{Ph}_3\text{P}}$$

In the second step of the mechanism, the alkene undergoes a syn addition to the Pd complex on either face of the alkene.

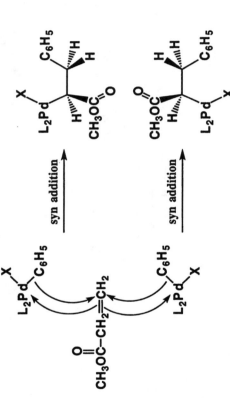

In the third step, the central bond of each enantiomer rotates to minimize torsional strain between the large -COOCH$_3$ and -C$_6$H$_5$ groups.

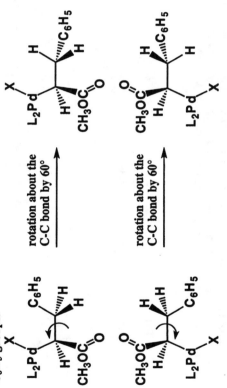

With the bond rotated in this conformation, the E product will be produced in the syn elimination step from both enantiomers. The key here is that the syn elimination occurs preferentially from the conformation that places the bulky -COOCH$_3$ and -C$_6$H$_5$ groups far apart from each other, because this is the lower energy conformation.

Problem 15.19 Heck reaction of bromobenzene and (E)-3-hexene gives a mixture of 3-phenyl-3-hexene and 4-phenyl-2-hexene in roughly equal amounts. Account for the formation of these two products.

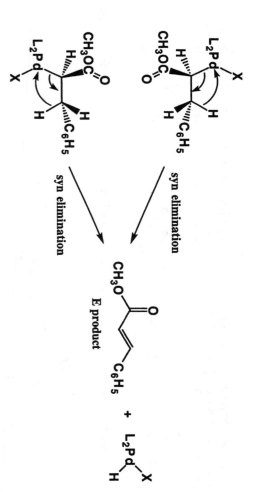

(E)-3-Hexene + C$_6$H$_5$Br $\xrightarrow[\text{(CH}_3\text{CH}_2)_3\text{N}]{\text{Pd(OAc)}_2,\ 2\text{Ph}_3\text{P}}$ (Z)-3-Phenyl-3-hexene + (E)-4-Phenyl-2-hexene

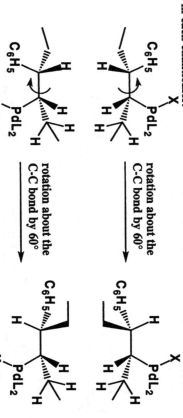

Following the syn addition step to either face of the double bond, the bond can rotate to set up the syn elimination in each enantiomer.

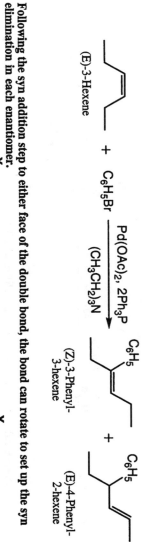

This intermediate can undergo syn elimination at either of the two hydrogens indicated to give the products. Note how the different enantiomeric intermediates give rise to different enantiomers of (E)-4-phenyl-2-hexene. (Z)-3-phenyl-3-hexene is achiral and is produced from both enantiomers of the intermediate.

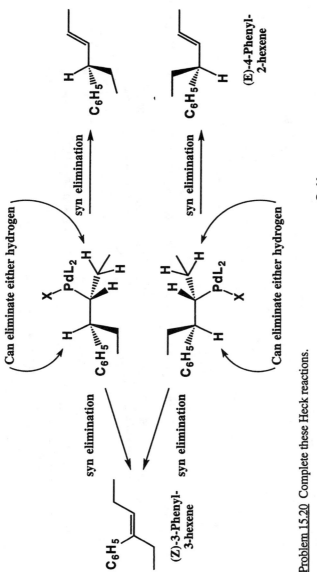

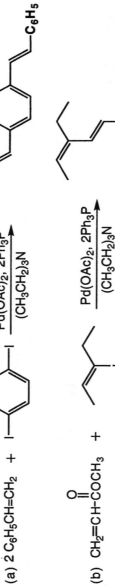

syn elimination

(E)-4-Phenyl-2-hexene

(Z)-3-Phenyl-3-hexene

Problem 15.20 Complete these Heck reactions.

(a) $2 \, C_6H_5CH=CH_2 \ + $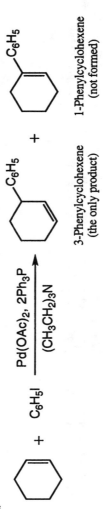

(b) $CH_2=CH-COCH_3 \ + $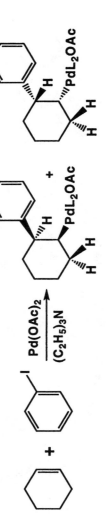

Problem 15.21 Treatment of cyclohexene with iodobenzene under the conditions of the Heck reaction might be expected to give 1-phenylcyclohexene. The exclusive product, however, is 3-phenylcyclohexene. Account for the formation of this product.

3-Phenylcyclohexene
(the only product)

1-Phenylcyclohexene
(not formed)

In the second step of the mechanism there is a syn addition to create the following Pd complex intermediate as two enantiomers.

In both enantiomers of the Pd complex intermediate, only one adjacent hydrogen atom is on the same side of the ring (syn) as the Pd atom as shown. Thus, this is the only hydrogen atom that can undergo syn elimination to generate product (3-phenylcyclohexene).

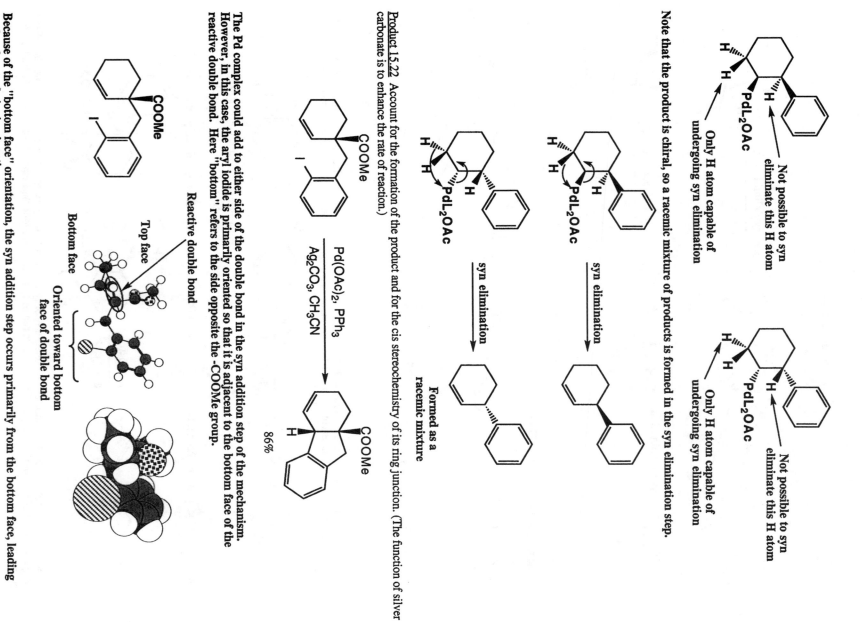

Note that the product is chiral, so a racemic mixture of products is formed in the syn elimination step.

Only H atom capable of undergoing syn elimination

Not possible to syn eliminate this H atom

PdL₂OAc

syn elimination

syn elimination

Formed as a racemic mixture

Product 15.22 Account for the formation of the product and for the cis stereochemistry of its ring junction. (The function of silver carbonate is to enhance the rate of reaction.)

$$ \xrightarrow[\text{Ag}_2\text{CO}_3,\ \text{CH}_3\text{CN}]{\text{Pd(OAc)}_2,\ \text{PPh}_3} $$

86%

The Pd complex could add to either side of the double bond in the syn addition step of the mechanism. However, in this case, the aryl iodide is primarily oriented so that it is adjacent to the bottom face of the reactive double bond. Here "bottom" refers to the side opposite the -COOMe group.

Reactive double bond

Top face

Bottom face

Oriented toward bottom face of double bond

Because of the "bottom face" orientation, the syn addition step occurs primarily from the bottom face, leading to the observed cis ring junction.

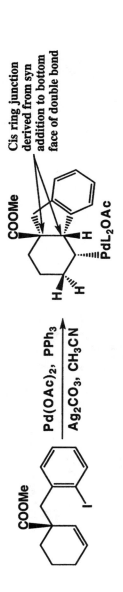

Syn elimination of this complex gives the predominant product.

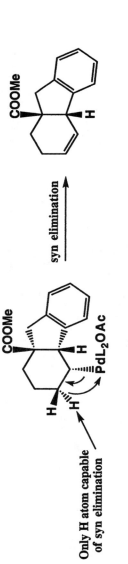

Cis ring junction derived from syn addition to bottom face of double bond

Only H atom capable of syn elimination

syn elimination

Problem 15.23 Account for the formation of the following product, including the cis stereochemistry at the ring junction. (Tf is an abbreviation for the trifluoromethanesulfonate group, $CF_3SO_2^-$, which can be used in the place of a halogen atom in a Heck reaction.)

Pd(OAc)₂, (R)-BINAP

K₂CO₃

The Pd complex could add to either side of the double bond in the syn addition step of the mechanism. However, in this case, the reactive TfO group is primarily oriented so that is adjacent to the top face of the reactive double bond. Here "top" refers to the side opposite the R group. Note that in the following molecular models a methyl group was used to illustrate the location of the "R" group in the problem.

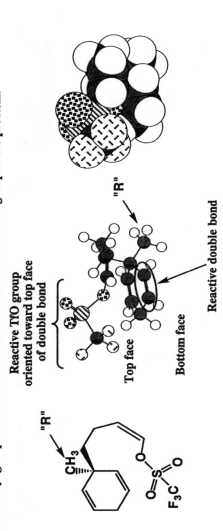

"R"

Reactive TfO group oriented toward top face of double bond

Top face

Bottom face

Reactive double bond

"R"

Because of the "top face" orientation, the syn addition step occurs primarily on the top face, leading to the observed cis ring junction.

Chapter 15: Organometallics

Syn elimination of this complex gives the predominant product.

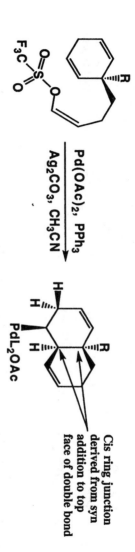

Syn elimination of this complex gives the predominant product.

Cis ring junction
derived from syn
addition to top
face of double bond

Only H atom capable
of syn elimination

syn elimination

<u>Problem 15. 24</u> The cyclic ester (lactone) Exaltolide (The Merck Index, 12th ed., #3956) has a musk-like fragrance and is used as a fixative in perfumery. Show how this compound could be synthesized from the indicated starting material. Give the structure of R.

Exaltolide

Like many syntheses, the key step is the carbon-carbon bond-forming step. Here propose an alkene metathesis reaction, dictating the structure shown. The numbers and letters were added to the structures to help you keep track of the atoms during the ring formin process.

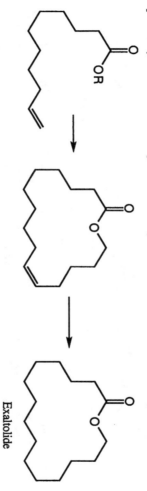

nucleophilic
carbene catalyst

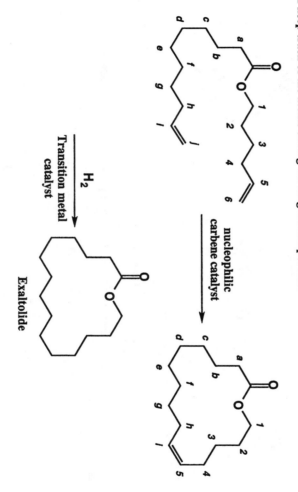

H$_2$
Transition metal
catalyst

Exaltolide

Problem 15.25 Show how the spiro[2.2]pentane can be prepared in one step from organic compounds containing three carbons or less and any necessary inorganic reagents or solvents.

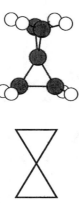

Spiro[2.2]pentane can be prepared in one step through the reaction of allene with the Simmons-Smith reagent.

$$H_2C=C=CH_2 \quad + \quad CH_2I_2 \quad \xrightarrow[\text{(excess)}]{Zn(Cu)}$$

Allene

Problem 15.26 Predict the product of each alkene metathesis reaction using a Ru-nucleophilic carbene catalyst.

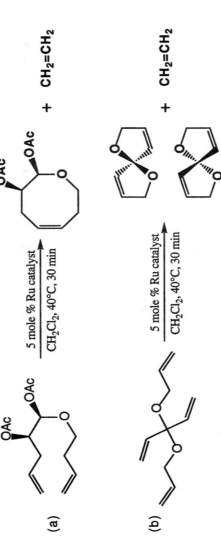

(a) $\xrightarrow[\text{CH}_2\text{Cl}_2,\ 40°C,\ 30\ \text{min}]{\text{5 mole \% Ru catalyst}}$ + $CH_2=CH_2$

(b) $\xrightarrow[\text{CH}_2\text{Cl}_2,\ 40°C,\ 30\ \text{min}]{\text{5 mole \% Ru catalyst}}$ + $CH_2=CH_2$

(formed as a racemic mixture)

Orangefin anemonefish *Amphiprion chrysopterus*
in a *Stichodactyla kenti* anenome
Bora Bora, French Polynesia

CHAPTER 16
Solutions to the Problems

Problem 16.1 Write IUPAC names for these compounds. Specify configuration in (c).

(a)

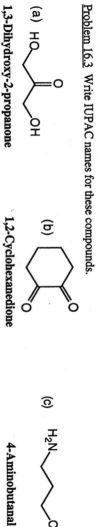

(b)

(c) C_6H_5—C$^{\text{\tiny ∥}}$—CHO with CH$_3$ and H

3,3-Dimethylbutanal   **1,3-Cyclohexanedione**   **(R)-2-Phenylpropanal**

Problem 16.2 Write structural formulas for all aldehydes of molecular formula $C_6H_{12}O$ and give each its IUPAC name. Which of these aldehydes are chiral?

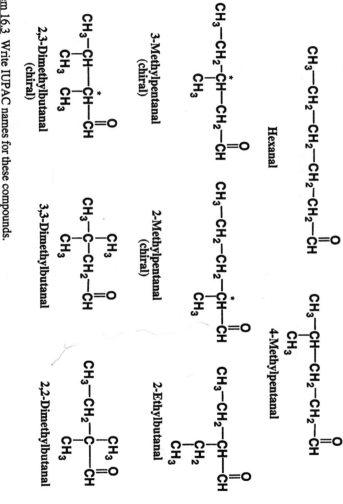

CH_3—CH_2—CH_2—CH_2—CH_2—CHO

Hexanal

CH_3—CH_2—CH_2—$\overset{*}{CH}$—CHO with CH_3

2-Methylpentanal (chiral)

CH_3—CH_2—$\overset{*}{CH}$—CH_2—CHO with CH_3

3-Methylpentanal (chiral)

CH_3—CH—CH_2—CH_2—CHO with CH_3

4-Methylpentanal

CH_3—CH_2—CH—CHO with CH_2—CH_3

2-Ethylbutanal

CH_3—CH—$\overset{*}{CH}$—CHO with CH_3 and CH_3

2,3-Dimethylbutanal (chiral)

CH_3—C—CH_2—CHO with CH_3 and CH_3

3,3-Dimethylbutanal

CH_3—CH_2—C—CHO with CH_3 and CH_3

2,2-Dimethylbutanal

Problem 16.3 Write IUPAC names for these compounds.

(a) HO—CH$_2$—C(=O)—CH$_2$—OH

1,3-Dihydroxy-2-propanone

(b)

1,2-Cyclohexanedione

(c) H_2N—CH$_2$CH$_2$CH$_2$—CHO

4-Aminobutanal

Problem 16.4 Show how these four products can be synthesized from the same Grignard reagent.

(a)

(b)

(c)

(d)

All of the products can be obtained using 2-cyclohexenylmagnesium bromide and the other reactants shown. Note that in each case at least one stereocenter is produced so that racemic mixtures will be generated in the reactions.

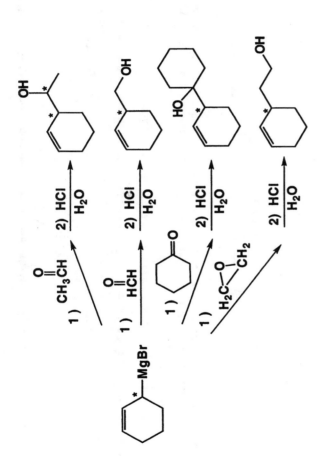

Problem 16.5 Show how each alkene can be synthesized by a Wittig reaction. There are two routes to each.

(a)

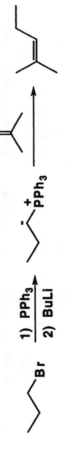

There are two combinations of Wittig reagent and carbonyl compound that could be used to make this alkene. The first uses 1-bromopropane to produce the Wittig reagent.

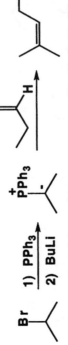

The other involves the Wittig reagent prepared from 2-bromopropane

The first synthesis is better because a more reactive 1° alkyl halide is used in the S_N2 reaction with PPh_3.

(b)

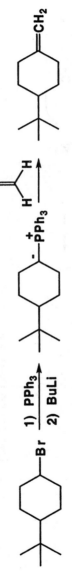

There are also two combinations of Wittig reagent and carbonyl compound that could be used to make this alkene. The first uses 1-bromo-4-*tert*-butylcyclohexane to produce the Wittig reagent.

The other involves the Wittig reagent prepared from bromomethane

$$CH_3Br \xrightarrow[\text{2) BuLi}]{\text{1) PPh}_3} H_2\overset{-}{C}\text{-}\overset{+}{P}Ph_3$$

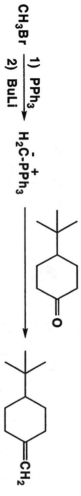

The second synthesis is better because the more reactive CH_3Br is used in the S_N2 reaction with PPh_3.

In the above examples, alkyl bromides were used. In fact, alkyl chlorides and iodides could have also been utilized.

Problem 16.6 Hydrolysis of an acetal in aqueous acid gives an aldehyde or ketone and two molecules of alcohol. Following are structural formulas for three acetals. Draw the structural formulas for the products of hydrolysis of each in aqueous acid.

(a)

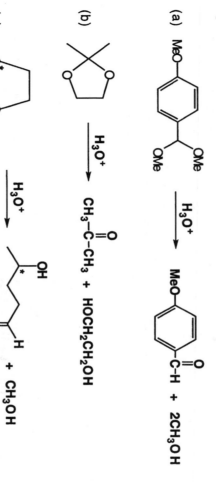

(b)

(c)

Problem 16.7 Write a mechanism for the acid-catalyzed hydrolysis of a THP ether to regenerate the original alcohol. Into what compound is the THP group converted?

The mechanism for acid-catalyzed THP ether hydrolysis is exactly the reverse of the acid-catalyzed THP ether formation.

Note that stereochemistry of this product will be that of the starting material, because the configuration of the stereocenter is not altered during the reaction.

Step 1:

Step 2:

Step 3:

A resonance-stabilized cation

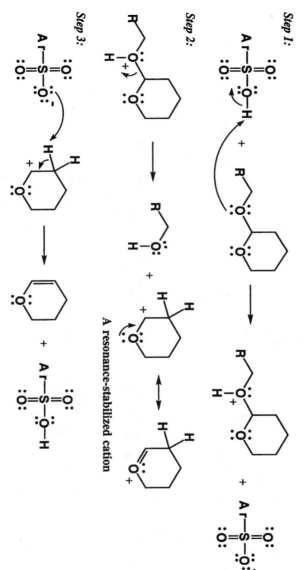

Problem 16.8 Show how to convert the given starting material into the indicated product using a 1,3–dithiane as an intermediate.

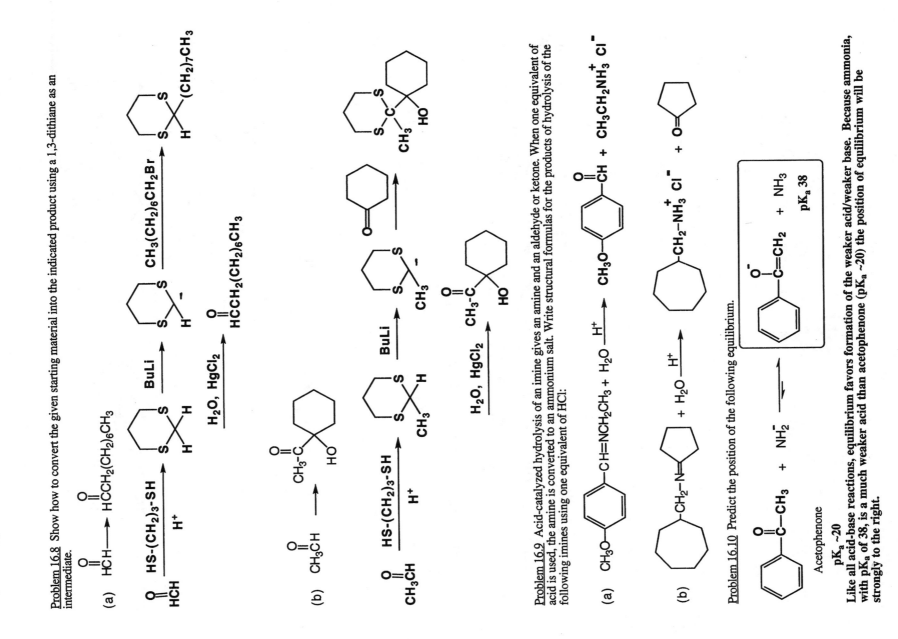

Problem 16.9 Acid-catalyzed hydrolysis of an imine gives an amine and an aldehyde or ketone. When one equivalent of acid is used, the amine is converted to an ammonium salt. Write structural formulas for the products of hydrolysis of the following imines using one equivalent of HCl:

Problem 16.10 Predict the position of the following equilibrium.

Like all acid-base reactions, equilibrium favors formation of the weaker acid/weaker base. Because ammonia, with pK$_a$ of 38, is a much weaker acid than acetophenone (pK$_a$ ~20) the position of equilibrium will be strongly to the right.

Problem 16.11 Draw the structural formula for the keto form of each enol.

(a)

(b)

(c)

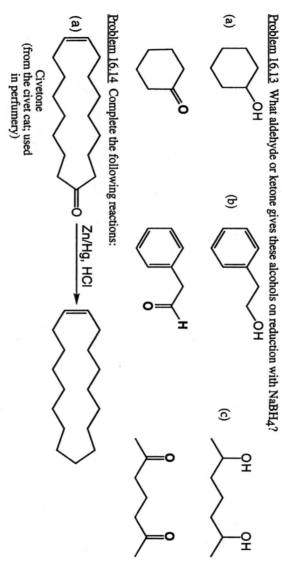

Problem 16.12 Complete these oxidations.

(a) Hexanal + H_2O_2 ⟶

$$CH_3(CH_2)_4\overset{O}{\overset{\|}{C}}H \;+\; H_2O_2 \longrightarrow CH_3(CH_2)_4\overset{O}{\overset{\|}{C}}OH \;+\; H_2O$$

(b) 3-Phenylpropanal + Tollens' reagent ⟶

$$CH_2CH_2\overset{O}{\overset{\|}{C}}H \;+\; 2\; Ag(NH_3)_2^+ \;+\; 2\; HO^- \;\xrightarrow[H_2O]{NH_3}$$

$$CH_2CH_2\overset{O}{\overset{\|}{C}}O^- \;+\; 2\; Ag \;+\; 3\; NH_3 \;+\; NH_4^+ \; H_2O$$

Problem 16.13 What aldehyde or ketone gives these alcohols on reduction with $NaBH_4$?

(a)

(b)

(c)

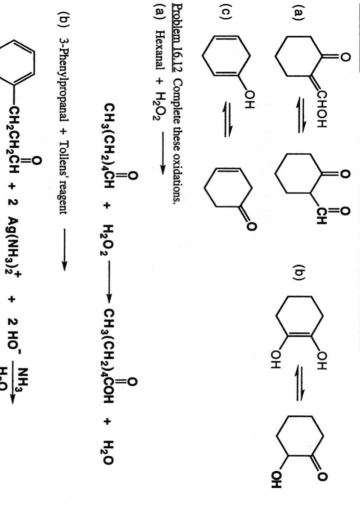

Problem 16.14 Complete the following reactions:

(a)

Civetone
(from the civet cat; used
in perfumery)

Zn/Hg, HCl ⟶

The above reaction is an example of a Clemmensen reduction.

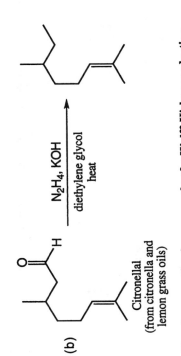

(b)

N$_2$H$_4$, KOH
diethylene glycol
heat

Citronellal
(from citronella and
lemon grass oils)

The above reaction is an example of a Wolff-Kishner reduction.

Structure and Nomenclature
Problem 16.15 Name these compounds.

(a)

**4-Nonanone
(dibutyl ketone)**

(b)

**(S)-2-Methylcyclo-
pentanone**

(c)

(Z)-2-Methyl-2-butenal

(d)

CHO
H······C
CH$_3$ CH$_2$OH

**(R)-3-Hydroxy-2-methyl-
propanal**

(e)

1-Phenyl-1-propanone

(f)

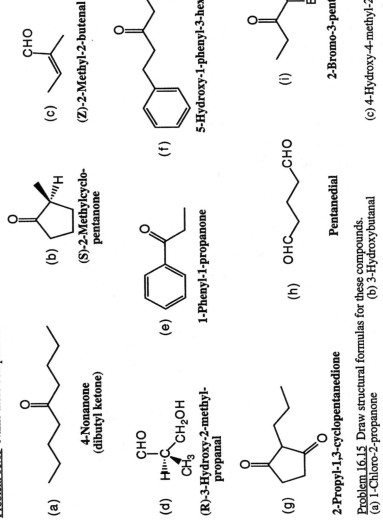

OH

5-Hydroxy-1-phenyl-3-hexanone

(g)

2-Propyl-1,3-cyclopentanedione

(h)

OHC   CHO

Pentanedial

(i)

Br

2-Bromo-3-pentanone

Problem 16.15 Draw structural formulas for these compounds.
(a) 1-Chloro-2-propanone   (b) 3-Hydroxybutanal

Cl

OH   H
*

(c) 4-Hydroxy-4-methyl-2-pentanone

OH

(d) 3-Methyl-3-phenylbutanal

H

(e) 1,3-Cyclohexanedione

(f) 3-Methyl-3-buten-2-one

(g) 5-Oxohexanal

(h) 2,2-Dimethylcyclohexanecarbaldehyde

or

(i) 3-Oxobutanoic acid

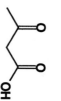

Spectroscopy

Problem 16.17 The infrared spectrum of Compound A, $C_6H_{12}O$, shows a strong, sharp peak at 1724 cm⁻¹. From this information and its ¹H-NMR spectrum, deduce the structure of compound A.

Compound A:

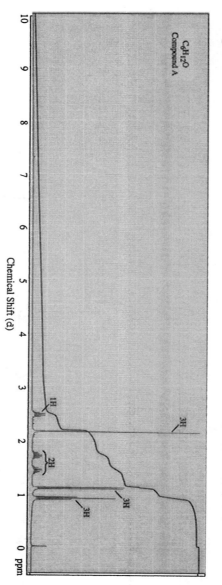

$C_6H_{12}O$
Compound A

Chemical Shift (δ)

Compound A has an aldehyde or ketone function judging from the peak at 1724 cm⁻¹ in the IR spectrum. The rest of the structure can be deduced from the ¹H-NMR, with assignments as listed below.

$$\underset{0.89}{CH_3}-\underset{1.42,1.68}{CH_2}-\underset{\underset{\underset{1.08}{CH_3}}{|}}{CH}-\overset{O}{\underset{\|}{C}}-\underset{2.1}{CH_3} \quad 2.45$$

The -CH₂- hydrogens with signals at δ 1.42 and δ 1.68 above are not equivalent because they are adjacent to a tetrahedral stereocenter in the molecule.

Problem 16.18 Following are ¹H-NMR spectra for compounds B, $C_6H_{12}O_2$, and C, $C_6H_{10}O$. On warming in dilute acid, compound B is converted to compound C. Deduce the structural formulas for compounds B and C.

Compound B:

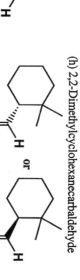

$C_6H_{12}O_2$
Compound B

Chemical Shift (δ)

Compound C:

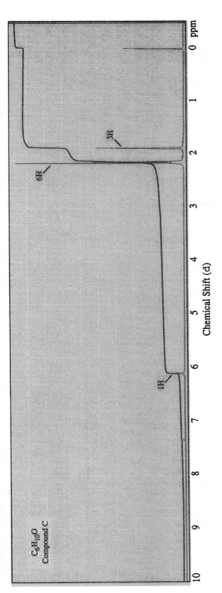

$C_6H_{10}O$
Compound C

6H

3H

1H

Chemical Shift (d)

10   9   8   7   6   5   4   3   2   1   0   ppm

From the molecular formula it is clear that compound B undergoes acid-catalyzed dehydration to create compound C. Thus, compound B must have an -OH group. Furthermore, because its index of hydrogen deficiency is one, compound B must have one ring or pi bond. However, compound C has an index of hydrogen deficiency of two, so it must have two pi bonds or rings. The ¹H-NMR spectrum of compound B shows all singlets. Especially helpful are the methyl group resonances; the singlet integrating to 6H at δ 1.22 and the singlet integrating to 3H at δ 2.18. This latter signal is assigned as a methyl ketone by comparison to the previous problem. The other two methyl groups are equivalent. There is the -OH hydrogen at δ 3.85 and a -CH₂- resonance at δ 2.62. The only structure consistent with these signals is 4-hydroxy-4-methyl-2-pentanone.

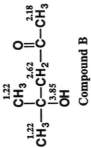

1.22
CH₃
| O   2.18
1.22 | 2.62 ||
CH₃—C—CH₂—C—CH₃
|
3.85
OH

Compound B

Upon dehydration, Compound B would be converted to 4-methyl-3-pentene-2-one.

2.15
CH₃
| O
| 6.10 || 1.86
2.15 | |
CH₃—C=CH—C—CH₃

Compound C

The structure of 4-methyl-3-pentene-2-one is entirely consistent with the ¹H-NMR spectrum, especially the presence of the signal at δ 6.10 (-CH=C) and the methyl singlets at δ 1.86 (integrating to 3H) and δ 2.15 (integrating to 6H).

Addition of Carbon Nucleophiles
Problem 16.19 Draw structural formulas for the product formed by treating each compound with propylmagnesium bromide followed by aqueous HCl.

The products after acid hydrolysis are given in bold.

(a) CH₂O

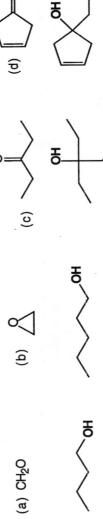

(b)

(c)

(d)

OH

OH

OH

OH

Problem 16.20 Suggest a synthesis for the following alcohols starting from an aldehyde or ketone and an appropriate Grignard reagent. Below each target molecule is the number of combinations of Grignard reagent and aldehyde or ketone that might be used.

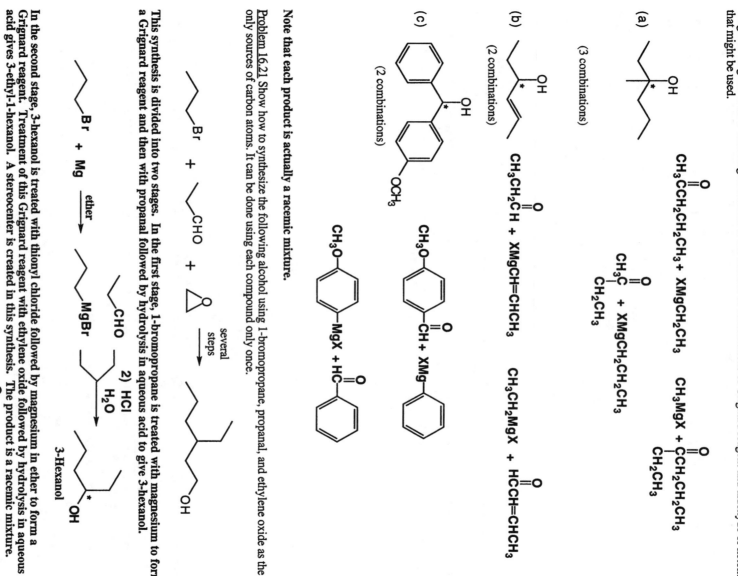

(a)

OH

$$CH_3\overset{O}{\overset{\|}{C}}CH_2CH_2CH_3 + XMgCH_2CH_3 \qquad CH_3MgX + \overset{O}{\overset{\|}{C}}CH_2CH_2CH_3$$

$$CH_3\overset{O}{\overset{\|}{C}} + XMgCH_2CH_2CH_3$$
$$CH_2CH_3$$

(3 combinations)

(b)

OH

(2 combinations)

$$CH_3CH_2\overset{O}{\overset{\|}{C}}H + XMgCH=CHCH_3 \qquad CH_3CH_2MgX + H\overset{O}{\overset{\|}{C}}CH=CHCH_3$$

(c)

OH

OCH₃

(2 combinations)

CH₃O——MgX + HC==O (phenyl)

CH₃O——CH==O + XMg (phenyl)

Note that each product is actually a racemic mixture.

Problem 16.21 Show how to synthesize the following alcohol using 1-bromopropane, propanal, and ethylene oxide as the only sources of carbon atoms. It can be done using each compound only once.

several
steps

Br

CHO

This synthesis is divided into two stages. In the first stage, 1-bromopropane is treated with magnesium to form a Grignard reagent and then with propanal followed by hydrolysis in aqueous acid to give 3-hexanol.

Br + Mg → ether → MgBr

CHO
2) HCl
H₂O

OH

3-Hexanol

In the second stage, 3-hexanol is treated with thionyl chloride followed by magnesium in ether to form a Grignard reagent. Treatment of this Grignard reagent with ethylene oxide followed by hydrolysis in aqueous acid gives 3-ethyl-1-hexanol. A stereocenter is created in this synthesis. The product is a racemic mixture.

OH
1) SOCl₂
2) Mg, ether
MgBr

1) O
2) HCl
H₂O

OH

3-Ethyl-1-hexanol

Problem 16.22 1-Phenyl-2-butanol is used in perfumery. Show how to synthesize this alcohol from bromobenzene, 1-butene, and any necessary inorganic reagents.

several steps →

Bromobenzene   +   1-Butene

1-Phenyl-2-butanol

(a) Bromobenzene is treated with magnesium in diethyl ether to form phenylmagnesium bromide, in preparation for part (c).

Br + Mg — ether → MgBr

(b) Treatment of 1-butene with a peroxycarboxylic acid gives 1,2-epoxybutane.

 + R—CO₃H → + RCOOH

(c) Treatment of phenylmagnesium bromide with 1,2-epoxybutane followed by hydrolysis in aqueous acid gives 1-phenyl-2-butanol.

 MgBr + O → 2) HCl / H₂O → OH

A stereocenter is created in this synthesis. The product is a racemic mixture. A Gilman reagent could also have been used in this step.

Problem 16.23 With organolithium and organomagnesium compounds, approach to the carbonyl carbon from the less hindered direction is generally preferred. Assuming this is the case, predict the structure of the major product formed by reaction of methylmagnesium bromide with 4-*tert*-butylcyclohexanone.

The bulky *tert*-butyl group lies in an equatorial position. Approach to the carbonyl carbon may be by way of a pseudo-axial direction or a pseudo-equatorial direction. The less hindered approach is from the pseudo-equatorial direction which then places the incoming group equatorial and the -OH axial.

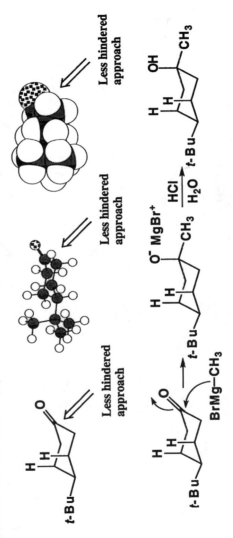

Less hindered approach

Less hindered approach

Less hindered approach

Less hindered approach

BrMg—CH₃

O⁻ MgBr⁺ / CH₃ — HCl / H₂O → t-Bu

CH₃ / OH

Problem 16.24 Draw structural formulas for (1) the triphenylphosphonium salt formed by treatment of each haloalkane with triphenylphosphine, (2) the phosphonium ylide formed by treatment of each phosphonium salt with butyllithium, and (3) the alkene formed by treatment of each phosphonium ylide with acetone.

In the following, the triphenylphosphonium salts are listed as (1), the phosphonium ylides as (2), and the alkene formed upon reaction with acetone as (3).

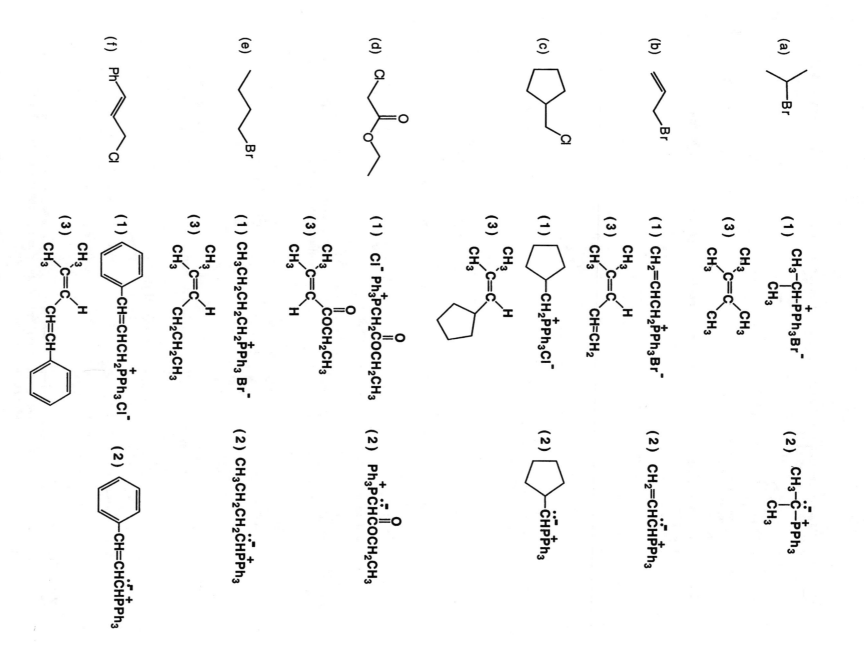

Problem 16.25 Show how to bring about the following conversions using a Wittig reaction.

(a)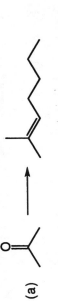

Start with 1-halopentane and, by treatment with triphenylphosphine followed by butyllithium, convert it to a Wittig ylide. Treatment of this ylide with acetone gives the desired alkene.

$$CH_3(CH_2)_3CH_2I \xrightarrow[\text{2) BuLi}]{\text{1) PPh}_3} CH_3(CH_2)_3\overset{-}{\underset{+}{\text{C}}}\text{HPPh}_3 \xrightarrow{CH_3\overset{O}{\overset{\|}{C}}CH_3}$$

$$\overset{CH_3}{\underset{}{CH_3\overset{|}{C}=CH(CH_2)_3CH_3}}$$

(b)

Treatment of acetophenone with the Wittig ylide derived from methyl iodide gives the desired alkene.

$$CH_3I \xrightarrow[\text{2) BuLi}]{\text{1) PPh}_3} \overset{..}{\underset{}{C}}H_2=\overset{+}{P}Ph_3$$

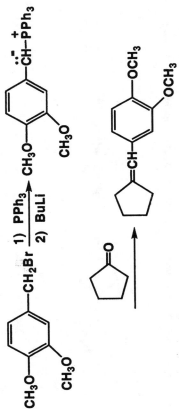

(c)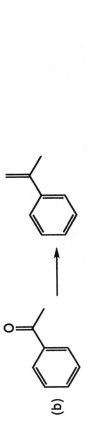

Treatment of cyclopentanone with the Wittig ylide derived from 3,4-dimethoxybenzyl bromide gives the desired alkene.

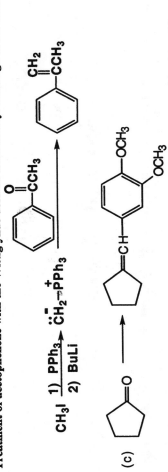

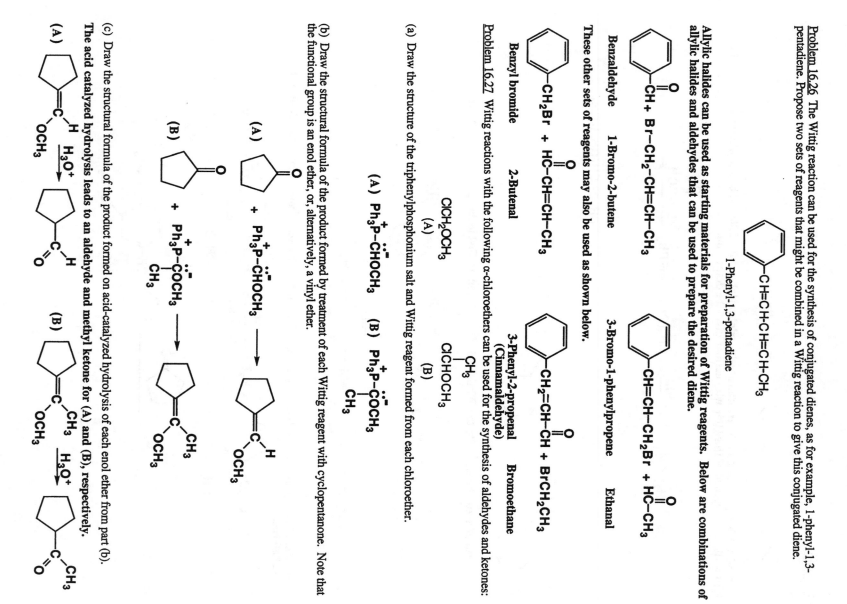

Problem 16.26 The Wittig reaction can be used for the synthesis of conjugated dienes, as for example, 1-phenyl-1,3-pentadiene. Propose two sets of reagents that might be combined in a Wittig reaction to give this conjugated diene.

1-Phenyl-1,3-pentadiene

Allylic halides can be used as starting materials for preparation of Wittig reagents. Below are combinations of allylic halides and aldehydes that can be used to prepare the desired diene.

Benzaldehyde   1-Bromo-2-butene

These other sets of reagents may also be used as shown below.

Benzyl bromide   2-Butenal   3-Bromo-1-phenylpropene   Ethanal

3-Phenyl-2-propenal (Cinnamaldehyde)   Bromoethane

Problem 16.27 Wittig reactions with the following α-chloroethers can be used for the synthesis of aldehydes and ketones:

(A) $ClCH_2OCH_3$ (B)

(a) Draw the structure of the triphenylphosphonium salt and Wittig reagent formed from each chloroether.

(A) $Ph_3\overset{+}{P}-\overset{..}{C}HOCH_3$ (B) $Ph_3\overset{+}{P}-\overset{..}{C}OCH_3$

(b) Draw the structural formula of the product formed by treatment of each Wittig reagent with cyclopentanone. Note that the functional group is an enol ether, or, alternatively, a vinyl ether.

(A)   + $Ph_3\overset{+}{P}-\overset{..}{C}HOCH_3$ ⟶

(B)   + $Ph_3\overset{+}{P}-\overset{..}{C}OCH_3$ ⟶

(c) Draw the structural formula of the product formed on acid-catalyzed hydrolysis of each enol ether from part (b). The acid catalyzed hydrolysis leads to an aldehyde and methyl ketone for (A) and (B), respectively.

(A) ⟶ H_3O^+

(B) ⟶ H_3O^+

Problem 16.28 It is possible to generate sulfur ylides in a manner similar to that used to produce phosphonium ylides. For example, treating a sulfonium salt with a strong base gives the sulfur ylide.

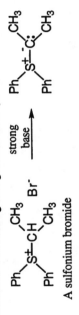

A sulfonium bromide

Sulfur ylides react with ketones to give epoxides. Suggest a mechanism for this reaction.

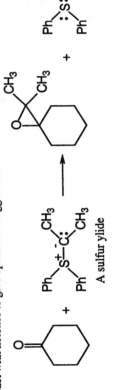

A sulfur ylide

Step 1:

Step 2:

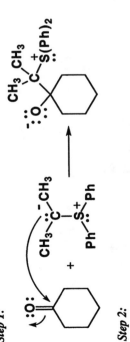

Problem 16.29 Propose a structural formula for Compound D and for the product, $C_9H_{14}O$ formed in this reaction sequence.

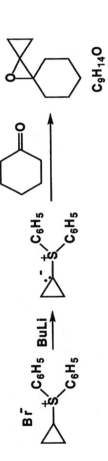

The product of this transformation is an epoxide with both a three-membered and a six-membered ring attached via so-called "spiro" attachments.

Addition of Oxygen Nucleophiles

Problem 16.30 5-Hydroxyhexanal forms a six-membered cyclic hemiacetal, which predominates at equilibrium in aqueous solution.

(a) Draw a structural formula for this cyclic hemiacetal.

5-Hydroxyhexanal forms a six-membered cyclic hemiacetal.

5-Hydroxyhexanal

$\xrightarrow{H^+}$ A cyclic hemiacetal

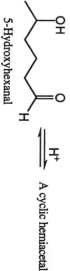

(b) How many stereoisomers are possible for 5-hydroxyhexanal?

Two stereoisomers are possible for 5-hydroxyhexanal; a pair of enantiomers.

(c) How many stereoisomers are possible for this cyclic hemiacetal?

Four stereoisomers are possible for the cyclic hemiacetal; two pairs of enantiomers. Following are planar hexagon formulas for each pair of enantiomers of the cyclic hemiacetal.

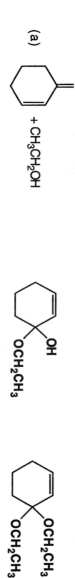

(A)　(B)

A pair of enantiomers

(C)　(D)

A pair of enantiomers

(d) Draw alternative chair conformations for each stereoisomer of the cyclic hemiacetal and label groups axial or equatorial. Also predict which of the alternative chair conformations for each stereoisomer is the more stable.

Alternative chair conformations are drawn for (A), one of the *cis* enantiomers, and for (C), one of the *trans* enantiomers. Methyl and hydroxyl groups are 7.28 kJ/mol and 3.9 kJ/mol more stable in the equatorial position, respectively. For (A), the diequatorial chair is the more stable by 7.28 + 3.9 = 11.18 kJ/mol. For (C), the chair with the methyl group is equatorial (structure on the left) is 7.28 - 3.9 = 3.38 kJ/mol more stable.

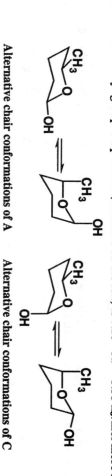

Alternative chair conformations of A

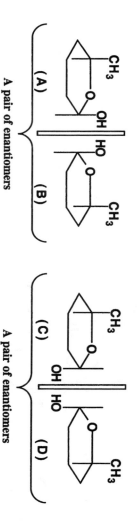

Alternative chair conformations of C

Problem 16.31 Draw structural formulas for the hemiacetal and then the acetal formed from each pair of reactants in the presence of an acid catalyst.

(a)

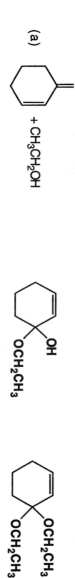

+ CH₃CH₂OH

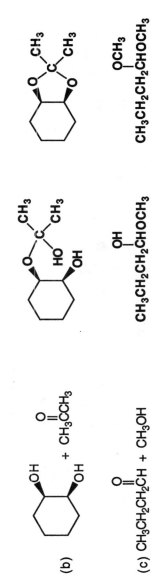

(b)

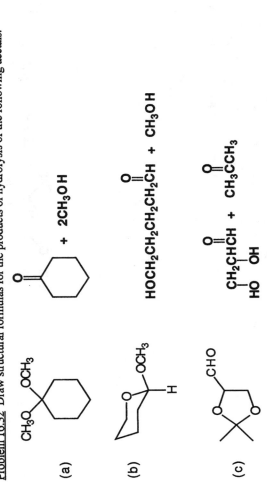

(c) $CH_3CH_2CH_2CH$ + CH_3OH $CH_3CH_2CH_2CHOCH_3$

<u>Problem 16.32</u> Draw structural formulas for the products of hydrolysis of the following acetals.

(a)

$+$ $2CH_3OH$

(b)

$HOCH_2CH_2CH_2CH_2CH$ + CH_3OH

(c)

CH_2CHCH + CH_3CCH_3

<u>Problem 16.33</u> Propose a mechanism to account for the formation of a cyclic acetal from 4-hydroxypentanal and one equivalent of methanol. If the carbonyl oxygen of 4-hydroxypentanal is enriched with oxygen-18, do you predict that the oxygen label appears in the cyclic acetal or in the water?

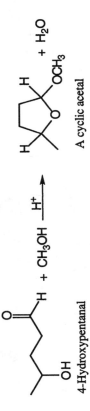

4-Hydroxypentanal   A cyclic acetal

Propose formation of a hemiacetal followed by protonation of the hemiacetal -OH and loss of water to form a resonance-stabilized cation. Then propose a Lewis acid-base reaction between this cation and methanol followed by loss of a proton to give the product. If the carbonyl group of 4-hydroxypentanal is enriched with oxygen-18, the oxygen-18 label appears in the water.

Step 1:

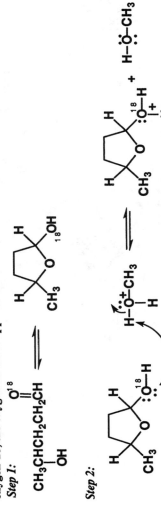

Step 2:

Step 3:

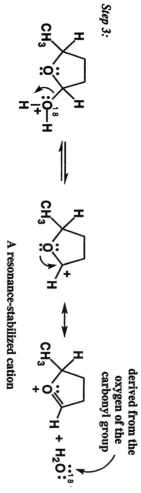

A resonance-stabilized cation

derived from the
oxygen of the
carbonyl group

Step 4:

Step 5:

Problem 16.34   Propose a mechanism for this acid-catalyzed hydrolysis.

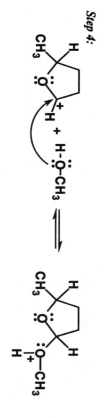

$+ \; H_2O \xrightarrow{\;H^+\;}$ $+ \; CH_3OH$

Step 1:

A reasonable mechanism for this reaction involves protonation of the pi bond of the carbon-carbon double bond to give a resonance stabilized cation intermediate.

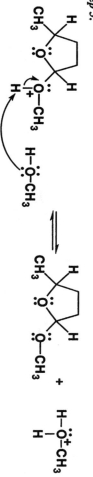

A resonance stabilized cation

Step 2:

In subsequent steps, water reacts with the cation to form a hemiacetal, that loses methanol to give the ketone.

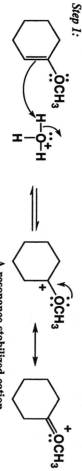

Step 3:

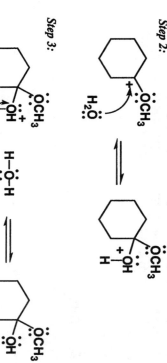

A hemiacetal

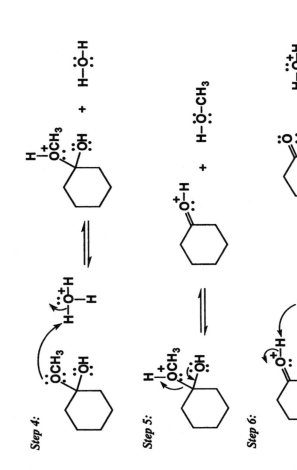

Step 4:

Step 5:

Step 6:

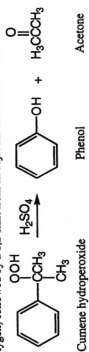

Problem 16.35 All rearrangements we have discussed so far involve generation of a positively charged electron-deficient carbon atom (a carbocation) followed by a 1,2-shift of an atom or group of atoms from an adjacent carbon to the carbocation. A mechanism that can be written for the following reaction also involves generation of an electron-deficient atom (in this case an oxygen) followed by a 1,2-shift from an adjacent carbon to the electron-deficient atom.

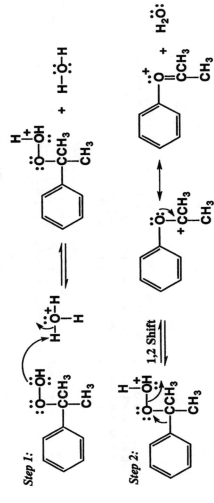

Cumene hydroperoxide   Phenol   Acetone

Propose a mechanism for the acid-catalyzed rearrangement of cumene hydroperoxide to phenol and acetone based on the previous rationale. In completing a mechanism, you will want to review the characteristic structural features of a hemiacetal (Section 16.8B) and its equilibration with a ketone by loss of an alcohol.

A reasonable mechanism for this reaction involves protonation of the terminal oxygen atom of the hydroperoxide group followed by a 1,2 shift of the benzene ring. Attack of the resulting carbocation by water and loss of a proton yields a hemiacetal, that decomposes to give phenol and acetone.

Step 1:

Step 2:

    1,2 Shift

Step 3:

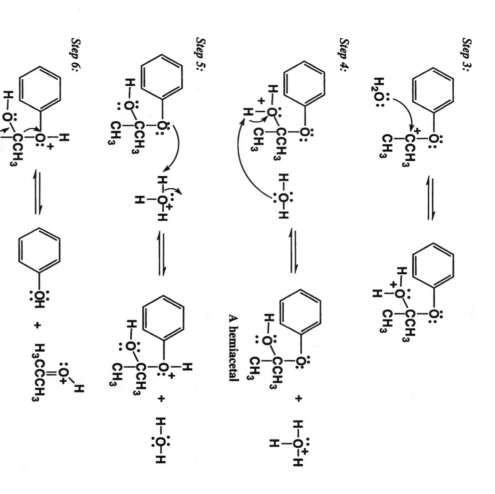

Step 4:

A hemiacetal

Step 5:

Step 6:

Step 7:

Problem 16.36 In Section 11.6A we saw that ethers, such as diethyl ether and tetrahydrofuran, are quite resistant to the action of dilute acids and require hot concentrated HI or HBr for cleavage. However, acetals in which two ether groups are linked to the same carbon, undergo hydrolysis readily even in dilute aqueous acid. How do you account for this marked difference in chemical reactivity toward dilute aqueous acid between ethers and acetals?

The first step of the cleavage reactions in acid for both ethers and acetals is protonation of an oxygen to form an oxonium ion. For acetals, the following step is cleavage of a carbon-oxygen bond to form a resonance-stabilized cation. For ethers, similar cleavage occurs, but the cation formed has no comparable resonance stabilization. Therefore, it is the resonance-stabilization of the cation intermediate formed during the cleavage of acetals that lowers the activation energy for their hydrolysis much below that for the hydrolysis of ethers.

Problem 16.37 Show how to bring about the following conversion:

The most convenient way to convert an alkene to a glycol is to oxidize the alkene with osmium tetroxide in the presence of hydrogen peroxide. These conditions, however, will also oxidize an aldehyde to a carboxylic acid.

Therefore, it is necessary to first protect the aldehyde by transformation to an acetal. In the following answer, ethylene glycol is the protecting agent.

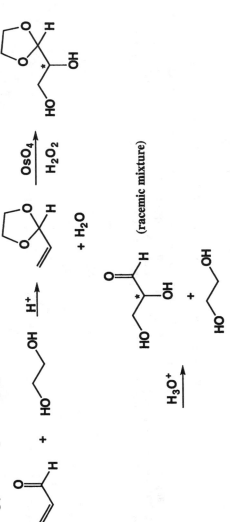

+ H₂O

(racemic mixture)

Problem 16.38 A primary or secondary alcohol can be protected by conversion to its tetrahydropyranyl ether. Why is formation of THP ethers by this reaction limited to primary and secondary alcohols?

It will help to refer to the mechanism given in Example 16.7. The concerns here are acid-catalyzed dehydration of the alcohol as a competing reaction and relative rates of nucleophilic reaction. Primary and secondary alcohols do not dehydrate nearly as readily as tertiary alcohols (due to relative carbocation stabilities). In addition, primary and secondary alcohols are relatively unhindered nucleophiles, while tertiary alcohols are so sterically hindered that they are sluggish nucleophiles. The net result is that tertiary alcohols dehydrate too readily, and react too slowly as nucleophiles, to be converted efficiently into tetrahydropyranyl ethers in acidic conditions. Primary and secondary alcohols, on the other hand, react fast enough as nucleophiles and dehyrate slowly enough to produce tetrahydropyranyl ethers in high yields with acid catalysis.

Problem 16.39 Which of these molecules will cyclize to give the insect pheromone frontalin?

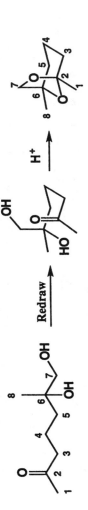

A   B   C

Frontalin

The answer is B. Using models may help with this answer.

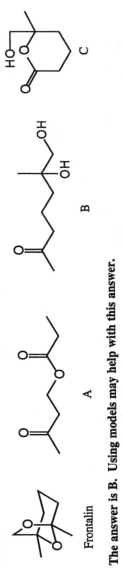

Redraw

Addition of Sulfur Nucleophiles

Problem 16.40 Draw a structural formula for the product of reaction of benzaldehyde with the following dithiols in the presence of an acid catalyst.

(a) 1,2-Ethanedithiol

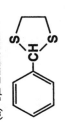

(b) 1,3-Propanedithiol

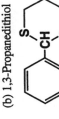

Problem 16.41 Draw a structural formula for the product formed by treating each of these compounds with (1) the lithium salt of the 1,3-dithiane derived from acetaldehyde and then (2) H_2O, $HgCl_2$.

(a)

$$\underset{CH_3}{\overset{HO}{\underset{*}{C}}}\overset{O}{\overset{\|}{C}}CH_3$$

(b)

$$\underset{CH_2CCH_3}{\overset{OH}{\underset{*}{CHCH_2CCH_3}}}$$

(c) $ClCH_2CH=CH_2$

$$CH_2=CHCH_2\overset{O}{\overset{\|}{C}}CH_3$$

In reaction of the 1,3-dithiane of acetaldehyde with butyllithium, the carbonyl carbon is in effect converted to a carbanion that then adds to the carbonyl carbon of benzaldehyde in (a), brings about an S_N2 opening of the epoxide in (b), and an S_N2 displacement of chlorine in (c). Hydrolysis in aqueous mercuric chloride regenerates the carbonyl group. In parts (a) and (b) a racemic mixture will be formed.

Problem 16.42 Show how to bring about the following conversions using a 1,3-dithiane:

(a)

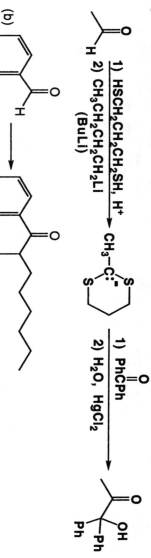

First convert acetaldehyde to a 1,3-dithiane and then to a lithium salt with butyllithium. Nucleophilic addition of this carbanion to the carbonyl group of benzophenone followed by hydrolysis of the 1,3-dithiane gives the desired ketoalcohol.

(b)

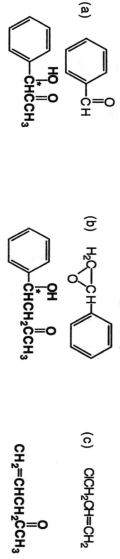

Convert benzaldehyde to a 1,3-dithiane and then to a lithium salt. Treatment of this nucleophile with 2-bromooctane via an S_N2 pathway followed by hydrolysis gives the desired product. Assuming the 2-bromooctane is racemic, the product will be a racemic mixture as well.

(c)

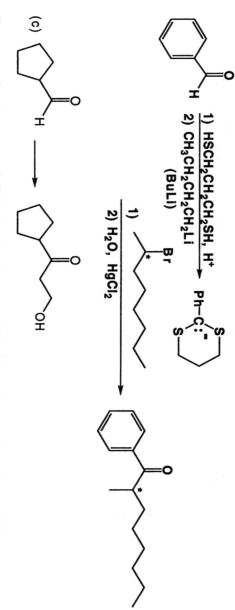

Treatment of cyclopentanecarbaldehyde with 1,3-propanedithiol followed by reaction with butyllithium gives a carbanion. Treatment of this carbanion with ethylene oxide followed by hydrolysis in aqueous mercuric chloride gives the product.

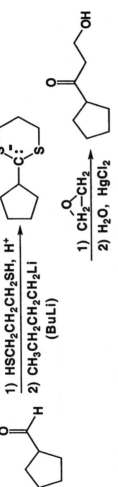

1) HSCH₂CH₂CH₂SH, H⁺

2) CH₃CH₂CH₂CH₂Li
(BuLi)

1) 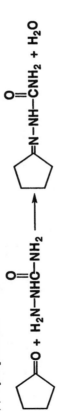 CH₂—CH₂

2) H₂O, HgCl₂

Addition of Nitrogen Nucleophiles

Problem 16.43 Draw structural formulas for the products of the following acid-catalyzed reactions:

(a) Phenylacetaldehyde + hydrazine ----->

PhCH₂CH + H₂N–NH₂ ⟶ PhCH₂CH=N–NH₂ + H₂O

(b) Cyclopentanone + semicarbazide ----->

 + H₂N–NHC–NH₂ ⟶ 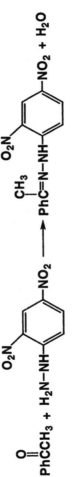 N–NH–CNH₂ + H₂O

(c) Acetophenone + 2,4-dinitrophenylhydrazine ----->

PhCCH₃ + H₂N–NH  NO₂ ⟶ PhC=N–NH 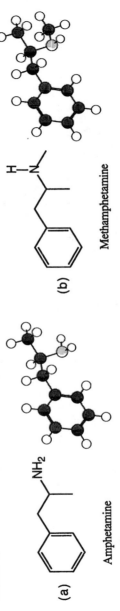 NO₂ + H₂O

(d) Benzaldehyde + hydroxylamine ----->

CH + H₂N–OH ⟶ C=NOH + H₂O

Problem 16.44 Following are structural formulas for amphetamine and methamphetamine (see The Merck Index, 12th ed., #623 and #6015). The major central nervous system effects of amphetamine and amphetamine-like drugs are locomotor stimulation, euphoria and excitement, stereotyped behavior, and anorexia. Show how each of these drugs can be synthesized by reductive amination of an appropriate aldehyde or ketone and amine. For the structural formulas of several more anorexics, see MC.42 and MC.43.

(a) Amphetamine

(b) Methamphetamine

Both molecules can be synthesized from the same ketone via reductive amination with either ammonia (amphetamine) or methylamine (methamphetamine). A new stereocenter is created in each case. Racemic mixtures will be produced.

Solutions

Problem 16.45 Following is the final step in the synthesis of the antiviral drug rimantadine. Describe experimental conditions to bring this conversion.

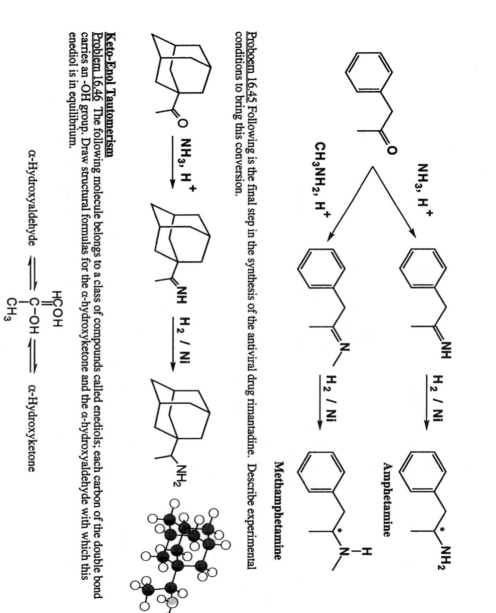

Amphetamine

Methamphetamine

Keto-Enol Tautomerism

Problem 16.46 The following molecule belongs to a class of compounds called enediols; each carbon of the double bond carries an -OH group. Draw structural formulas for the α-hydroxyketone and the α-hydroxyaldehyde with which this enediol is in equilibrium.

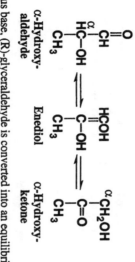

α-Hydroxyaldehyde ⇌ An enediol ⇌ α-Hydroxyketone

Following are formulas for the α-hydroxyaldehyde and α-hydroxyketone in equilibrium by way of the enediol intermediate.

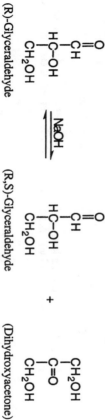

α-Hydroxy-aldehyde   Enediol   α-Hydroxy-ketone

Problem 16.47 In dilute aqueous base, (R)-glyceraldehyde is converted into an equilibrium mixture of (R,S)-glyceraldehyde and dihydroxyacetone. Propose a mechanism for this isomerization.

(R)-Glyceraldehyde   (R,S)-Glyceraldehyde   (Dihydroxyacetone)

The key is keto-enol tautomerism. In the presence of base, (R)-glyceraldehyde undergoes base-catalyzed keto-enol tautomerism to form an enediol in which carbon-2 is achiral. This enediol is in turn in equilibrium with (S)-glyceraldehyde and dihydroxyacetone.

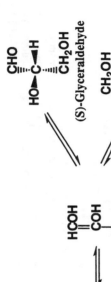

(R)-Glyceraldehyde

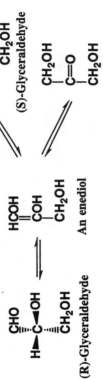

CHO
HCOH
COH
CH₂OH

An enediol

CHO
HO—C—H
CH₂OH

(S)-Glyceraldehyde

CH₂OH
C=O
CH₂OH

Dihydroxyacetone

Problem 16.48 When *cis-2*-decalone is dissolved in ether containing a trace of HCl, an equilibrium is established with *trans-2*-decalone. The latter ketone predominates in the equilibrium mixture. Propose a mechanism for this isomerization and account for the fact that the *trans* isomer predominates at equilibrium.

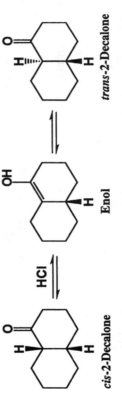

The *cis-2*-decalone and *trans-2*-decalone are equilibrating via the enol intermediate.

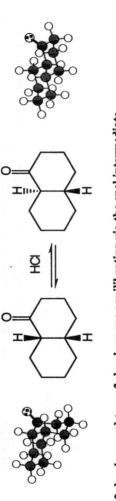

cis-2-Decalone   Enol   *trans-2*-Decalone

The *trans-2*-decalone is more stable because all of the carbons of the ring fusions are in the more stable equatorial arrangement. In *cis-2*-decalone, there is one equatorial and one less-favored axial arrangement at the ring fusion. Note that in the models of *cis-2*-decalone, the carbonyl oxygen is hidden from view because it is behind the structure.

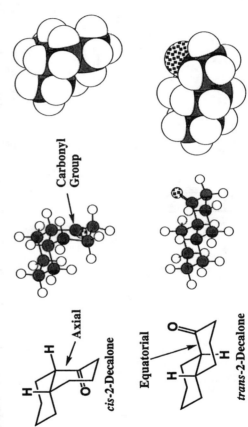

Reactions at the α-Carbon

Problem 16.49 The following bicyclic ketone has two α-carbons and three α-hydrogens. When this molecule is treated with D₂O in the presence of an acid catalyst, only two α-hydrogens exchange with deuterium. The α-hydrogen at the bridgehead does not exchange. How do you account for the fact that two α-hydrogens do exchange but the third does not? You will find it helpful to build a model of this molecule and of the fact that two α-hydrogens do exchange but the third does not?

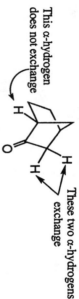

This α-hydrogen does not exchange

These two α-hydrogens exchange

Exchange of α-hydrogens is through keto-enol tautomerism and an enol intermediate. The key to this problem centers on the possibility of placing a carbon-carbon double bond between carbons 2-3 of the bicyclic ring and between carbons 1-2 of the ring. Enolization in the first direction is possible with the result that the two α-hydrogens on carbon-3 exchange. Enolization toward the bridgehead carbon is not possible because of the geometry of the bicyclo[2.2.1]heptane ring. The energy required to force the bridgehead carbon and the three other carbons attached to it into a planar conformation with bond angles of 120° is prohibitively high.

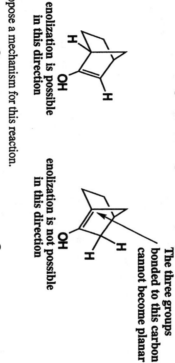

enolization is possible in this direction

enolization is not possible in this direction

The three groups bonded to this carbon cannot become planar

Problem 16.50 Propose a mechanism for this reaction.

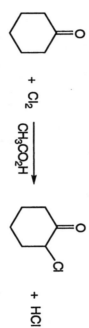

+ Cl₂ → $\xrightarrow{CH_3CO_2H}$

+ HCl

A reasonable mechanism for this reaction involves tautomerization to the enol form, that then reacts with the Cl₂. A final proton transfer completes the reaction.

Step 1:

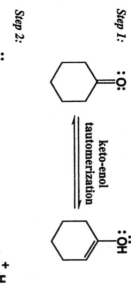

:O:

keto-enol tautomerization

:ÖH

Step 2:

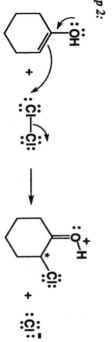

Step 3:

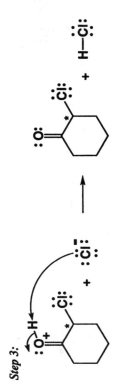

A new stereocenter is created in the product. A racemic mixture is produced.

Problem 16.51 The base-promoted rearrangement of an α-haloketone to a carboxylic acid, known as the Favorskii rearrangement, is illustrated by the conversion of 2-chlorocyclohexanone to cyclopentanecarboxylic acid. It is proposed that NaOH first converts the α-haloketone to the substituted cyclopropanone shown in brackets, and then to the sodium salt of cyclopentanecarboxylic acid.

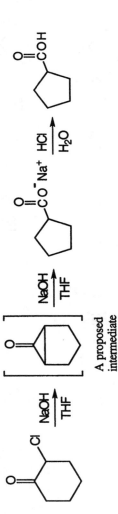

(a) Propose a mechanism for base-promoted conversion of 2-chlorocyclohexanone to the proposed intermediate.

A reasonable mechanism involves formation of an enolate anion that then displaces chloride in an intramolecular step that produces the three-membered ring intermediate.

Step 1:

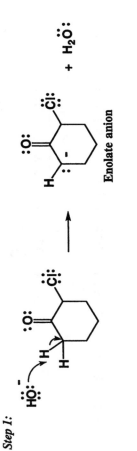

Step 2:

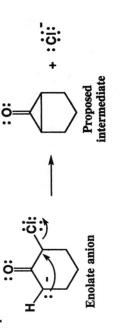

(b) Propose a mechanism for base-promoted conversion of the bracketed intermediate to sodium cyclopentanecarboxylate.

As stated in the hint, a reasonable mechanism can be written that starts with formation of a tetrahedral carbonyl addition intermediate, followed by collapse of the intermediate and breaking of one of the cyclopropane ring bonds. A proton transfer completes the reaction.

Step 1:

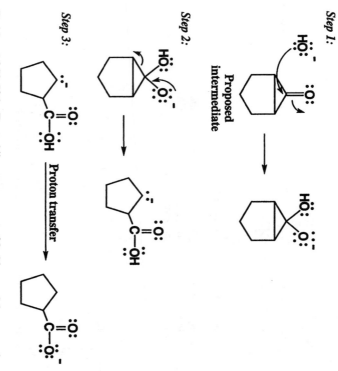

Proposed
intermediate

Step 2:

Step 3:

Proton transfer

Problem 16.52 If the Favorskii rearrangement of 2-chlorocyclohexanone is carried out using sodium ethoxide in ethanol, the product is ethyl cyclopentanecarboxylate. Propose a mechanism for this reaction.

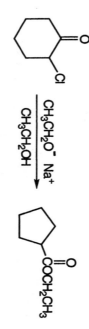

$$\xrightarrow[\text{CH}_3\text{CH}_2\text{OH}]{\text{CH}_3\text{CH}_2\text{O}^- \text{ Na}^+}$$

This mechanism is entirely analogous to that of the previous problem, except this time ethoxide is the base and nucleophile, not hydroxide.

Step 1:

Step 2:

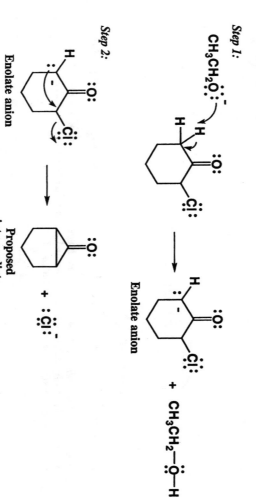

Enolate anion

Proposed
intermediate

Step 3:

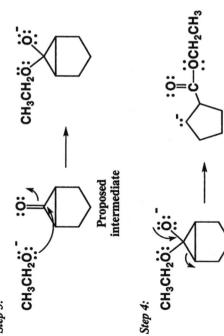

Proposed
intermediate

Step 4:

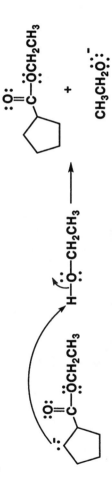

Step 5:

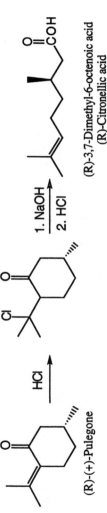

Problem 16.53 (R)-(+)-Pulegone is converted to (R)-citronellic acid by addition of HCl followed by treatment with NaOH. Propose a mechanism for each step in this transformation, including the regioselectivity of HCl addition.

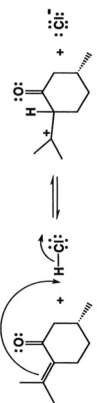

(R)-(+)-Pulegone

$\xrightarrow[\text{2. HCl}]{\text{1. NaOH}}$

(R)-3,7-Dimethyl-6-octenoic acid
(R)-Citronellic acid

The first step of the process involves protonation of the carbon-carbon double bond to create the tertiary carbocation shown below. Chloride reacts with this carbocation to give the chlorinated intermediate. This regiochemistry of hydrochlorination reflects the most stable carbocation that is possible. The other possible carbocation produced upon protonation would also have been a tertiary carbocation, but one that is α to the carbonyl group. Carbonyl groups are electron withdrawing and thus destabilizing to an adjacent carbocation.

Step 1:

(R)-(+)-Pulegone

Step 2:

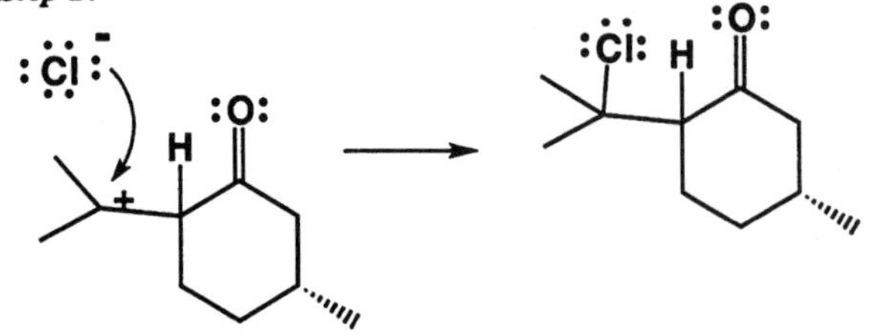

Hydroxide attacks the carbonyl group to give a tetrahedral carbonyl addition intermediate that then decomposes in Step 4 to break one bond of the six-membered ring while displacing chloride to create a new carbon-carbon double bond.

Step 3:

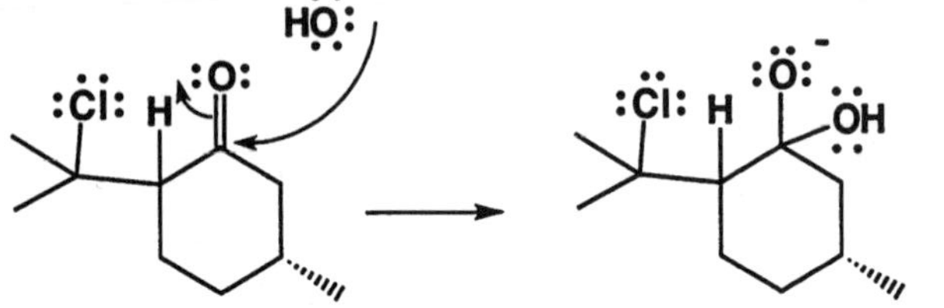

Step 4:

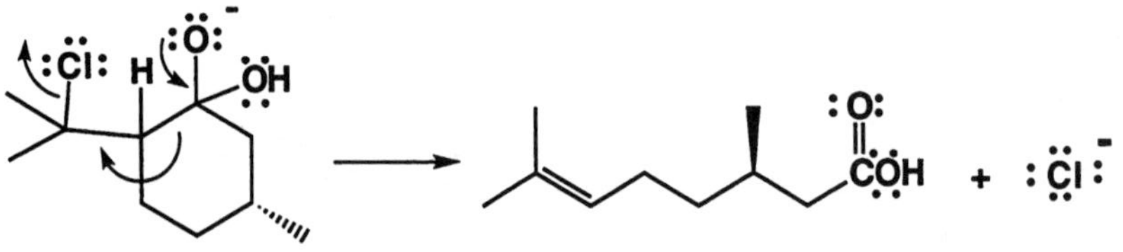

Oxidation/Reduction of Aldehydes and Ketones

Problem 16.54 Draw structural formulas for the products formed by treatment of butanal with the following reagents.

(a) LiAlH₄ followed by H₂O   (b) NaBH₄ in CH₃OH/H₂O   (c) H₂/Pt

CH₃CH₂CH₂CH₂OH   CH₃CH₂CH₂CH₂OH   CH₃CH₂CH₂CH₂OH

(d) Ag(NH₃)₂⁺ in NH₃/H₂O   (e) H₂CrO₄, heat   (f) HOCH₂CH₂OH, HCl

CH₃CH₂CH₂COO⁻NH₄⁺   CH₃CH₂CH₂COOH

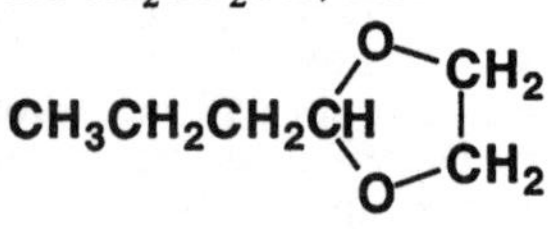

                                                                  + H₂O

(g) Zn(Hg)/HCl   (h) N₂H₄, KOH at 250°C   (i) C₆H₅NH₂

CH₃CH₂CH₂CH₃   CH₃CH₂CH₂CH₃

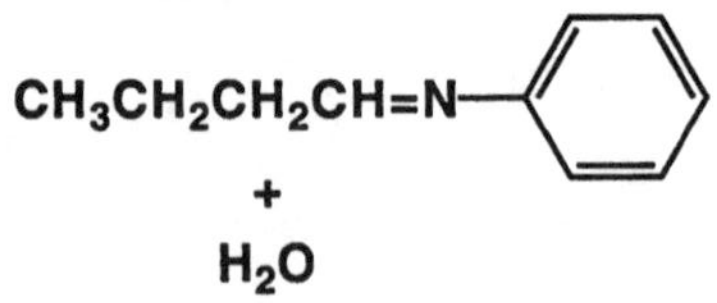

                                                                  + H₂O

(j) C₆H₅NHNH₂

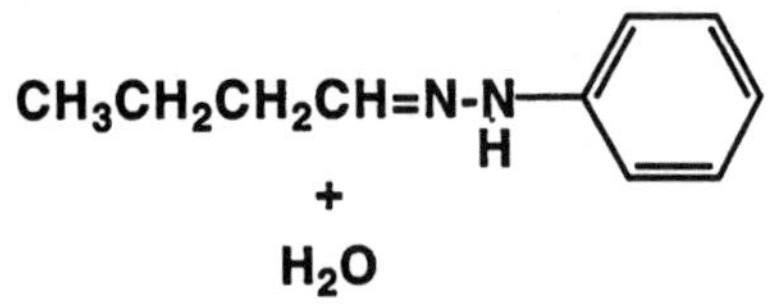

+ H₂O

Problem 16.55 Draw structural formulas for the products of the reaction of acetophenone with the reagents given in Problem 16.54.

Following are structural formulas for each product. Note that in parts (a), (b), and (c) a new stereocenter is created in the reaction so the products will actually be racemic mixtures.

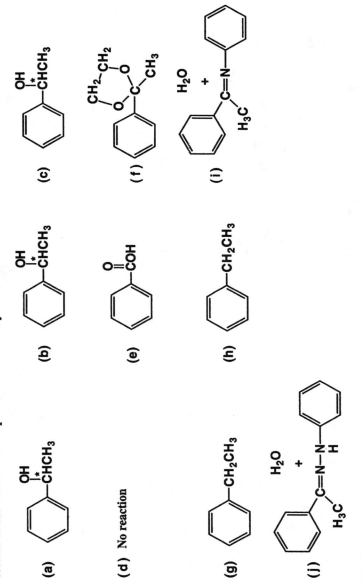

(a)

(b)

(c)

(d) No reaction

(e)

(f)

(g)

(h)

(i)

(j)

Synthesis
Problem 16.56 Starting with cyclohexanone, show how to prepare these compounds. In addition to the given starting material, use any other organic or inorganic reagents as necessary.
(a) Cyclohexanol

This transformation can be accomplished with any of three different sets of reagents:

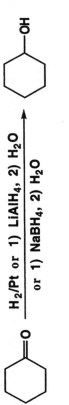

$$\xrightarrow[\text{or 1) NaBH}_4, \text{ 2) H}_2\text{O}]{\text{H}_2/\text{Pt or 1) LiAlH}_4, \text{ 2) H}_2\text{O}}$$

(b) Cyclohexene

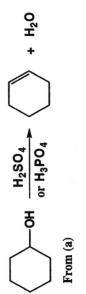

$$\xrightarrow[\text{or H}_3\text{PO}_4]{\text{H}_2\text{SO}_4}$$

+ H$_2$O

From (a)

(c) cis-1,2-Cyclohexanediol

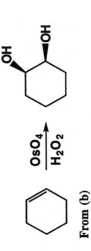

$$\xrightarrow[\text{H}_2\text{O}_2]{\text{OsO}_4}$$

From (b)

(d) 1-Methylcyclohexanol

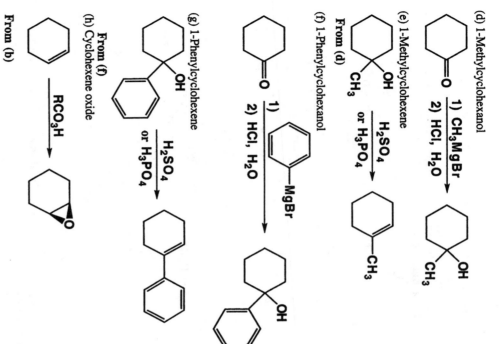

$$\text{(ketone)} \xrightarrow[\text{2) HCl, H}_2\text{O}]{\text{1) CH}_3\text{MgBr}} \text{(1-methylcyclohexanol)}$$

(e) 1-Methylcyclohexene

From (d)

$$\text{(1-methylcyclohexanol)} \xrightarrow[\text{or H}_3\text{PO}_4]{\text{H}_2\text{SO}_4} \text{(1-methylcyclohexene)}$$

(f) 1-Phenylcyclohexanol

$$\text{(ketone)} \xrightarrow[\text{2) HCl, H}_2\text{O}]{\text{1) PhMgBr}} \text{(1-phenylcyclohexanol)}$$

(g) 1-Phenylcyclohexene

From (f)

$$\text{(1-phenylcyclohexanol)} \xrightarrow[\text{or H}_3\text{PO}_4]{\text{H}_2\text{SO}_4} \text{(1-phenylcyclohexene)}$$

(h) Cyclohexene oxide

From (b)

$$\text{(cyclohexene)} \xrightarrow{\text{RCO}_3\text{H}} \text{(cyclohexene oxide)}$$

(i) trans-1,2-Cyclohexanediol

From (h)

Recall that ring-opening of an epoxide in acid gives the desired *trans* product. Compare this to part (c) of this problem in which the *cis* product is desired.

$$\text{(epoxide)} + \text{H}_2\text{O} \xrightarrow{\text{H}^+} \text{(trans-diol)} + \text{(trans-diol)}$$

(racemic mixture)

Problem 16.57 Show how to convert cyclopentanone to these compounds. In addition to cyclopentanone, use other organic or inorganic reagents as necessary.

(a)

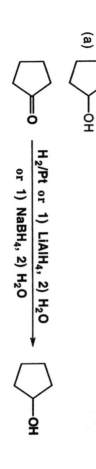
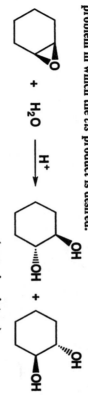

$$\text{(cyclopentanone)} \xrightarrow[\text{or 1) NaBH}_4, \text{ 2) H}_2\text{O}]{\text{H}_2/\text{Pt or 1) LiAlH}_4, \text{ 2) H}_2\text{O}} \text{(cyclopentanol)}$$

(b)

$\xrightarrow[\text{Pyridine}]{\text{SOCl}_2}$

from (a)

(c)

$\xrightarrow[\text{Ph}_3\overset{+}{\text{P}}-\overset{-}{\text{CH}}-\text{CH}=\text{CH}_2]{}$

(d)

$\xrightarrow[\text{Pyridine}]{\text{SOCl}_2}$

from (a)

1) Mg/ether

2)

3) HCl, H₂O

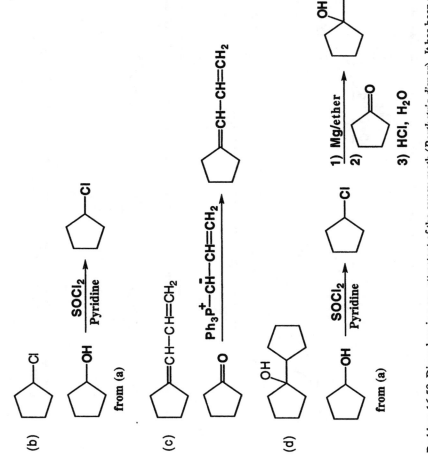

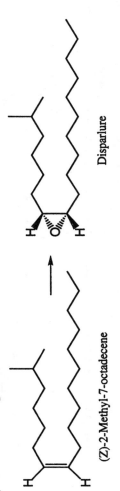

Problem 16.58 Disparlure is a sex attractant of the gypsy moth (*Portheria dispar*). It has been synthesized in the laboratory from the following (Z)-alkene.

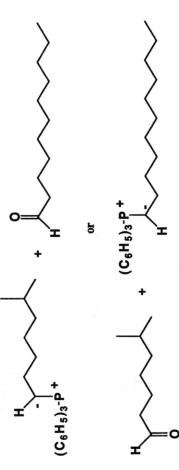

(Z)-2-Methyl-7-octadecene   Disparlure

(a) Propose two sets of reagents that might be combined in a Wittig reaction to give the indicated (Z)-alkene. Note that at least for simple phosphonium ylides, the product of a Wittig reaction is predominantly the (Z)-alkene.

(b) How might the (Z)-alkene be converted to disparlure?

The (Z)-alkene is converted to disparlure by reaction with a peracid.

(c) How many stereoisomers are possible for disparlure? How many are formed in the sequence you have chosen?

Disparlure has two stereocenters, so there are a total of 4 stereoisomers possible. Starting with the (Z)-alkene means that only (Z) epoxides are formed. Thus, there will only be two stereoisomers formed as shown below.

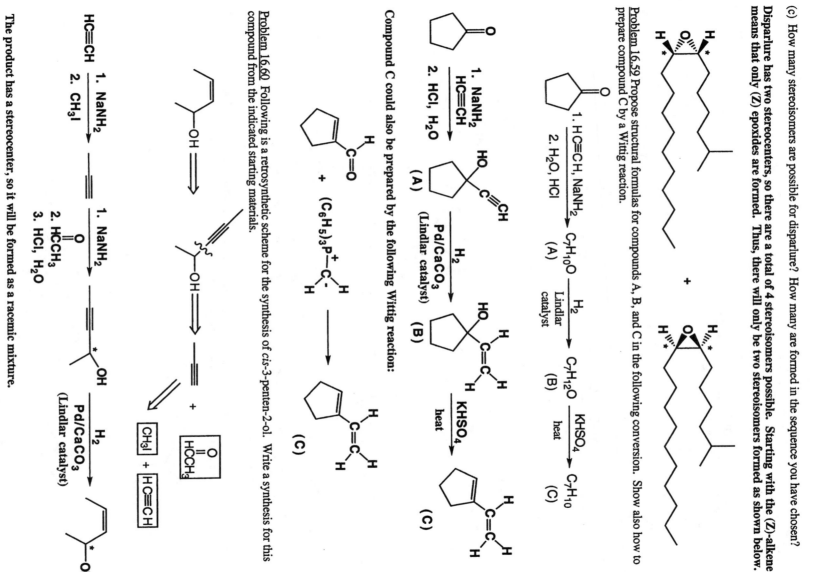

Problem 16.59 Propose structural formulas for compounds A, B, and C in the following conversion. Show also how to prepare compound C by a Wittig reaction.

The product has a stereocenter, so it will be formed as a racemic mixture.

Problem 16.61 Following is the structural formula of the tranquilizer Oblivon (meparfynol, The Merck Index, 12th ed., #5890). Propose a synthesis for this molecule starting with acetylene and a ketone.

Oblivon

Oblivon is prepared by reaction of sodium acetylide with 2-butanone followed by hydrolysis of the resulting sodium alkoxide in aqueous acid. It has a stereocenter, so it will be produced as a racemic mixture.

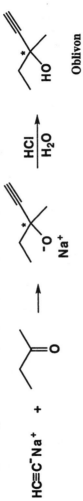

Oblivon

Problem 16.62 Following is the structural formula of surfynol, a defoaming surfactant. Describe the synthesis of this molecule from acetylene and a ketone.

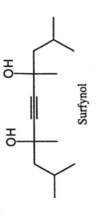

Surfynol

Surfynol (surfactant containing alkyne and alcohol functional groups) is synthesized by reaction of the sodium salt of acetylene in the presence of excess sodium amide with two moles 4-methyl-2-pentanone. Surfynol has two stereocenters, so a racemic mixture of stereoisomers is produced in this synthesis.

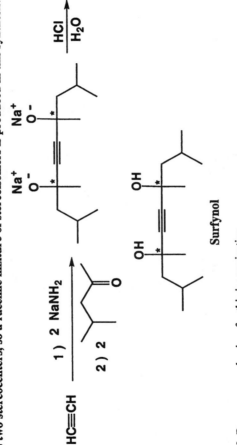

Surfynol

Problem 16.63 Propose a mechanism for this isomerization.

A reasonable mechanism for this transformation involves initial protonation of the -OH group, followed by loss of water to generate a resonance stabilized carbocation. This carbocation is attacked by water to give an intermediate that loses a proton to give an enol that tautomerizes to produce the final product.

Step 1:

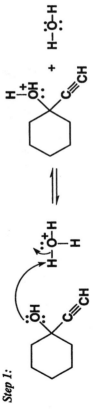

Step 2:

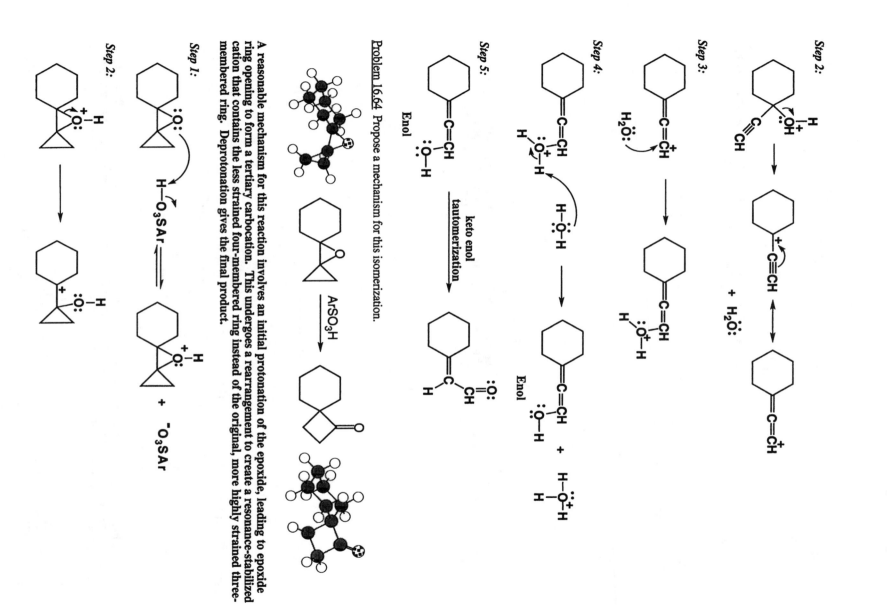

Step 3:

Step 4:

Enol

Step 5:

Enol

keto enol
tautomerization

Problem 16.64 Propose a mechanism for this isomerization.

ArSO₃H

Step 1:

H—O₃SAr ⟷

+ ⁻O₃SAr

Step 2:

A reasonable mechanism for this reaction involves an initial protonation of the epoxide, leading to epoxide ring opening to form a tertiary carbocation. This undergoes a rearrangement to create a resonance-stabilized cation that contains the less strained four-membered ring instead of the original, more highly strained three-membered ring. Deprotonation gives the final product.

Step 3:

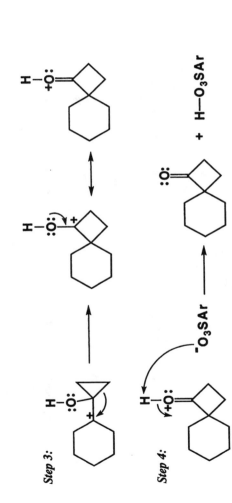

Step 4:

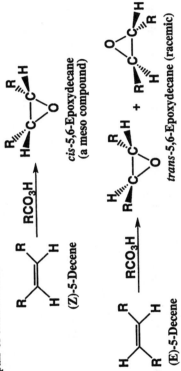

Problem 16.65 Starting with acetylene and 1-bromobutane as the only sources of carbon atoms, show how to synthesize the following:

All molecules in this problem are decanes. The way to build this carbon skeleton using the compounds given is by dialkylation of acetylene to give 5-decyne.

$$HC \equiv CH \xrightarrow[\text{2) } 2CH_3CH_2CH_2CH_2Br]{\text{1) } 2NaNH_2} CH_3(CH_2)_3C \equiv C(CH_2)_3CH_3$$

5-Decyne

Catalytic reduction of 5-decyne using hydrogen over a Lindlar catalyst gives (Z)-5-decene. Chemical reduction of 5-decyne using sodium or lithium metal in liquid ammonia gives (E)-5-decene.

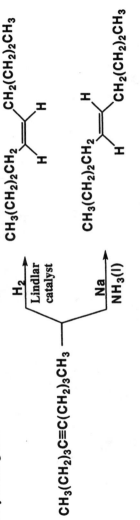

In the following formulas, the CH$_3$(CH$_2$)$_2$CH$_2$- group is represented as R-. Oxidation of (Z)-5-decene with a peracid gives *cis*-5,6-epoxydecane as a meso compound. Oxidation of (E)-5-decene by a peracid gives *trans*-5,6-epoxydecane as a pair of enantiomers.

(a) meso-5,6-Decanediol

Oxidation of (Z)-5-decene by osmium tetroxide in the presence of hydrogen peroxide gives meso-5,6-decanediol.

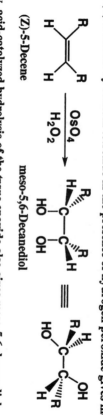

(Z)-5-Decene   meso-5,6-Decanediol

Alternatively, acid-catalyzed hydrolysis of the *trans* epoxide also gives meso-5,6-decanediol.

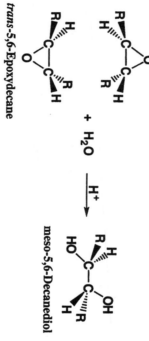

trans-5,6-Epoxydecane   meso-5,6-Decanediol

(b) Racemic 5,6-decanediol

Oxidation of (E)-5-decene by osmium tetroxide in the presence of hydrogen peroxide gives racemic 5,6-decanediol.

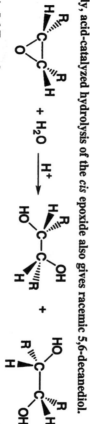

(E)-5-Decene   5,6-Decanediol (racemic)

Alternatively, acid-catalyzed hydrolysis of the *cis* epoxide also gives racemic 5,6-decanediol.

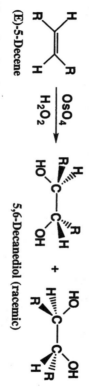

cis-5,6-Epoxydecane   5,6-Decanediol (racemic)

(c) 5-Decanone

Acid-catalyzed hydration of 5-decyne gives 5-decanone.

$$CH_3(CH_2)_3C{\equiv}C(CH_2)_3CH_3 + H_2O \xrightarrow[\text{HgSO}_4]{\text{H}_2\text{SO}_4} CH_3(CH_2)_3\overset{\displaystyle O}{\overset{\|}{C}}CH_2(CH_2)_3CH_3$$

5-Decyne   5-Decanone

(d) 5,6-Epoxydecane

See the beginning of this solution for methods to prepare both the *cis*-epoxide and the *trans*-epoxide.

(e) 5-Decanol

Catalytic reduction of 5-decanone using hydrogen over a transition metal catalyst, or chemical reduction using NaBH₄ or LiAlH₄ gives 5-decanol as a racemic mixture.

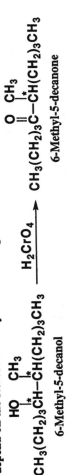

$$CH_3(CH_2)_3\overset{\overset{\displaystyle O}{\|}}{C}(CH_2)_4CH_3 + NaBH_4 \longrightarrow CH_3(CH_2)_3\overset{\overset{\displaystyle OH}{|}}{\underset{*}{C}H}(CH_2)_4CH_3$$

5-Decanone   5-Decanol (racemic)

(f) Decane

Catalytic reduction of 5-decyne using hydrogen over a transition metal catalyst gives decane.

$$CH_3(CH_2)_3C\equiv C(CH_2)_3CH_3 + 2H_2 \xrightarrow{\text{Pt}} CH_3(CH_2)_8CH_3$$

5-Decyne   Decane

(g) 6-Methyl-5-decanol

Treatment of either the *cis*-epoxide or the *trans*-epoxide from part (d) with methylmagnesium iodide gives 6-methyl-5-decanol as a racemic mixture, the exact stereochemistry of which depends on whether the cos or trans epoxide is used as starting material.

$$CH_3(CH_2)_3\overset{\overset{\displaystyle O}{\diagup\hspace{-0.5em}\diagdown}}{CH-CH}(CH_2)_3CH_3 \xrightarrow[\text{2) HCl, H}_2\text{O}]{\text{1) CH}_3\text{MgI}} CH_3(CH_2)_3\overset{\overset{\displaystyle HO}{|}}{\underset{*}{C}H}\overset{\overset{\displaystyle CH_3}{|}}{\underset{*}{C}H}-CH(CH_2)_3CH_3$$

cis or trans 5,6-Epoxydecane   6-Methyl-5-decanol

(h) 6-Methyl-5-decanone

Chromic acid oxidation of 6-methyl-5-decanol from part (g) gives 6-methyl-5-decanone, the stereochemistry of which will depend on the stereochemistry of the starting material.

$$CH_3(CH_2)_3\overset{\overset{\displaystyle HO}{|}}{\underset{*}{C}H}\overset{\overset{\displaystyle CH_3}{|}}{\underset{*}{C}H}-CH(CH_2)_3CH_3 \xrightarrow{\text{H}_2\text{CrO}_4} CH_3(CH_2)_3\overset{\overset{\displaystyle O}{\|}}{C}\overset{\overset{\displaystyle CH_3}{|}}{\underset{*}{C}H}-CH(CH_2)_3CH_3$$

6-Methyl-5-decanol   6-Methyl-5-decanone

<u>Problem 16.66</u> Following are the final steps in one industrial synthesis of vitamin A acetate:

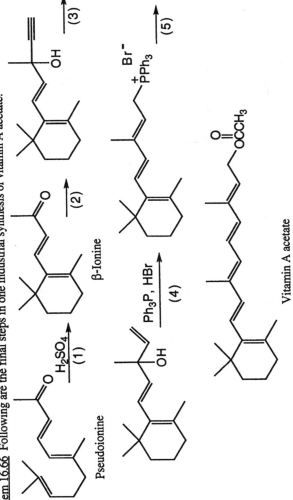

(a) Propose a mechanism for the acid-catalyzed cyclization in Step 1.

Step 1:

Step 2:

Step 3:

(b) Propose reagents to bring about Step 2.

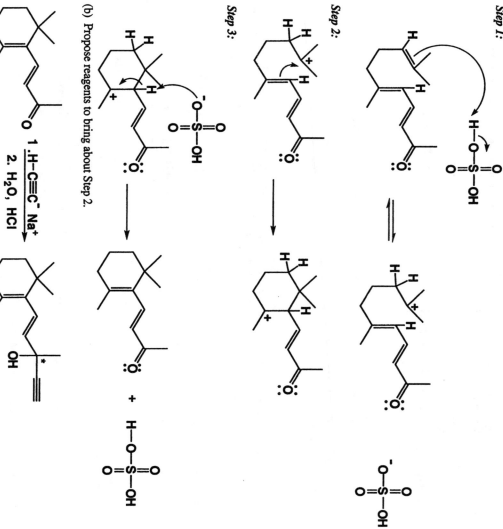

(c) Propose reagents to bring about Step 3

1. H–C≡C⁻ Na⁺
2. H₂O, HCl

(racemic mixture)

H₂

Lindlar
Catalyst

(racemic mixture assuming racemic
starting material)

(d) Propose a mechanism for formation of the phosphonium salt in Step 4.

Step 1:

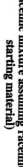
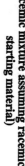
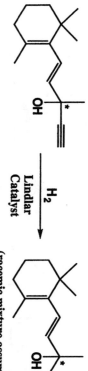

Step 2:

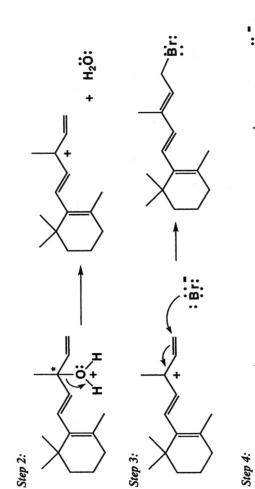

Step 3:

Step 4:

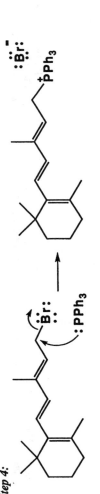

Note it is also possible the PPh₃ might attack the allylic cation directly in Step 3, without the intermediate alkyl bromide being formed.

(e) Show how Step 5 can be completed by a Wittig reaction.

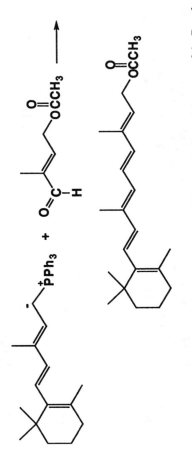

<u>Problem 16.67</u> Following is the structural formula of the principal sex pheromone of the Douglas fir tussock moth (*Orgyia pseudotsugata*), a severe defoliant of the fir trees of western North America.

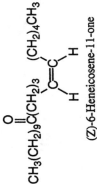

$$CH_3(CH_2)_9C(CH_2)_3 \quad (CH_2)_4CH_3$$

(Z)-6-Heneicosene-11-one

Several syntheses of this compound have been reported in the literature, starting materials for three of which are given here. Show a series of steps by which each set of starting materials could be converted into the target molecule.

When analyzing difficult synthesis problems such as these, it is important to 1) work backwards from the product and 2) focus on the key carbon-carbon bond forming steps. For example, in both (a) and (b), an alkyne is present that must be converted to a Z alkene. Working backwards, propose that hydrogenation using Lindlar's catalyst will likely be the last step in both of these cases. The question now becomes, what is the nature of the key carbon-carbon bond forming reaction?

(a) $CH_3(CH_2)_9CH=O$ + $CH_3(CH_2)_4C\equiv C(CH_2)_3OH$

The key to deducing the appropriate synthesis strategy is to recognize elements in the starting materials that suggest the carbon-carbon bond forming step. The key element to recognize in the above reagents is the aldehyde, that can be turned into a strong nucleophile by conversion to a dithiane.

$CH_3(CH_2)_9CH=O$
$\xrightarrow[\text{2) } CH_3CH_2CH_2CH_2Li]{\text{1) } H^+ / HS\text{~~}SH}$

[dithiane structure] $CH_3(CH_2)_9C^-\!\!\!<\!\!S\,S\!\!>$ Li$^+$

The alcohol can be converted into an alkyl chloride using thionyl chloride

$CH_3(CH_2)_4C\equiv C(CH_2)_3OH$ $\xrightarrow{SOCl_2}$ $CH_3(CH_2)_4C\equiv C(CH_2)_3Cl$

The carbon-carbon bond can now be formed in an alkylation reaction followed by removal of the dithiane group then hydrogenation using Lindlar's catalyst completes the synthesis. For the original reference, see R. Smith and D. Daves, Journal of Organic Chemistry, 40, 1593 (1975).

$CH_3(CH_2)_9C^-\!\!\!<\!\!S\,S\!\!>$ Li$^+$ + $CH_3(CH_2)_4C\equiv C(CH_2)_3Cl$ $\longrightarrow$ [dithiane] $CH_3(CH_2)_9C\!<\!S\,S\!>\!(CH_2)_3C\equiv C(CH_2)_4CH_3$

$\xrightarrow{H_2O,\ HgCl_2}$ $CH_3(CH_2)_9C(\!=\!O)(CH_2)_3C\equiv C(CH_2)_4CH_3$

$\xrightarrow[\substack{Pd/CaCO_3 \\ \text{(Lindlar catalyst)}}]{H_2}$ $CH_3(CH_2)_9C(\!=\!O)(CH_2)_3$ C=C $(CH_2)_4CH_3$ (Z)-6-Heneicosene-11-one

(b) $CH_3(CH_2)_9C(\!=\!O)(CH_2)_3C\equiv CH$ + $CH_3(CH_2)_4Br$

The key carbon-carbon bond forming step can be deduced from the presence of the terminal alkyne and alkyl halide. Here, alkylation of an acetylide anion with the alkyl halide will be used to construct the carbon backbone of the molecule. However, the ketone group must first be protected as an acetal, since ketones also react with acetylide anions. Reduction using Lindlar's catalyst completes the synthesis. For the original reference, see K. Mori, M. Uchida, and M. Matsui, Tetrahedron, 33, 385 (1977).

$CH_3(CH_2)_9C(\!=\!O)(CH_2)_3C\equiv CH$ $\xrightarrow{H^+ / HO\text{~~}OH}$ $CH_3(CH_2)_9C\!<\!O\,O\!>\!(CH_2)_3C\equiv CH$ $\xrightarrow{NaNH_2}$

$CH_3(CH_2)_9C\!<\!O\,O\!>\!(CH_2)_3C\equiv C^-\ Na^+$ $\xrightarrow{CH_3(CH_2)_4Br}$ $CH_3(CH_2)_9C\!<\!O\,O\!>\!(CH_2)_3C\equiv C(CH_2)_4CH_3$

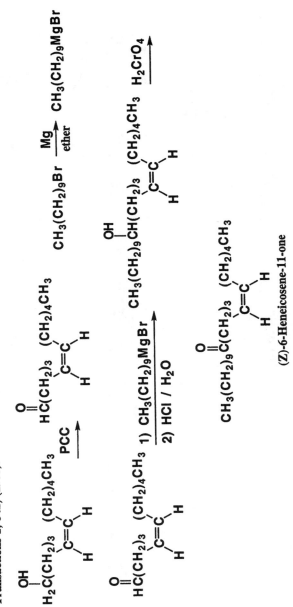

H_3O^+

$CH_3(CH_2)_9\overset{\text{O}}{\overset{\|}{C}}(CH_2)_3C\equiv C(CH_2)_4CH_3$

$\xrightarrow[\text{(Lindlar catalyst)}]{\begin{array}{c}H_2\\ \text{Pd/CaCO}_3\end{array}}$

(Z)-6-Heneicosene-11-one

(c)

$+ \; CH_3(CH_2)_5Br \; + \; CH_3(CH_2)_9Br$

This part is by far the most challenging since it is difficult to deduce the nature of the carbon-carbon bond forming steps. The key element to recognize is that the ring structure is actually a cyclic hemiacetal, in other words the cyclized form of 5-hydroxypentanal.

$\rightleftharpoons \; + \; H_2O$

As a result, the cyclic hemiacetal can take part in a Wittig reaction like other aldehyes. The first key carbon-carbon bond forming step uses a Wittig reagent prepared from hexyl bromide. The authors report that the desired Z isomer was the major product.

$CH_3(CH_2)_5Br \xrightarrow[\text{2) } CH_3CH_2CH_2CH_2Li]{\text{1) } PPh_3} CH_3(CH_2)_4\overset{+}{\overset{-}{C}H}{-}PPh_3$

The next carbon-carbon bond forming step must add the remaining ten carbon atoms onto the end of the molecule containing the primary hydroxyl group. Propose to carry out a Grignard reaction after first transforming the primary alcohol into an aldehyde group using PCC. Oxidation of the resulting alcohol to the ketone completes the synthesis. See M. Fetizon and C. Lazare, Journal of the Chemical Society, Perkin Transactions 1, 842, (1978).

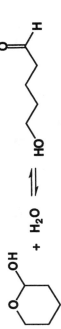

(Z)-6-Heneicosene-11-one

Note how three very different approaches all produce the same molecule in high yield. One of the fascinating aspects of organic synthesis is that there are usually several good ways to make a molecule, proving that organic chemistry involves not only knowlegde and experience, but considerable scientific creativity as well.

Problem16.68 Both (S)-citronellal and isopulegol are naturally occurring terpenes (Section 5.4). When (S)-citronellal is treated with tin(IV) chloride (a Lewis acid) followed by neutralization with aqueous ammonium chloride, isopulegol is obtained in 85% yield.

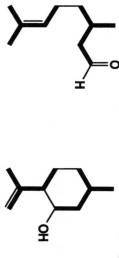

(S)-Citronellal
(C₁₀H₁₈O)

$$\xrightarrow[\text{2. NH}_4\text{Cl}]{\text{1. SnCl}_4,\ \text{CH}_2\text{Cl}_2}$$

Isopulegol
(C₁₀H₁₈O)

(a) Show that both compounds are terpenes.

The five carbon units characteristic of terpenes are shown in bold in the following representations.

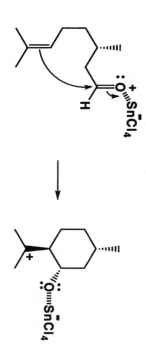

(b) Propose a mechanism for the conversion of (S)-citronellal to isopulegol.

In the first step of the reaction, the very strong Lewis acid SnCl₄ associates with the most Lewis basic site in the molecule, namely the carbonyl oxygen atom.

Step 1:

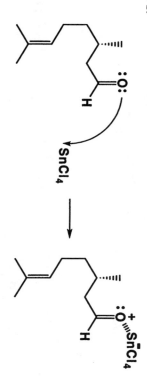

Association of the carbonyl oxygen atom with the strong Lewis acid SnCl₄ increases the electrophilic character of the carbonyl carbon atom, facilitating nucleophilic attack by the adjacent pi bond. Note that a six-membered ring is formed in the process, the relative stability of six-membered rings assists in the reaction process. Note also that the Lewis acid helps stabilize the oxygen anion produced.

Step 2:

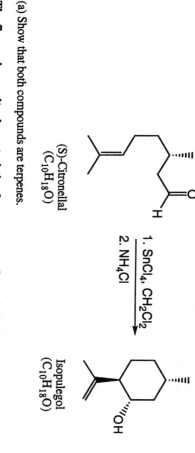

Step 3:

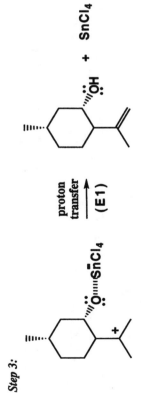

proton
transfer
(E1)

+ SnCl₄

The NH₄Cl removes the SnCl₄ from the reaction.

(c) How many stereocenters are present in isopulegol? How many stereoisomers are possible for a molecule with this number of stereocenters?

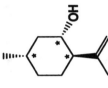

Isopulegol has three stereocenters so there are 2 x 2 x 2 = 8 possible stereoisomers.

(d) Isopulegol is formed as a single, enantiomerically pure stereoisomer. Account for the fact that only a single stereoisomer is formed.

The key step that sets the stereochemistry of the final product is Step 2. In this step, the pi bond of the alkene acts as a nucleophile to attack the carbonyl carbon atom. A six-membered ring is formed during the step. As a result, we can approximate the transition state leading to product as a six-membered ring. Of the two possible chair conformations that could form as transition states, the most stable one places the methyl group equatorial. During the reaction, the two new stereocenters will form such that both substituents will also be equatorial. The net result is a single enantiomer, derived from the most stable chair transition state.

This methyl group will
be equatorial in the most
stable chair conformation
of the transition state

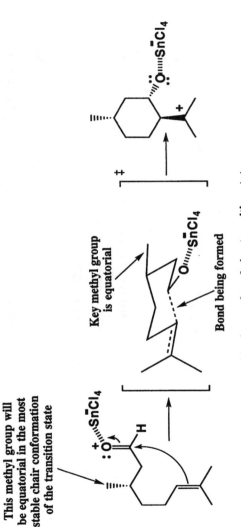

Key methyl group
is equatorial

Bond being formed

More stable six-membered ring transition state.

A careful evaluation of the stereocenters being created in the equatorial positions of the transition state shown above will reveal that they do accurately predict the observed stereochemistry in the isopulegol product.

CHAPTER 17
Solutions to the Problems

Problem 17.1 Each of these carboxylic acids has a well-recognized common name. A derivative of glyceric acid is an intermediate in glycolysis. Maleic acid is an intermediate in the tricarboxylic acid (TCA) cycle. Mevalonic acid is an intermediate in the biosynthesis of steroids. Write the IUPAC name for each compound. Be certain to specify configuration.

(a)

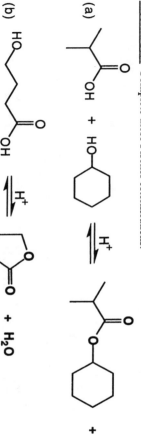

Glyceric acid

(R)-2,3-Dihydroxy-
propanoic acid

(b)

Maleic acid

(Z)-2-Butenedioic acid

(c)

Mevalonic acid

(R)-3,5-Dihydroxy-3-
methylpentanoic acid

Problem 17.2 Which is the stronger acid in each pair?

(a) CH_3COOH or $\boxed{CH_3SO_3H}$

Acetic acid   Methanesulfonic
                acid

$pK_a = 4.8$ $pK_a = -1.8$

(b)

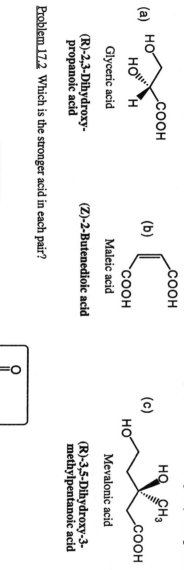

2-Oxopropanoic acid
(Pyruvic acid)

$pK_a = 2.4$

or

Propanoic acid

$pK_a = 4.8$

In (a), the third oxygen atom on sulfur increases acidity by both inductive and resonance effects compared to the carboxylic acid. In (b), the electron withdrawing carbonyl group in the 2 position of pyruvic acid increases acidity due to inductive effects.

Problem 17.3 Write equations for the reaction of each acid in Example 17.3 with ammonia, and name the carboxylic salt formed.

(a)

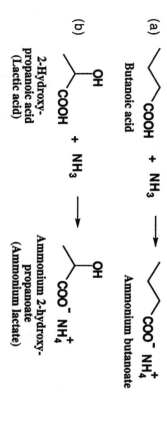

Butanoic acid

$+ NH_3 \longrightarrow$

Ammonium butanoate

(b)

2-Hydroxy-
propanoic acid
(Lactic acid)

$+ NH_3 \longrightarrow$

Ammonium 2-hydroxy-
propanoic
(Ammonium lactate)

Problem 17.4 Complete these Fischer esterifications.

(a)

(b)

Problem 17.5 Complete the following equations.

(a)

+ SOCl₂ ⟶ + SO₂ + HCl

(b)

+ SOCl₂ ⟶ + SO₂ + HCl

Problem 17.6 Account for the observation that the following β-ketoacid does not undergo thermal decarboxylation. It can be heated for extended periods at temperatures above its melting point without noticeable decomposition.

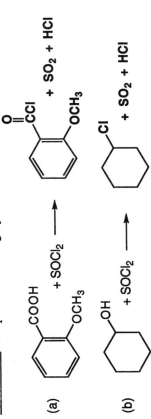

2-Oxobicyclo[2.2.1]heptane-1-carboxylic acid

The mechanism we have proposed for decarboxylation of a β-ketoacid involves formation of an enol intermediate and equilibrium of the enol, via keto-enol tautomerism, with the keto-form. Given the geometry of a bicyclo[2.2.1]alkane, it is not possible to have a carbon-carbon double bond to a bridgehead carbon because of the prohibitively high angle strain. Therefore, because the enol intermediate cannot be formed, this β-ketoacid does not undergo thermal decarboxylation.

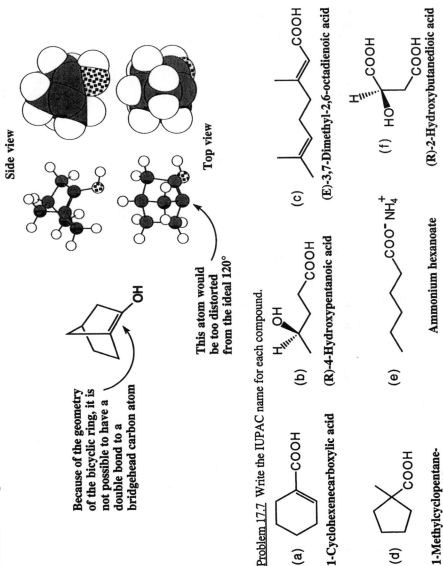

Side view

Top view

This atom would be too distorted from the ideal 120°

Because of the geometry of the bicyclic ring, it is not possible to have a double bond to a bridgehead carbon atom

Problem 17.7 Write the IUPAC name for each compound.

(a)

1-Cyclohexenecarboxylic acid

(b)

(R)-4-Hydroxypentanoic acid

(c)

(E)-3,7-Dimethyl-2,6-octadienoic acid

(d)

1-Methylcyclopentane-carboxylic acid

(e)

Ammonium hexanoate

(f)

(R)-2-Hydroxybutanedioic acid

Problem 17.8 Draw structural formulas for the following carboxylic acids.

(a) Phenylacetic acid

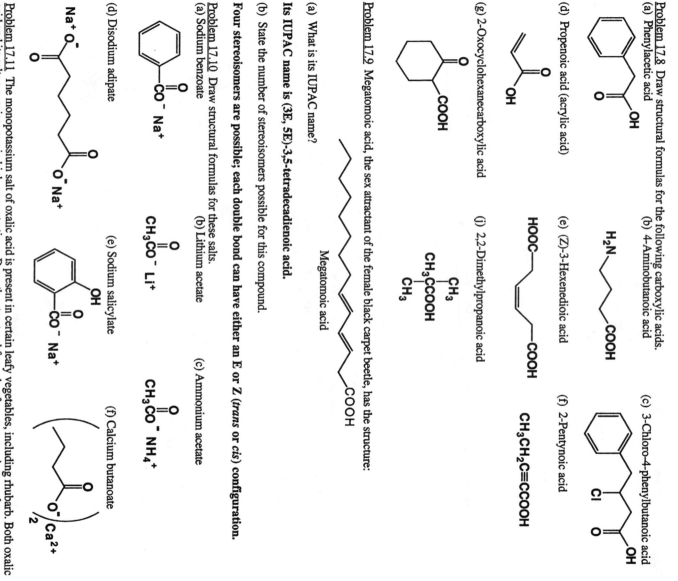

(b) 4-Aminobutanoic acid

H$_2$N——COOH

(c) 3-Chloro-4-phenylbutanoic acid

(d) Propenoic acid (acrylic acid)

(e) (Z)-3-Hexenedioic acid

HOOC——COOH

(f) 2-Pentynoic acid

CH$_3$CH$_2$C≡CCOOH

(g) 2-Oxocyclohexanecarboxylic acid

(j) 2,2-Dimethylpropanoic acid

CH$_3$
CH$_3$CCOOH
CH$_3$

Problem 17.9 Megatomoic acid, the sex attractant of the female black carpet beetle, has the structure:

Megatomoic acid

(a) What is its IUPAC name?

Its IUPAC name is (3E, 5E)-3,5-tetradecadienoic acid.

(b) State the number of stereoisomers possible for this compound.

Four stereoisomers are possible; each double bond can have either an E or Z (*trans* or *cis*) configuration.

Problem 17.10 Draw structural formulas for these salts.

(a) Sodium benzoate

CO——Na$^+$

(b) Lithium acetate

CH$_3$CO——Li$^+$

(c) Ammonium acetate

CH$_3$CO——NH$_4$$^+$

(d) Disodium adipate

Na$^+$O——O——Na$^+$

(e) Sodium salicylate

OH
CO——Na$^+$

(f) Calcium butanoate

(O——)$_2$Ca^{2+}

Problem 17.11 The monopotassium salt of oxalic acid is present in certain leafy vegetables, including rhubarb. Both oxalic acid and its salts are poisonous in high concentrations. Draw the structural formula of monopotassium oxalate.

HOC—CO——K$^+$

Monopotassium oxalate

Problem 17.12 Potassium sorbate is added as a preservative to certain foods to prevent bacteria and molds from causing food spoilage and to extend the foods' shelf life. The IUPAC name of potassium sorbate is potassium (2E,4E)-2,4-hexadienoate. Draw the structural formula of potassium sorbate.

O——K$^+$

Potassium sorbate

Problem 17.13 Zinc 10-undecenoate, the zinc salt of 10-undecenoic acid, is used to treat certain fungal infections, particularly *tinea pedis* (athlete's foot). Draw the structural formula of this zinc salt.

$$\left(CH_2{=}CH(CH_2)_8\overset{\overset{\displaystyle O}{\|}}{C}O^{-} \right)_2 Zn^{2+}$$

Zinc 10-undecenoate

Problem 17.14 On a cyclohexane ring, an axial carboxyl group has a conformational energy of 5.9 kJ (1.4 kcal)/mol relative to an equatorial carboxyl group. Consider the equilibrium for the alternative chair conformations of *trans*-1,4-cyclohexanedicarboxylic acid. Draw the less stable chair conformation on the left of the equilibrium arrows and the more stable chair on the right. Calculate ΔG° for the equilibrium as written, and calculate the ratio of more stable chair to less stable chair at 25°C.

$$\Delta G° = -11.8 \text{ kJ/mol}$$

As written, the conformation on the right is (2 x 5.9) = 11.8 kJ/mol more stable than the conformation on the left.

At equilibrium the relative amounts of each form are given by the equation:

$$\Delta G° = -RT \ln K_{eq}$$

Here K_{eq} refers to the ratio of the alternative chair conformations. Converting to base 10 and rearranging gives

$$\log K_{eq} = \frac{-\Delta G°}{(2.303)RT}$$

Taking the antilog of both sides gives:

$$K_{eq} = 10^{\left(\frac{-\Delta G°}{(2.303)RT}\right)}$$

Plugging in the values for ΔG°, R, and 298 K gives the answer:

$$\boxed{K_{eq} = 10^{\left(\frac{-(-11.8 \text{ kJ/mol})}{(2.303)(8.314 \times 10^{-3} \text{ kJ/mol K})(298 \text{ K})}\right)} = 10^{2.06} = 1.17 \times 10^2}$$

Physical Properties
Problem 17.15 Arrange the compounds in each set in order of increasing boiling point:

The better the hydrogen bond capability, the higher the boiling point.

(a) $CH_3(CH_2)_5COOH$ $CH_3(CH_2)_6CH$ $CH_3(CH_2)_6CH_2OH$

$\overset{\overset{\displaystyle O}{\|}}{}$

The following are listed in order of increasing boiling point:

$CH_3(CH_2)_6\overset{\overset{\displaystyle O}{\|}}{C}H$ $CH_3(CH_2)_6CH_2OH$ $CH_3(CH_2)_5\overset{\overset{\displaystyle O}{\|}}{C}OH$

bp 171°C   bp 195°C   bp 223°C

(b)

The following are listed in order of increasing boiling point:

bp 35°C   bp 117°C   bp 141°C

Problem 17.16 Acetic acid has a boiling point of 118°C, whereas its methyl ester has a boiling point of 57°C. Account for the fact that the boiling point of acetic acid is higher than that of its methyl ester, even though acetic acid has a lower molecular weight.

Acetic acid can make strong hydrogen bonds, but the methyl ester lacks a hydrogen bonding hydrogen atom. Thus, acetic acid will have the much higher boiling point compared to the methyl ester.

Can form hydrogen bonds with this H atom

Acetic acid
bp 118°C

Methyl acetate
bp 57°C

Spectroscopy

Problem 17.17 Given here are ^{1}H-NMR and ^{13}C-NMR spectral data for nine compounds. Each compound shows strong absorption between 1720 and 1700 cm^{-1}, and strong, broad absorption over the region 2500 - 3500 cm^{-1}. Propose a structural formula for each compound. Refer to Appendices 4, 5, and 6 for spectral correlation tables.

(a) $C_5H_{10}O_2$

^{1}H-NMR	^{13}C-NMR
0.94 (t, 3H)	180.71
1.39 (m, 2H)	33.89
1.62 (m, 2H)	26.76
2.35 (t, 2H)	22.21
12.0 (s, 1H)	13.69

0.94 1.39 1.62 2.35 12.0
CH₃CH₂CH₂CH₂COOH

(b) $C_6H_{12}O_2$

^{1}H-NMR	^{13}C-NMR
1.08 (s, 9H)	179.29
2.23 (s, 2H)	47.82
12.1 (s, 1H)	30.62
	29.57

1.08
CH₃
1.08
CH₃CCH₂COOH 2.23 12.1
CH₃
1.08

(c) $C_5H_8O_4$

^{1}H-NMR	^{13}C-NMR
0.93 (t, 3H)	170.94
1.80 (m, 2H)	53.28
3.10 (t, 1H)	21.90
12.7 (s, 2H)	11.81

12.7
HOOCCHCOOH 12.7
CH₂CH₃ 3.10 1.80 0.93

(d) $C_5H_8O_4$

^{1}H-NMR	^{13}C-NMR
1.29 (s, 6H)	174.01
12.8 (s, 2H)	48.77
	22.56

1.29
CH₃
12.8 **HOOCCCOOH** 12.8
CH₃
1.29

(e) $C_4H_6O_2$

^{1}H-NMR	^{13}C-NMR
1.91 (d, 3H)	172.26
5.86 (d, 1H)	147.53
7.10 (m, 1H)	122.24
12.4 (s, 1H)	18.11

1.91 7.10 5.86 12.4
CH₃CH=CHCOOH

(f) $C_3H_4Cl_2O_2$

^{1}H-NMR	^{13}C-NMR
2.34 (s, 3H)	171.82
11.3 (s, 1H)	79.36
	34.02

Cl
CH₃CCOOH 11.3
Cl 2.34

(g) $C_5H_8Cl_2O_2$

^1H-NMR	^{13}C-NMR
1.42 (s, 6H)	180.15
6.10 (s, 1H)	77.78
12.4 (s, 1H)	51.88
	20.71

$$\overset{1.42}{C}H_3$$
$$\overset{6.10}{Cl_2CHC}\overset{}{C}\overset{12.4}{COOH}$$
$$\overset{1.42}{C}H_3$$

(h) $C_5H_9BrO_2$

^1H-NMR	^{13}C-NMR
0.97 (t, 3H)	176.36
1.50 (m, 2H)	45.08
2.05 (m, 2H)	36.49
4.25 (t, 1H)	20.48
12.1 (s, 1H)	13.24

$$\overset{}{B\,r}$$
$$\overset{0.97}{C}H_3\overset{1.50}{C}H_2\overset{2.05}{C}H_2\overset{4.25}{C}H\overset{12.1}{C}HCOOH$$

(i) $C_4H_8O_3$

^1H-NMR	^{13}C-NMR
2.62 (t, 2H)	177.33
3.38 (s, 3H)	67.55
3.68 (t, 2H)	58.72
11.5 (s, 1H)	34.75

$$\overset{3.38}{C}H_3O\overset{3.68}{C}H_2\overset{2.62}{C}H_2\overset{11.5}{C}OOH$$

Preparation of Carboxylic Acids
Problem 17.18 Complete these reactions:

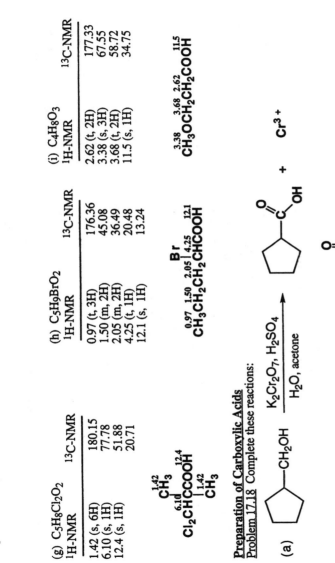

(a)   cyclopentyl-CH_2OH → $K_2Cr_2O_7, H_2SO_4$ / H_2O, acetone → cyclopentyl-COOH + Cr^{3+}

(b) HO— *—CHO with OH → 1. $Ag(NH_3)_2^+$ 2. H_2O, HCl → HO—*—$COOH$ with OH + $Ag°$

The stereochemistry of the product will be the same as the stereochemistry of the starting material.

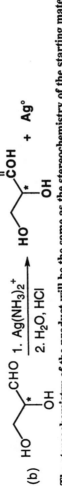

(c)   1. Cl_2, KOH in water/dioxane 2. HCl, H_2O →

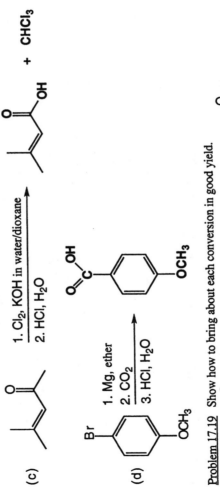

(d) Br—(benzene)—OCH_3 → 1. Mg, ether 2. CO_2 3. HCl, H_2O → HO_2C—(benzene)—OCH_3

Problem 17.19 Show how to bring about each conversion in good yield.

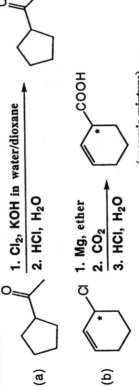

(a)   cyclopentyl-C(=O)-CH_3 → 1. Cl_2, KOH in water/dioxane 2. HCl, H_2O → cyclopentyl-COOH

(b)   Cl—(cyclohexene)* → 1. Mg, ether 2. CO_2 3. HCl, H_2O → (cyclohexene)*—COOH (racemic mixture)

(c)

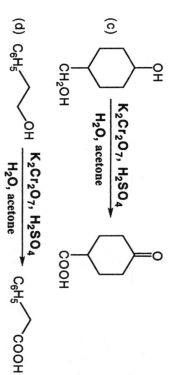

$$\text{OH} \xrightarrow[\text{H}_2\text{O, acetone}]{\text{K}_2\text{Cr}_2\text{O}_7, \text{H}_2\text{SO}_4} \text{COOH}$$

(d) C_6H_5 \quad $\xrightarrow[\text{H}_2\text{O, acetone}]{\text{K}_2\text{Cr}_2\text{O}_7, \text{H}_2\text{SO}_4}$ \quad C_6H_5 \quad COOH

<u>Problem 17.20</u> Show how to prepare pentanoic acid from these compounds:

(a) 1-Pentanol

Oxidation of 1-pentanol by chromic acid gives pentanoic acid.

$$\text{CH}_3(\text{CH}_2)_3\text{CH}_2\text{OH} \xrightarrow[\text{H}_2\text{O, acetone}]{\text{K}_2\text{Cr}_2\text{O}_7, \text{H}_2\text{SO}_4} \text{CH}_3(\text{CH}_2)_3\text{COOH} + \text{Cr}^{3+}$$

(b) Pentanal

Oxidation of pentanal by chromic acid, Tollens' solution, or molecular oxygen gives pentanoic acid.

$$\text{CH}_3(\text{CH}_2)_3\overset{\text{O}}{\overset{\|}{\text{CH}}} + \text{O}_2 \longrightarrow \text{CH}_3(\text{CH}_2)_3\overset{\text{O}}{\overset{\|}{\text{COH}}}$$

(c) 1-Pentene

Hydroboration/oxidation of 1-pentene gives 1-pentanol. Oxidation of 1-pentanol as in part (a) gives pentanoic acid.

$$\text{CH}_3\text{CH}_2\text{CH}_2\text{CH}=\text{CH}_2 \xrightarrow[\text{2) H}_2\text{O}_2, \text{ NaOH}]{\text{1) BH}_3 \cdot \text{THF}} \text{CH}_3(\text{CH}_2)_3\text{CH}_2\text{OH} \text{ then as in part (a)}$$

(d) 1-Butanol

Conversion of 1-butanol to 1-bromobutane, then to butyllithium or butylmagnesium bromide followed by carbonation and then acidification gives pentanoic acid.

$$\text{CH}_3(\text{CH}_2)_3\text{OH} \xrightarrow[\text{2) Mg,ether}]{\text{1) PBr}_3} \text{CH}_3(\text{CH}_2)_3\text{MgBr} \xrightarrow[\text{2) HCl,H}_2\text{O}]{\text{1) CO}_2} \text{CH}_3(\text{CH}_2)_3\text{COOH}$$

(e) 1-Bromopropane

Treatment of 1-bromopropane with magnesium in ether followed by treatment of the Grignard reagent with ethylene oxide gives 1-pentanol. Oxidation of 1-pentanol as in part (a) gives pentanoic acid.

$$\text{CH}_3\text{CH}_2\text{CH}_2\text{Br} \xrightarrow[]{\text{Mg, ether}} \text{CH}_3\text{CH}_2\text{CH}_2\text{MgBr} \xrightarrow[\text{2) HCl, H}_2\text{O}]{\text{1) H}_2\text{C}-\text{CH}_2 \text{ (O)}} \text{CH}_3(\text{CH}_2)_3\text{CH}_2\text{OH}$$

$$\xrightarrow[]{\text{Oxidize as in part (a)}} \text{CH}_3(\text{CH}_2)_3\text{COOH}$$

(f) 2-Hexanone

Haloform reaction of 2-hexanone gives pentanoic acid and a haloform.

$$CH_3(CH_2)_3\overset{O}{\overset{\|}{C}}CH_3 \xrightarrow[\text{2) HCl, H}_2\text{O}]{\text{1) Br}_2, \text{ NaOH}} CH_3(CH_2)_3\overset{O}{\overset{\|}{C}}OH + CHBr_3$$

(g) 1-Hexene

Acid-catalyzed hydration or oxymercuration/reduction of 1-hexene gives 2-hexanol. Oxidation of 2-hexanol as in part (f) gives pentanoic acid.

$$CH_3(CH_2)_3CH{=}CH_2 + H_2O \xrightarrow{\text{H}_2\text{SO}_4} CH_3(CH_2)_3\overset{\overset{*}{OH}}{\overset{|}{C}}HCH_3$$

$$\xrightarrow[\substack{K_2Cr_2O_7, \ H_2SO_4 \\ H_2O, \ \text{acetone}}]{\text{PCC or}} CH_3(CH_2)_3\overset{O}{\overset{\|}{C}}CH_3 \xrightarrow[\text{2) HCl, H}_2\text{O}]{\text{1) Br}_2, \text{ NaOH}} CH_3(CH_2)_3\overset{O}{\overset{\|}{C}}OH$$

Problem 17.21 Draw the structural formula of a compound of the given molecular formula that, on oxidation by potassium dichromate in aqueous sulfuric acid, gives the carboxylic acid or dicarboxylic acid shown.

(a) $C_6H_{14}O$ $\xrightarrow{\text{oxidation}}$

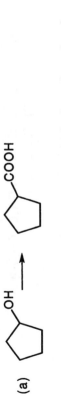

(b) $C_6H_{12}O$ $\xrightarrow{\text{oxidation}}$

(c) $C_6H_{14}O_2$ $\xrightarrow{\text{oxidation}}$

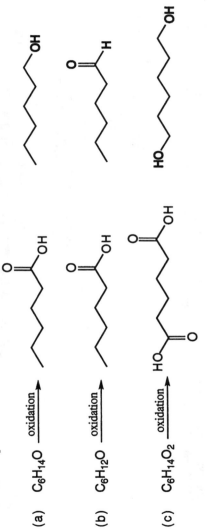

Problem 17.22 Show the reagents and experimental conditions necessary to bring about each conversion in good yield.

(a)

The most convenient way to add a carbon atom in the form of a carboxyl group is carbonation of an organolithium or organomagnesium compound.

$$\text{(cyclopentyl)-OH} \xrightarrow[\text{2) Mg, ether}]{\text{1) SOCl}_2} \text{(cyclopentyl)-MgCl} \xrightarrow[\text{2) HCl, H}_2\text{O}]{\text{1) CO}_2} \text{(cyclopentyl)-COOH}$$

(b)

$$CH_3\text{-}\overset{\displaystyle CH_3}{\underset{\displaystyle CH_3}{C}}\text{-}OH \longrightarrow CH_3\text{-}\overset{\displaystyle CH_3}{\underset{\displaystyle CH_3}{C}}\text{-}COOH$$

Use the same set of reactions in this part as you used in part (a), except because the starting material is a tertiary alcohol, HCl is used in place of $SOCl_2$.

$$CH_3\overset{\displaystyle CH_3}{\underset{\displaystyle CH_3}{C}}OH \xrightarrow{HCl} CH_3\overset{\displaystyle CH_3}{\underset{\displaystyle CH_3}{C}}Cl \xrightarrow[\substack{\text{2) } CO_2 \\ \text{3) } HCl,\ H_2O}]{\text{1) Mg, ether}} CH_3\overset{\displaystyle CH_3}{\underset{\displaystyle CH_3}{C}}COOH$$

(c)

$$CH_3\overset{\displaystyle CH_3}{\underset{\displaystyle CH_3}{C}}OH \longrightarrow CH_3\overset{\displaystyle CH_3}{CH}COOH$$

This conversion is best accomplished by acid-catalyzed dehydration of the tertiary alcohol to an alkene followed by hydroboration/oxidation. These two reactions in sequence shift the -OH group from the tertiary carbon to a primary carbon. Oxidation of this primary alcohol with chromic acid gives 2-methylpropanoic acid.

$$CH_3\overset{\displaystyle CH_3}{\underset{\displaystyle CH_3}{C}}OH \xrightarrow[(-H_2O)]{H_2SO_4} CH_3\overset{\displaystyle CH_3}{C}{=}CH_2 \xrightarrow[\substack{\text{2) } H_2O_2,\ NaOH}]{\text{1) } BH_3 \cdot THF} CH_3\overset{\displaystyle CH_3}{CH}CH_2OH$$

$$\xrightarrow[H_2O,\ \text{acetone}]{K_2Cr_2O_7,\ H_2SO_4} CH_3\overset{\displaystyle CH_3}{CH}COOH$$

(d)

$$CH_3\overset{\displaystyle CH_3}{\underset{\displaystyle CH_3}{C}}OH \longrightarrow CH_3\overset{\displaystyle CH_3}{CH}CH_2COOH$$

From part (c)

Repeat the first set of steps as in part (c) to convert *tert*-butyl alcohol to isobutyl alcohol. Then use the sequence of steps as in part (a) and (b) to convert this alcohol to a carboxylic acid containing one more carbon atom.

$$CH_3\overset{\displaystyle CH_3}{CH}CH_2OH \xrightarrow[\substack{\text{2) Mg, ether}}]{\text{1) } SOCl_2} CH_3\overset{\displaystyle CH_3}{CH}CH_2Cl \xrightarrow[\substack{\text{2) } HCl,\ H_2O}]{\text{1) } CO_2} CH_3\overset{\displaystyle CH_3}{CH}CH_2COOH$$

(e) $CH_3CH{=}CHCH_3 \longrightarrow CH_3CH{=}CHCH_2COOH$

This problem is analogous to part (a) except an allylic halogenation of the alkene starting material is the first step.

$$CH_3CH{=}CHCH_3 \xrightarrow[\text{Light or peroxides}]{NBS} CH_3CH{=}CHCH_2Br \xrightarrow[\substack{\text{2) } CO_2 \\ \text{3) } HCl,\ H_2O}]{\text{1) Mg, ether}} CH_3CH{=}CHCH_2COOH$$

Problem 17.23 Succinic acid can be synthesized by the following series of reactions from acetylene. Show the reagents and experimental conditions necessary to carry out this synthesis in good yield.

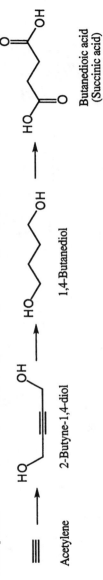

Acetylene → 2-Butyne-1,4-diol → 1,4-Butanediol → Butanedioic acid (Succinic acid)

Two one-carbon fragments in the form of formaldehyde are added to the carbon skeleton of acetylene. To bring about this double addition, acetylene is treated with two moles of sodamide followed by two moles of formaldehyde. Acidification of the resulting disodium salt gives 2-butyne-1,4-diol.

HC≡CH

1. 2 NaNH₂
2. 2 H–C–H (O)

→ Na⁺O⁻ ... O⁻Na⁺ + 2 NH₃

Na⁺O⁻ ... O⁻Na⁺ + 2HCl → HO ... OH

2-Butyne-1,4-diol

Catalytic reduction of the carbon-carbon triple bond in 2-butyne-1,4-diol over a transition metal catalyst gives 1,4-butanediol.

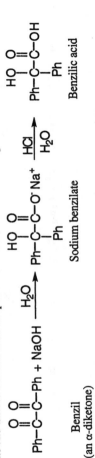

HO ... OH + 2 H₂ → (Pt or other transition metal catalyst) HO ... OH

2-Butyne-1,4-diol   1,4-Butanediol

Oxidation of 1,4-butanediol by chromic acid gives succinic acid.

HO ... OH → (K₂Cr₂O₇, H₂SO₄ / H₂O, acetone) → Butanedioic acid (Succinic acid)

1,4-Butanediol

Problem 17.24 The reaction of an α–diketone with concentrated sodium or potassium hydroxide to give the salt of an α–hydroxyacid is given the general name benzil-benzilic acid rearrangement. It is illustrated by the conversion of benzil to sodium benzilate and then to benzilic acid. Propose a mechanism for this rearrangement.

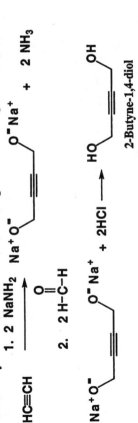

Ph–C–C–Ph + NaOH →(H₂O) Ph–C–C–O⁻Na⁺ →(HCl / H₂O) Ph–C–C–OH

Benzil (an α–diketone)   Sodium benzilate   Benzilic acid

Addition of hydroxide ion to one of the carbonyl groups to form a tetrahedral carbonyl addition intermediate followed by collapse of this intermediate with simultaneous regeneration of the carbonyl group and migration of a phenyl group gives sodium benzilate.

Step 1:

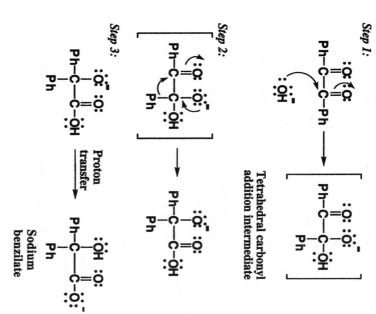

Tetrahedral carbonyl
addition intermediate

Step 2:

Proton
transfer

Step 3:

Sodium
benzilate

Acidity of Carboxylic Acids
Problem 17.25 Select the stronger acid in each set.
(a) Phenol (pK$_a$ 9.95) and benzoic acid (pK$_a$ 4.17)

Recall that pK$_a$ is the negative log$_{10}$ of K$_a$. The smaller the pK$_a$, the stronger the acid, so benzoic acid is the stronger acid.

(b) Lactic acid (K$_a$ 8.4 x 10^{-4}) and ascorbic acid (K$_a$ 7.9 x 10^{-5})

The larger the value of K$_a$, the stronger the acid, so lactic acid is the stronger acid.

Problem 17.26 Assign the acid in each set its appropriate pK$_a$.

(a)

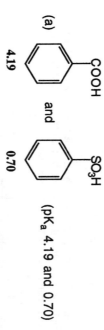

COOH   and   SO$_3$H

4.19   0.70   (pK$_a$ 4.19 and 0.70)

The third oxygen on sulfur increases acidity compared to an analogous carboxylic acid.

(b) CH$_3$CCH$_2$COOH   and   CH$_3$CH$_2$CCOOH   (pK$_a$ 3.58 and 2.49)

3.58   2.49

The electron withdrawing carbonyl group increases acidity through inductive effects, so the closer the carbonyl group to the carboxylic acid, the greater the effect and the stronger the acid.

(c) CH$_3$CH$_2$COOH   and   N≡CCH$_2$COOH   (pK$_a$ 4.78 and 2.45)

4.78   2.45

The nitrile is electron withdrawing, so it increases acidity through inductive effects.

Problem 17.27 Low-molecular-weight dicarboxylic acids normally exhibit two different pK_a values. Ionization of the first carboxyl group is easier than the second. This effect diminishes with molecular size, and, for adipic acid and longer chain dicarboxylic acids, the two acid ionization constants differ by about one pK unit.

Dicarboxylic Acid	Structural Formula	pK_{a1}	pK_{a2}
Oxalic	HOOCCOOH	1.23	4.19
Malonic	HOOCCH$_2$COOH	2.83	5.69
Succinic	HOOC(CH$_2$)$_2$COOH	4.16	5.61
Glutaric	HOOC(CH$_2$)$_3$COOH	4.31	5.41
Adipic	HOOC(CH$_2$)$_4$COOH	4.43	5.41

Why do the two pK_a values differ more for the shorter chain dicarboxylic acids than for the longer chain dicarboxylic acids?

For these dicarboxylic acids, in going from the first dissociation to the second, the molecule is changing from a monoanion to a dianion. Electrostatic repulsion hinders formation of two negative charges in nearby regions of space. Thus, the second pK_a values are higher than the first, and the effect is more pronounced the shorter the chain between the two carboxyl groups.

Problem 17.28 Complete the following acid-base reactions:

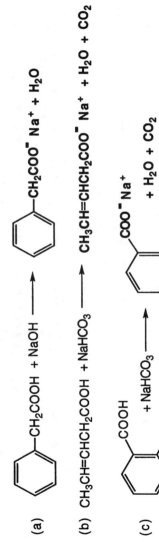

(a) [structure]—CH$_2$COOH + NaOH $\longrightarrow$ [structure]—CH$_2$COO$^-$ Na$^+$ + H$_2$O

(b) CH$_3$CH=CHCH$_2$COOH + NaHCO$_3$ $\longrightarrow$ CH$_3$CH=CHCH$_2$COO$^-$ Na$^+$ + H$_2$O + CO$_2$

(c) [structure] COOH, OCH$_3$ + NaHCO$_3$ $\longrightarrow$ [structure] COO$^-$ Na$^+$, OCH$_3$ + H$_2$O + CO$_2$

(d) CH$_3$CHCOOH (OH) + H$_2$NCH$_2$CH$_2$OH $\longrightarrow$ CH$_3$CHCOO$^-$ (OH) (H$_3$NCH$_2$CH$_2$OH)

(e) CH$_3$CH=CHCH$_2$COO$^-$Na$^+$ + HCl $\longrightarrow$ CH$_3$CH=CHCH$_2$COOH + NaCl

(f) CH$_3$CH$_2$CH$_2$CH$_2$Li + CH$_3$COOH $\longrightarrow$ CH$_3$CH$_2$CH$_2$CH$_3$ + CH$_3$COO$^-$ Li$^+$

(g) CH$_3$CH$_2$CH$_2$CH$_2$MgBr + CH$_3$CH$_2$OH $\longrightarrow$ CH$_3$CH$_2$CH$_2$CH$_3$ + CH$_3$CH$_2$O$^-$ MgBr$^+$

Problem 17.29 The normal pH range for blood plasma is 7.35 - 7.45. Under these conditions, would you expect the carboxyl group of lactic acid (pK_a 3.07) to exist primarily as a carboxyl group or as a carboxylate anion? Explain.

Recall from the definition of K_a that:

$$K_a = \frac{[A^-][H^+]}{[H-A]} \qquad \text{so dividing both sides by } [H^+] \text{ gives} \qquad \frac{K_a}{[H^+]} = \frac{[A^-]}{[H-A]}$$

Here, [H$^+$] is concentration of H$^+$, [H-A] is concentration of protonated acid (lactic acid in this case) and [A$^-$] is the concentration of deprotonated acid (lactic acid carboxylate anion in this case). Therefore, if the ratio of K_a / [H$^+$] is greater than 1, [A$^-$] will be the predominant form, and if the ratio of K_a / [H$^+$] is less than 1, then [H-A] will be the predominant form. Recall that pH = -log₁₀ [H⁺], so a pH of 7.4 corresponds to a [H⁺] of 10⁻(pH) = 10⁻(7.4) = 4.0 x 10⁻⁸. Similarly, pKₐ = -log₁₀ Kₐ, so for lactic acid Kₐ = 10⁻(pKa) = 10⁻(3.07) = 8.5 x 10⁻⁴. Using these numbers:

$$\frac{[A^-]}{[H\text{-}A]} = \frac{K_a}{[H^+]} = \frac{8.5 \times 10^{-4}}{4.0 \times 10^{-8}} = 2.1 \times 10^4$$

Therefore, lactic acid will exist primarily as the carboxylate anion in blood plasma.

Problem 17.30 The K_{a1} of ascorbic acid is 7.94 x 10⁻⁵. Would you expect ascorbic acid dissolved in blood plasma to exist primarily as ascorbic acid or as ascorbate anion? Explain.

Using the same reasoning described in the answer to Problem 17.29:

$$\frac{[A^-]}{[H\text{-}A]} = \frac{K_a}{[H^+]} = \frac{7.9 \times 10^{-5}}{4.0 \times 10^{-8}} = 2.0 \times 10^3$$

Therefore, ascorbic acid will exist primarily as the ascorbate anion in blood plasma.

Problem 17.31 Excess ascorbic acid is excreted in the urine, the pH of which is normally in the range 4.8 - 8.4. What form of ascorbic acid would you expect to be present in urine of pH 8.4, free ascorbic acid or ascorbate anion? Explain.

At pH 8.4, [H⁺] = 4.0 x 10⁻⁹, therefore using the same reasoning as described in the answer to Problem 17.29 and 17.30:

$$\frac{[A^-]}{[H\text{-}A]} = \frac{K_a}{[H^+]} = \frac{7.9 \times 10^{-5}}{4.0 \times 10^{-9}} = 2.0 \times 10^4$$

Ascorbic acid will exist primarily as the ascorbate anion in urine of pH 8.4.

Reactions of Carboxylic Acids

Problem 17.32 Give the expected organic product when phenylacetic acid, PhCH₂COOH, is treated with each reagent.

(a) SOCl₂

$$\underset{\text{O}}{\overset{\text{O}}{PhCH_2\overset{\|}{C}OH}} + SOCl_2 \longrightarrow \underset{\text{O}}{\overset{\text{O}}{PhCH_2\overset{\|}{C}Cl}} + SO_2 + HCl$$

(b) NaHCO₃, H₂O

$$PhCH_2\overset{\text{O}}{\overset{\|}{C}}OH + NaHCO_3 \longrightarrow PhCH_2\overset{\text{O}}{\overset{\|}{C}}O^-Na^+ + H_2O + CO_2$$

(c) NaOH, H₂O

$$PhCH_2\overset{\text{O}}{\overset{\|}{C}}OH + NaOH \longrightarrow PhCH_2\overset{\text{O}}{\overset{\|}{C}}O^-Na^+ + H_2O$$

(d) CH₃MgBr (1 equivalent)

$$PhCH_2\overset{\text{O}}{\overset{\|}{C}}OH + CH_3MgBr \longrightarrow PhCH_2\overset{\text{O}}{\overset{\|}{C}}O^-MgBr^+ + CH_4$$

(e) LiAlH₄ followed by H₂O

$$PhCH_2\overset{\text{O}}{\overset{\|}{C}}OH \xrightarrow[\text{2) H}_2\text{O}]{\text{1) LiAlH}_4} PhCH_2CH_2OH$$

(f) CH$_2$N$_2$

$$PhCH_2\overset{\displaystyle O}{\overset{\|}{C}}OH + CH_2N_2 \longrightarrow PhCH_2\overset{\displaystyle O}{\overset{\|}{C}}OCH_3 + N_2$$

(g) CH$_3$OH + H$_2$SO$_4$ (catalyst)

$$PhCH_2\overset{\displaystyle O}{\overset{\|}{C}}OH + CH_3OH \xrightarrow{H_2SO_4} PhCH_2\overset{\displaystyle O}{\overset{\|}{C}}OCH_3 + H_2O$$

Problem 17.33 Show how to convert *trans*-3-phenyl-2-propenoic acid (cinnamic acid) to these compounds.

(a)

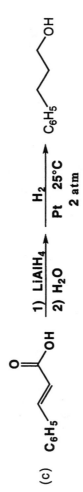

$$\xrightarrow[\text{2) H}_2\text{O}]{\text{1) LiAlH}_4}$$

(b)

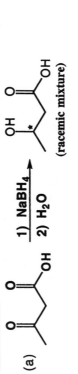

$$\xrightarrow[\text{2 atm}]{\text{H}_2 \quad \text{25°C}}{\text{Pt}}$$

(c)

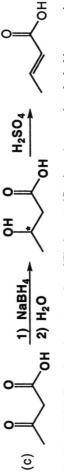

$$\xrightarrow[\text{2) H}_2\text{O}]{\text{1) LiAlH}_4} \xrightarrow[\text{2 atm}]{\text{H}_2 \quad \text{25°C}}{\text{Pt}}$$

Problem 17.34 Show how to convert 3-oxobutanoic acid (acetoacetic acid) to these compounds.

(a)

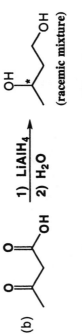

$$\xrightarrow[\text{2) H}_2\text{O}]{\text{1) NaBH}_4}$$

(racemic mixture)

(b)

$$\xrightarrow[\text{2) H}_2\text{O}]{\text{1) LiAlH}_4}$$

(racemic mixture)

(c)

$$\xrightarrow[\text{2) H}_2\text{O}]{\text{1) NaBH}_4} \xrightarrow{\text{H}_2\text{SO}_4}$$

Problem 17.35 Complete these examples of Fischer esterification. Assume that alcohol is present in excess.

(a)

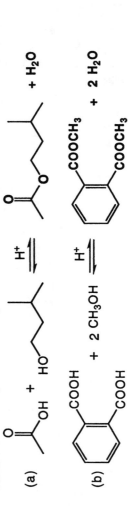

$$\rightleftharpoons^{H^+}$$

+ H$_2$O

(b)

+ 2 CH$_3$OH $\rightleftharpoons^{H^+}$ + 2 H$_2$O

(c)

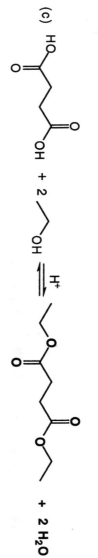

+ 2 H₂O

Problem 17.36 Benzocaine, a topical anesthetic, is prepared by treatment of 4-aminobenzoic acid with ethanol in the presence of an acid catalyst followed by neutralization. Draw the structural formula of benzocaine.

H₂N—⟨benzene⟩—COOH + CH₃CH₂OH
$\xrightarrow[\text{2) Mild base to}]{\text{1) H}_2\text{SO}_4}$
deprotonate
amino group

H₂N—⟨benzene⟩—COCH₂CH₃ + H₂O

4-Aminobenzoic acid

Benzocaine
(a topical anesthetic)

Problem 17.37 Name the carboxylic acid and alcohol from which each ester is derived.

(a)

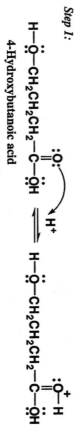

+ CH₃OH $\overset{\text{H}^+}{\rightleftharpoons}$

⟨cyclohexane-CO-O-CH₃⟩ + H₂O

Cyclohexyl-carboxylic acid　　**Methanol**

(b)

HO—⟨cyclohexane⟩—OH + 2 ⟨CH₃COOH⟩ $\overset{\text{H}^+}{\rightleftharpoons}$

⟨diester⟩ + 2 H₂O

1,4-Cyclohexanediol　　**Acetic acid**

(c)

⟨pentenoic acid⟩ + ⟨isopropanol⟩ $\overset{\text{H}^+}{\rightleftharpoons}$ ⟨ester⟩ + H₂O

trans-2-Pentenoic acid　　**2-Propanol**

(d)

HO—⟨butanedioic acid⟩—OH + 2 CH₃OH $\overset{\text{H}^+}{\rightleftharpoons}$ ⟨dimethyl ester⟩ + 2 H₂O

Butanedioic acid
(Succinic acid)　　**Methanol**

Problem 17.38 When 4-hydroxybutanoic acid is treated with an acid catalyst, it forms a lactone (a cyclic ester). Draw the structural formula of this lactone and propose a mechanism for its formation.

Step 1:

H—Ö—CH₂CH₂CH₂—C—ÖH $\rightleftharpoons$ H—Ö—CH₂CH₂CH₂—C—ÖH⁺—H

＋ H⁺

4-Hydroxybutanoic acid

Step 2:

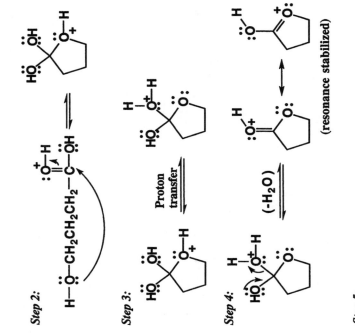

Step 3:

Proton transfer

Step 4:

(-H₂O)

(resonance stabilized)

Step 5:

(-H⁺)

Lactone

Problem 17.39 Fischer esterification cannot be used to prepare *tert*-butyl esters. Instead, carboxylic acids are treated with 2-methylpropene and an acidic catalyst to generate them.
(a) Why does the Fischer esterification fail for the synthesis of *tert*-butyl esters?

The Fischer esterification does not work for at least three reasons. First, the *tert*-butyl alcohol is not very nucleophilic because of steric hindrance. Second, the *tert*-butyl alcohol dehydrates in acid. Third, any *tert*-butyl ester that forms falls apart to give the carboxylic acid and *tert*-butyl cation in acid as shown in the two part mechanism below.

Step 1:

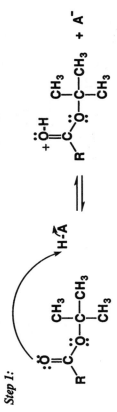

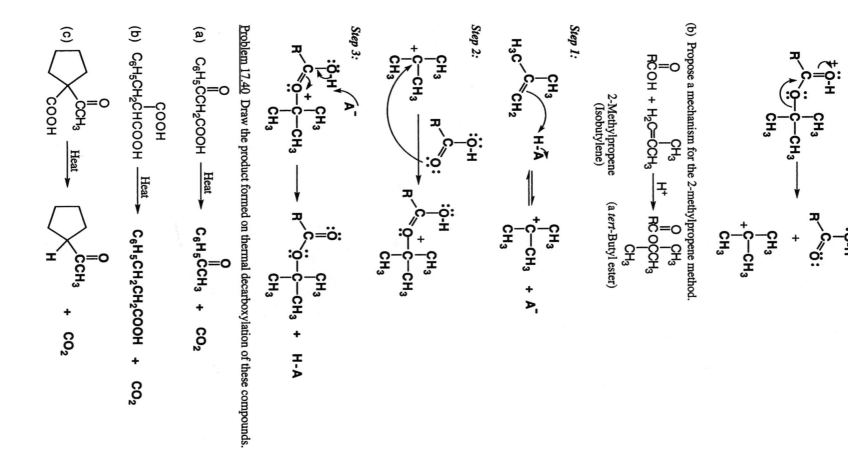

Step 2:

(b) Propose a mechanism for the 2-methylpropene method.

$$RCOH + H_2C=CCH_3 \xrightarrow{H^+} RCOCCH_3$$

2-Methylpropene   (a *tert*-Butyl ester)
(Isobutylene)

Step 1:

Step 2:

Step 3:

Problem 17.40 Draw the product formed on thermal decarboxylation of these compounds.

(a) $C_6H_5CCH_2COOH \xrightarrow{\text{Heat}} C_6H_5CCH_3 + CO_2$

(b) $C_6H_5CH_2CHCOOH \xrightarrow{\text{Heat}} C_6H_5CH_2CH_2COOH + CO_2$

(c)

$\xrightarrow{\text{Heat}}$

$+ CO_2$

Problem 17.41 When heated, carboxylic salts in which there is a good leaving group on the carbon beta to the carboxylate group undergo decarboxylation/elimination to give an alkene. Propose a mechanism for this type of decarboxylation/elimination. Compare the mechanism of these decarboxylations with the mechanism for decarboxylation of β-ketoacids; in what way(s) are the mechanisms similar?

The mechanisms are similar in that, in both cases, CO_2 is generated and a carbon-carbon bond is broken. As can be seen in the mechanisms shown below, that is where the similarity ends. In the case of the decarboxylation/elimination reaction, the starting material is the carboxylate anion and a bromide anion is lost in the decarboxylation step to give the product alkene. In the case of decarboxylation of β-ketoacids, the protonated carboxylic acid group is usually used. A proton is transferred from the carboxylic acid to the ketone to give an enol in the decarboxylation step, that tautomerizes to produce the product ketone.

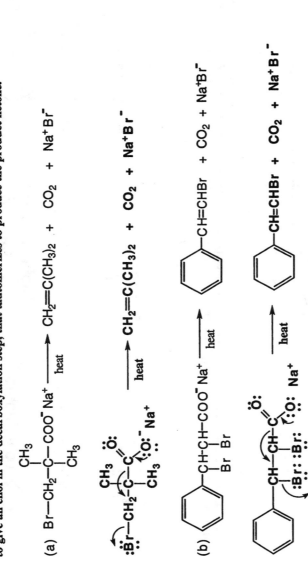

Problem 17.42 A mnemonic phrase for remembering the common names for the dicarboxylic acids oxalic through adipic is "Oh my, such good apples". Explain.

The first letter of each each work in the phrase corresponds to the first letter in the names of the dicarboxylic acids ordered in increasing numbers of carbon atoms.

	Dicarboxylic Acid	Structural Formula
Oh	**O**xalic	HOOCCOOH
My	**M**alonic	HOOCCH₂COOH
Such	**S**uccinic	HOOC(CH₂)₂COOH
Good	**G**lutaric	HOOC(CH₂)₃COOH
Apples	**A**dipic	HOOC(CH₂)₄COOH

Problem 17.43 Show how cyclohexanecarboxylic acid could be synthesized from cyclohexane in good yield.

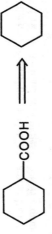

Propose a synthesis that begins with radical halogenation. Note that this is the only synthetic reaction you have learned that uses an alkane as a starting material. Next create a Grignard reagent, followed by reaction with CO_2 then an acid workup to give cyclohexanecarboxylic acid in high yield.

CHAPTER 18
Solutions to the Problems

Problem 18.1 Draw structural formulas for the following compounds.

(a) *N*-Cyclohexylacetamide.

(b) 1-Methylpropyl methanoate

(c) Cyclobutyl butanoate

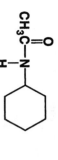

(d) *N*-(1-Methylheptyl)succinate

(e) Diethyl adipate

(f) 2-Aminopropanamide

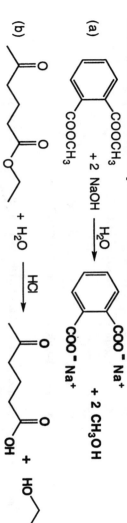

$C_2H_5OC(CH_2)_4COC_2H_5$

Problem 18.2 Will phthalimide dissolve in aqueous sodium bicarbonate?

In order to dissolve in water, the phthalimide must be deprotonated to give the negatively-charged phthalimide anion. The question becomes whether sodium bicarbonate is a strong enough base to deprotonate phthalimide according to the following equilibrium:

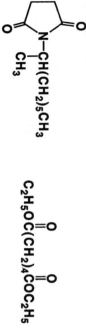

$$NH + NaHCO_3 \rightleftharpoons N^- \ Na^+ + H_2O + CO_2$$

$$pK_a = 8.3$$

The pK_a of phthalimide is 8.3, while the pK_a of carbonic acid H_2CO_3, which dissociates into H_2O and CO_2, is 6.4. Because equilibria in acid-base reactions favor formation of the weaker acid (phthalimide), the equilibrium will be to the left and the phthalimide will not dissolve.

Problem 18.3 Complete and balance equations for the hydrolysis of each ester in aqueous solution. Show each product as it is ionized under the indicated experimental conditions.

(a)

COOCH₃
COOCH₃
+ 2 NaOH →(H₂O)→ COO⁻ Na⁺
COO⁻ Na⁺ + 2 CH₃OH

(b)

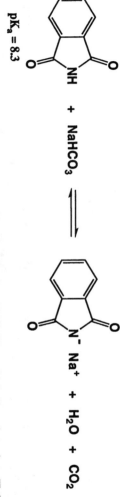

+ H₂O →(HCl)→   OH + HO

Problem 18.4 Complete equations for the hydrolysis of the amides in Example 18.4 in concentrated aqueous NaOH. Show all products as they exist in aqueous NaOH, and the number of moles of NaOH required for hydrolysis of each amide.

Each product is shown as it would exist in aqueous NaOH.

(a) $CH_3\overset{O}{\overset{\|}{C}}-\overset{CH_3}{\overset{|}{N}}-CH_3 + NaOH \xrightarrow{H_2O} CH_3CO^-Na^+ + (CH_3)_2NH$

(b)

NH + NaOH $\xrightarrow{\text{H}_2\text{O}}$ $\text{H}_2\text{NCH}_2\text{CH}_2\text{CH}_2\text{CH}_2\text{CO}^-\ \text{Na}^+$

Problem 18.5 Synthesis of nitriles by nucleophilic displacement of halide from an alkyl halide is practical only with primary and secondary alkyl halides. It fails with tertiary alkyl halides. Why? What is the major product of the following reaction?

$\xrightarrow[\text{water}]{\text{KCN, ethanol,}}$

CH_3 + HCN + KCl

Cyanide ion is both a base and a nucleophile. With a tertiary halide and moderate base moderate nucleophile such as CN⁻, E2 is the principal reaction. Thus, the major product in this instance is 1-methylcyclopentene.

Problem 18.6 Complete the following transesterification reaction. The stoichiometry is given in the equation.

2 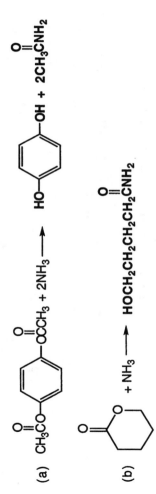 COCH_3 + HOCH$_2$CH$_2$OH $\xrightarrow{\text{H}^+}$ 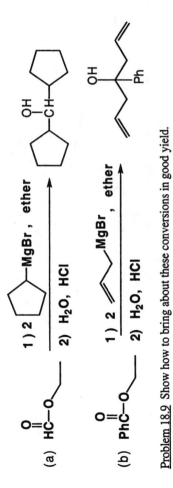 $\text{COCH}_2\text{CH}_2\text{OC}$ + 2CH$_3$OH

Problem 18.7 Complete and balance equations for the following reactions. The stoichiometry of each reaction is given in the equation.

Each is an example of aminolysis of an ester.

(a) CH$_3$CO—⟨⟩—OCCH$_3$ + 2NH$_3$ ⟶ HO—⟨⟩—OH + 2CH$_3$CNH$_2$

(b) ⟨lactone⟩ + NH$_3$ ⟶ HOCH$_2$CH$_2$CH$_2$CH$_2$CNH$_2$

Problem 18.8 Show how to prepare these alcohols by treatment of an ester with a Grignard reagent.

Each can be prepared by treatment of an ester with two moles of an organomagnesium reagent. In these solutions, the ester chosen is the ethyl ester.

(a) HC—O—⟨⟩ $\xrightarrow[\text{2) H}_2\text{O, HCl}]{\text{1) 2 ⟨⟩—MgBr , ether}}$ ⟨⟩—CH(OH)—⟨⟩

(b) PhC—O—⟨⟩ $\xrightarrow[\text{2) H}_2\text{O, HCl}]{\text{1) 2 ⟨⟩—MgBr , ether}}$ ⟨C(OH)(Ph)⟩

Problem 18.9 Show how to bring about these conversions in good yield.

Both of these conversions can be brought about using a lithium diorganocopper reagent.

(a) PhCOOH $\xrightarrow{SOCl_2}$ PhCCl $\xrightarrow[\text{2) } H_2O]{\text{1) } [CH_3(CH_2)_4CH_2]_2CuLi}$

(b) CH$_2$=CHCl $\xrightarrow[\text{3. HCl, } H_2O]{\substack{\text{1. Mg, ether} \\ \text{2. } CO_2}}$ CH$_2$=CHCOH $\xrightarrow{SOCl_2}$ CH$_2$=CHCCl $\xrightarrow[\text{2) } H_2O]{\text{1) } [CH_3(CH_2)_3CH_2]_2CuLi}$

Problem 18.10 Show how to convert hexanoic acid to each amine in good yield.

In both cases, the conversions can be carried out through formation of an amide, followed by reduction to give the desired amine.

(a)

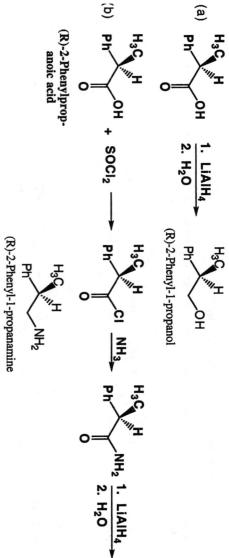

Hexanoic acid

(b)

Hexanoic acid

Problem 18.11 Show how to convert (R)-2-phenylpropanoic acid to these compounds:

(a)

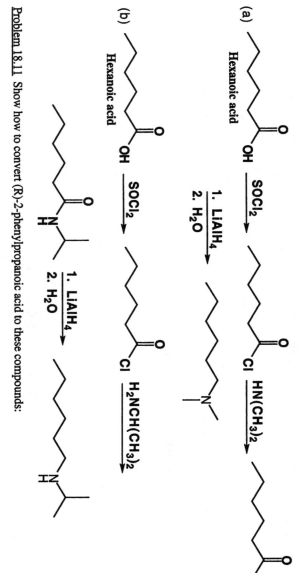

(R)-2-Phenyl-1-propanol

(b)

(R)-2-Phenylprop-
anoic acid

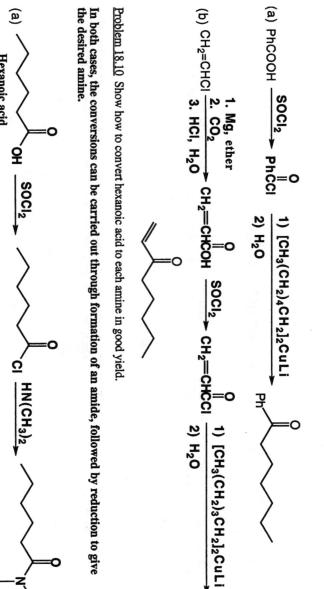

(R)-2-Phenyl-1-propanamine

Problem 18.12 Show how to convert phenylacetic acid into the following in good yield.

(a) PhCH₂COH $\xrightarrow{\text{1. SOCl}_2}$ PhCH₂CNH₂ $\xrightarrow[\text{2. H}_2\text{O}]{\text{1. LiAlH}_4}$ PhCH₂CH₂NH₂

Phenylacetic acid

(b) PhCH₂CNH₂ $\xrightarrow{\text{Br}_2/\text{NaOH}}$ PhCH₂NH₂

From part (a)

PROBLEMS
Structure and Nomenclature
Problem 18.13 Draw structural formulas for these compounds:

(a) Dimethyl carbonate   (b) Benzonitrile   (c) Isopropyl 3-methylhexanoate

(d) Diethyl oxalate   (e) Ethyl (Z)-2-pentenoate   (f) Butanoic anhydride

(g) Dodecanamide   (h) Ethyl 3-hydroxybutanoate

(i) Octanoyl chloride   (j) Diethyl cis 1,2-cyclohexane-dicarboxylate   (k) Methanesulfonyl chloride

(l) p-Toluenesulfonyl chloride

Problem 18.14 Write the IUPAC name for each compound.

(a) Benzoic anhydride   (b) Benzenesulfonamide   (c) N-Methylhexanamide

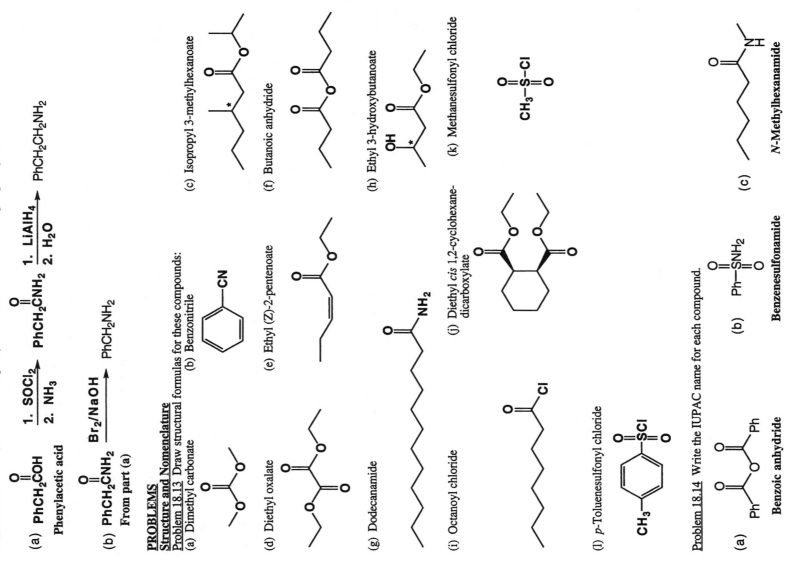

352

Solutions

Chapter 18: Derivatives of Carboxylic Acids

(d)

Octanamide

(e)

**Diethyl propanedioate
(Diethyl malonate)**

(f) CH₃O—S—OCH₃

Dimethyl sulfate

(g)

**Methyl 2-methyl-3-oxo-
4-phenylbutanoate**

(h)

**Hexanedioyl chloride
(Adipoyl chloride)**

(i) CH₃(CH₂)₅CN

Heptanenitrile

Physical Properties

Problem 18.15   Both the melting point and boiling point of acetamide are higher than those of its *N,N*-dimethyl derivative.

CH₃CNH₂	CH₃CN(CH₃)₂
Acetamide	*N,N*-Dimethylacetamide
mp 82.3°C, bp 221°C	mp -20°C, bp 165°C

How do you account for these differences?

Acetamide has two amide N-H atoms that can take part in hydrogen bonding with carbonyl oxygen atoms, while *N,N*-dimethylacetamide does not. Thus, due to this ability to hydrogen bond, acetamide molecules can associate with each other more strongly, leading to higher melting and boiling points compared to *N,N*-dimethylacetamide.

Problem 18.16   Each hydrogen of a primary amide typically has a separate ¹H-NMR resonance, as illustrated by the separate signals for the two amide hydrogens of propanamide, which fall at δ 6.22 and δ 6.58. Furthermore, each methyl group of *N,N*-dimethylformamide has a separate resonance (δ 3.88 and δ 3.98). How do you account for these observations?

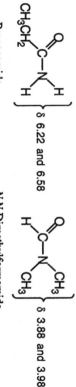

CH₃CH₂ } δ 6.22 and 6.58

Propanamide

} δ 3.88 and 3.98

***N,N*-Dimethylformamide**

Amide resonance causes a rotation barrier around the C(O)-N bond that prevents interconversion of the two nitrogen substituents on the NMR time scale. For this reason, the two hydrogens and two methyl groups, respectively, are in different chemical environments and thus give different signals.

Problem 18.17 Propose a structural formula for compound A, C$_7$H$_{14}$O$_2$, consistent with its ^{1}H-NMR and infrared spectra.

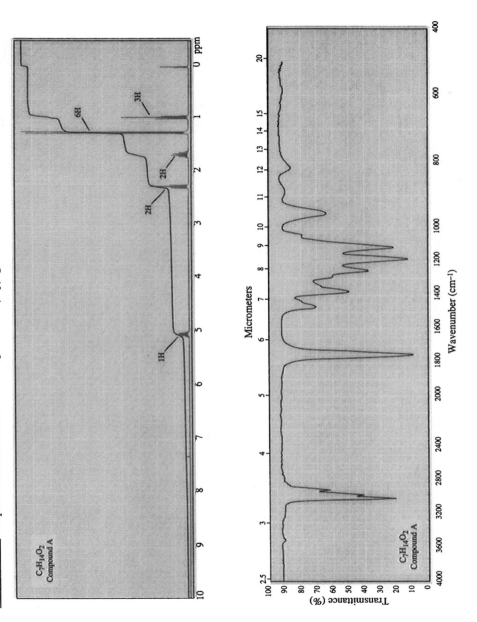

C$_7$H$_{14}$O$_2$
Compound A

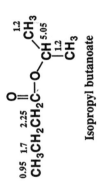

Isopropyl butanoate

Problem 18.18 Propose a structural formula for compound B, $C_6H_{13}NO$, consistent with its ^1H-NMR and infrared spectra.

$C_6H_{13}NO$
Compound B

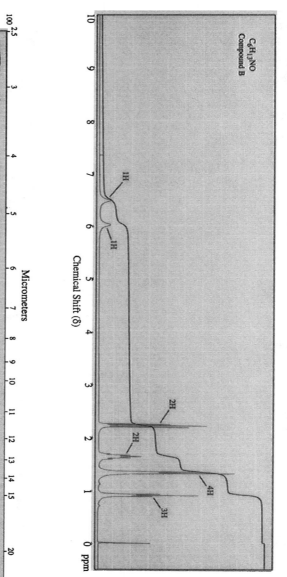

$C_6H_{13}NO$
Compound B

Problem 18.19 Propose a structural formula for each compound consistent with its ^1H-NMR and ^{13}C-NMR spectra.

(a) $C_5H_{10}O_2$

^1H-NMR	^{13}C-NMR
0.96 (d, 6H)	161.11
1.96 (m, 1H)	70.01
3.95 (d, 2H)	27.71
8.08 (s, 1H)	19.00

$\underset{0.96}{(CH_3)_2}\underset{1.96}{CH}\underset{3.95}{CH_2}\underset{\overset{O}{\|}}{C}\underset{8.08}{OH}$

(b) $C_7H_{14}O_2$

^1H-NMR	^{13}C-NMR
0.92 (d, 6H)	171.15
1.52 (m, 2H)	63.12
1.70 (m, 1H)	37.31
2.09 (s, 3H)	25.05
4.10 (t, 2H)	22.45
	21.06

$\underset{2.09}{CH_3}\underset{\overset{O}{\|}}{C}\underset{4.10}{OCH_2}\underset{1.52}{CH_2}\underset{1.70}{CH}\underset{0.92}{(CH_3)_2}$

(c) $C_6H_{12}O_2$

^1H-NMR	^{13}C-NMR
1.18 (d, 6H)	177.16
1.26 (t, 3H)	60.17
2.51 (m, 1H)	34.04
4.13 (q, 2H)	19.01
	14.25

$\underset{1.18}{(CH_3)_2}\underset{2.51}{CH}\underset{\overset{O}{\|}}{C}\underset{4.13}{OCH_2}\underset{1.26}{CH_3}$

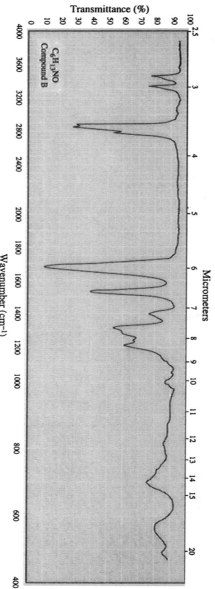

$\underset{0.95}{CH_3}\underset{1.35}{CH_2}\underset{1.35}{CH_2}\underset{1.7}{CH_2}\underset{2.2}{CH_2}\underset{\overset{O}{\|}}{C}\underset{6.0,6.5}{NH_2}$

Hexanamide

(d) $C_7H_{12}O_4$

^{1}H-NMR	^{13}C-NMR
1.28 (t, 6H)	166.52
3.36 (s, 2H)	61.43
4.21 (q, 4H)	41.69
	14.07

CH₃CH₂OCCH₂COCH₂CH₃
1.28   4.21   O(3.36) O   4.21   1.28

(e) $C_4H_7ClO_2$

^{1}H-NMR	^{13}C-NMR
1.68 (d, 3H)	170.51
3.80 (s, 3H)	52.92
4.42 (q, 1H)	52.32
	21.52

CH₃CHClCOCH₃
1.68   4.42   O   3.80

(f) $C_4H_6O_2$

^{1}H-NMR	^{13}C-NMR
2.29 (m, 2H)	177.81
2.50 (t, 2H)	68.58
4.36 (t, 2H)	27.79
	22.17

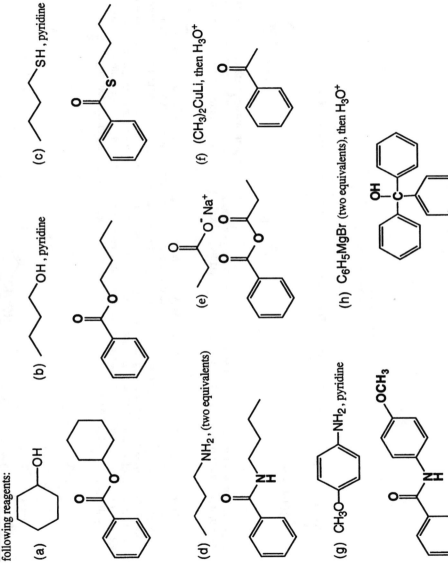

Problem 18.20 Write the structural formula of the principal product formed when benzoyl chloride is treated with the following reagents:

(a)   —OH , pyridine

(b)   —OH , pyridine

(c)   —SH , pyridine

(d)   —NH₂ , (two equivalents)

(e)   O⁻ Na⁺ (propanoate)

(f)   (CH₃)₂CuLi , then H₃O⁺

(g)   CH₃O— —NH₂ , pyridine

(h)   C₆H₅MgBr (two equivalents), then H₃O⁺

Problem 18.21 Write the structural formula of the principal product formed when ethyl benzoate is treated with the following reagents:

(a) H_2O, NaOH, heat

(b) H_2O, H_2SO_4, heat

(c) $CH_3CH_2CH_2CH_2NH_2$

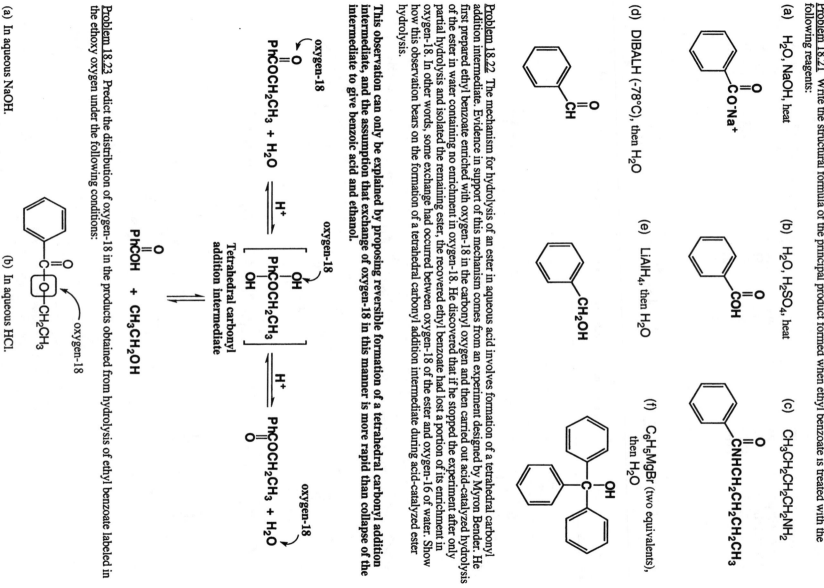

(d) DIBALH (-78°C), then H_2O

(e) LiAlH$_4$, then H_2O

(f) C_6H_5MgBr (two equivalents), then H_2O

Problem 18.22 The mechanism for hydrolysis of an ester in aqueous acid involves formation of a tetrahedral carbonyl addition intermediate. Evidence in support of this mechanism comes from an experiment designed by Myron Bender. He first prepared ethyl benzoate enriched with oxygen-18 in the carbonyl oxygen and then carried out acid-catalyzed hydrolysis of the ester in water containing no enrichment in oxygen-18. He discovered that if he stopped the experiment after only partial hydrolysis and isolated the remaining ester, the recovered ethyl benzoate had lost a portion of its enrichment in oxygen-18. In other words, some exchange had occurred between oxygen-18 of the ester and oxygen-16 of water. Show how this observation bears on the formation of a tetrahedral carbonyl addition intermediate during acid-catalyzed ester hydrolysis.

This observation can only be explained by proposing reversible formation of a tetrahedral carbonyl addition intermediate, and the assumption that exchange of oxygen-18 in this manner is more rapid than collapse of the intermediate to give benzoic acid and ethanol.

oxygen-18

PhCOCH$_2$CH$_3$ + H$_2$O $\underset{H^+}{\rightleftharpoons}$

oxygen-18

$\left[\text{PhCOCH}_2\text{CH}_3 \right]$ with OH / OH

Tetrahedral carbonyl
addition intermediate

$\underset{H^+}{\rightleftharpoons}$

oxygen-18

PhCOCH$_2$CH$_3$ + H$_2$O

$\rightleftharpoons$

PhCOH + CH$_3$CH$_2$OH

Problem 18.23 Predict the distribution of oxygen-18 in the products obtained from hydrolysis of ethyl benzoate labeled in the ethoxy oxygen under the following conditions:

oxygen-18

(a) In aqueous NaOH.

(b) In aqueous HCl.

Under each set of experimental conditions, oxygen-18 will appear in ethanol.

Problem 18.24 Write the structural formula of the principal product formed when benzamide is treated with the following reagents:

(a) H_2O, HCl, heat

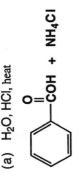
C_6H_5—COH + NH_4Cl

(b) NaOH, H_2O, heat

C_6H_5—CO^-Na^+ + NH_3

(c) $LiAlH_4$, then H_2O

C_6H_5—CH_2NH_2

(d) Br_2, NaOH, heat

C_6H_5—NH_2 + CO_2

Problem 18.25 Write the structural formula of the principal product formed when benzonitrile is treated with the following reagents.

(a) H_2O (one equivalent), H_2SO_4, heat

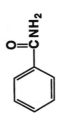
C_6H_5—CNH_2

(b) H_2O (excess), H_2SO_4, heat

C_6H_5—COH + NH_4^+ HSO_4^-

(c) NaOH, H_2O, heat

C_6H_5—CO^- Na^+ + NH_3

(d) $LiAlH_4$, then H_2O

C_6H_5—CH_2NH_2

Problem 18.26 Show the product expected when the following unsaturated δ-ketoester is treated with each reagent.

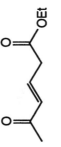

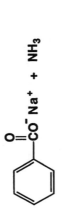

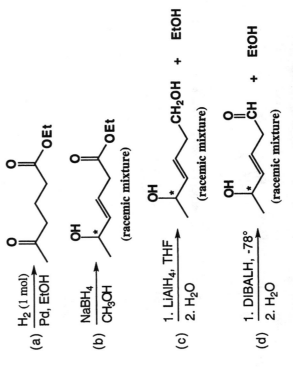

(a) $\dfrac{H_2 \text{ (1 mol)}}{\text{Pd, EtOH}}$

(b) $\dfrac{NaBH_4}{CH_3OH}$

(c) $\dfrac{1.\ LiAlH_4,\ THF}{2.\ H_2O}$

(d) $\dfrac{1.\ DIBALH,\ -78°}{2.\ H_2O}$

Problem 18.27 The reagent diisobutylaluminum hydride (DIBALH) reduces esters to aldehydes. When nitriles are treated with DIBALH, followed by mild acid hydrolysis, the product is also an aldehyde. Give a mechanism for this reaction.

A reasonable mechanism for this reaction involves an initial reaction of the hydride with the nitrile carbon atom, followed by hydrolysis of the resulting imine anion.

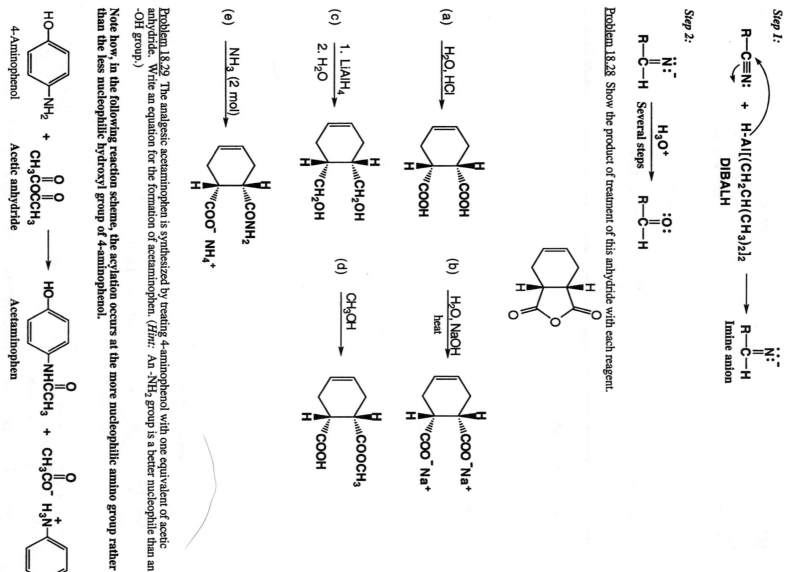

Step 1:

$$R-C \equiv N: \quad + \quad H-Al[(CH_2CH(CH_3)_2]_2 \longrightarrow R-\overset{\cdot\cdot}{C}-H$$

DIBALH

Imine anion

Step 2:

$$R-\overset{\cdot\cdot}{\underset{}{C}}-H \xrightarrow[\text{Several steps}]{H_3O^+} R-\overset{\cdot\cdot}{C}-H$$

Problem 18.28 Show the product of treatment of this anhydride with each reagent.

(a) H₂O, HCl

(b) H₂O, NaOH / heat

(c) 1. LiAlH₄ / 2. H₂O

(d) CH₃OH

(e) NH₃ (2 mol)

Problem 18.29 The analgesic acetaminophen is synthesized by treating 4-aminophenol with one equivalent of acetic anhydride. Write an equation for the formation of acetaminophen. (*Hint:* An -NH₂ group is a better nucleophile than an -OH group.)

Note how, in the following reaction scheme, the acylation occurs at the more nucleophilic amino group rather than the less nucleophilic hydroxyl group of 4-aminophenol.

4-Aminophenol + Acetic anhydride (CH₃COCCH₃) → Acetaminophen + CH₃CO⁻ H₃N⁺

Problem 18.30 Treatment of choline with acetic anhydride gives acetylcholine, a neurotransmitter. Write an equation for the formation of acetylcholine.

$$(CH_3)_3\overset{+}{N}CH_2CH_2OH + CH_3\overset{O}{\overset{\|}{C}}OCCH_3 \longrightarrow (CH_3)_3\overset{+}{N}CH_2CH_2O\overset{O}{\overset{\|}{C}}CH_3 + CH_3\overset{O}{\overset{\|}{C}}OH$$

Choline   Acetylcholine

Problem 18.31 Nicotinic acid, more commonly named niacin, is one of the B vitamins. Show how nicotinic acid can be converted to (a) ethyl nicotinate and then to (b) nicotinamide.

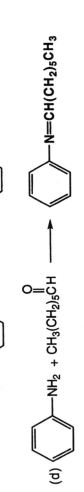

Nicotinic acid
(Niacin)

1) CH₃CH₂OH/H₂SO₄

2) Mild base (to deprotonate the pyridine N atom)

Ethyl nicotinate
(A)

NH₃

Nicotinamide
(B)

Problem 18.32 Complete these reactions.

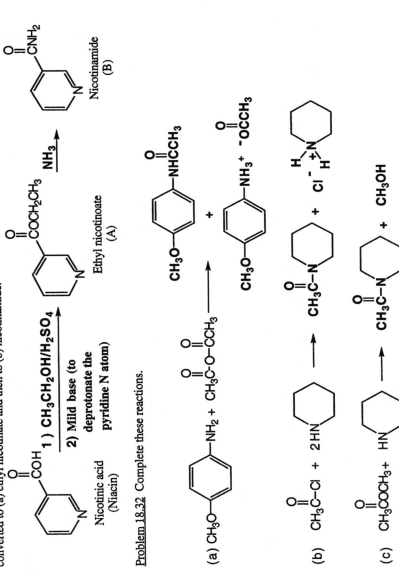

(a)

(b) CH₃C—Cl + 2HN

(c) CH₃OOCH₃ + HN

(d)

Problem 18.33 Show the product of treatment of γ-butyrolactone with each reagent.

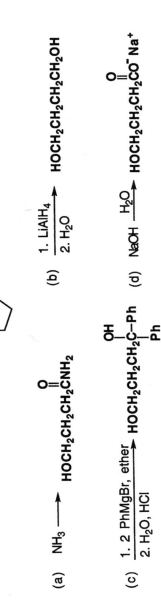

(a) NH₃ ⟶ HOCH₂CH₂CH₂CNH₂

(b) 1. LiAlH₄ / 2. H₂O ⟶ HOCH₂CH₂CH₂CH₂OH

(c) 1. 2 PhMgBr, ether / 2. H₂O, HCl ⟶ HOCH₂CH₂CH₂C(OH)(Ph)Ph

(d) NaOH, H₂O ⟶ HOCH₂CH₂CH₂CO⁻ Na⁺

(e) $\dfrac{\text{1. 2 CH}_3\text{Li, ether}}{\text{2. H}_2\text{O, HCl}}$ HOCH$_2$CH$_2$CH$_2$C—CH$_3$

(f) $\dfrac{\text{1. DIBALH, ether, -78°C}}{\text{2. H}_2\text{O, HCl}}$ HOCH$_2$CH$_2$CH$_2$CH

Problem 18.34 Show the product of treatment of the following γ-lactam with each reagent.

(a) $\dfrac{\text{H}_2\text{O, HCl}}{\text{heat}}$

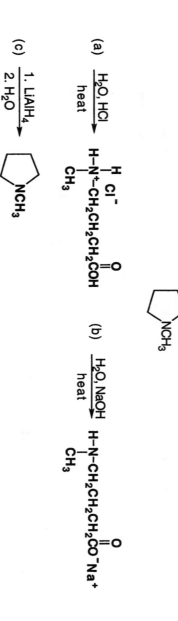

(b) $\dfrac{\text{H}_2\text{O, NaOH}}{\text{heat}}$

(c) $\dfrac{\text{1. LiAlH}_4}{\text{2. H}_2\text{O}}$

Problem 18.35 Draw structural formulas for the products of complete hydrolysis of meprobamate, phenobarbital, and pentobarbital in hot aqueous acid. Meprobamate (see The Merck Index, 12th ed., #5908) is a tranquilizer prescribed under 58 different trade names, including Equanil and Miltown. Phenobarbital (see The Merck Index, 12th ed., #7386) is a long-acting sedative, hypnotic, and anticonvulsant. Luminal is one of over a dozen names under which it is prescribed. Pentobarbital (see The Merck Index, 12th ed., #7272) is a short-acting sedative, hypnotic, and anticonvulsant. Nembutal is one of several trade names under which it is prescribed.

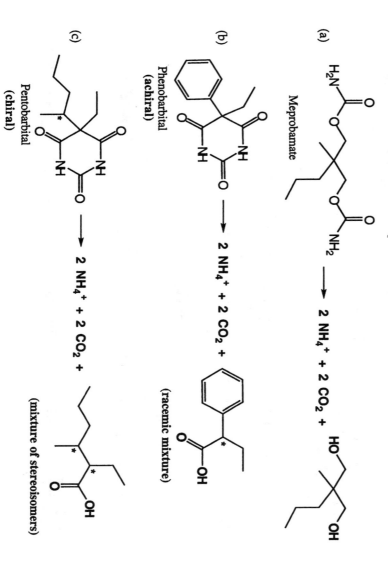

(a)

Meprobamate

2 NH$_4$$^+$ + 2 CO$_2$ +

(b)

Phenobarbital (achiral)

2 NH$_4$$^+$ + 2 CO$_2$ +

(racemic mixture)

(c)

Pentobarbital (chiral)

2 NH$_4$$^+$ + 2 CO$_2$ +

(mixture of stereoisomers)

In parts (b) and (c) above there is a decarboxylation that occurs from a β-diacid that accompanies hydrolysis.

The Hofmann Rearrangement

Problem 18.36 Following are steps in a synthesis of anthranilic acid from phthalic anhydride. Describe how you could bring about each step.

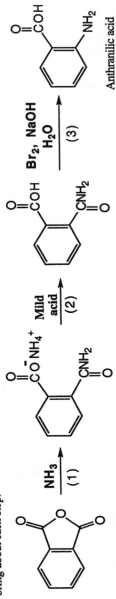

Phthalic anhydride

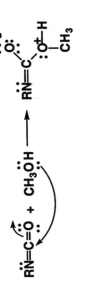

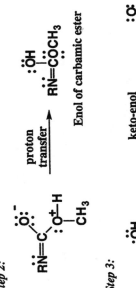

Anthranilic acid

Problem 18.37 Hofmann rearrangements of lower molecular-weight primary amides can be brought about using bromine in aqueous NaOH. Primary amides larger than about seven or eight carbon atoms are not sufficiently soluble in aqueous solution to react. Instead, they are dissolved in methanol or ethanol, and the corresponding sodium alkoxide is used as the base. Under these conditions, the isocyanate intermediate reacts with the alcohol to form a carbamate.

$$RCNH_2 + Br_2 \xrightarrow[CH_3OH]{CH_3O^-Na^+} RN=C=O \xrightarrow{CH_3OH} RNHCOCH_3$$

A primary amide   An isocyanate   A carbamate

Propose a mechanism for the reaction of an isocyanate with methanol to form a methyl carbamate.

A reasonable mechanism for this reaction involves addition of methanol to the carbonyl group to give an enol of a carbamic ester, followed by keto-enol tautomerism to give the carbamic ester product.

Step 1:

$$RN=C=O + CH_3OH \longrightarrow RN=C\overset{O^-}{\underset{O^+-H}{|}}CH_3$$

Step 2:

proton transfer

Enol of carbamic ester

Step 3:

keto-enol tautomerization

Enol of carbamic ester   A carbamic ester

Problem 18.38 (R)-(+)-Pulegone, readily available from pennyroyal oil (see The Merck Index, 12th ed., #8124), is an important enantiopure building block for organic syntheses. Propose a mechanism for each step in this transformation of pulegone.

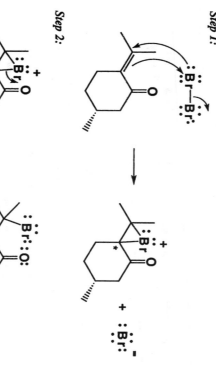

(R)-(+)-Pulegone

The first stage is a simple electrophilic addition to an alkene.

Step 1:

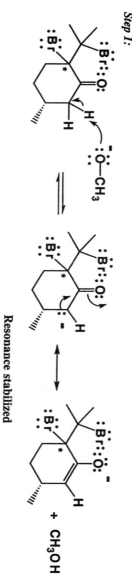

Step 2:

For the second stage, a rearrangement of the carbon skeleton occured. The first step of the process involves deprotonation α to the ketone. The hydrogen atoms on the α carbon are relatively acidic because the resulting anion is resonance stabilized as shown.

Step 1:

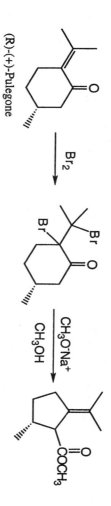

Resonance stabilized
enolate anion

The enolate then attacks the carbon-halogen bond to create the three-membered ring.

Step 2:

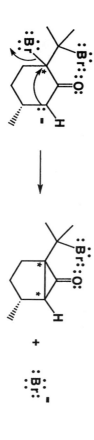

Methoxide then reacts with the electrophilic carbonyl group to create a tetrahedral addition intermediate.

Step 3:

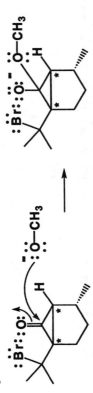

The reaction is completed when the tetrahedral addition intermediate collapses by relieving the ring strain in the three-membered ring while simultaneously eliminating a bromide leaving group.

Step 4:

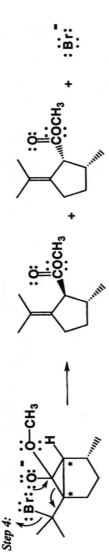

Synthesis

Problem 18.39 *N,N*-Diethyl *m*-toluamide (DEET) is the active ingredient in several common insect repellents. Propose a synthesis for DEET from 3-methylbenzoic acid.

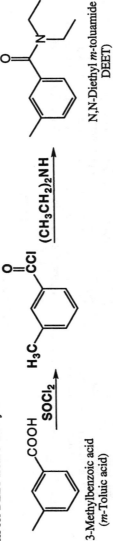

3-Methylbenzoic acid
(*m*-Toluic acid)

N,N-Diethyl *m*-toluamide
(DEET)

Problem 18.40 Following is the structural formula of isoniazid (see The Merck Index, 12th ed., #5203), a drug used to treat tuberculosis. It is estimated that one-third of the world's population is infected with tuberculosis, which results in approximately 3 million TB-related deaths per year. Show how to prepare isoniazid from pyridine-4-carboxylic acid.

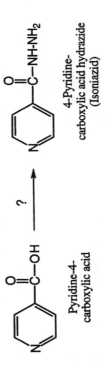

Pyridine-4-
carboxylic acid

4-Pyridine-
carboxylic acid hydrazide
(Isoniazid)

Isoniazid can be prepared by reaction of hydrazine (NH₂-NH₂) with a suitable carboxylic acid derivative, such as an ester.

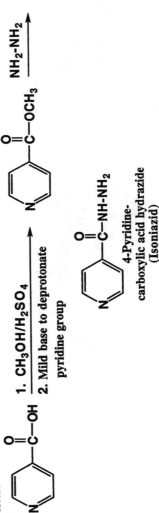

1. CH₃OH/H₂SO₄
2. Mild base to deprotonate pyridine group

4-Pyridine-
carboxylic acid hydrazide
(Isoniazid)

Chapter 18: Derivatives of Carboxylic Acids

Problem 18.41 Show how to convert phenylacetylene to allyl phenylacetate.

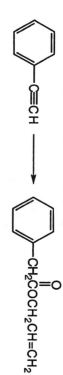

The key to this synthesis is the first step in which hydroboration / oxidation is used to make the aldehyde. Oxidation to the carboxylic acid, conversion to the acid chloride, and reaction with allyl alcohol gives the desired product.

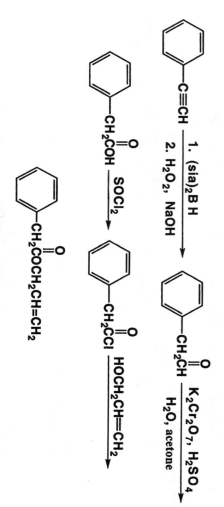

Problem 18.42 A step in a synthesis of PGE₁ (prostaglandin E₁, alprostadil; The Merck Index, 12th ed., #8063) is the reaction of a trisubstituted cyclohexene with bromine to form a bromolactone. Alprostadil is used as a temporary therapy for infants born with congenital heart defects that restrict pulmonary blood flow. It brings about dilation of the ductus arteriosus, which in turn increases blood flow in the lungs and blood oxygenation. Propose a mechanism for formation of this bromolactone and account for the observed stereochemistry of each substituent on the cyclohexane ring.

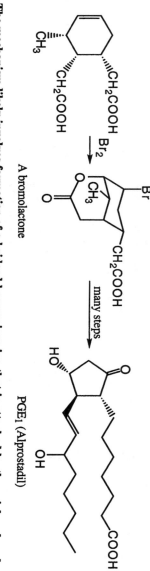

The mechanism likely involves formation of a bridged bromonium ion that is attacked by the axial carboxyl group. It is the axial carboxyl group that is in the proper location to attack the bromonium ion as shown:

Step 1:

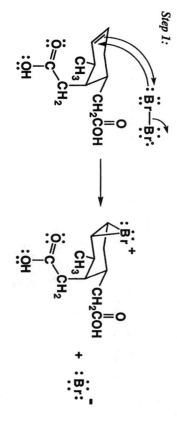

Step 2:

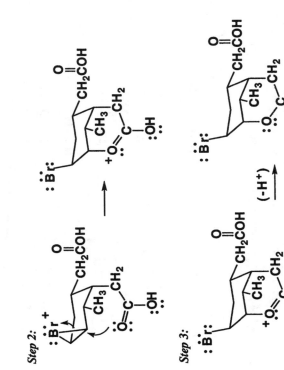

Step 3:

(-H⁺)

The stereochemistry is fixed by the bicyclic structure. The bromine atom is axial, while the -CH₂COOH and methyl group are equatorial.

<u>Problem 18.43</u> Barbiturates are prepared by treatment of a derivative of diethyl malonate with urea in the presence of sodium ethoxide as a catalyst. Following is an equation for the preparation of barbital, a long-duration hypnotic and sedative, from diethyl diethylmalonate and urea. Barbital is prescribed under one of a dozen or more trade names.

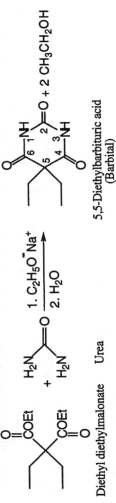

| Diethyl diethylmalonate | Urea | | 5,5-Diethylbarbituric acid (Barbital) |

(a) Propose a mechanism for this reaction.

Treatment of urea with ethoxide ion gives an anion that then attacks a carbonyl group of the malonic ester to displace ethoxide ion. This second reaction is an example of nucleophilic displacement at a carbonyl carbon.

Step 1:

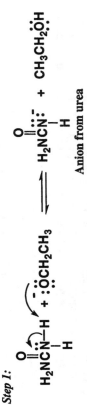

Note that each of the following carbonyl addition reactions involves formation of a tetrahedral addition intermediate that is not drawn because you should be very familiar with this type of mechanism by now.

Step 2:

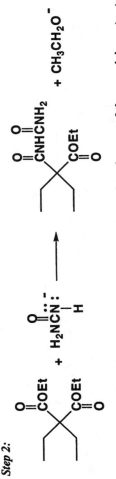

This is followed by nucleophilic displacement at the carbonyl group of the remaining ester to complete formation of the six-membered ring.

Step 3:

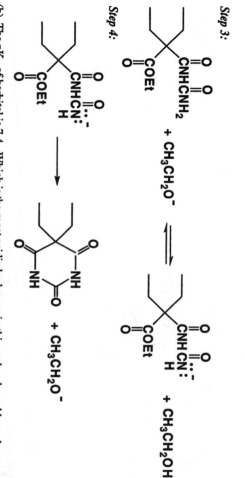

Step 4:

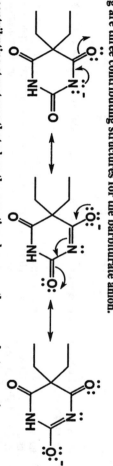

(b) The pKa of barbital is 7.4. Which is the most acidic hydrogen in this molecule and how do you account for its acidity?

The most acidic hydrogen is the imide hydrogen. Acidity results from the inductive effects of the adjacent carbonyl groups and stabilization of the deprotonated anion by resonance interaction with the carbonyl groups. Following are three contributing structures for the barbiturate anion.

The two contributing structures that place the negative charge on the more electronegative oxygen atoms make the greater contribution to the resonance hybrid.

Problem 18.44   The following compound is one of a group of β-chloroamines, many of which have antitumor activity. Describe a synthesis of this compound from anthranilic acid and ethylene oxide. This compound is a nitrogen mustard (Section 8.5).

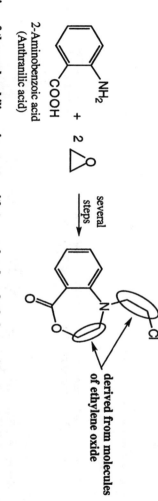

derived from molecules
of ethylene oxide

Reaction of the nucleophilic amino group with two molecules of ethylene oxide gives the diol intermediate that is converted to the dichloride by treatment with HCl. The dichloride intermediate is actually a nitrogen mustard derivative that will react with the carboxylate function to give the desired lactone, presumably via the three-membered ring intermediate characteristic of nitrogen mustards (Section 8.5).

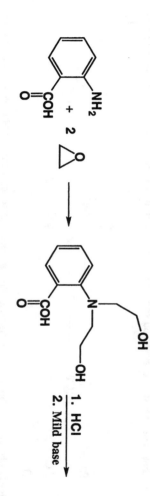

2-Aminobenzoic acid
(Anthranilic acid)

+ 2 ▷O

several
steps

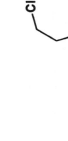

Reaction with carboxylate via three-membered ring intermediate

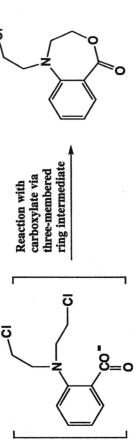

Problem 18.45 Following is a retrosynthetic scheme for the synthesis of 5-nonanone from 1-bromobutane as the only organic starting material. Show reagents and experimental conditions to bring about this synthesis.

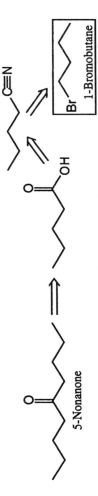

5-Nonanone

1-Bromobutane

1-Bromobutane is converted to pentanoic acid via the nitrile, which is hydrolyzed to the carboxylic acid.

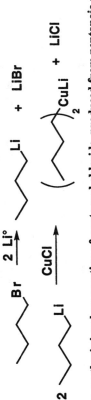

1-Bromobutane is also converted to a Gilman reagent via the organolithium species.

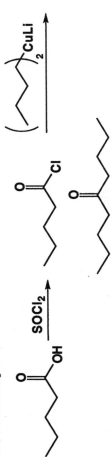

The key step in the synthesis involves reaction of pentanoyl chloride, produced from pentanoic acid by reaction with SOCl₂, with the Gilman reagent to give the product 5-nonanone. This is an example of a so-called "convergent synthesis" in which two pieces, themselves the product of several synthetic steps, are combined in a late step in the reaction sequence.

Problem 18.46 Procaine (its hydrochloride is marketed as Novocaine) was one of the first local anesthetics for infiltration and regional anesthesia. According to this retrosynthetic scheme, procaine can be synthesized from 4-aminobenzoic acid, ethylene oxide, and diethylamine as sources of carbon atoms. Provide reagents and experimental conditions to carry out the synthesis of procaine from these three compounds. For the synthesis of several more members of the "caine" family of local anesthetics, see MC.42-MC.46. See also "Chemistry in Action: From Cocaine to Procaine and Beyond"

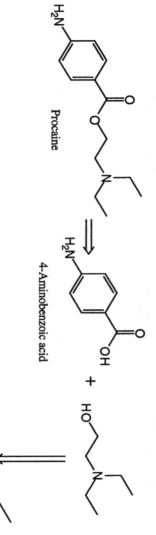

Procaine

4-Aminobenzoic acid   Ethylene oxide   Diethylamine

A reasonable synthetic scheme is shown below. First, the amine is reacted with ethylene oxide to create the corresponding β-aminoalcohol:

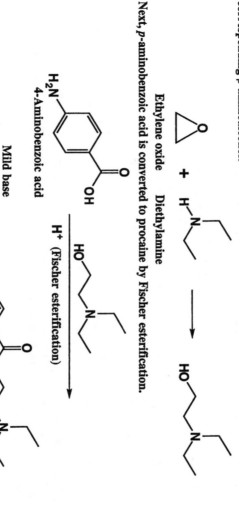

Ethylene oxide   Diethylamine

Next, *p*-aminobenzoic acid is converted to procaine by Fischer esterification.

H₂N
4-Aminobenzoic acid

H⁺ (Fischer esterification)

Mild base
(to deprotonate amino groups)

Procaine

Problem 18.47 The following sequence of steps converts (R)-2-octanol to (S)-2-octanol. Propose structural formulas for intermediates A and B, specify the configuration of each, and account for the inversion of configuration in this sequence.

(R)-2-Octanol

p-TsCl / pyridine → A $\xrightarrow{CH_3COO^-Na^+}$ B $\xrightarrow[2.\ H_2O]{1.\ LiAlH_4}$

(S)-2-Octanol

Formation of the tosylate ester (A) occurs with retention of configuration; reaction takes place on oxygen of the alcohol and does not involve the tetrahedral stereocenter. Nucleophilic substitution to form (B) occurs by an S$_N$2 pathway with inversion of configuration at the stereocenter undergoing reaction. Reduction of the ester with LAH occurs at the carbonyl carbon, not at the stereocenter. These relationships are shown in the following structural formulas.

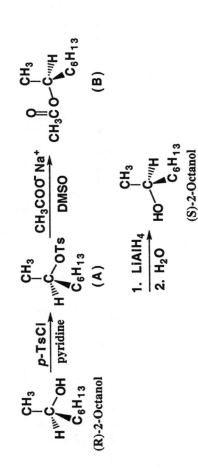

(R)-2-Octanol

(A)

(B)

(S)-2-Octanol

__Problem 18.48__ Reaction of a primary or secondary amine with diethyl carbonate under controlled conditions gives a carbamic ester. Propose a mechanism for this reaction.

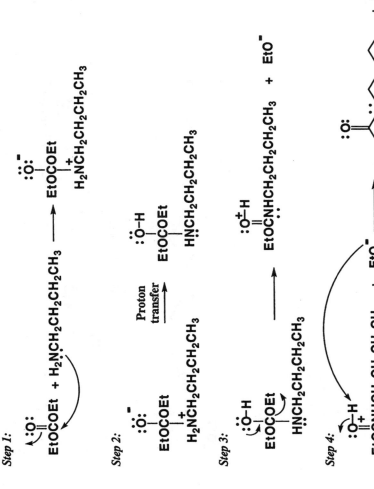

Diethyl carbonate   Butylamine

Ethyl *N*-butylcarbamate

A reasonable mechanism for this reaction involves initial attack by the nucleophilic nitrogen atom of butylamine on the carbonyl carbon atom of diethyl carbonate to create a tetrahedral carbonyl addition intermediate that leads to the product carbamate after a proton transfer then elimination of ethoxide.

Step 1:

Step 2:

Proton
transfer

Step 3:

Step 4:

Problem 18.49 Several sulfonylureas, a class of compounds containing RSO2NHCONHR, are useful drugs as orally active replacements for injected insulin in patients with adult-onset diabetes. It was discovered in 1942 that certain members of this class cause hypoglycemia in laboratory animals. Clinical trials of tolbutamide were begun in the early 1950s, and, since that time, more than 20 sulfonylureas have been introduced into clinical medicine. The sulfonylureas decrease blood glucose concentrations by stimulating β cells of the pancreas to release insulin and by increasing the sensitivity of insulin receptors in peripheral tissues to insulin stimulation.

Tolbutamide is synthesized by the reaction of the sodium salt of *p*-toluenesulfonamide and ethyl *N*-butylcarbamate (see Problem 18.48 for the synthesis of this carbamic ester). Propose a mechanism for the following step in the synthesis of tolbutamide:

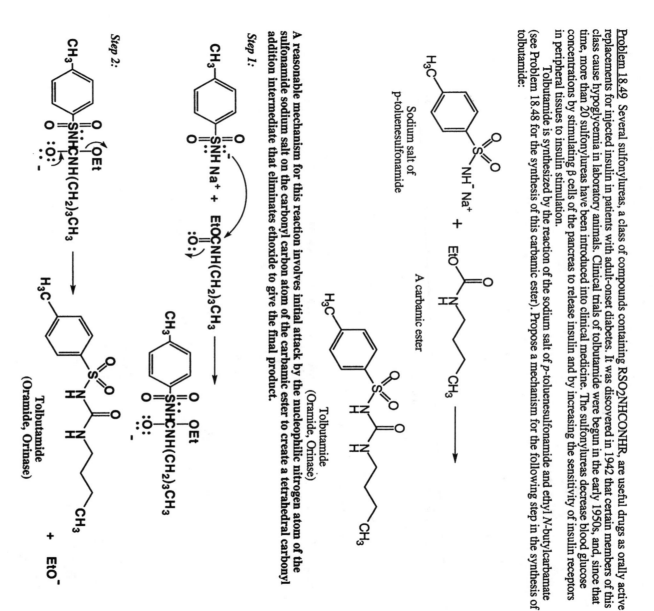

Sodium salt of
p-toluenesulfonamide

A carbamic ester

Tolbutamide
(Oramide, Orinase)

A reasonable mechanism for this reaction involves initial attack by the nucleophilic nitrogen atom of the sulfonamide sodium salt on the carbonyl carbon atom of the carbamic ester to create a tetrahedral carbonyl addition intermediate that eliminates ethoxide to give the final product.

Step 1:

Step 2:

Tolbutamide
(Oramide, Orinase)

Problem 18.50 Following are structural formulas for two more widely used sulfonylurea hypoglycemic agents. Show how each might be synthesized by converting an appropriate amine to a carbamic ester and then treating the carbamate with the sodium salt of a substituted benzenesulfonamide.

(a)

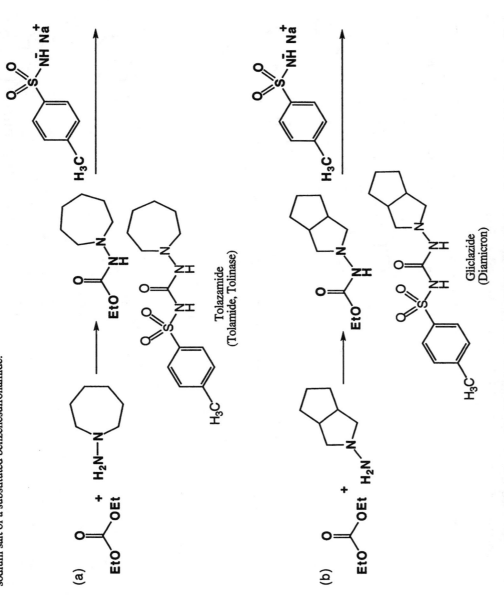

Tolazamide
(Tolamide, Tolinase)

(b)

Gliclazide
(Diamicron)

Problem 18.51 Amantadine (see The Merck Index, 12th ed., #389) is one of the very few available antiviral agents and is effective in preventing infections caused by the influenza A virus and in treating established illnesses. It is thought to block a late stage in the assembly of the virus. Amantadine is synthesized as follows. Treatment of 1-bromoadamantane with acetonitrile in sulfuric acid gives N-adamantylacetamide, which is then converted to amantadine.

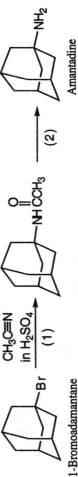

1-Bromoadamantane

Amantadine

(a) Propose a mechanism for the transformation in Step 1.

A reasonable mechanism is shown below. The last proton transfer may occur at the same time as formation of the carbonyl.

Step 1:

Step 2:

Step 3:

Step 4:

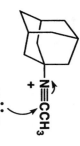

(-H⁺)

Step 5:

proton
transfer

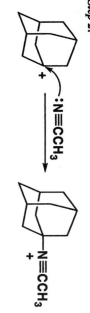

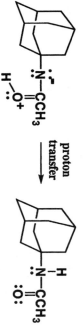

(b) Describe experimental conditions to bring about Step 2.

Step 2 is a simple hydrolysis carried out with HCl and H₂O.

Problem 18.52 In a series of seven steps, (S)-malic acid is converted to the bromoepoxide shown on the right in 50% overall yield. (S)-Malic acid occurs in apples and many other fruits. It is also available in enantiopure form from microbiological fermentation. This synthesis is enantioselective: of the stereoisomers possible for the bromoepoxide, only one is formed. In thinking about the chemistry of these steps, you will want to review the use of dihydropyran as an -OH protecting group (Section 16.8C) and the use the *p*-toluenesulfonyl chloride to convert the -OH, a poor leaving group, into a tosylate, a good leaving group (Section 9.5D).

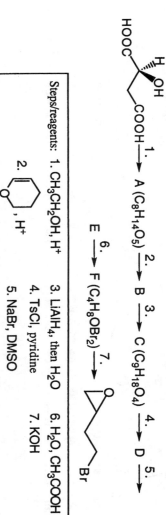

HOOC [H, OH] COOH $\xrightarrow{1.}$ A (C₈H₁₄O₅) $\xrightarrow{2.}$ B $\xrightarrow{3.}$ C (C₉H₁₈O₄) $\xrightarrow{4.}$ D $\xrightarrow{5.}$

E $\xrightarrow{6.}$ F (C₄H₈OBr₂) $\xrightarrow{7.}$ [epoxide-Br structure]

Steps/reagents:	1. CH₃CH₂OH, H⁺	3. LiAlH₄, then H₂O	6. H₂O, CH₃COOH
	2. [dihydropyran], H⁺	4. TsCl, pyridine	7. KOH
		5. NaBr, DMSO	

(a) Propose structural formulas for intermediates A through F. Also specify configuration at each stereocenter.

Transformation 2. deserves special comment. This reaction produces a tetrahydropyranyl ether (intermediate B) that serves as an -OH protecting group. Note how a new stereocenter is created as shown, but that it will be a mixture of stereoisomers (R,S). This center has no influence over the stereochemistry of the product because the tetrahydropyranyl ether is cleaved in acid during the second to last step of the sequence.

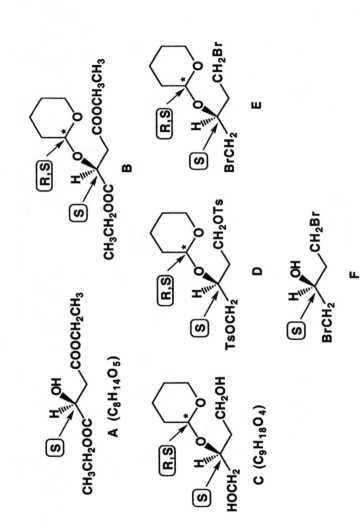

A (C₈H₁₄O₅)

B

C (C₉H₁₈O₄)

D

E

F

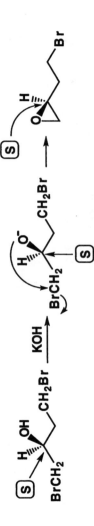

(b) What is the configuration of the stereocenter in the bromoepoxide and how do you account for the stereoselectivity of this seven-step conversion?

As shown on the above structures, the configuration at the 2 position during the synthesis is S. This stereochemistry is set in the starting material and maintained throughout the synthetic sequence, because none of the reactions involves a change or loss of stereochemistry at this position.

Problem 18.56 Following is a retrosynthetic analysis for the synthesis of the herbicide (S)-Metolachlor. According to this retrosynthetic analysis, starting materials are 2-ethyl-6-methylaniline, chloroacetic acid, acetone, and methanol. Show reagents and experimental conditions for the synthesis of Metolachlor from these four organic starting materials. Your synthesis will most likely give a racemic mixture. The chiral catalyst used by Novartis for Step 2 gives 80% enantiomeric excess of the S enantiomer.

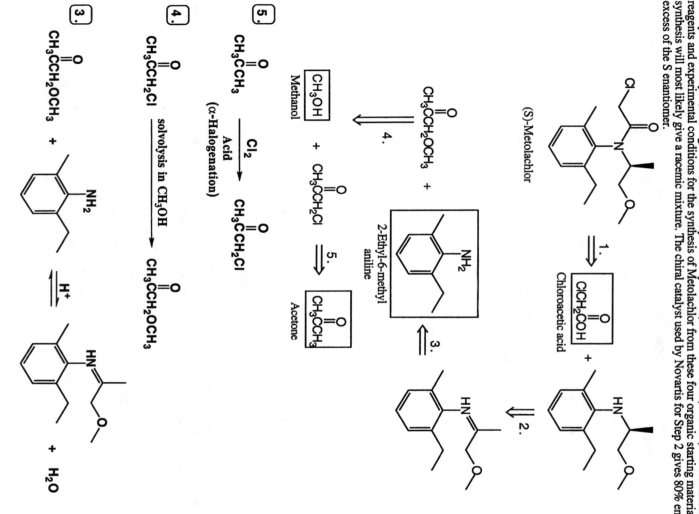

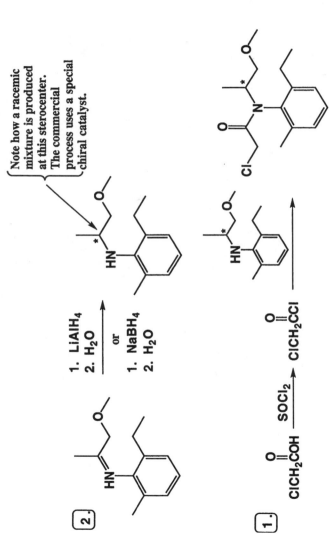

Note how a racemic mixture is produced at this stereocenter. The commercial process uses a special chiral catalyst.

2.

1. LiAlH₄
2. H₂O

or

1. NaBH₄
2. H₂O

1.

ClCH₂COH $\xrightarrow{\text{SOCl}_2}$ ClCH₂CCl

(racemic)-Metolachlor

Raccoon butterflyfish *Chaetodon lunula*
Kauai, Hawaii

CHAPTER 19
Solutions to the Problems

Problem 19.1 Draw the product of the base-catalyzed aldol reaction of these compounds.
(a) Phenylacetaldehyde   (b) Cyclopentanone

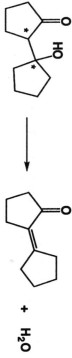

2 C₆H₅CH₂CH → C₆H₅CH₂CHCHCH
(with O, OH, O groups, C₆H₅ substituent, stereocenters marked *)

2 [cyclopentanone] → [aldol product with cyclopentanone rings, HO, O, stereocenters *]

Two new stereocenters are created in each reaction, so racemic mixtures of stereoisomers will be produced.

Problem 19.2 Draw the product of dehydration of each aldol product in Problem 19.1.

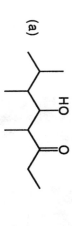

C₆H₅CH₂CHCHCH → C₆H₅CH₂CH=CCH + H₂O
   | | |
  OH   C₆H₅   C₆H₅

[cyclopentanone aldol] → [cyclopentanone with exocyclic alkene to cyclopentane] + H₂O

Problem 19.3 Draw the product of the base-catalyzed crossed aldol reaction between benzaldehyde and 3-pentanone and the product formed by its dehydration.

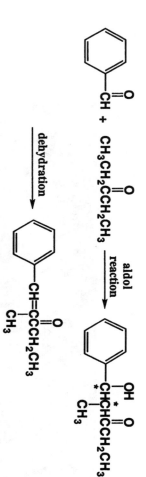

[benzaldehyde] + CH₃CH₂CCH₂CH₃ →(aldol reaction) [product: C₆H₅-CHCHCCH₂CH₃ with OH, CH₃, O, stereocenters *]

→(dehydration) [product: C₆H₅-CH=CCCH₂CH₃ with CH₃, O] + H₂O

Problem 19.4 Show how you might prepare these compounds by directed aldol reactions.

(a)

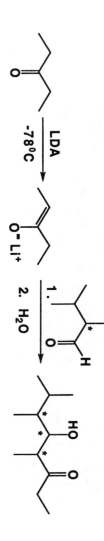

An enolate is made from 3-pentanone that is used to carry out an aldol reaction with 2,3-dimethylbutanal. Two new stereocenters will be created in the reaction and one stereocenter is present in the aldehyde starting material, so a mixture of stereoisomers will be produced.

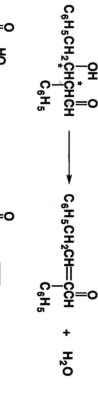

[3-pentanone] →(LDA, -78°C)→ [enolate, O⁻ Li⁺] →(1. 2,3-dimethylbutanal; 2. H₂O)→ [aldol product]

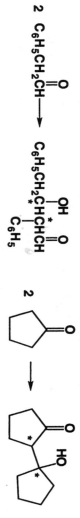

(b)

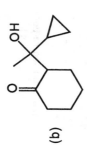

An enolate is made from cyclohexanone that is used to carry out an aldol reaction with cyclopropyl methyl ketone. Two new stereocenters are created in the reaction, so a racemic mixture of stereoisomers will be produced.

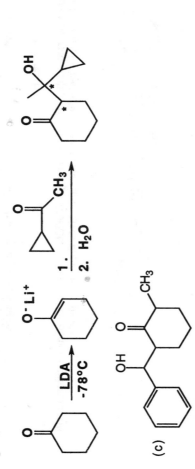

(c)

An enolate is made from 2-methylcyclohexanone. Note that in this case it is the kinetic product, made with an excess of LDA, that is used to carry out an aldol reaction with benzaldehyde. One new stereocenter will be created in the reaction and one stereocenter is present in the ketone starting material, so a mixture of stereoisomers will be produced.

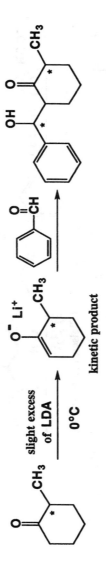

kinetic product

<u>Problem 19.5</u> Show the product of Claisen condensation of ethyl 3-methylbutanoate in the presence of sodium ethoxide followed by acidification with aqueous HCl.

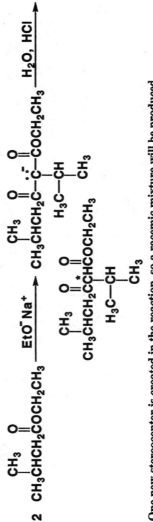

One new stereocenter is created in the reaction, so a racemic mixture will be produced.

<u>Problem 19.6</u> Complete the equation for this crossed Claisen condensation:

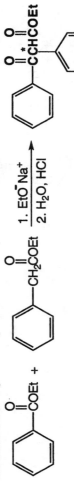

One new stereocenter is created in the reaction, so a racemic mixture will be produced.

Problem 19.7 Show how to convert benzoic acid to 3-methyl-1-phenyl-1-butanone (isobutyl phenyl ketone) by the following synthetic strategies, each of which uses a different type of reaction to form the new carbon-carbon bond to the carbonyl group of benzoic acid.

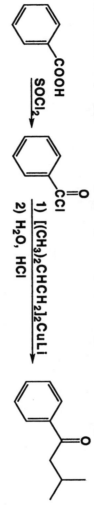

Benzoic acid   3-Methyl-1-phenyl-1-butanone

(a) A lithium diorganocopper (Gilman) reagent

Treatment of benzoic acid with thionyl chloride gives benzoyl chloride. Treatment of benzoyl chloride with lithium diisobutylcopper followed by hydrolysis in aqueous acid gives the desired product.

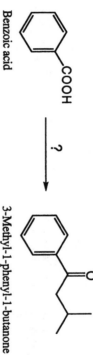

(b) A Claisen condensation

Crossed Claisen condensation of ethyl benzoate and ethyl 3-methylbutanoate gives a β-ketoester. Saponification of the ester followed by acidification gives the β-ketoacid. Heating causes decarboxylation and gives the desired product.

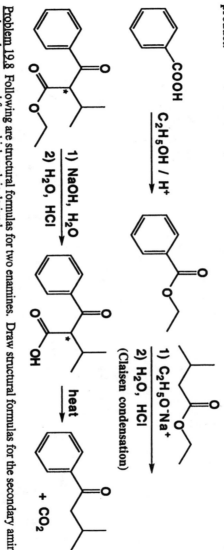

Problem 19.8 Following are structural formulas for two enamines. Draw structural formulas for the secondary amine and carbonyl compound from which each is derived.

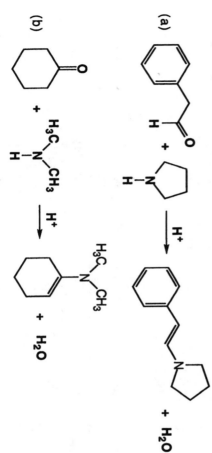

Problem 19.9 Write a mechanism for the hydrolysis of the following iminium chloride in aqueous HCl:

+ HCl/H₂O

The positively charged iminium ion is attacked by water acting as a nucleophile. A proton is transferred to the nitrogen, followed by departure of the amine and formation of the carbon-oxygen double bond. The reaction is completed with a proton transfer step to produce the product ketone and protonated morpholine. Under the acidic conditions in this reaction, an enamine is not formed, so hydrolysis predominates.

Step 1:

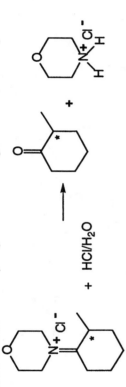

Step 2:

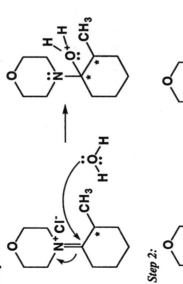

proton transfer

Step 3:

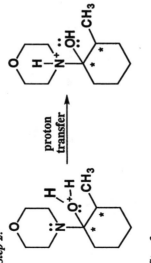

Step 4:

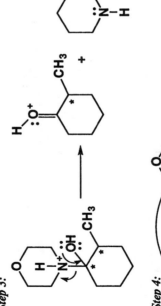

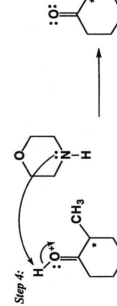

Problem 19.10 Show how to use alkylation or acylation of an enamine to convert acetophenone to the following compounds.

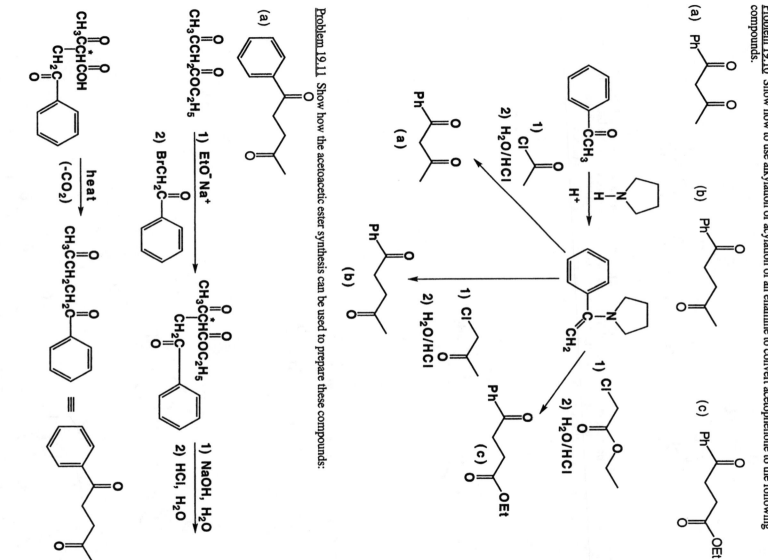

Problem 19.11 Show how the acetoacetic ester synthesis can be used to prepare these compounds:

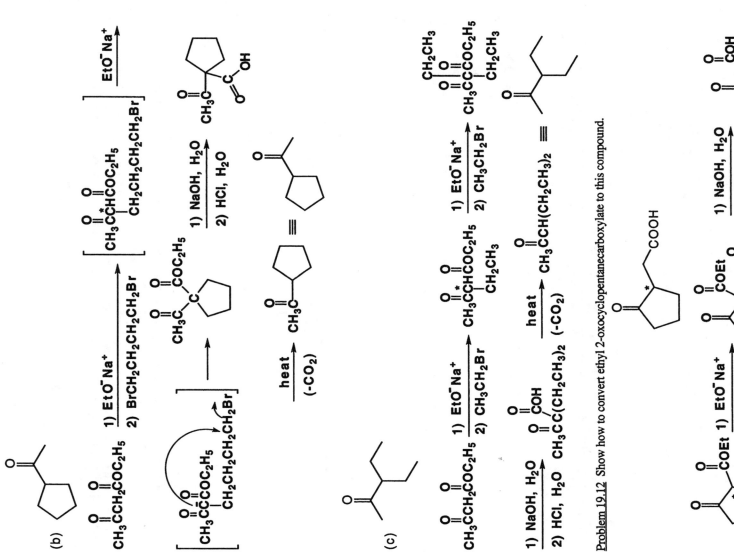

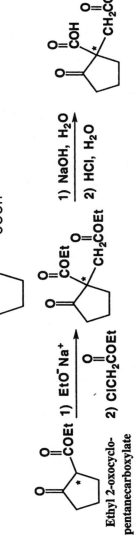

Problem 19.12 Show how to convert ethyl 2-oxocyclopentanecarboxylate to this compound.

Ethyl 2-oxocyclo-
pentanecarboxylate

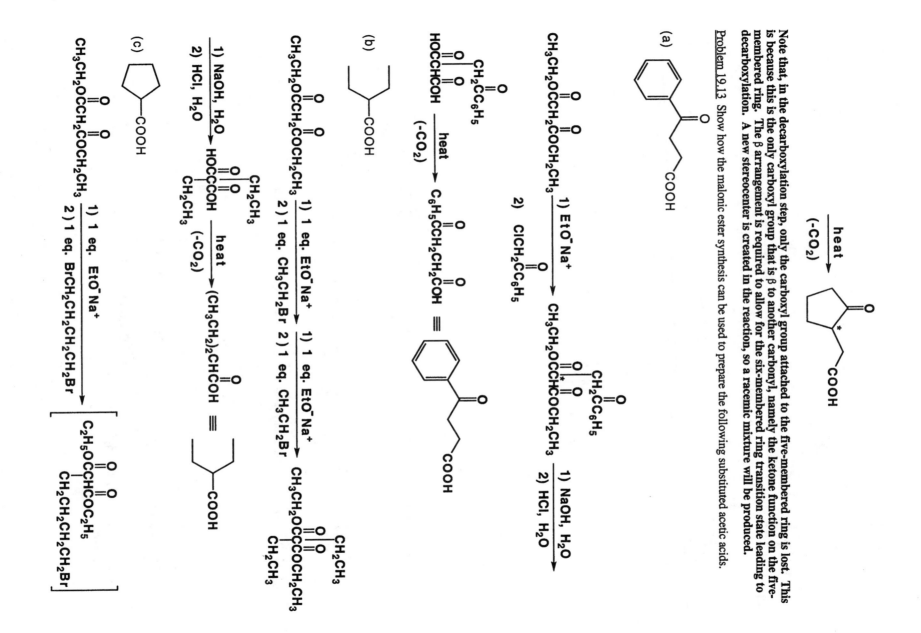

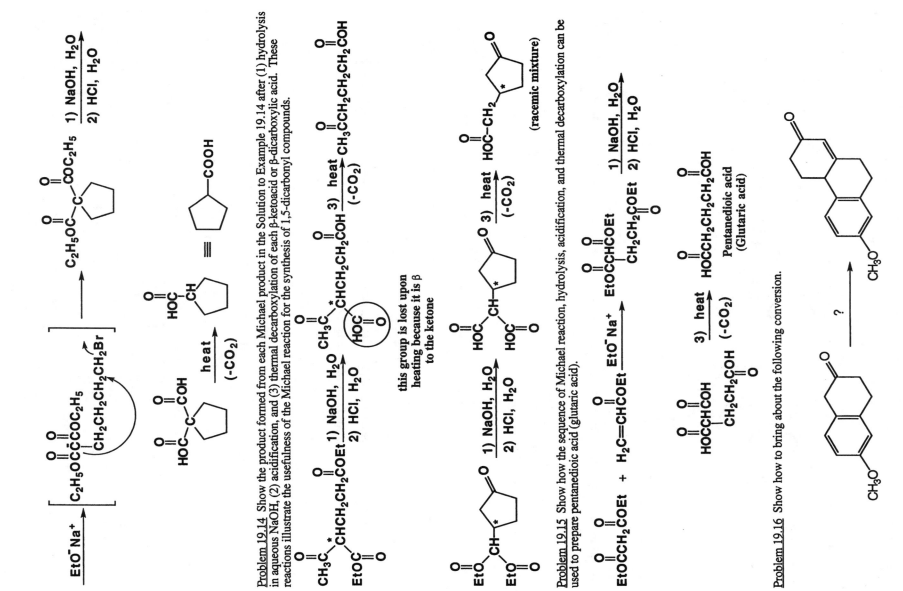

Problem 19.14 Show the product formed from each Michael product in the Solution to Example 19.14 after (1) hydrolysis in aqueous NaOH, (2) acidification, and (3) thermal decarboxylation of each β-ketoacid or β-dicarboxylic acid. These reactions illustrate the usefulness of the Michael reaction for the synthesis of 1,5-dicarbonyl compounds.

this group is lost upon heating because it is β to the ketone

(racemic mixture)

Problem 19.15 Show how the sequence of Michael reaction, hydrolysis, acidification, and thermal decarboxylation can be used to prepare pentanedioic acid (glutaric acid).

Pentanedioic acid (Glutaric acid)

Problem 19.16 Show how to bring about the following conversion.

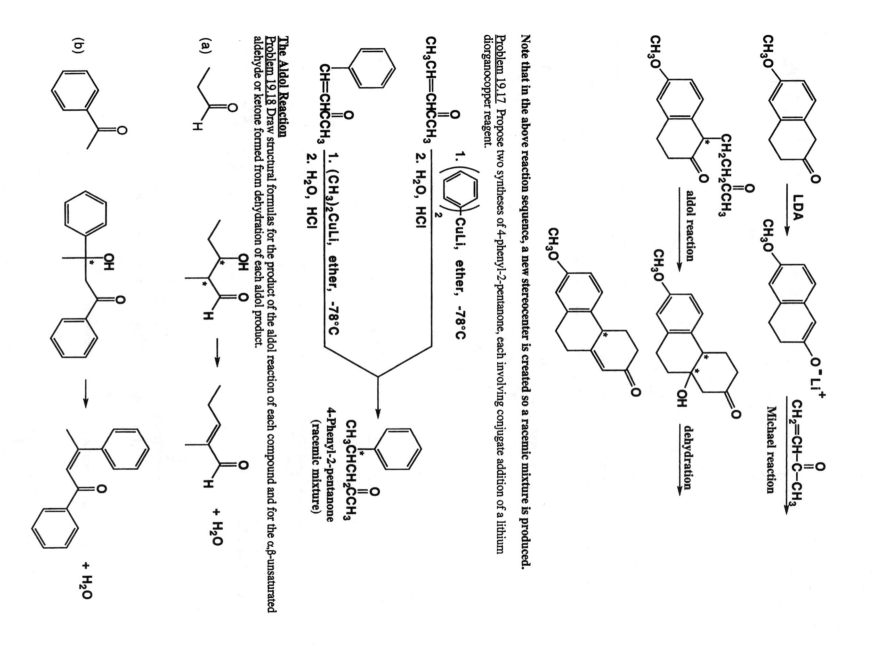

Note that in the above reaction sequence, a new stereocenter is created so a racemic mixture is produced.

Problem 19.17 Propose two syntheses of 4-phenyl-2-pentanone, each involving conjugate addition of a lithium diorganocopper reagent.

(c)

→

+ H₂O

Note that in this problem the relatively stable five-membered ring was formed instead of the highly strained three-membered ring that could have formed. Note also that all the dehydrations take place so that the C=C is in conjugation with the carbonyl.

Problem 19.19 Draw structural formulas for the product of each crossed aldol reaction and for the compound formed by dehydration of each aldol product.

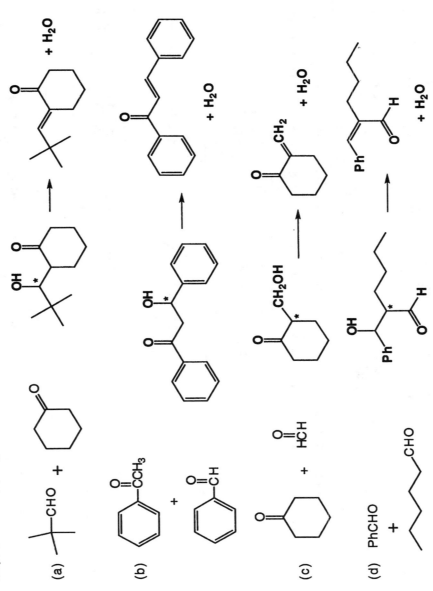

Problem 19.20 When a 1:1 mixture of acetone and 2-butanone is treated with base, six aldol products are possible. Draw structural formulas for these six aldol products.

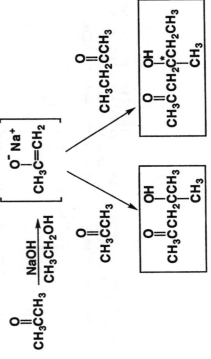

$$CH_3CH_2\overset{O}{\underset{\|}{C}}CH_3 \xrightarrow[CH_3CH_2OH]{NaOH} \left[CH_3CH_2\overset{O^-Na^+}{\underset{|}{C}}=CH_2 \right]$$

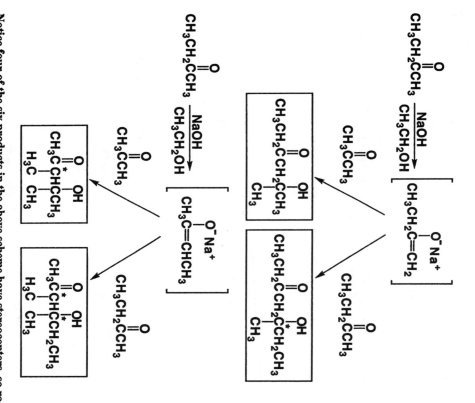

$$CH_3CH_2\overset{O}{\underset{\|}{C}}CH_3$$

$$CH_3CH_2\overset{O}{\underset{\|}{C}}CH_3$$

CH₃CH₂CCH₃   CH₃CH₂CCH₃

Boxed:
$$CH_3CH_2\overset{O}{\underset{\|}{C}}\overset{OH}{\underset{|}{C}}CH_2\overset{O}{\underset{\|}{C}}CH_3 \atop CH_3$$

$$CH_3CH_2\overset{O}{\underset{\|}{C}}\overset{OH}{\underset{|}{C}}CH_2\overset{O}{\underset{\|}{C}}CH_2CH_3 \atop CH_3$$

$$CH_3CH_2\overset{O}{\underset{\|}{C}}CH_3 \xrightarrow[CH_3CH_2OH]{NaOH} \left[CH_3\overset{O^-Na^+}{\underset{|}{C}}=CHCH_3 \right]$$

$$CH_3CCH_3$$ (with =O)

$$CH_3CH_2CCH_3$$ (with =O)

Boxed:
$$CH_3\overset{O}{\underset{\|}{C}}\overset{OH}{\underset{|}{\overset{*}{C}}H}\overset{}{\underset{}{C}}CH_3 \atop H_3C \quad CH_3$$

$$CH_3\overset{O}{\underset{\|}{C}}\overset{OH}{\underset{|}{\overset{*}{C}}H}\overset{O}{\underset{\|}{\overset{*}{C}}H}CH_2CH_3 \atop H_3C \quad CH_3$$

Notice four of the six products in the above scheme have stereocenters, so racemic mixtures will be formed.

Problem 19.21   Show how to prepare these α,β-unsaturated ketones by an aldol reaction followed by dehydration of the aldol product.

(a)
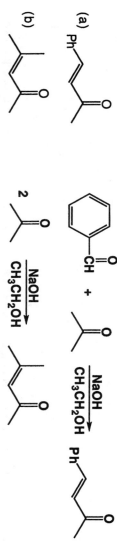

Ph —CH=O + CH₃CCH₃ (with =O) $\xrightarrow[CH_3CH_2OH]{NaOH}$ Ph —CH=CH—CCH₃ (with =O)

(b)

2 CH₃CCH₃ (with =O) $\xrightarrow[CH_3CH_2OH]{NaOH}$

Problem 19.22   Show how to prepare these α,β-unsaturated aldehydes by an aldol reaction followed by dehydration of the aldol product.

(a)
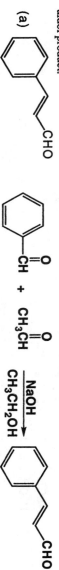

—CH=O + CH₃CH=O $\xrightarrow[CH_3CH_2OH]{NaOH}$ —CH=CH—CHO

Note that in the reaction above, some CH₃CH=CHCHO may form, but the product shown should predominate because conjugation with the aromatic ring makes it the more stable product. The way in which the reaction is run can also have a major influence on the products here. Formation of the desired product is favored by adding the ethanal dropwise to a solution of benzaldehyde and NaOH.

(b)

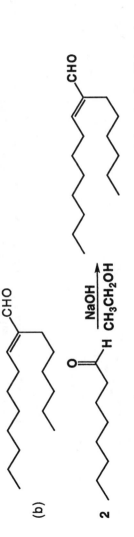

H   CH₃CH₂OH $\xrightarrow{\text{NaOH}}$

2

$\xrightarrow{\text{base}}$ $C_{10}H_{14}O + H_2O$

Problem 19.23   When treated with base, the following compound undergoes an intramolecular aldol reaction to give a product containing a ring (yield 78%). Propose a structural formula for this product.

Analyze this problem in the following way. There are three α-carbons which might form an anion and then condense with either of the other carbonyl groups. Two of these condensations lead to three-membered rings and, therefore, are not feasible. The third anion leads to formation of a five-membered ring, so this is the pathway followed.

An intramolecular aldol condensation by this α-carbon gives a five-membered ring

An intramolecular aldol condensation by these α-carbons would give a three- membered rings

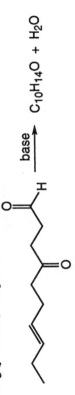

CH₃CH₂CH=CHCH₂CH₂CH₂CH $\longrightarrow$ **CH₃CH₂CH=CHCH₂**

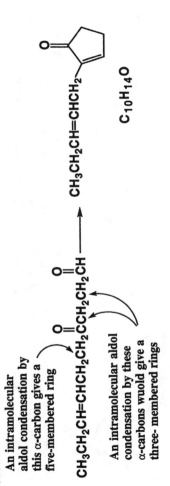

$C_{10}H_{14}O$

Problem 19.24   Cyclohexene can be converted to 1-cyclopentenecarbaldehyde by the following series of reactions. Propose a structural formula for each intermediate compound.

$\xrightarrow[\text{H}_2\text{O}_2]{\text{OsO}_4}$ $C_6H_{12}O_2$ $\xrightarrow{\text{HIO}_4}$ $C_6H_{10}O_2$ $\xrightarrow{\text{base}}$

1-Cyclopentenecarbaldehyde

Oxidation of cyclohexene by osmium tetroxide gives a *cis*-glycol, which is then oxidized by periodic acid to hexanedial. Base-catalyzed aldol reaction of this dialdehyde followed by dehydration of the aldol product gives 1-cyclopentenecarbaldehyde.

Cyclo-hexene $\xrightarrow[\text{H}_2\text{O}_2]{\text{OsO}_4}$ *cis*-1,2-Cyclo-hexanediol $\xrightarrow{\text{HIO}_4}$ Hexanedial $\xrightarrow{\text{base}}$ 1-Cyclopentene-carbaldehyde

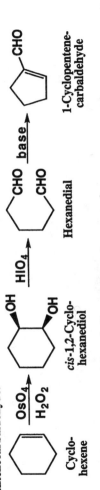

Problem 19.25   Propose a structural formula for each lettered compound.

$\xrightarrow{\text{CrO}_3, \text{pyridine}}$ A ($C_{11}H_{18}O_2$) $\xrightarrow[\text{EtOH}]{\text{EtO}^-\text{Na}^+}$ B ($C_{11}H_{16}O$)

Oxidation of the starting material by CrO₃ gives the diketone, compound A. Base-catalyzed aldol reaction of this diketone could potentially give cyclic products with either a four-membered or six-membered ring. The six-

membered ring is much more stable, so this product predominates. Dehydration of the aldol product gives compound B. One new stereocenter is created, so assuming racemic starting material, a racemic mixture will be produced.

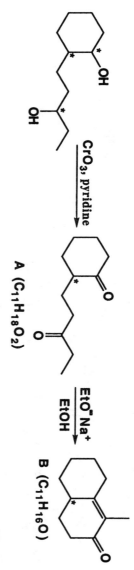

$A\ (C_{11}H_{18}O_2)$

$B\ (C_{11}H_{16}O)$

Problem 19.26 How might you bring about the following conversions?

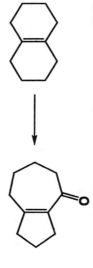

The alkene is oxidized to a diketone by OsO₄/H₂O₂ followed by treatment with HIO₄. Note how ozonolysis could have been used for this step as well. Sodium ethoxide or other base catalyzes an intramolecular aldol reaction of the diketone to give a β-hydroxyketone. Dehydration of this aldol product under the conditions of the aldol reaction gives the desired α,β-unsaturated ketone.

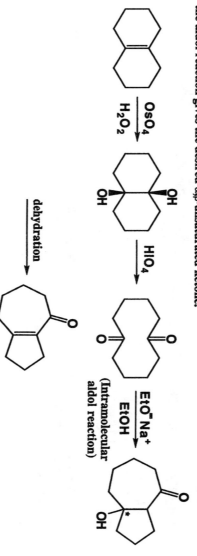

(Intramolecular aldol reaction)

dehydration

Problem 19.27 Pulegone, $C_{10}H_{16}O$, a compound from oil of pennyroyal, has a pleasant odor midway between peppermint and camphor. (The Merck Index, 12th ed., #8124.) Treatment of pulegone with steam produces acetone and 3-methylcyclohexanone.

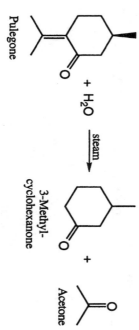

Pulegone   3-Methyl-
                                           cyclohexanone

(a) Natural pulegone has the configuration shown. Assign an R,S configuration to its stereocenter.

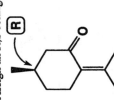

(b) Propose a mechanism for the steam hydrolysis of pulegone to the compounds shown.

An aldol reaction is reversible, and this steam hydrolysis of pulegone is formally the reverse of an aldol reaction. Thus, a reasonable mechanism is exactly the reverse of what would be written for an aldol reaction between acetone and 3-methylcyclohexanone.

Step 1:

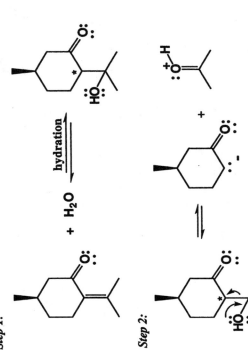

Step 2:

Step 3:

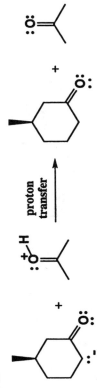

(c) In what way does this steam hydrolysis affect the configuration of the stereocenter in pulegone? Assign an R,S configuration to the 3-methylcyclohexanone formed in this reaction.

This reaction mechanism does not involve the stereocenter; it remains in the R configuration in the 3-methylcyclohexanone product.

Problem 19.28 Propose a mechanism for this acid-catalyzed aldol reaction and the dehydration of the resulting aldol product.

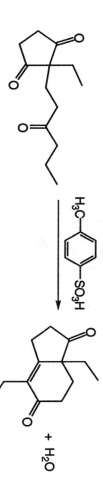

A reasonable mechanism for this acid-catalyzed aldol reaction involves attack of the enol on the protonated carbonyl group, followed by acid-catalyzed dehydration. Note how the first two steps could occur in either order.

Step 1:

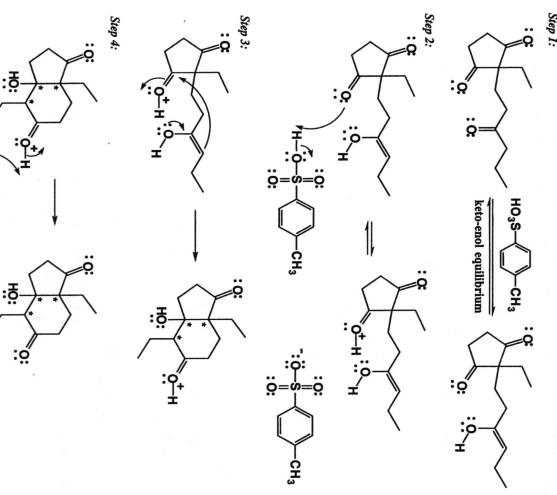

Step 2:

Step 3:

Step 4:

Step 5:

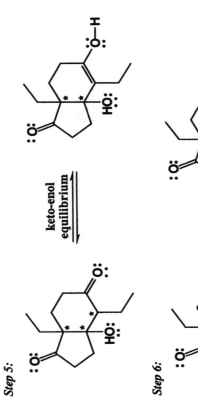

keto-enol
equilibrium

Step 6:

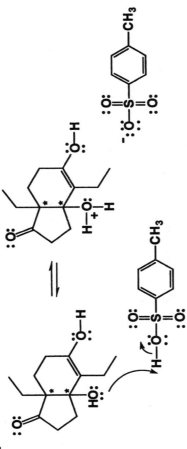

Step 7:

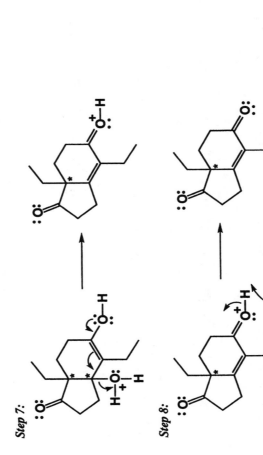

Step 8:

Note that steps 7 and 8 may take place more or less simultaneously. One stereocenter is created, so a racemic mixture is produced in the overall reaction.

Directed Aldol Reactions

Problem 19.29 In Section 19.2B, it was stated that four possible aldol products are formed when phenylacetaldehyde and acetone are mixed in the presence of base. Draw structural formulas for each of these aldol products.

$$C_6H_5CCH_3 \;+\; CH_3CCH_3 \xrightarrow[CH_3CH_2OH]{NaOH}\; \text{a mixture of four aldol products}$$

Phenyl-   Acetone
acetaldehyde

For the following scheme, in the cases where one stereocenter is created, a racemic mixture is produced.

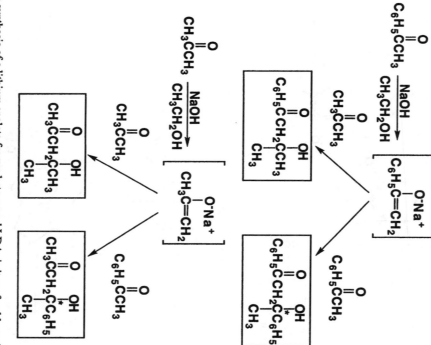

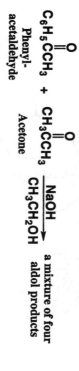

Problem 19.30 In the synthesis of a lithium enolate from a ketone and LDA, is it preferable to (a) add a solution of LDA to a solution of the ketone or (b) add a solution of the ketone to a solution of LDA or (c) conclude that the order in which the solutions are mixed makes no difference? Explain your answer.

It is important to (b), add a solution of the ketone to the LDA. This ensures that excess base is available to convert all of the ketone to the enolate form as quickly as possible during the addition. This prevents any ketone from being available to take part in an unwanted aldol reaction.

The Claisen Condensation

Problem 19.31 Show the product of Claisen condensation of these esters.
(a) Ethyl phenylacetate in the presence of sodium ethoxide.

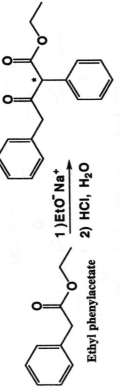

Ethyl phenylacetate

1)EtO⁻Na⁺
2) HCl, H₂O

One stereocenter is created, so a racemic mixture is produced.

(b) Methyl hexanoate in the presence of sodium methoxide.

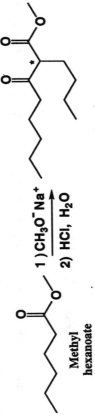

**Methyl
hexanoate**

1)CH₃O⁻Na⁺
2) HCl, H₂O

One stereocenter is created, so a racemic mixture is produced.

Problem 19.32 When a 1:1 mixture of ethyl propanoate and ethyl butanoate is treated with sodium ethoxide, four Claisen condensation products are possible. Draw a structural formula for each product.

For the following scheme, in the cases where one stereocenter is created, a racemic mixture is produced.

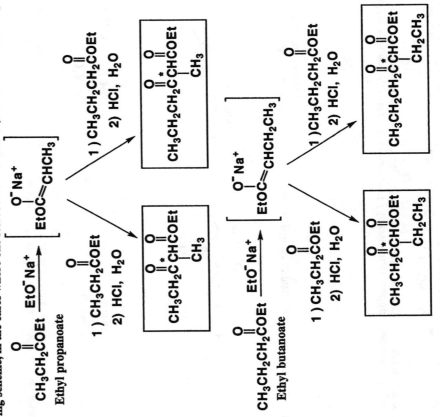

Problem 19.33 Draw structural formulas for the β-ketoesters formed by Claisen condensation of ethyl propanoate with each ester:

(a) EtOC—COEt

(b) PhCOEt

(c) HCOEt

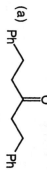

In all three of the above reactions, one stereocenter is created so racemic mixtures are produced.

Problem 19.34 Draw a structural formula for the product of saponification, acidification, and decarboxylation of each β-ketoester formed in Problem 19.33.

Following are the structures for the products of saponification and decarboxylation of each β-ketoester from the previous problem.

(a) HOC—CCH₂CH₃

(b) PhCCH₂CH₃

(c) HCCH₂CH₃

Problem 19.35 The Claisen condensation can be used as one step in the synthesis of ketones, as illustrated by this reaction sequence. Propose structural formulas for compounds A, B, and the ketone formed in this sequence.

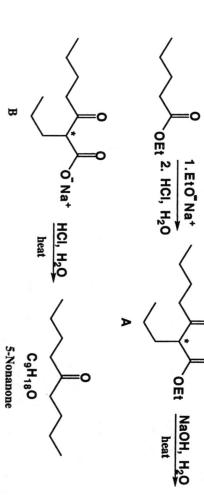

Compound (A) is a β-ketoester, compound (B) is the sodium salt of a β-ketoacid, and the final ketone is 5-nonanone. Compounds A and B are racemic mixtures, while the final product is not chiral.

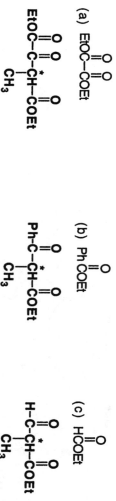

Problem 19.36 Propose a synthesis for each ketone, using as one step in the sequence a Claisen condensation and the reaction sequence illustrated in Problem 19.35.

(a)

Each target molecule is synthesized using the sequence of Claisen condensation, saponification, and decarboxylation as outlined in the previous problem. The starting ester is ethyl 3-phenylpropanoate.

2 PhCH₂CH₂COEt → PhCH₂CH₂CCHCOEt →
CH₂Ph

1. EtO⁻ Na⁺
2. HCl, H₂O

1. NaOH, H₂O
2. HCl, H₂O
3. Heat

(b)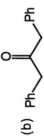

The starting ester is ethyl phenylacetate.

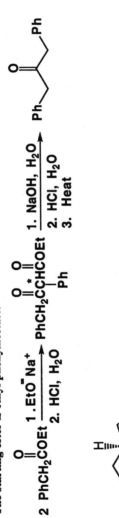

$$2 \text{ PhCH}_2\text{COEt} \xrightarrow[\text{2. HCl, H}_2\text{O}]{\text{1. EtO}^-\text{Na}^+} \text{PhCH}_2\overset{*}{\text{C}}\text{CHCOEt} \xrightarrow[\substack{\text{2. HCl, H}_2\text{O} \\ \text{3. Heat}}]{\text{1. NaOH, H}_2\text{O}}$$

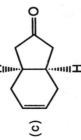

(c)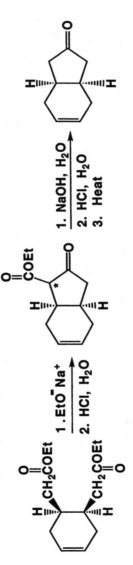

The starting diester is derived from a *cis*-4,5-disubstituted cyclohexene.

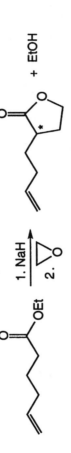

Problem 19.37 Propose a mechanism for the following conversion.

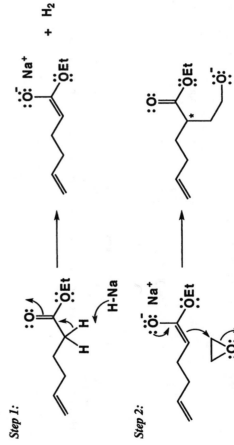

A reasonable mechanism for this conversion involves an initial reaction of the ester with the strong base NaH to give an enolate, that attacks the ethylene oxide.

Step 1:

Step 2:

Step 3: The alkoxide intermediate then undergoes an intramolecular reaction with the ester function to give the product.

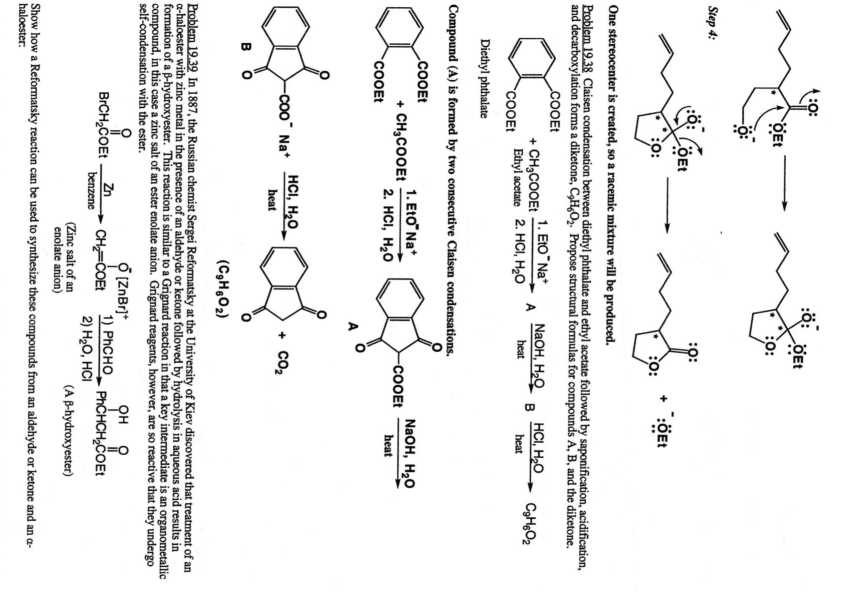

Step 4:

One stereocenter is created, so a racemic mixture will be produced.

Problem 19.38 Claisen condensation between diethyl phthalate and ethyl acetate followed by saponification, acidification, and decarboxylation forms a diketone, $C_9H_6O_2$. Propose structural formulas for compounds A, B, and the diketone.

Diethyl phthalate

Compound (A) is formed by two consecutive Claisen condensations.

$(C_9H_6O_2)$

Problem 19.39 In 1887, the Russian chemist Sergei Reformatsky at the University of Kiev discovered that treatment of an α-haloester with zinc metal in the presence of an aldehyde or ketone followed by hydrolysis in aqueous acid results in formation of a β-hydroxyester. This reaction is similar to a Grignard reaction in that a key intermediate is an organometallic compound, in this case a zinc salt of an ester enolate anion. Grignard reagents, however, are so reactive that they undergo self-condensation with the ester.

(Zinc salt of an enolate anion)

(A β-hydroxyester)

Show how a Reformatsky reaction can be used to synthesize these compounds from an aldehyde or ketone and an α-haloester:

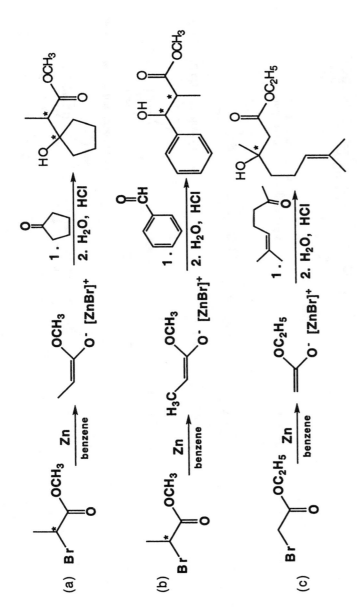

(a)

(b)

(c)

In parts (a) and (b), one new stereocenter is created and one stereocenter is present in the ester starting material, so a mixture of stereoisomers will be produced. In part (c), one new stereocenter is created, so a racemic mixture will be produced.

Problem 19.40 Many types of carbonyl condensation reactions have acquired specialized names, after the 19th century organic chemists who first studied them. Propose mechanisms for the following named condensations.

(a) Perkin condensation: Condensation of an aromatic aldehyde with a carboxylic acid anhydride.

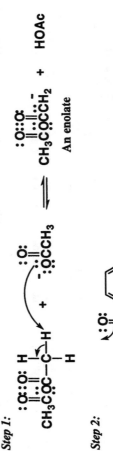

Cinnamic acid

In the Perkin reaction, an enolate of the anhydride is formed with acetate acting as the base. This enolate then attacks benzaldehyde reminiscent of an aldol reaction. Hydrolysis of the anhydride and dehydration complete the mechanism.

Step 1:

Step 2:

An enolate

Step 3:

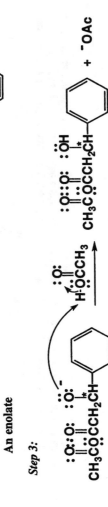

PDT is missing a carbon

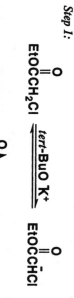

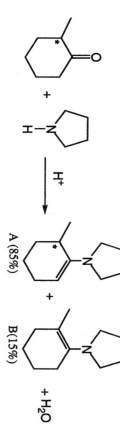

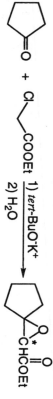

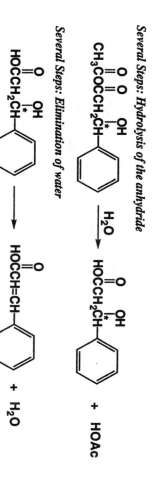

Several Steps: Hydrolysis of the anhydride

CH₃COCH₂CH → H₂O → HOCH₂CH + HOAc

Several Steps: Elimination of water

HOCH₂CH → HOCH=CH + H₂O

(b) Darzens condensation: Condensation of an α-haloester with a ketone or an aromatic aldehyde.

In the Darzens condensation, the enolate of the α-haloester is formed, which then attacks the ketone carbonyl group reminiscent of an aldol reaction.

Step 1:

EtOCCH₂Cl ⇌ *tert-BuO⁻K⁺* EtOCCHCl

Step 2:

EtOCCHCl

Step 3:

The oxyanion displaces chloride via backside attack to create an epoxide.

—CHCOEt + Cl⁻

One new stereocenter is created, so a racemic mixture will be produced.

Enamines

Problem 19.41 When 2-methylcyclohexanone is treated with pyrrolidine, two isomeric enamines are formed. Why is enamine A with the less substituted double bond the thermodynamically favored product? You will find it helpful to examine the models on the CD of these two enamines.

A (85%)   +   B(15%)   + H₂O

Remember that in the six atom system of an enamine, the nitrogen is sp² hybridized, so there is overlap of the 2p orbital of nitrogen atom with the 2p orbitals of the carbon-carbon double bond. This overlap allows electron delocalization and thus stabilization of the enamine.

this six-atom sytem is planar

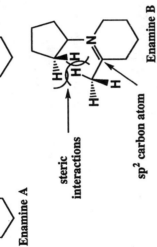

Enamine A

Enamine B

steric interactions

sp² carbon atom

In enamine B, non-bonded interactions force rotation about the C-N bond and reduction in planarity of the enamine system, resulting in loss of stability. In enamine A, there is no such non-bonded interaction, so the enamine system is planar providing for maximum stabilization. Thus, enamine A predominates.

Problem 19.42 Enamines normally react with methyl iodide to give two products: one arising from alkylation at nitrogen, the second arising from alkylation at carbon. For example:

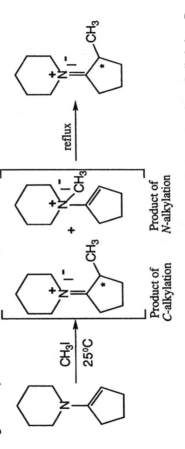

Product of C-alkylation

Product of N-alkylation

Heating the mixture of C-alkylation and N-alkylation products gives only the product from C-alkylation. Propose a mechanism for this isomerization.

The I⁻ that is in the solution can act as a nucleophile to regenerate CH₃I from the N-alkylated product. This then reacts with the enamine to make the C-alkylated product.

Step 1:

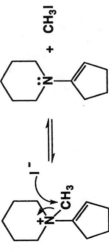

Solutions

Step 2:

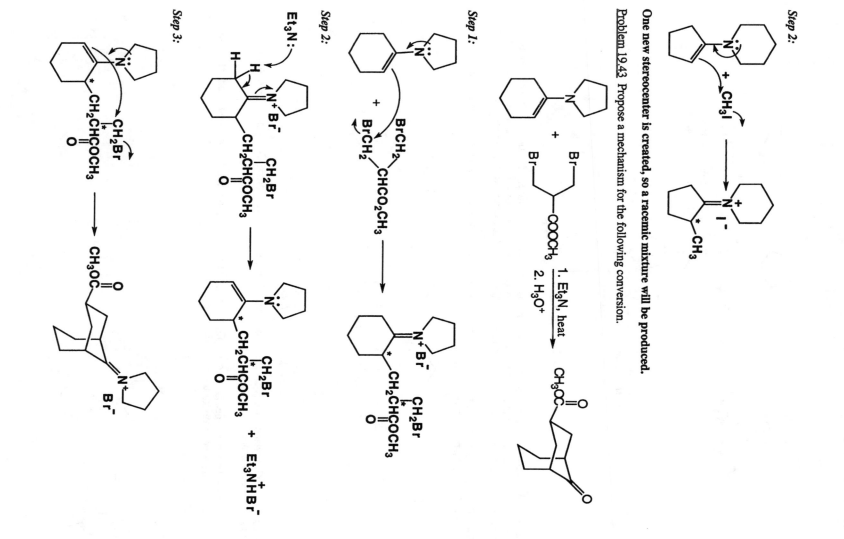

One new stereocenter is created, so a racemic mixture will be produced.

Problem 19.43 Propose a mechanism for the following conversion.

Step 1:

Step 2:

Step 3:

Step 4:

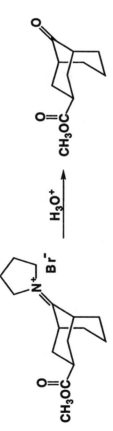

Due to symmetry, the product is not chiral.

Problem 19.44 The following intermediate was needed for the synthesis of tamoxifen, a widely used antiestrogen drug. For the structural formula and synthesis of tamoifen, see MedChem Problem MC.26. Propose a synthesis for this intermediate from compound A.

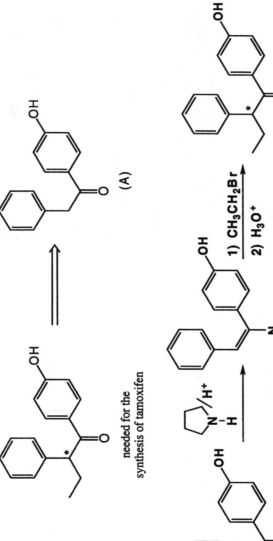

(A)

needed for the synthesis of tamoxifen

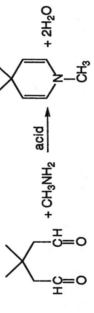

One new stereocenter is created, so a racemic mixture will be produced.

Problem 19.45 Propose a mechanism for the following reaction.

Note that in the following mechanism familiar steps (amino alcohol formation, acid catalyzed dehydration) are summarized rather than shown explicitly.

Step 1:

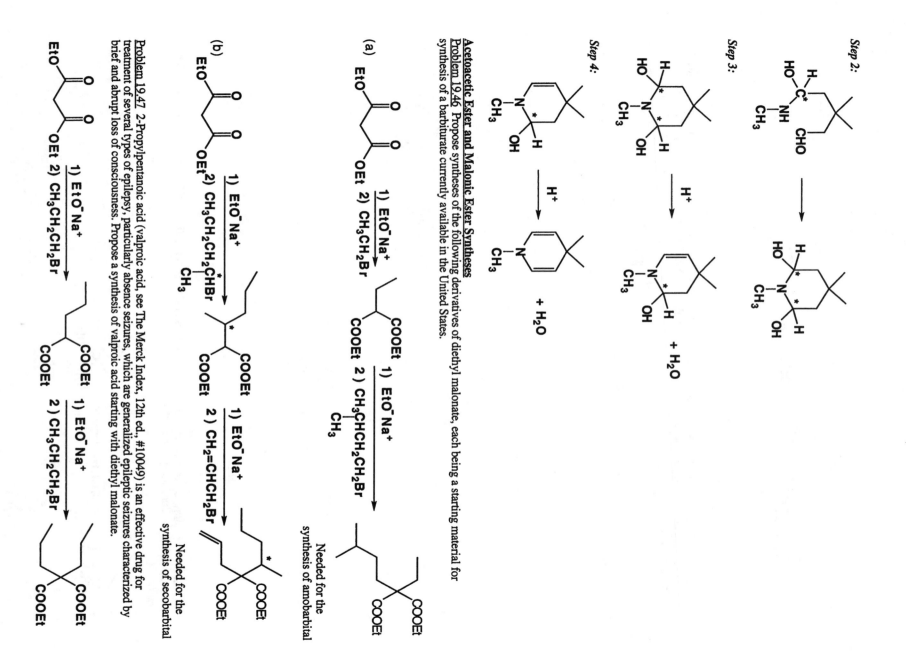

Step 2:

Step 3:

Step 4:

Acetoacetic Ester and Malonic Ester Syntheses

Problem 19.46 Propose syntheses of the following derivatives of diethyl malonate, each being a starting material for synthesis of a barbiturate currently available in the United States.

(a)

1) EtO⁻Na⁺
2) CH₃CH₂CH₂Br

1) EtO⁻Na⁺
2) CH₃CHCH₂CH₂Br
      |
      CH₃

Needed for the synthesis of amobarbital

(b)

1) EtO⁻Na⁺
2) CH₃CH₂CH₂CHBr
                  |
                  CH₃

1) EtO⁻Na⁺
2) CH₂=CHCH₂Br

Needed for the synthesis of secobarbital

Problem 19.47 2-Propylpentanoic acid (valproic acid, see The Merck Index, 12th ed., #10049) is an effective drug for treatment of several types of epilepsy, particularly absence seizures, which are generalized epileptic seizures characterized by brief and abrupt loss of consciousness. Propose a synthesis of valproic acid starting with diethyl malonate.

1) EtO⁻Na⁺
2) CH₃CH₂CH₂Br

1) EtO⁻Na⁺
2) CH₃CH₂CH₂Br

1) NaOH, H₂O
2) HCl, H₂O

heat
(-CO₂)

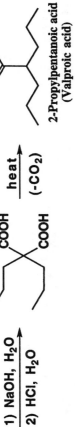

2-Propylpentanoic acid
(Valproic acid)

Problem 19.48 Show how to synthesize the following compounds using either the malonic ester synthesis or the acetoacetic ester synthesis.

a) 4-Phenyl-2-butanone

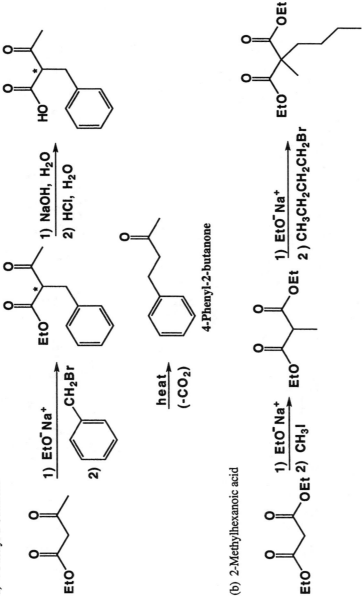

1) EtO⁻ Na⁺

2) CH₂Br

1) NaOH, H₂O
2) HCl, H₂O

heat
(-CO₂)

4-Phenyl-2-butanone

(b) 2-Methylhexanoic acid

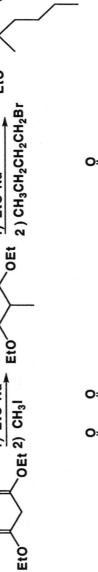

1) EtO⁻ Na⁺ 2) CH₃I

1) EtO⁻ Na⁺
2) CH₃CH₂CH₂CH₂Br

1) NaOH, H₂O
2) HCl, H₂O

heat
(-CO₂)

2-Methylhexanoic acid

One new stereocenter is created, so a racemic mixture will be produced.

(c) 3-Ethyl-2-pentanone

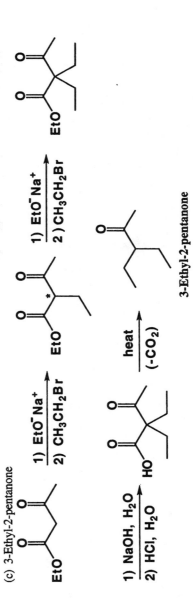

1) EtO⁻ Na⁺
2) CH₃CH₂Br

1) EtO⁻ Na⁺
2) CH₃CH₂Br

1) NaOH, H₂O
2) HCl, H₂O

heat
(-CO₂)

3-Ethyl-2-pentanone

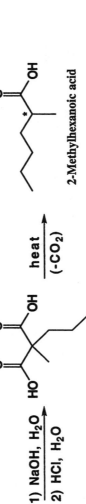

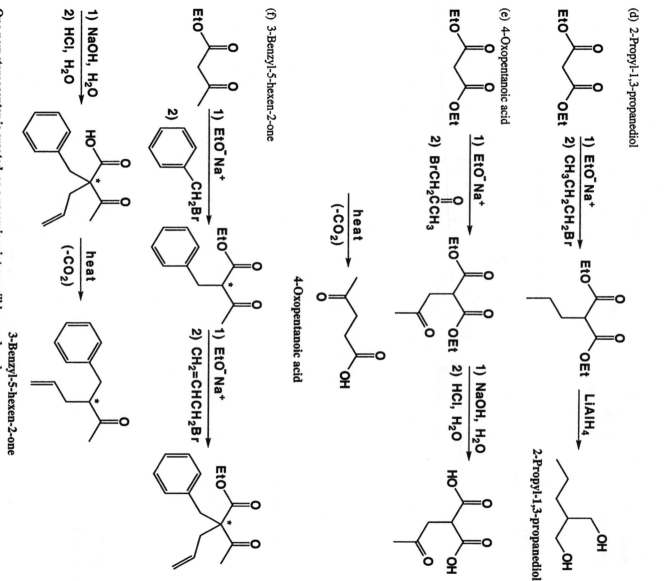

(d) 2-Propyl-1,3-propanediol

(e) 4-Oxopentanoic acid

4-Oxopentanoic acid

(f) 3-Benzyl-5-hexen-2-one

3-Benzyl-5-hexen-2-one

One new stereocenter is created, so a racemic mixture will be produced.

(g) Cyclopropanecarboxylic acid

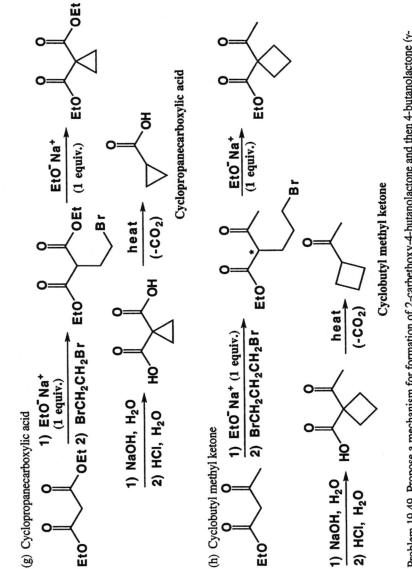

(h) Cyclobutyl methyl ketone

Cyclobutyl methyl ketone

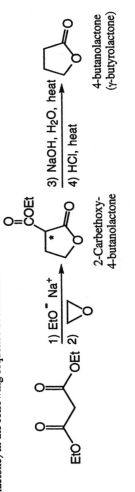

2-Carbethoxy-
4-butanolactone

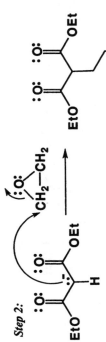

4-butanolactone
(γ-butyrolactone)

<u>Problem 19.49</u> Propose a mechanism for formation of 2-carbethoxy-4-butanolactone and then 4-butanolactone (γ-butyrolactone) in the following sequence of reactions.

Step 1:

Step 2:

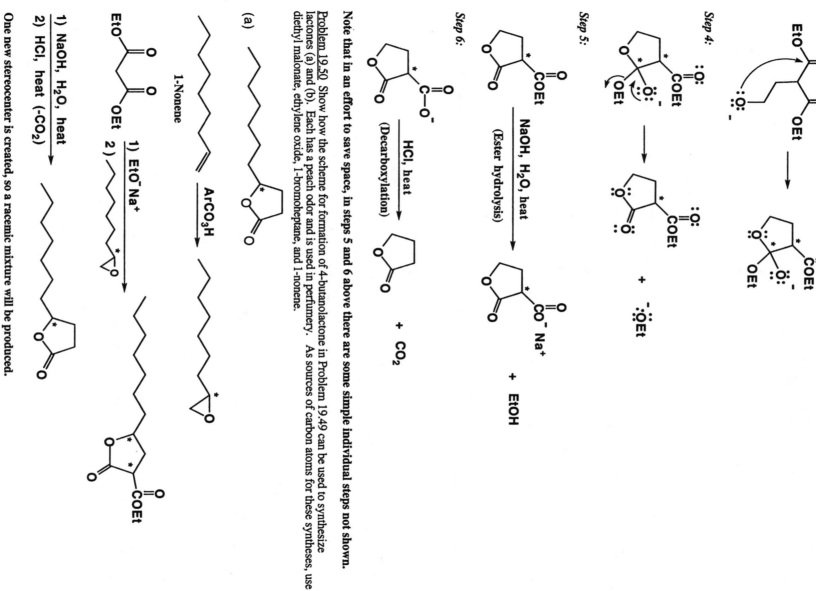

Step 3:

Step 4:

Step 5:

NaOH, H₂O, heat

(Ester hydrolysis)

+ EtOH

Step 6:

HCl, heat

(Decarboxylation)

+ CO₂

Note that in an effort to save space, in steps 5 and 6 above there are some simple individual steps not shown.

Problem 19.50 Show how the scheme for formation of 4-butanolactone in Problem 19.49 can be used to synthesize lactones (a) and (b). Each has a peach odor and is used in perfumery. As sources of carbon atoms for these syntheses, use diethyl malonate, ethylene oxide, 1-bromoheptane, and 1-nonene.

(a)

1-Nonene

ArCO₃H

EtO malonate

1) EtO⁻ Na⁺

2)

1) NaOH, H₂O, heat

2) HCl, heat (-CO₂)

One new stereocenter is created, so a racemic mixture will be produced.

(b)

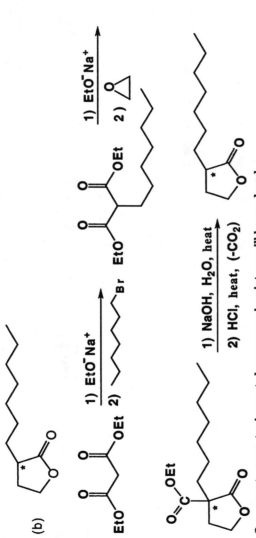

One new stereocenter is created, so a racemic mixture will be produced.

Michael Reactions

Problem 19.51 The following synthetic route is used to prepare an intermediate in the total synthesis of the anticholinergic drug benzilonium bromide. Write structural formulas for intermediates (A), (B), (C), and (D) in this synthesis.

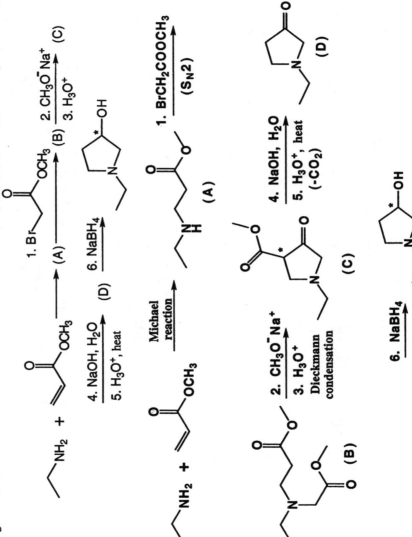

One new stereocenter is created, so a racemic mixture will be produced.

Problem 19.52 Propose a mechanism for formation of the bracketed intermediate, and for the bicyclic ketone formed in the following reaction sequence.

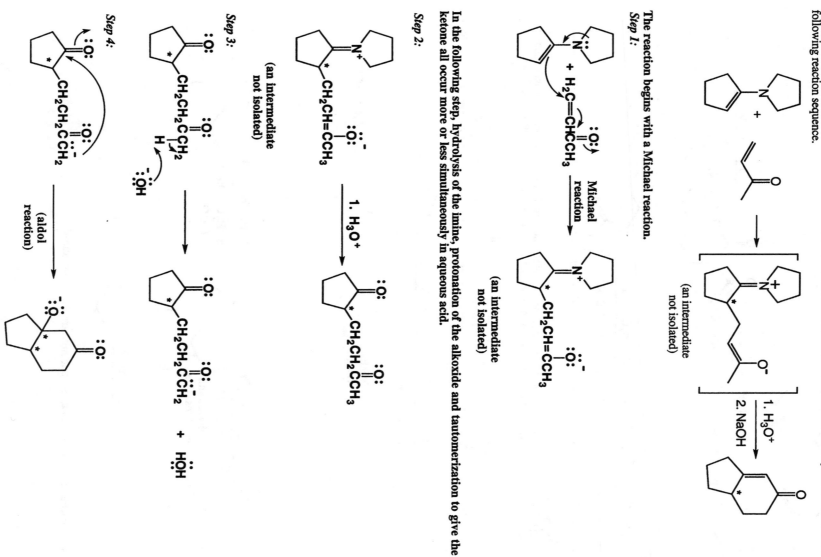

The reaction begins with a Michael reaction.

Step 1:

Step 2:

In the following step, hydrolysis of the imine, protonation of the alkoxide and tautomerization to give the ketone all occur more or less simultaneously in aqueous acid.

Step 3:

Step 4:

Step 5:

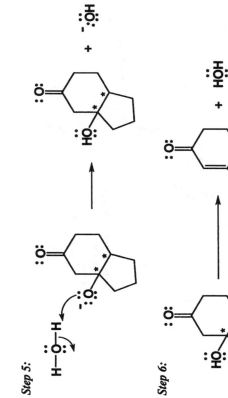

Step 6:

One new stereocenter is created, so a racemic mixture will be produced.

Synthesis

<u>Problem 19.53</u>   Show experimental conditions by which to carry out the following synthesis starting with benzaldehyde and methyl acetoacetate.

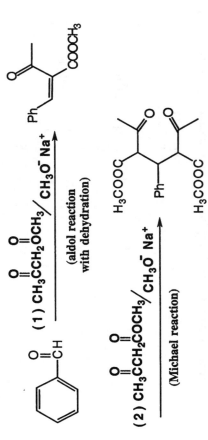

<u>Problem 19.54</u>   Nifedipine (Procardia, Adalat, see The Merck Index, 12 ed., #6617) belongs to a class of drugs called calcium channel blockers and is effective in the treatment of various types of angina, including that induced by exercise. Show how nifedipine can be synthesized from 2-nitrobenzaldehyde, methyl acetoacetate, and ammonia. (Hint: Review the chemistry of your answers to Problems 19.45 and 19.53, and then combine that chemistry to solve this problem.)

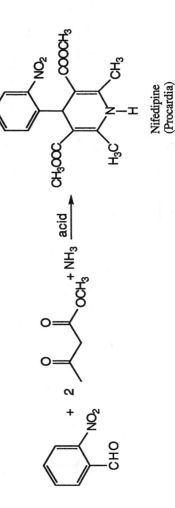

Solutions

Problem 19.55 The compound 3,5,5-trimethyl-2-cyclohexenone can be synthesized using acetone and ethyl acetoacetate as sources of carbon atoms. New carbon-carbon bonds in this synthesis are formed by a combination of aldol reactions and Michael reactions. Show reagents and conditions by which this synthesis might be accomplished.

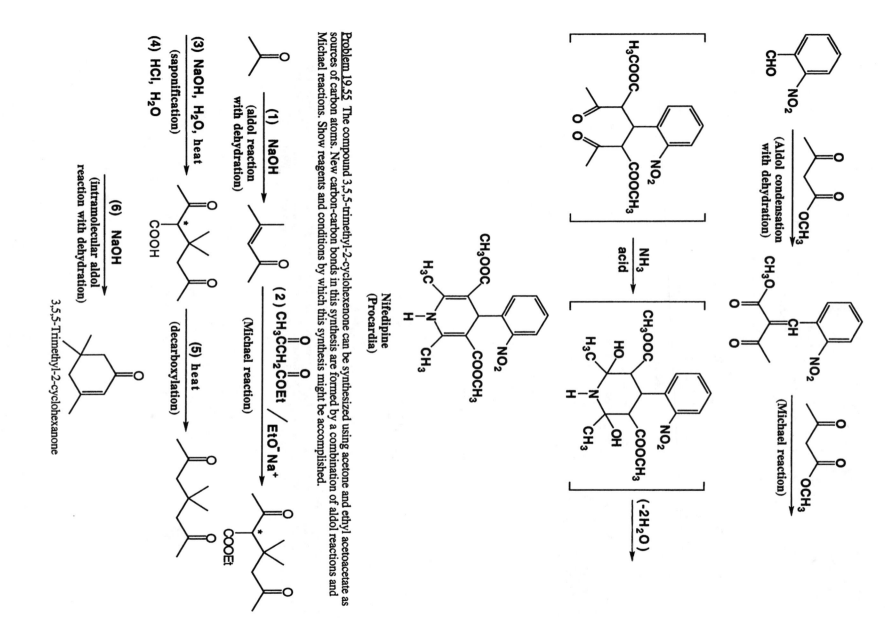

Nifedipine
(Procardia)

(1) NaOH
(aldol reaction
with dehydration)

(2) CH₃CCH₂COEt / EtO⁻Na⁺
(Michael reaction)

(3) NaOH, H₂O, heat
(saponification)

(4) HCl, H₂O

(5) heat
(decarboxylation)

(6) NaOH
(intramolecular aldol
reaction with dehydration)

3,5,5-Trimethyl-2-cyclohexanone

Problem 19.56 The Weiss reaction, discovered in 1968 by Dr. Ulrich Weiss at the National Institutes of Health, is a route to fused five-membered rings. An example of a Weiss reaction is treating dimethyl 3-oxopentanedioate with ethanedial (glyoxal) in aqueous base under carefully controlled conditions. The bicyclo[3.3.0]octane derivative (A) is formed in 90% yield.

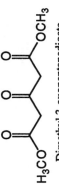

Dimethyl 3-oxopentanedioate

The mechanism of the Weiss reaction has been investigated, and the overall steps, as presently understood, involve a combination of aldol, Michael, and dehydration reactions. The molecule shown in brackets is assumed to be an intermediate, but it is not isolated.

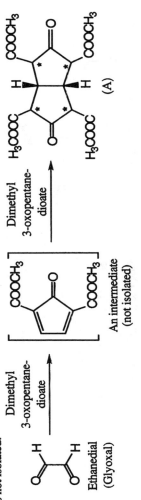

Propose a mechanism for the formation Compound A.

The reaction starts with a "double" aldol reaction between glyoxal and dimethyl 3-oxopentanedioate.
Step 1:

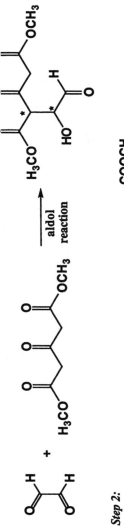

Step 2:

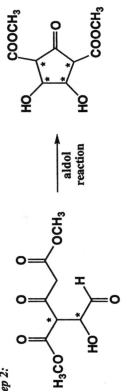

Dehydration gives the intermediate shown above.
Step 3:

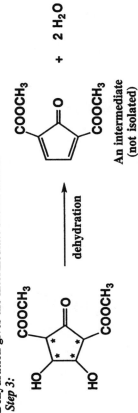

Two Michael reaction steps completes the mechanism.

Step 4:

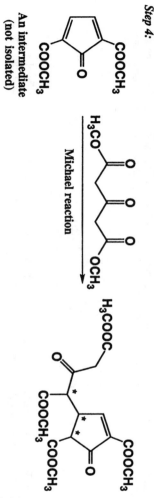

An intermediate
(not isolated)

Step 5:

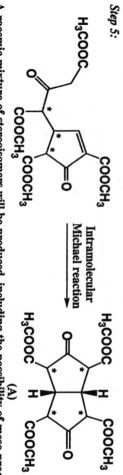

A racemic mixture of stereoisomers will be produced, including the possibility of meso products.

<u>Problem 19.57</u> The following β-diketone (A) can be synthesized from cyclopentanone and an acid chloride using an enamine reaction.

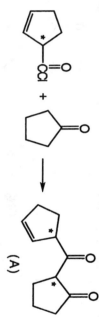

(A)

(a) Propose a synthesis of the starting acid chloride from cyclopentene.

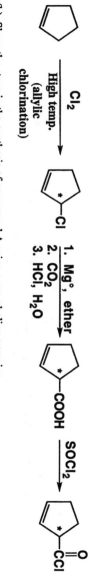

(b) Show the steps in the synthesis of compound A using a morpholine enamine.

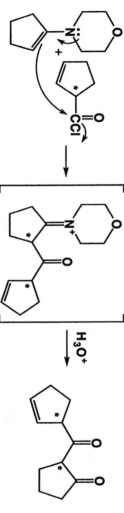

A racemic mixture of stereoisomers will be produced.

Problem 19.58 Cisplatin (see The Merck Index, 12th ed., #2378) was first prepared in 1844, but it was not until 1964 that its value as an anticancer drug was realized. In that year, Barnett Rosenberg and coworkers at Michigan State University observed that when platinum electrodes are inserted into a growing bacterial culture and an electric current passed through the culture, all cell division ceased within 1 to 2 hours. The result was surprising. Equally surprising was their finding that cell division was inhibited by *cis*-diamminedichloroplatinum(II), more commonly named cisplatin, a platinum complex formed in the presence of ammonia and chloride ion. Cisplatin has a broad spectrum of anticancer activity and is particularly useful for treatment of epithelial malignancies. Evidence suggests that platinum(II) in the complex bonds to DNA and forms intrachain and interchain cross linkages. More than 1000 platinum complexes have since been prepared and tested in attempts to discover even more active cytotoxic drugs. In spiroplatin, the two NH_3 groups are replaced by primary amino groups. This drug showed excellent antileukemic activity in animal models but was disappointing in human trials. In carboplatin, the two chloride ions are replaced by carboxylate groups (see The Merck Index, 12 ed., #1870). In 1989, carboplatin was approved by the FDA for treatment of ovarian cancers.

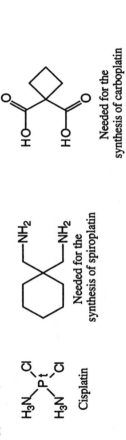

Cisplatin

Needed for the
synthesis of spiroplatin

Needed for the
synthesis of carboplatin

(a) Devise a synthesis for the diamine needed in the synthesis of spiroplatin starting with diethyl malonate and 1,5-dibromopentane as the sources of carbon atoms.

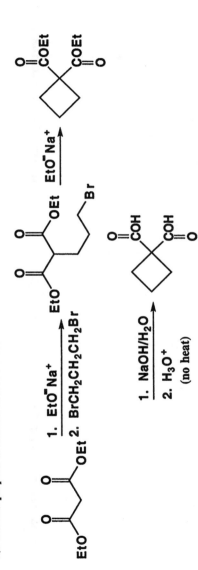

(b) Devise a synthesis for the dicarboxylic acid needed in the synthesis of carboplatin starting with diethyl malonate and 1,3-dibromopropane as sources of carbon atoms.

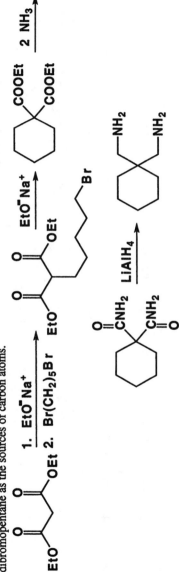

Problem 19.59 Oxanamide (see The Merck Index, 12th ed., #7053) is a mild sedative belonging to a class of molecules called oxanamides; (it contains an oxirane (epoxide) group and an amide group. As seen in this retrosynthetic scheme, the source of carbon atoms for the synthesis of oxanamide is butanal.

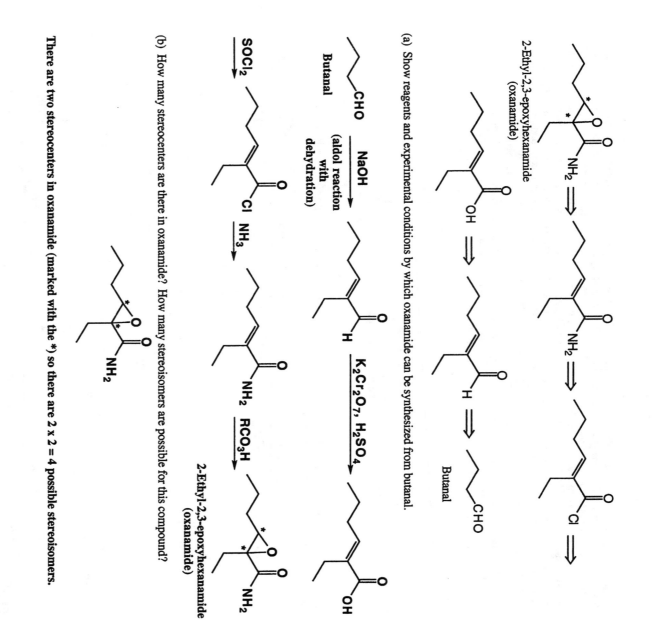

(a) Show reagents and experimental conditions by which oxanamide can be synthesized from butanal.

(b) How many stereocenters are there in oxanamide? How many stereoisomers are possible for this compound?

There are two stereocenters in oxanamide (marked with the *) so there are 2 x 2 = 4 possible stereoisomers.

Problem 19.60 The widely used anticoagulant warfarin (see Chapter 18, "Chemistry in Action: From Moldy Clover to a Blood Thinner") is synthesized from 4-hydroxycoumarin, benzaldehyde, and acetone as shown in this retrosynthesis. Show how warfarin is synthesized from these reagents.

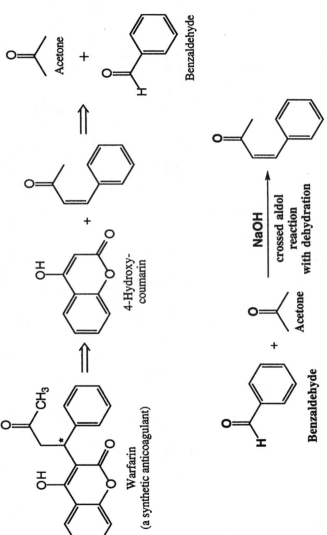

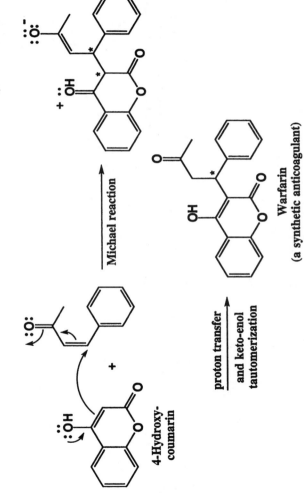

One new stereocenter is created, so a racemic mixture will be produced.

Problem 19.61 Following is a retrosynthetic analysis for an intermediate in the industrial synthesis of vitamin A.

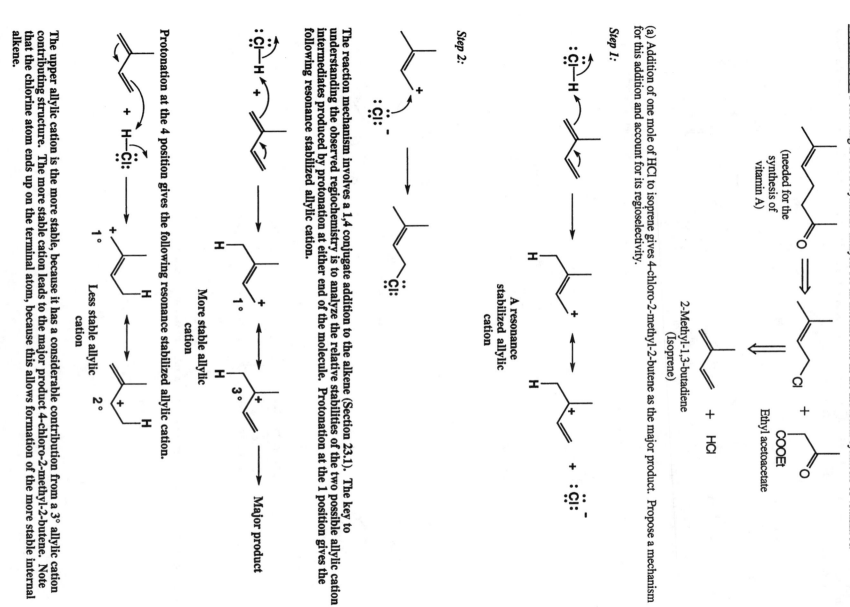

(a) Addition of one mole of HCl to isoprene gives 4-chloro-2-methyl-2-butene as the major product. Propose a mechanism for this addition and account for its regioselectivity.

Step 1:

A resonance stabilized allylic cation

The reaction mechanism involves a 1,4 conjugate addition to the alkene (Section 23.1). The key to understanding the observed regiochemistry is to analyze the relative stabilities of the two possible allylic cation intermediates produced by protonation at either end of the molecule. Protonation at the 1 position gives the following resonance stabilized allylic cation.

More stable allylic cation

Step 2:

Protonation at the 4 position gives the following resonance stabilized allylic cation.

Less stable allylic cation

Major product

The upper allylic cation is the more stable, because it has a considerable contribution from a 3° allylic cation contributing structure. The more stable cation leads to the major product 4-chloro-2-methyl-2-butene. Note that the chlorine atom ends up on the terminal atom, because this allows formation of the more stable internal alkene.

(b) Propose a synthesis of the vitamin A precursor from this allylic chloride and ethyl acetoacetate.

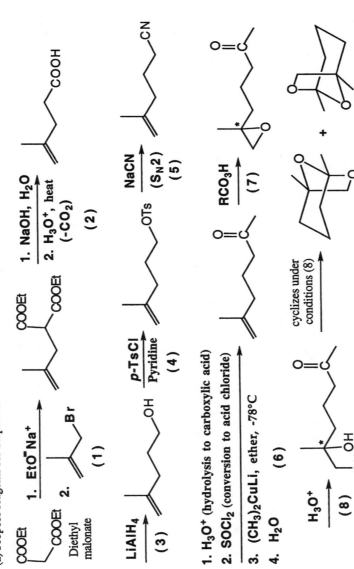

1. EtO⁻Na⁺

2.

1. NaOH, H₂O

2. H₃O⁺, heat (-CO₂)

(needed for the synthesis of vitamin A)

Problem 19.62 Following are the steps in one of the several published synthesis of frontalin, a pheromone of the western pine beetle. See Mori, K., Kobayashi, S., and Matsui, Agric. Biol. Chem. 39: 1889 (1975).

(a) Propose reagents for Steps 1-8.

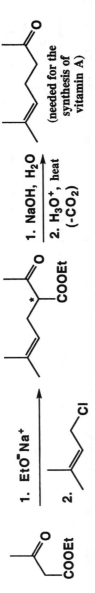

Diethyl malonate

1. EtO⁻Na⁺

2.

(1)

1. NaOH, H₂O

2. H₃O⁺, heat (-CO₂)

(2)

LiAlH₄

(3)

p-TsCl Pyridine

(4)

NaCN (Sₙ2)

(5)

1. H₃O⁺ (hydrolysis to carboxylic acid)

2. SOCl₂ (conversion to acid chloride)

3. (CH₃)₂CuLi, ether, –78°C

4. H₂O

(6)

RCO₃H

(7)

H₃O⁺

(8)

cyclizes under conditions (8)

+

Two new stereocenters are created, but because of the bicyclic structure, there are only two possible stereoisomers of the product. These stereoisomers are enantiomers, so a racemic mixture of these enantiomers will be produced. Both enantiomers are drawn above.

(b) Propose a mechanism for the cyclization of the ketodiol from Step 8 to frontalin.

This reaction is an intramolecular acetal formation. In the following steps, the atoms are labeled to help keep track of them during the cyclization steps. It may be helpful to create molecular models to follow this reaction.

Step 1:

intramolecular hemiacetal formation

Step 2:

acetal formation ⟶

The other enantiomer is formed by attack of the carbonyl from the other face in Step 1.

Problem 19.63 2-Ethyl-1-hexanol was needed for the synthesis of the sunscreen octyl *p*-methylcinnamate (See Chemistry in Action: Sunscreens and Sunblocks). Show how this alcohol could be synthesized.

(a) by an aldol condensation of butanal

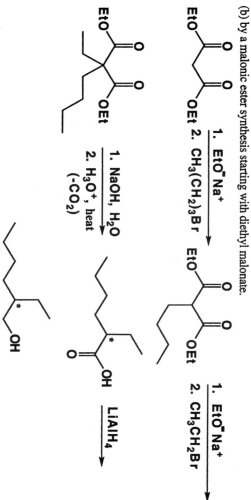

(b) by a malonic ester synthesis starting with diethyl malonate.

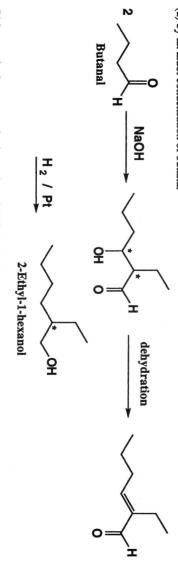

Because a new stereocenter is created, reactions in parts (a) and (b) both create a racemic mixture.

Chapter 20: Aromatics I

Solutions

419

CHAPTER 20
Solutions to the Problems

Problem 20.1 Construct a Frost circle for a planar eight-membered ring with one 2p orbital on each atom of the ring and show the relative energies of its eight pi molecular orbitals. Which are bonding, which are antibonding, and which are nonbonding?

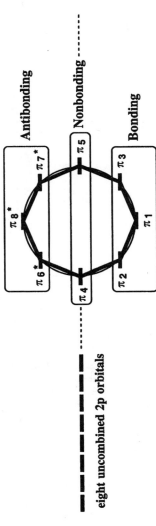

Problem 20.2 Describe the ground-state electron configuration of the cyclopentadienyl cation and radical. Assuming each species is planar, would you expect it to be aromatic or antiaromatic?

Cyclopentadienyl cation:

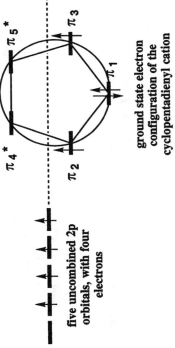

Cyclopentadienyl radical:

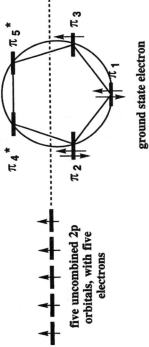

The cyclopentadienyl cation has 4 pi electrons so it is antiaromatic (4n + 2) nor antiaromatic (4n). The cyclopentadienyl radical has five pi electrons, so it is neither aromatic (4n + 2) nor antiaromatic (4n).

Problem 20.3 Describe the ground-state electron configuration of the cycloheptatrienyl radical and anion. Assuming that each species is planar, would you expect it to be aromatic or antiaromatic?

Cycloheptatrienyl radical:

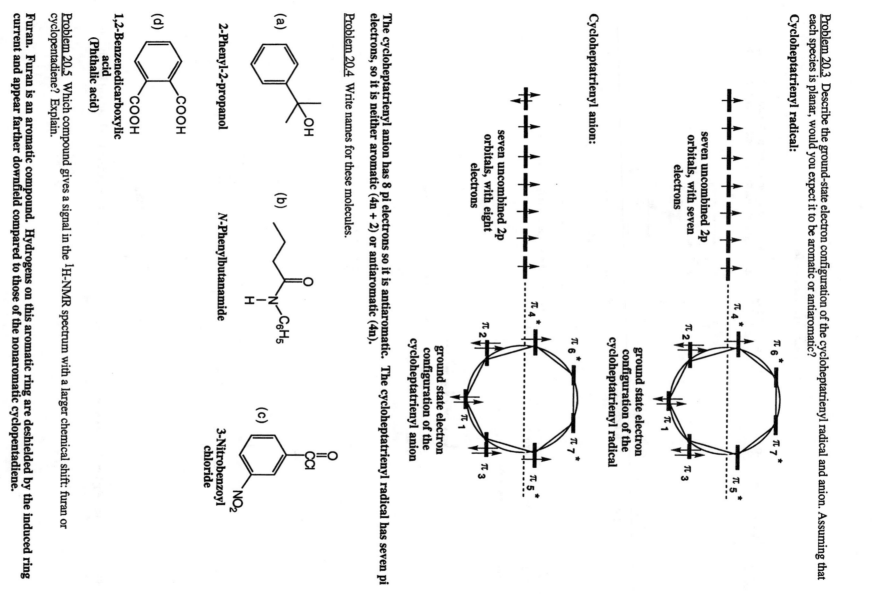

seven uncombined 2p
orbitals, with seven
electrons

ground state electron
configuration of the
cycloheptatrienyl radical

Cycloheptatrienyl anion:

seven uncombined 2p
orbitals, with eight
electrons

ground state electron
configuration of the
cycloheptatrienyl anion

The cycloheptatrienyl anion has 8 pi electrons so it is antiaromatic. The cycloheptatrienyl radical has seven pi electrons, so it is neither aromatic (4n + 2) or antiaromatic (4n).

Problem 20.4 Write names for these molecules.

(a)

2-Phenyl-2-propanol

(b)

N-Phenylbutanamide

(c)

3-Nitrobenzoyl
chloride

(d)

1,2-Benzenedicarboxylic
acid
(Phthalic acid)

Problem 20.5 Which compound gives a signal in the ^{1}H-NMR spectrum with a larger chemical shift: furan or cyclopentadiene? Explain.

Furan. Furan is an aromatic compound. Hydrogens on this aromatic ring are deshielded by the induced ring current and appear farther downfield compared to those of the nonaromatic cyclopentadiene.

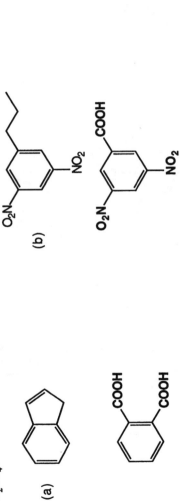

Furan　　　Cyclopentadiene

Problem 20.6　Arrange these compounds in order of increasing acidity: 2,4-dichlorophenol, phenol, cyclohexanol.

The following compounds are ranked from least to most acidic:

Cyclohexanol　<　Phenol　<　2,4-Dichlorophenol
pKa ~18　　　pKa 9.95　　　pKa ~7.5

A good way to predict relative acidities between related compounds is to keep track of the anionic conjugate bases produced upon deprotonation. In general, the more stable the conjugate base anion, the stronger the acid. Anions become increasingly stabilized as the negative charge is more delocalized around the molecule. Thus, phenol is more acidic than an aliphatic alcohol like cyclohexanol, because resonance involving the aromatic ring of phenol leads to increased charge delocalization and thus stabilization of the phenoxide anion compared to the cyclohexylalkoxide anion. The electronegative chlorine atoms of 2,4-dichlorophenol withdraw electron density from the aromatic ring and thus help to stabilize the 2,4-dichlorophenoxide anion even further than what is seen with phenoxide (inductive effect).

Problem 20.7　Predict the products resulting from vigorous oxidation of these compounds by K₂Cr₂O₇ in aqueous H₂SO₄.

(a)

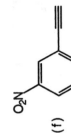

(b)

(c)

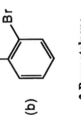

3-Phenyl-1-propanol

Nomenclature and Structural Formulas
Problem 20.8　Name the following molecules and ions.

(a)

1-Chloro-4-nitrobenzene

(b)

2-Bromotoluene
(o-Bromotoluene)

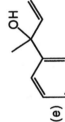

(d)

1,5-Dinitronaphthalene

(e)

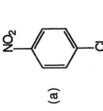

2-Phenyl-3-buten-2-ol

(f)

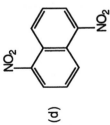

3-Nitrophenylethyne
(m-Nitrophenylacetylene)

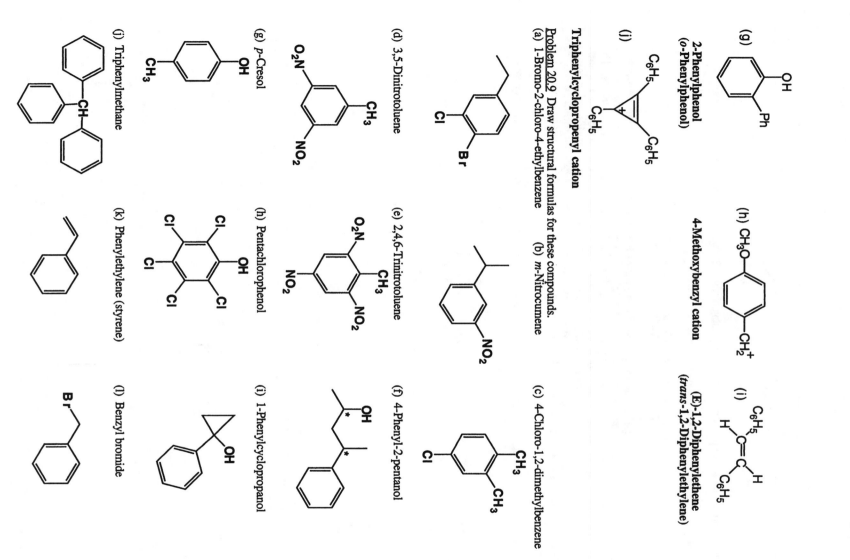

(g)

OH
Ph

2-Phenylphenol
(o-Phenylphenol)

(h) CH₃O

CH₂⁺

4-Methoxybenzyl cation

(i) C₆H₅   H

H   C₆H₅

(E)-1,2-Diphenylethene
(trans-1,2-Diphenylethylene)

(j)

C₆H₅
+
C₆H₅   C₆H₅

Triphenylcyclopropenyl cation

<u>Problem 20.9</u> Draw structural formulas for these compounds.
(a) 1-Bromo-2-chloro-4-ethylbenzene   **(b)** *m*-Nitrocumene

(d) 3,5-Dinitrotoluene

O₂N

CH₃

NO₂

(c) 4-Chloro-1,2-dimethylbenzene

Cl

CH₃

CH₃

Cl

Br

(e) 2,4,6-Trinitrotoluene

O₂N

CH₃

NO₂

NO₂

(f) 4-Phenyl-2-pentanol

OH

*

*

(g) *p*-Cresol

CH₃

OH

(h) Pentachlorophenol

Cl   Cl

Cl   OH

Cl   Cl

(i) 1-Phenylcyclopropanol

OH

(j) Triphenylmethane

CH

(k) Phenylethylene (styrene)

(l) Benzyl bromide

Br

(m) 1-Phenyl-1-butyne

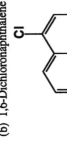

(n) 3-Phenyl-2-propen-1-ol

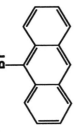

Problem 20.10 Draw structural formulas for these molecules.

(a) 1-Nitronaphthalene

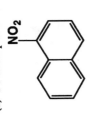

(b) 1,6-Dichloronaphthalene

(c) 9-Bromoanthracene

(d) 2-Methylphenanthrene

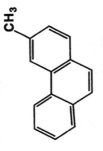

Problem 20.11 Molecules of 6,6'-dinitrobiphenyl-2,2'-dicarboxylic acid have no tetrahedral stereocenter, and yet they can be resolved to a pair of enantiomers. Account for this chirality. You will find it helpful to examine the model on the CD of this compound.

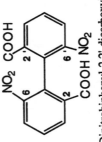

6,6'-Dinitrobiphenyl-2,2'-dicarboxylic acid

The key here is that the central bond between the benzene rings cannot rotate freely at room temperature due to the steric hindrance provided by the nitro and carboxyl groups. In other words, these groups run into each other as the molecule attempts to rotate around the central bond, so it is prevented from rotating. As a result, the molecule is chiral because, like a propeller, there are two different orientations possible. The two orientations represent non-superposable mirror images so they are a pair of enantiomers.

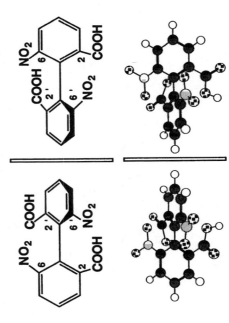

Resonance in Aromatic Compounds

Problem 20.12 Following each name is the number of Kekulé structures that can be drawn for it. Draw these Kekulé structures, and show, using curved arrows how the first contributing structure for each molecule is converted to the second and so forth.

(a) Naphthalene (3)

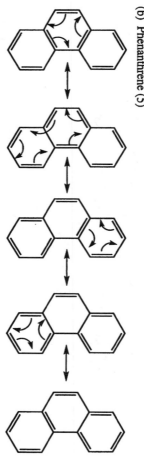

(b) Phenanthrene (5)

Problem 20.13 Each molecule can be drawn as a hybrid of five contributing structures: two Kekulé structures and three that involve creation and separation of unlike charges. For (a) and (b), the creation and separation of unlike charges places a positive charge on the substituent and a negative charge on the ring. For (c), a positive charge is placed on the ring and an additional negative charge is placed on the -NO₂ group. Draw these five contributing structures for each

(a) Chlorobenzene

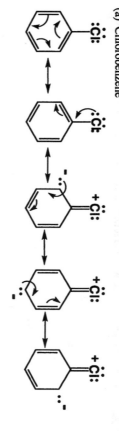

(b) Phenol

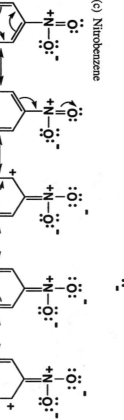

(c) Nitrobenzene

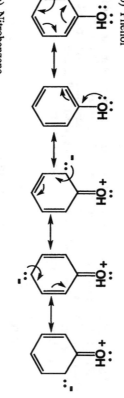

Problem 20.14 Following are structural formulas for furan and pyridine.

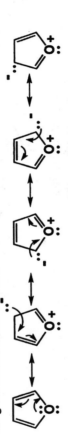

Furan

Pyridine

(a) Write four contributing structures for the furan hybrid that place a positive charge on oxygen and a negative charge first on carbon 3 of the ring and then on each other carbon of the ring.

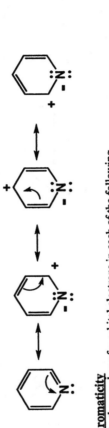

(b) Write three contributing structures for the pyridine hybrid that place a negative charge on nitrogen and a positive charge first on carbon 2, then on carbon 4, and finally carbon 6.

The Concept of Aromaticity

Problem 20.15 State the number of p orbital electrons in each of the following.

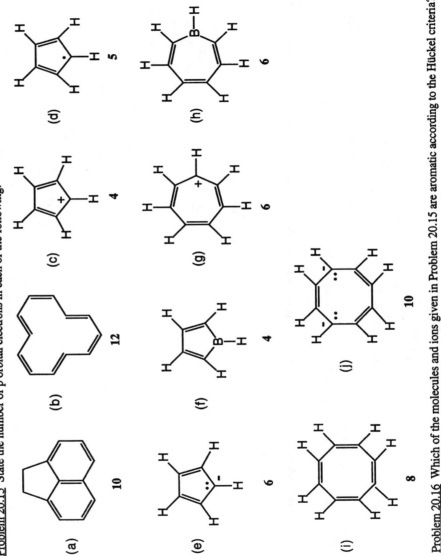

Problem 20.16 Which of the molecules and ions given in Problem 20.15 are aromatic according to the Hückel criteria? Which, if planar, would be antiaromatic?

The following molecules are aromatic because they have 4n + 2 π electrons: a, e, g, h, and j.
The following molecules would be antiaromatic if planar because they have 4n π electrons: b, c, f, and i.

Problem 20.17 Construct MO energy diagrams for the cyclopropenyl cation, radical, and anion. Which of these species is aromatic according to the Hückel criteria?

Cyclopropenyl cation:

three uncombined 2p orbitals, with two electrons

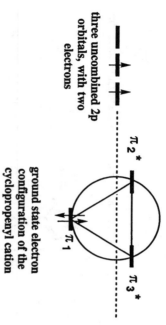

ground state electron configuration of the cyclopropenyl cation

Cyclopropenyl radical:

three uncombined 2p orbitals, with three electrons

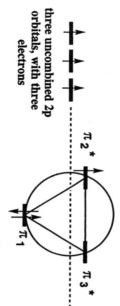

ground state electron configuration of the cyclopropenyl radical

Cyclopropenyl anion:

three uncombined 2p orbitals, with four electrons

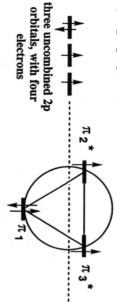

ground state electron configuration of the cyclopropenyl anion

Of these species, the only one that satisfies the Hückel criteria is the first one, namely the cyclopropenyl cation with 2 pi electrons, a Hückel number (4n + 2, here n = 0).

Problem 20.18 Naphthalene and azulene are constitutional isomers of molecular formula $C_{10}H_8$. Naphthalene is a colorless solid with a dipole moment of zero. Azulene is a solid with an intense blue color and a dipole moment of 1.0 D. Account for the difference in dipole moments of these constitutional isomers.

Naphthalene   Azulene

Naphthalene has no permanent dipole moment because it possesses a high degree of symmetry. Azulene has a remarkably large permanent dipole moment (1.0 D) for a hydrocarbon. The dipole moment of azulene can be explained using the contributing structures such as the one shown below on the right:

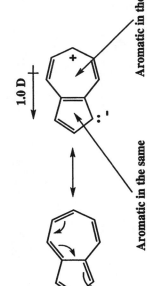

1.0 D

Aromatic in the same way the cycloheptatrienyl cation is aromatic

Aromatic in the same way the cyclopentadienyl anion is aromatic

Aromatic in the same way the cyclopentadienyl anion is aromatic

Neither a cyclopentadiene or cycloheptatriene ring is aromatic in the neutral state. However, by transferring electron density from the seven-membered ring pi system to the five-membered ring pi system of azulene, the aromatic cyclopentadienyl anion and cycloheptatrienyl cation pi systems are formed. This aromatic stabilization explains why the resonance structure on the right makes an important contribution to the overall resonance hybrid, resulting in the large observed permanent dipole moment.

Spectroscopy
Problem 20.19 Compound A, molecular formula C_9H_{12}, shows prominent peaks in its mass spectrum at m/z 120 and 105. Compound B, also molecular formula C_9H_{12}, shows prominent peaks at m/z 120 and 91. On vigorous oxidation with chromic acid, both compounds give benzoic acid. From this information, deduce the structural formulas of compounds A and B.

A   **B**

The compounds can be identified by the major fragments in the mass spectrum, both of which are benzylic cations. For compound A, the peak at m/z 105 corresponds to the cation shown below:

Peak at m/z 105 for compound A

For compound B, the peak at m/z 91 corresponds to the tropylium ion that was produced by rearrangement of the benzyl cation (Section 15.4)

+

Peak at m/z 91 for compound B

Both of the compounds will produce benzoic acid upon oxidation with chromic acid.

Problem 20.20 Compound C shows a molecular ion at *m/z* 148, and other prominent peaks at *m/z* 105, and 77. Following are its infrared and ^{1}H-NMR spectra.

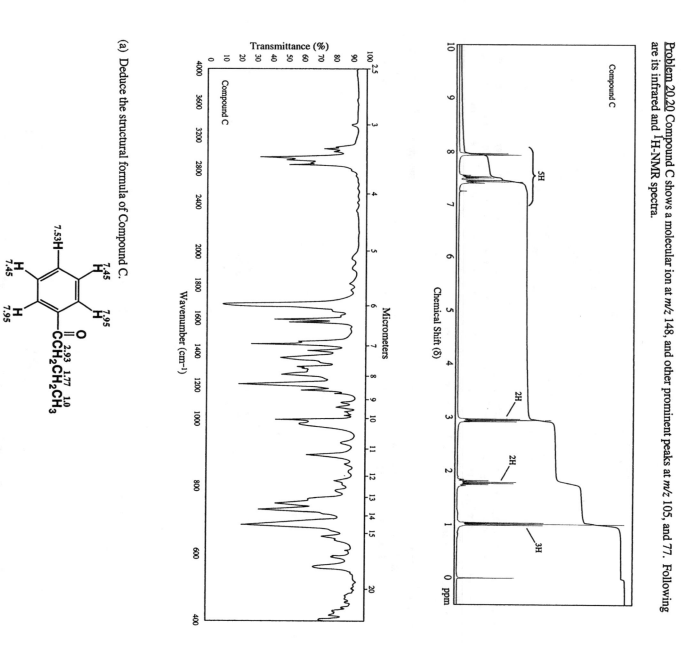

(a) Deduce the structural formula of Compound C.

This compound, $C_{10}H_{12}O$, has the correct molecular formula of 148. This compound also has a carbonyl corresponding to the peak at 1680 cm^{-1} in the IR spectrum and the correct pattern of hydrogens to explain the ^{1}H-NMR spectrum.

(b) Account for the appearance of peaks in its mass spectrum at *m/z* 105 and 77.

The peaks at *m/z* 105 and 77 correspond to the following fragments produced by α-cleavage (Section 17.4) on either side of the carbonyl group.

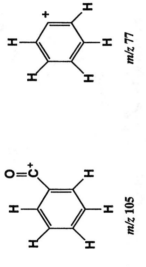

m/z 105   *m/z* 77

Problem 20.21 Following are IR and ^{1}H-NMR spectra of compound D. The mass spectrum of compound D shows a molecular ion peak at m/z 136, a base peak at m/z 107, and other prominent peaks at m/z 118 and 59.

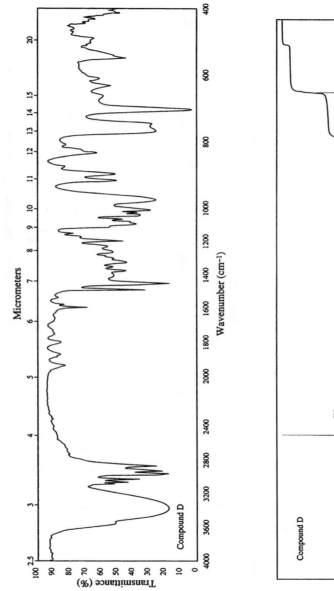

Compound D

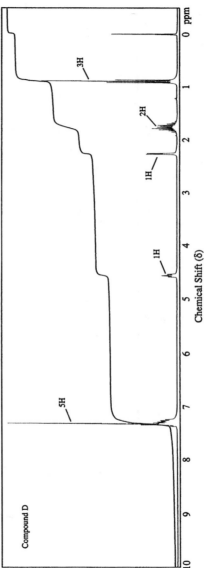

Compound D

(a) Propose a structural formula for compound D based on this information.

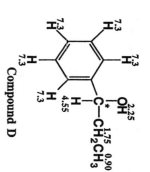

Compound D

This compound, C$_9$H$_{12}$O, has the correct molecular weight of 136. The hydroxyl group gives the broad peak at 3300 cm^{-1} in the IR spectrum and the pattern of hydrogens present explains the ^{1}H-NMR spectrum.

(b) Propose structural formulas for ions in the mass spectrum at m/z 118, 107 and 59.

The peaks at m/z 118, 107 and 77 correspond to the following fragments:

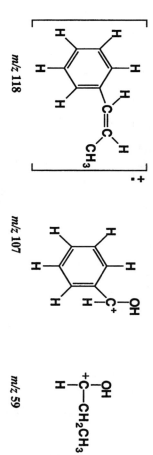

m/z 118   m/z 107   m/z 59

Problem 20.22   Compound E is a neutral solid of molecular formula C$_8$H$_{10}$O$_2$. Its mass spectrum shows a molecular ion at m/z 138 and prominent peaks at M-1 and M-17. Following are IR and ^{1}H-NMR spectra of compound E. Deduce the structure of compound E.

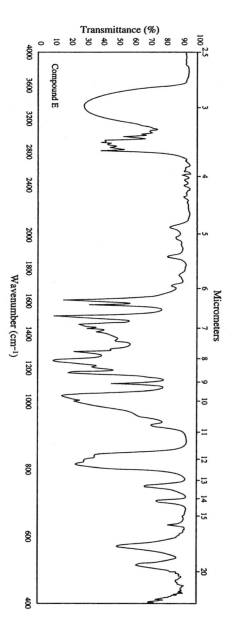

Compound E

Transmittance (%)

Wavenumber (cm^{-1})

Micrometers

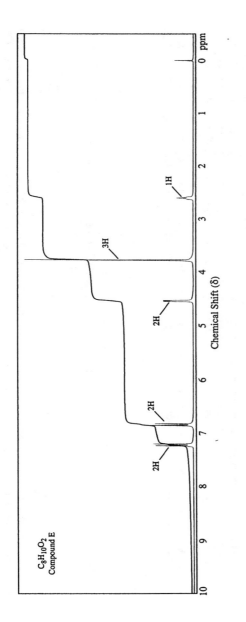

C₈H₁₀O₂
Compound E

Chemical Shift (δ)

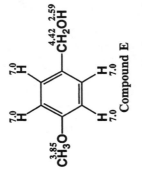

Compound E

A molecule of molecular formula $C_8H_{10}O_2$ has an index of hydrogen deficiency of 4, which is accounted for by the three double bonds and ring of the benzene ring of compound E. Compound E also has an alcohol function corresponding to the broad peak at 3350 cm⁻¹ in the IR spectrum, and the correct pattern of hydrogens to explain the ¹H-NMR spectrum. Notice especially the distinctive *para* pattern near δ 7.0 and the methyl group signal at δ 3.85.

Problem 20.23 Following are ¹H-NMR and ¹³C-NMR spectral data for compound F, $C_{12}H_{16}O$. From this information, deduce the structure of compound F.

¹H-NMR	¹³C-NMR
0.83 d (6H)	207.82 50.88
2.11 m (1H)	134.24 50.57
2.30 d (2H)	129.36 24.43
3.64 s (2H)	128.60 22.48
7.2 - 7.4 m (5H)	126.86

The signal in the ¹³C-NMR spectrum at δ 207.82 indicates the presence of a carbonyl group. The signals around δ 130 indicate there is an aromatic ring. The doublet in the ¹H-NMR at δ 0.83 that integrates to 6H indicates two methyl groups adjacent to a -CH- group. There are also two -CH₂- groups; one that is not adjacent to other hydrogens (the singlet at δ 3.64) and one next to a -CH- group (the doublet at δ 2.30). The multiplet at δ 2.11 must be this -CH- group that is also adjacent to the two methyl groups. Five aromatic hydrogens are found in the complex set of signals at δ 7.2-7.4. The only structure that is consistent with all of these facts is 4-methyl-1-phenyl-2-pentanone.

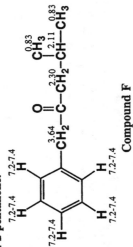

Compound F

A molecule of molecular formula $C_{12}H_{16}O$ has an index of hydrogen deficiency of 5, which is accounted for by the three double bonds and ring of the benzene ring plus the pi bond of the carbonyl group in compound F.

Problem 20.24 Following are ^{1}H-NMR and ^{13}C-NMR spectral data for compound G, $C_{10}H_{10}O$. From this information, deduce the structure of compound G.

^{1}H-NMR	^{13}C-NMR	
2.50 t (2H)	210.19	126.82
3.05 t (2H)	136.64	126.75
3.58 s (2H)	133.25	45.02
7.1 to 7.3 m (4H)	128.14	38.11
	127.75	28.34

The signal at δ 210.19 in the ^{13}C-NMR indicate the presence of a carbonyl species. The six signals between δ 126 and δ 137 indicate that there is a phenyl ring and the three signals between δ 28 and δ 45 indicate there are three more sp^3 carbon atoms. The signals between δ 7.1 and δ 7.3 in the ^{1}H-NMR integrate to 4H so the phenyl ring must have two hydrogens replaced by other atoms. The three signals between δ 2.6 and δ 3.6 integrate to 2H each so these must be three -CH$_2$- groups. Furthermore, the splitting pattern indicates that two of these are adjacent to each other (the two triplets) while one is not adjacent to any carbons with hydrogen atoms attached. The only structure fully consistent with these spectra is β-tetralone.

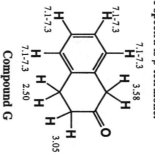

Compound G

Problem 20.25 Compound H, $C_8H_6O_3$, gives a precipitate when treated with hydroxylamine in aqueous ethanol, and a silver mirror when treated with Tollens' solution. Following is its ^{1}H-NMR spectrum. Deduce the structure of compound H.

A molecule of molecular formula $C_{10}H_{10}O$ has an index of hydrogen deficiency of 6, which is accounted for by the three double bonds and ring of the benzene ring, the pi bond of the carbonyl group, and the cyclohexane ring in compound G.

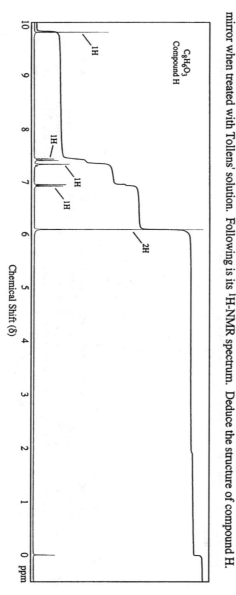

$C_8H_6O_3$
Compound H

Chemical Shift (δ)

The hydroxylamine reaction indicates the presence of a carbonyl species, and the positive Tollens' test confirms the presence of an aldehyde. The aldehyde is also indicated by the singlet in the ^{1}H-NMR at δ 9.8. The splitting pattern of the aromatic hydrogen signals between δ 6.95 and 7.5 indicates that the phenyl ring only has three hydrogens, two are adjacent to each other (the two doublets), while the third is not adjacent to any hydrogens. The singlet at δ 6.1, integrating to 2H, indicates the presence of a -CH$_2$- group bound to two very electronegative atoms (oxygen). Given the molecular formula, the only structure that agrees with this information is piperonal.

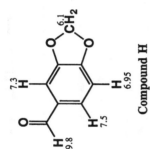

Compound H

A molecule of molecular formula $C_8H_6O_3$ has an index of hydrogen deficiency of 6, which is accounted for by the three double bonds and ring of the benzene ring, the pi bond of the carbonyl group, and the five-membered ring in compound H.

<u>Problem 20.26</u> Compound I, $C_{11}H_{14}O_2$, is insoluble in water, aqueous acid, and aqueous $NaHCO_3$ but dissolves readily in 10% Na_2CO_3 and 10% NaOH. When these alkaline solutions are acidified with 10% HCl, compound I is recovered unchanged. Given this information and its ^1H-NMR spectrum, deduce the structure of compound I.

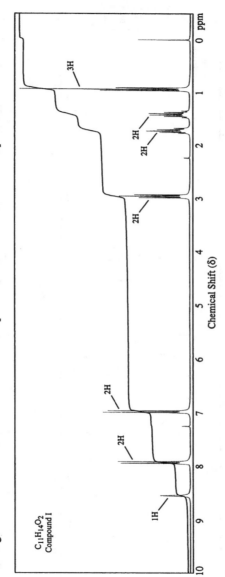

$C_{11}H_{14}O_2$
Compound I

Compound I dissolves in Na_2CO_3 or NaOH so it must be deprotonated by these bases. This behavior is expected for a weakly acidic species like a phenol. The ^1H-NMR is also consistent with the presence of a phenol since there is a broad singlet at δ 8.54 that integrates to 1H. There is no aldehyde hydrogen signal and the molecule does not undergo an aldol reaction since it is recovered unchanged from alkaline solution, therefore there is no aldehyde in the molecule. The two doublets at δ 6.9 and δ 7.9, each integrating to 2H, indicate a 1,4 disubstituted phenyl ring. The signals between δ 0.9 and δ 3.0 indicate a $-CH_2-CH_2-CH_2-CH_3$ group. Finally, the molecular formula indicates an index of hydrogen deficiency of 5. The phenyl ring accounts for 4 so there must be one other π bond (or ring) in the molecule. The only structure consistent with the molecular formula and the spectrum is 1-(4-hydroxyphenyl)-1-pentanone (4-hydroxy-valerophenone).

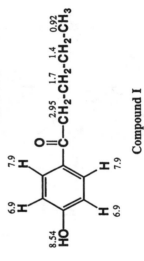

Compound I

Problem 20.27 Propose a structural formula for compound J, $C_{11}H_{14}O_3$, consistent with its ^{1}H-NMR and infrared spectra.

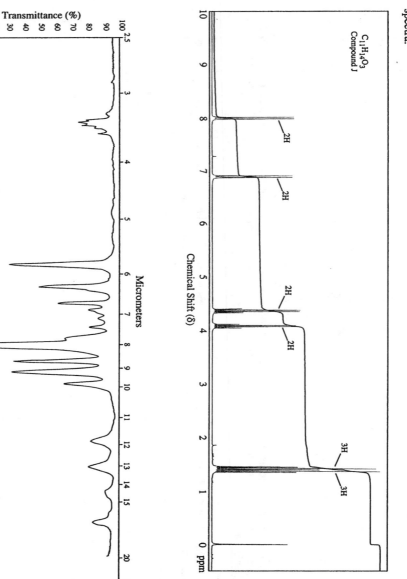

C_{11}H_{14}O_3
Compound J

Chemical Shift (δ)

2H   2H   2H   2H   3H   3H

10   9   8   7   6   5   4   3   2   1   0   ppm

Transmittance (%)

C_{11}H_{14}O_3
Compound J

100
90
80
70
60
50
40
30
20
10
0

4000 3600 3200 2800 2400 2000 1800 1600 1400 1200 1000 800 600 400

Wavenumber (cm^{-1})

2.5   3   4   5   6   7   8   9   10   11   12   13   14   15   20

Micrometers

The strong peak at 1720 cm^{-1} in the IR indicates the presence of a carbonyl group. The two sets of doublets integrating to 2H each between δ 6.9 and 8.0 indicate the presence of a 1,4 disubstituted benzene ring. The two quartets integrating to 2H each at δ 4.35 and δ 4.05 indicate there are two -CH$_2$- groups attached to oxygen atoms as well methyl groups. The molecular formula has three oxygen atoms, so the -OCH$_2$- groups are likely part of one ester and one ether. The two triplets integrating to a total of 6H at δ 1.4 are the signals from the two methyl groups, confirming the presence of two ethyl groups. The only structure that is consistent with the molecular formula $C_{11}H_{14}O_3$ and the spectra is ethyl 4-ethoxybenzoate.

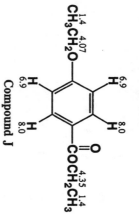

$\overset{1.4}{CH_3}\overset{4.07}{CH_2}O$

H 6.9   H 6.9

H 8.0   H 8.0

$\overset{O}{\underset{\parallel}{C}}\overset{4.35}{OCH_2}\overset{1.4}{CH_3}$

Compound J

A molecule of molecular formula $C_{11}H_{14}O_3$ has an index of hydrogen deficiency of 5, which is accounted for by the three double bonds and ring of the benzene ring plus the pi bond of the carbonyl group in compound J.

Problem 20.28 Propose a structural formula for the analgesic phenacetin, molecular formula $C_{10}H_{13}NO_2$, based on its ^1H-NMR spectrum.

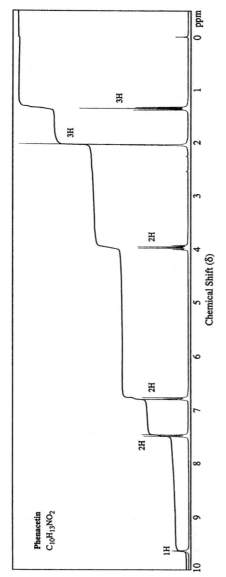

Phenacetin
$C_{10}H_{13}NO_2$

1H

2H

2H

2H

3H

3H

10   9   8   7   6   5   4   3   2   1   0 ppm

Chemical Shift (δ)

This structure is not only consistent with the molecular formula, but also with the ^1H-NMR spectrum. The characteristic two doublets centered at δ 6.8 and 7.5 indicate the presence of a 1,4-disubstituted benzene ring. The singlet at δ 9.65 integrating to 1H indicates a primary amide, and the singlet integrating to 3H at δ 2.05 indicates this is an acetamide. Finally, the typical ethyl splitting pattern for the signals at δ 1.32 and δ 3.95 indicates the presence of an ethyl group. These signals are shifted so far downfield that they must be part of an ethoxy group.

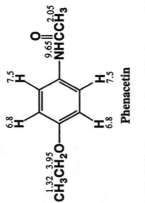

Phenacetin

A molecule of molecular formula $C_{10}H_{13}NO_2$ has an index of hydrogen deficiency of 5, which is accounted for by the three double bonds and ring of the benzene ring plus the pi bond of the carbonyl group in phenacetin.

Problem 20.29 Compound K, $C_{10}H_{12}O_2$, is insoluble in water, 10% NaOH, and 10% HCl. Given this information and the following ^1H-NMR and ^{13}C-NMR spectral information, deduce the structural formula of Compound K.

^1H-NMR		^{13}C-NMR	
2.10 (s 3H)		206.51	114.17
3.61 (s 2H)		158.67	55.21
3.77 (s 3H)		130.33	50.07
6.86 (d 2H)		126.31	29.03
7.12 (d 2H)			

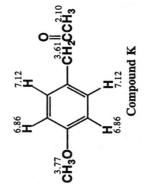

Compound K

Problem 20.30 Propose a structural formula for each compound given these NMR data.

(a) C₉H₉BrO₂

^{1}H-NMR   **^{13}C-NMR**
1.39 (t, 3H)   165.73
4.38 (q, 2H)   131.56
7.57 (d, 2H)   131.01
7.90 (d, 2H)   129.34
                   127.81
                   61.18
                   14.18

(b) C₈H₉NO

^{1}H-NMR   **^{13}C-NMR**
2.06 (s, 3H)   168.14
7.01 (t, 1H)   139.24
7.30 (m, 2H)   128.51
7.59 (d, 2H)   122.83
9.90 (s, 1H)   118.90
                   23.93

(c) C₉H₉NO₃

^{1}H-NMR   **^{13}C-NMR**
2.10(s, 3H)   168.74
7.72 (d, 2H)   166.85
7.91 (d, 2H)   143.23
10.3 (s, 1H)   130.28
12.7 (s, 1H)   124.80
                   118.09
                   24.09

Problem 20.31 Given here are ^{1}H-NMR and ^{13}C-NMR spectral data for two compounds. Each shows strong, sharp absorption between 1700 and 1720 cm^{-1}, and strong, broad absorption over the region 2500 - 3000 cm^{-1}. Propose a structural formula for each compound.

(a) C₁₀H₁₂O₃

^{1}H-NMR   **^{13}C-NMR**
2.49 (t, 2H)   173.89
2.80 (t, 2H)   157.57
3.72 (s, 3H)   132.62
6.78 (d, 2H)   128.99
7.11 (d, 2H)   113.55
12.4 (s 1H)   54.84
                   35.75
                   29.20

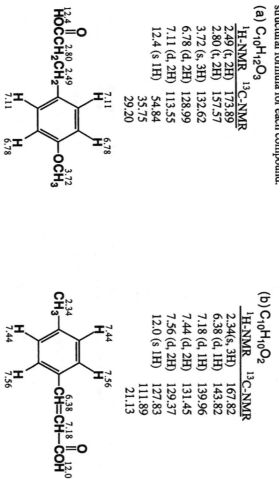

(b) C₁₀H₁₀O₂

^{1}H-NMR   **^{13}C-NMR**
2.34(s, 3H)   167.82
6.38 (d, 1H)   143.82
7.18 (d, 1H)   139.96
7.44 (d, 2H)   131.45
7.56 (d, 2H)   129.37
12.0 (s 1H)   127.83
                   111.89
                   21.13

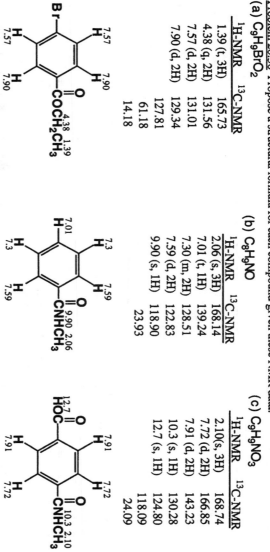

Acidity of Phenols

Problem 20.32 Account for the fact that *p*-nitrophenol is a stronger acid than phenol. Consider both the resonance and inductive effects of the nitro group.

OH   O₂N—⟨ ⟩—OH

K$_a$ = 1.0 x 10^{-10} K$_a$ = 7.0 x 10^{-8}

As seen from the acid ionization constants, *p*-nitrophenol is the stronger acid. To account for this fact, draw contributing structures for each anion and compare the degree of delocalization of negative charge (i.e., the resonance stabilization of each anion). Phenoxide ion is a resonance hybrid of five important contributing structures, two of which place the negative charge on the phenoxide oxygen, and three of which place the negative charge on the atoms of the ring.

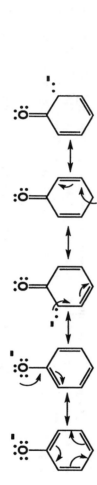

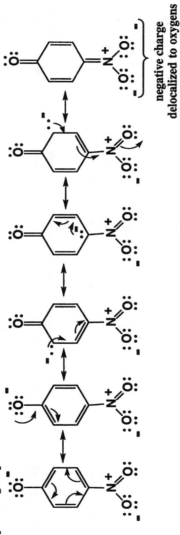

The *p*-nitrophenoxide ion is a hybrid of six important contributing structures. In addition to the five similar to those drawn above for the phenoxide ion, there is a sixth that places the negative charge on the oxygen atoms of the *p*-nitro group.

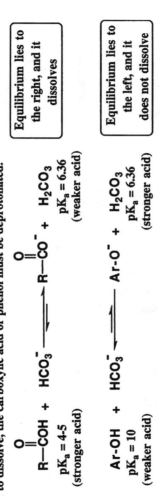

negative charge delocalized to oxygens of nitro group

Thus, because of the greater delocalization of the negative charge onto the more electronegative oxygen atoms of the nitro group, *p*-nitrophenol is a stronger acid than phenol.

In addition to the resonance effect discussed above, the nitro group is highly electron-withdrawing since nitrogen and oxygen are both more electronegative than carbon. The inductive effect also helps stabilize the phenoxide anion by pulling electron density away from the anionic oxygen atom.

Problem 20.33   Account for the fact that water-insoluble carboxylic acids (pK$_a$ 4-5) dissolve in 10% aqueous sodium bicarbonate (pH 8.5) with the evolution of a gas but water-insoluble phenols (pK$_a$ 9.5-10.5) do not dissolve in 10% sodium bicarbonate.

In order to dissolve, the carboxylic acid or phenol must be deprotonated.

$$R\!-\!\overset{\displaystyle O}{\overset{\|}{C}}\!OH \; + \; HCO_3^- \; \longrightarrow \; R\!-\!\overset{\displaystyle O}{\overset{\|}{C}}\!O^- \; + \; H_2CO_3$$

pK$_a$ = 4-5   pK$_a$ = 6.36
(stronger acid)   (weaker acid)

Equilibrium lies to the right, and it dissolves

$$Ar\!-\!OH \; + \; HCO_3^- \; \longrightarrow \; Ar\!-\!O^- \; + \; H_2CO_3$$

pK$_a$ = 10   pK$_a$ = 6.36
(weaker acid)   (stronger acid)

Equilibrium lies to the left, and it does not dissolve

Problem 20.34   Match each compound with its appropriate pK$_a$ value.
(a) 4-Nitrobenzoic acid, benzoic acid, 4-chlorobenzoic acid   pK$_a$ = 4.19, 3.98, and 3.41
pK$_a$ 3.41   4.19   3.98

Electron-withdrawing groups increase acidity through a combination of resonance and inductive effects. The nitro group is more withdrawing than the chloro group, explaining the observed trend.

(b) Benzoic acid, cyclohexanol, phenol   pK$_a$ = 18.0, 9.95, and 4.19
pK$_a$ 4.19   18.0   9.95

Acidity increases in the order of aliphatic alcohols, phenols, and carboxylic acids.

(c) 4-Nitrobenzoic acid, 4-nitrophenol, 4-nitrophenylacetic acid

pK_a	3.41	7.15	3.85

From part (a) it is clear that 4-nitrobenzoic acid has a pK_a of 3.41, meaning that the other carboxylic acid, 4-nitrophenylacetic acid, must be slightly less acidic with a pK_a of 3.85. This makes sense since the electron-withdrawing nitro group is farther away from the carboxylic acid group in 4-nitrophenylacetic acid compared to 4-nitrobenzoic acid. The 4-nitrophenol is significantly less acidic with a pK_a of 7.15.

Problem 20.35 Arrange the molecules and ions in each set in order of increasing acidity (from least acidic to most acidic).

(a)

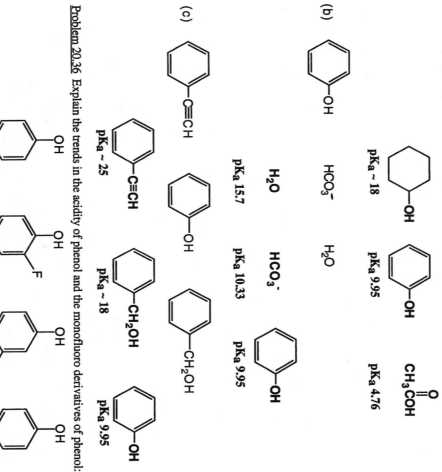

To arrange these in order of increasing acidity, refer to Table 4.2. For those compounds not listed in Table 3.2, estimate pK_a using values for compounds that are given in the table.

(a)

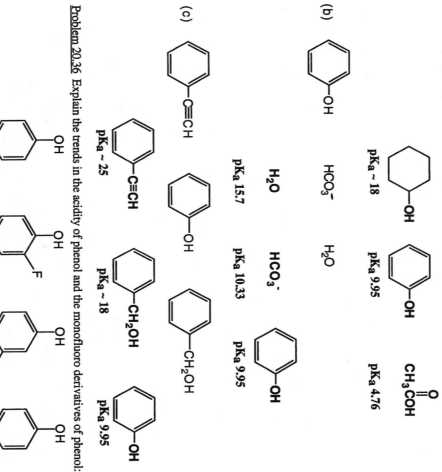

pKₐ ~ 18   pKₐ 9.95   pKₐ 4.76

(b)

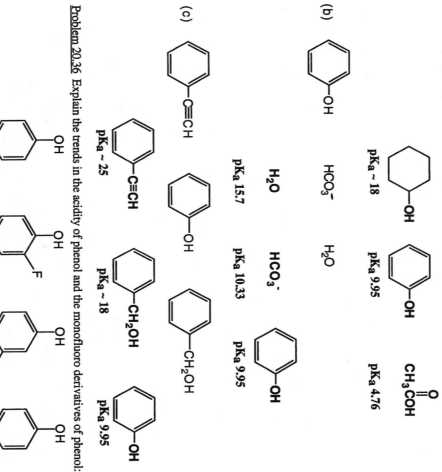

OH   HCO₃⁻   H₂O

pKₐ 15.7

(c)

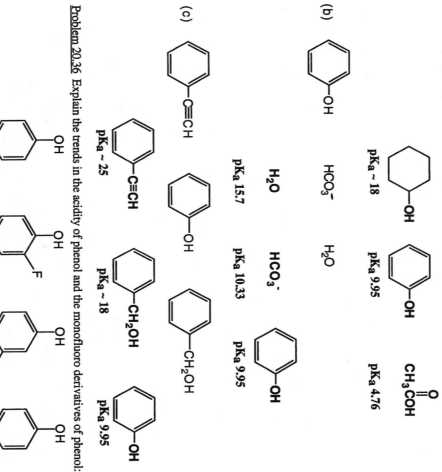

C≡CH   H₂O   HCO₃⁻

pKₐ ~ 25   pKₐ 10.33   pKₐ 9.95

Problem 20.36 Explain the trends in the acidity of phenol and the monofluoro derivatives of phenol:

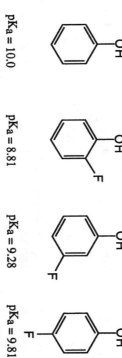

pKₐ = 10.0   pKₐ = 8.81   pKₐ = 9.28   pKₐ = 9.81

The electronegative fluoro substituent increases the acidity of the phenol through an inductive effect, so all of the monofluoro isomers of phenol are more acidic than phenol itself. Because this is an inductive effect, the closer the fluorine atom to the phenolic OH group, the stronger the effect and the greater the acidity. Thus, the ortho fluoro derivative is most acidic, followed by the meta fluoro derivative, then finally the para fluoro derivative.

Problem 20.37 You wish to determine the inductive effects of a series of functional groups, for example Cl, Br, CN, COOH, and C_6H_5. Is it best to use a series of ortho-, meta-, or para-substituted phenols? Explain your answer.

The question to be addressed involves inductive effects only. It would be best to use the derivatives with the substituents in the *meta* position, because this would minimize any contributions from resonance effects. Resonance effects are maximal when substituents are in the *ortho* and *para* positions.

Problem 20.38 From each pair, select the stronger base.

To estimate which is the stronger base, first determine which conjugate acid is the weaker acid. The weaker the acid, the stronger its conjugate base.

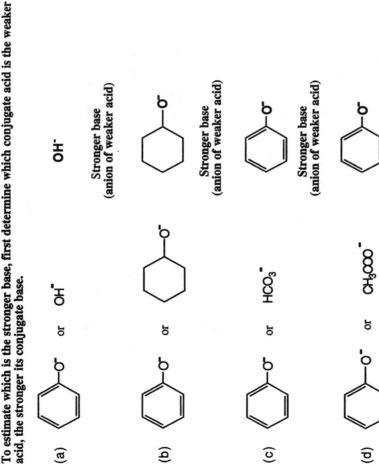

(a)

$C_6H_5O^-$ or   **OH⁻**

OH⁻
Stronger base
(anion of weaker acid)

(b)

$C_6H_5O^-$ or   (cyclohexoxide)

(cyclohexoxide)
Stronger base
(anion of weaker acid)

(c)

$C_6H_5O^-$ or   **HCO₃⁻**

$C_6H_5O^-$
Stronger base
(anion of weaker acid)

(d)

$C_6H_5O^-$ or   **CH₃COO⁻**

$C_6H_5O^-$
Stronger base
(anion of weaker acid)

Problem 20.39 Describe a chemical procedure to separate a mixture of benzyl alcohol and *o*-cresol and recover each in pure form.

CH₂OH

Benzyl alcohol
(bp 205°C)

CH₃ / OH

o-Cresol
(bp 191°C)

Following is a flow chart for an experimental method for separating these two compounds. Separation is based on the facts that each is insoluble in water, soluble in diethyl ether, and that *o*-cresol reacts with 10% NaOH to form a water-soluble phenoxide salt while benzyl alcohol does not.

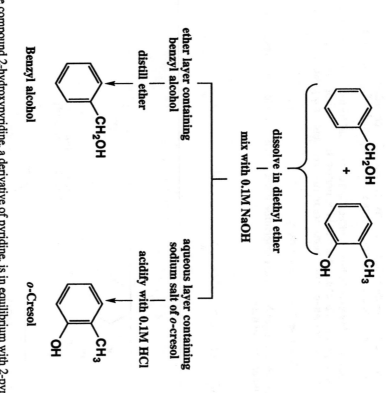

dissolve in diethyl ether

mix with 0.1M NaOH

ether layer containing
benzyl alcohol

distill ether

aqueous layer containing
sodium salt of o-cresol

acidify with 0.1M HCl

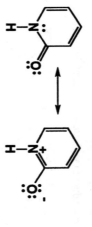

Benzyl alcohol

o-Cresol

Problem 20.40 The compound 2-hydroxypyridine, a derivative of pyridine, is in equilibrium with 2-pyridone. 2-Hydroxypyridine is aromatic. Does 2-pyridone have comparable aromatic character? Explain.

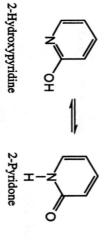

2-Hydroxypyridine   2-Pyridone

2-Pyridones have aromatic character because of the contributing structure shown on the right.

Reactions at the Benzylic Position

Problem 20.41 Write a balanced equation for the oxidation of *p*-xylene to 1,4-benzenedicarboxylic acid (terephthalic acid) using potassium dichromate in aqueous sulfuric acid. How many milligrams of $K_2Cr_2O_7$ is required to oxidize 250 mg of *p*-xylene to terephthalic acid?

Following are balanced equations for the oxidation half-reaction and the reduction half-reaction. Because oxidation takes place in aqueous acid, the reactions are balanced with H_2O and H^+.

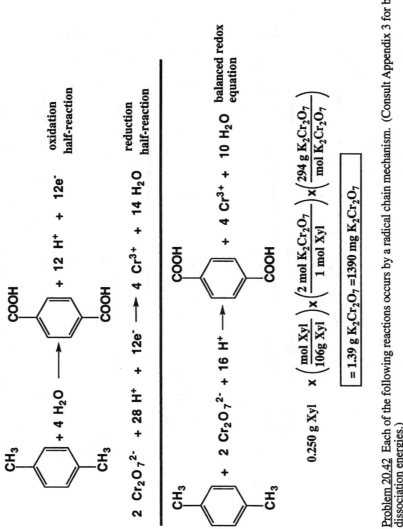

$$2\ Cr_2O_7{}^{2-} + 28\ H^+ + 12e^- \longrightarrow 4\ Cr^{3+} + 14\ H_2O$$

reduction
half-reaction

$$+ 2\ Cr_2O_7{}^{2-} + 16\ H^+ \longrightarrow \qquad + 4\ Cr^{3+} + 10\ H_2O$$

balanced redox
equation

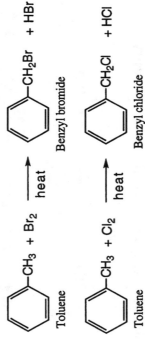

$$0.250\ g\ Xyl \quad \times \left(\frac{mol\ Xyl}{106g\ Xyl}\right) \times \left(\frac{2\ mol\ K_2Cr_2O_7}{1\ mol\ Xyl}\right) \times \left(\frac{294\ g\ K_2Cr_2O_7}{mol\ K_2Cr_2O_7}\right)$$

$$\boxed{= 1.39\ g\ K_2Cr_2O_7 = 1390\ mg\ K_2Cr_2O_7}$$

Problem 20.42 Each of the following reactions occurs by a radical chain mechanism. (Consult Appendix 3 for bond dissociation energies.)

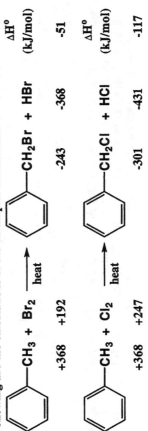

(a) Calculate the heat of reaction, $\Delta H°$, in kilocalories per mol for each reaction.

Following are the calculations of the enthalpies of reaction.

	$CH_3 + Br_2$ $\xrightarrow{heat}$	$CH_2Br + HBr$	$\Delta H°$ (kJ/mol)
	+368	-368	
	+192	-243	-51

	$CH_3 + Cl_2$ $\xrightarrow{heat}$	$CH_2Cl + HCl$	$\Delta H°$ (kJ/mol)
	+368	-301	
	+247	-431	-117

Solutions

(b) Write a pair of chain propagation steps for each mechanism, and show that the net result of the chain propagation steps is the observed reaction.

(c) Calculate ΔH° for each chain propagation step, and show that the sum for each pair of chain propagation steps is identical with the ΔH° value calculated in part (a).

Shown below are pairs of chain propagation steps for each reaction. Each pair adds up to the observed reaction and the observed enthalpy of reaction. This answers both (b) and (c).

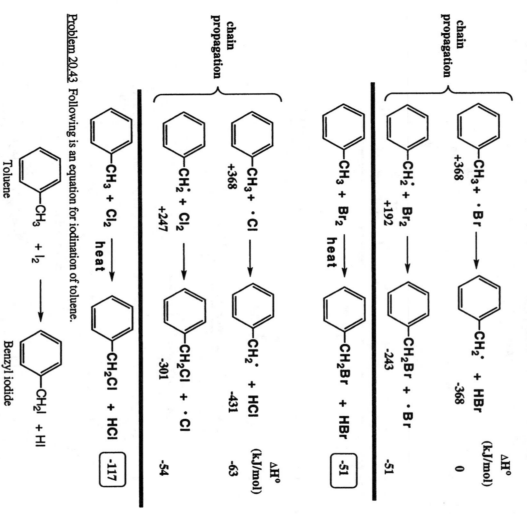

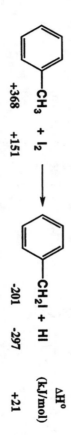

| | ΔH° (kJ/mol) |

Problem 20.43 Following is an equation for iodination of toluene.

Reaction of toluene and iodine to form benzyl iodide and HI is endothermic.

This reaction does not take place. All that happens under experimental conditions for the formation of radicals is initiation to form iodine radicals, I•, followed by termination to reform I₂. How do you account for these observations?

Using values for bond dissociation energies, calculate the enthalpy for each of the most likely chain propagation steps. Abstraction of hydrogen by an iodine radical (an iodine atom) is endothermic by 71 kJ/mol. The activation energy for this step is approximately a few kJ/mol greater than 71 kJ/mol. Given this large activation energy and the fact that the overall reaction is endothermic, it does not occur as written.

chain
propagation

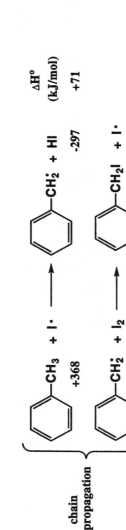

$$\Delta H^{\circ}$$
(kJ/mol)

—CH$_3$ + I·   +368

—CH$_2^{\cdot}$ + HI   -297   +71

—CH$_2^{\cdot}$ + I$_2$ +151

—CH$_2$I + I·   -201   -50

Problem 20.44 Although most alkanes react with chlorine by a radical chain mechanism when reaction is initiated by light or heat, benzene fails to react under the same conditions. Benzene cannot be converted to chlorobenzene by treatment with chlorine in the presence of light or heat.

—H + Cl$_2$ $\xrightarrow[\text{heat}]{\text{light or}}$ —Cl + HCl

(a) Explain why benzene fails to react under these conditions. Consult Appendix 3 for relevant bond dissociation energies.

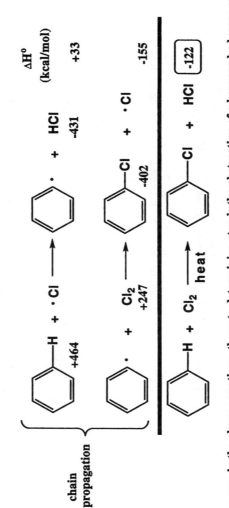

chain
propagation

$$\Delta H^{\circ}$$
(kcal/mol)

—H + ·Cl ⟶ · + HCl
+464   -431   +33

· + Cl$_2$ ⟶ —Cl + ·Cl
+247   -402   -155

—H + Cl$_2$ $\xrightarrow{\text{heat}}$ —Cl + HCl   -122

As can be seen in the above equations, the rate-determining step is the abstraction of a benzene hydrogen atom by a chlorine radical. This process is endothermic by 33 kJ/mol because the C-H bonds are relatively strong (464 kJ/mol). Therefore, even though the entire process is exothermic by 122 kJ/mol, the reaction does not proceed because of the endothermic rate-determining step.

(b) Explain why the bond dissociation energy of a C-H bond in benzene is significantly greater than that in alkanes. (*Hint:* Think of the orbitals forming the C-H bond in alkanes compared with those in benzene.)

The benzene carbon atoms are sp^2 hybridized, so the C-H bonds are derived from sp^2 hybridized orbitals. Compared with alkanes that use sp^3 hybridized orbitals, the sp^2 orbitals of benzene have greater s character and thus the electrons are held closer to the nucleus. This has the effect of increasing C-H bond strength in benzene relative to alkanes.

Problem 20.45 Following is an equation for hydroperoxidation of cumene. Assume that initiation is by an unspecified radical, R·.

—CH(CH$_3$)$_2$ + O$_2$ $\xrightarrow{\text{light}}$

Cumene

OOH
—C(CH$_3$)$_2$

Cumene hydroperoxide

Propose a radical chain mechanism for this reaction. Assume that initiation is by an unspecified radical, R·.

The stability of the benzyl radical, especially with the added methyl groups, facilitates the reaction with the radical initiator according to the following mechanism.

Step 1: **Initiation**

$\cdot$ R

H–R

Step 2: **Propagation**

Step 3: **2nd Propagation Step**

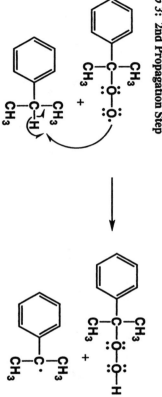

Problem 20.46 Para-substituted benzyl halides undergo reaction with methanol by an S_N1 mechanism to give a benzyl ether. Account for the following order of reactivity under these conditions.

R—CH₂Br + CH₃OH $\xrightarrow{\text{methanol}}$ R—CH₂OCH₃ + HBr

A reasonable S_N1 mechanism is shown below:

Step 1: **Rate-determining step**

Step 2:

Step 3:

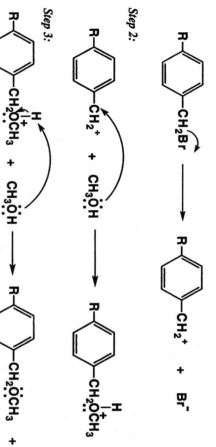

Rate of S_N1 reaction: $R = CH_3O^- > CH_3^- > H^- > NO_2^-$

The rate-determining step in this process is formation of the benzylic cation, so anything that stabilizes the benzylic cation will speed up the S_N1 reaction. Electron-donating groups such as the methoxy group are stabilizing to a benzylic cation. Electron-withdrawing groups such as the nitro group are destabilizing to a benzylic cation. Thus, the groups are listed in order from most electron-donating to most electron-withdrawing.

<u>Problem 20.47</u> When warmed in dilute sulfuric acid, 1-phenyl-1,2-propanediol undergoes dehydration and rearrangement to give 2-phenylpropanal.

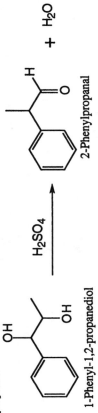

1-Phenyl-1,2-propanediol

2-Phenylpropanal

(a) Propose a mechanism for this example of a pinacol rearrangement (Section 9.7).

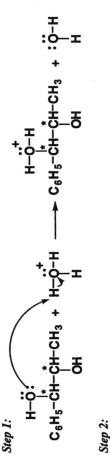

Step 1:

Step 2:

Step 3: Note that the intermediate produced in this step is resonance stabilized

Step 4:

(b) Account for the fact that 2-phenylpropanal is formed rather than its constitutional isomer, 1-phenyl-1-propanone.

1-Phenyl-1-propanone

2-Phenylpropanal

In the observed reaction that leads to the aldehyde product, protonation of the benzylic hydroxyl followed by loss of H_2O gives a benzylic carbocation (Step 2). 1-Phenyl-1-propanone would result from protonation and loss of the other -OH group, but that process involves a less stable secondary carbocation so it is not observed.

Problem 20.48 In the chemical synthesis of DNA and RNA, hydroxyl groups are normally converted to triphenylmethyl (trityl) ethers to protect the hydroxyl group from reaction with other reagents.

$$RCH_2OH + Ph_3CCl \xrightarrow[\text{amine}]{\text{tertiary}} RCH_2OCPh_3 + HCl \quad \text{(neutralized by the tertiary amine)}$$

triphenylmethyl chloride   a triphenylmethyl ether
(trityl chloride)   (a trityl ether)

Triphenylmethyl ethers are stable to aqueous base, but are rapidly cleaved in aqueous acid.

$$RCH_2OCPh_3 + H_2O \xrightarrow{H^+} RCH_2OH + Ph_3COH$$

(a) Why are triphenylmethyl ethers so readily hydrolyzed by aqueous acid?

The triphenylmethyl ethers are hydrolyzed according to the following mechanism.

Step 1:

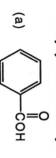

Step 2:

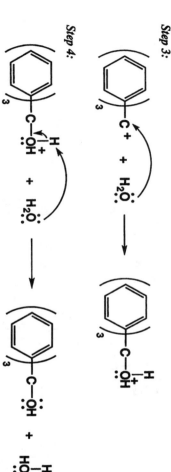

Note that the cation produced in this step is a highly resonance-stabilized benzylic cation due to the three adjacent phenyl rings.

Step 3:

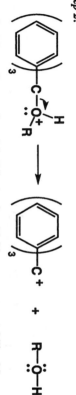

Step 4:

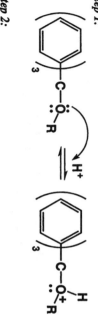

(b) How might the structure of the triphenylmethyl group be modified in order to increase or decrease its acid sensitivity?

Electron-releasing substituents like methoxy groups stabilize the triphenylmethyl cation and thereby increase the sensitivity of these triphenyl methyl ethers to acid. On the other hand, electron-withdrawing groups like nitro groups destabilize the triphenyl methyl cation and thereby decrease the sensitivity of these triphenylmethyl ethers to acid.

Synthesis

Problem 20.49 Using ethylbenzene as the only aromatic starting material, show how to synthesize the following compounds. In addition to ethylbenzene, use any other necessary organic or inorganic chemicals. Note that any compound already synthesized in one part of this problem may then be used to make any other compound in the problem.

(a)

Oxidation of ethylbenzene using $K_2Cr_2O_7$ in H_2SO_4 gives benzoic acid.

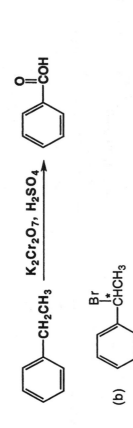

$$CH_2CH_3 \xrightarrow{\text{K}_2\text{Cr}_2\text{O}_7, \text{ H}_2\text{SO}_4} \text{COOH}$$

(b)

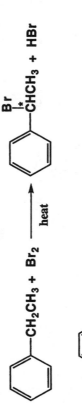

Br
|
*CHCH₃

Bromination of the benzylic position using bromine at elevated temperature or NBS in the presence of peroxide. The reaction involves a radical chain mechanism. A racemic mixture is produced because the product is chiral.

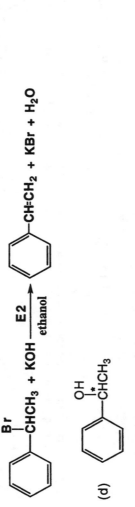

$$CH_2CH_3 + Br_2 \xrightarrow{\text{heat}} \begin{array}{c}\text{Br}\\|\\ \text{*CHCH}_3\end{array} + HBr$$

(c)

CH=CH₂

Dehydrohalogenation of the alkyl bromide from (b), brought about by a strong base such as KOH.

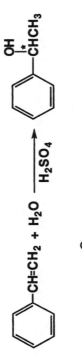

Br
|
*CHCH₃ + KOH $\xrightarrow[\text{ethanol}]{\text{E2}}$ CH=CH₂ + KBr + H₂O

(d)

OH
|
*CHCH₃

Acid-catalyzed hydration of a carbon-carbon double bond of (c). The same reaction may also be brought about by oxymercuration followed by reduction with NaBH₄. A racemic mixture is produced because the product is chiral.

CH=CH₂ + H₂O $\xrightarrow{\text{H}_2\text{SO}_4}$

OH
|
*CHCH₃

(e)

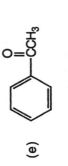

O
||
CCH₃

This oxidation is best brought about using the more selective oxidizing agents chromium trioxide in pyridine or pyridinium chlorochromate to avoid over oxidation to benzoic acid as in part (a).

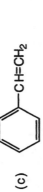

OH
|
CHCH₃ $\xrightarrow{\text{PCC}}$

O
||
CCH₃

(f)

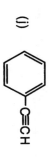

—CH₂CH₂OH

Hydroboration/oxidation of styrene from part (c).

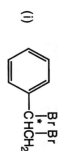

CH=CH₂ → —CH₂CH₂OH

1) B₂H₆
2) H₂O₂, NaOH

(g)

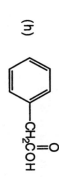

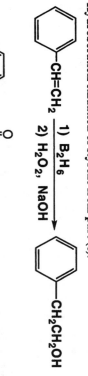

—CH₂CH₂OH + PCC → —CH₂CH

CH₂Cl₂

O
║

Oxidation of the primary alcohol of (f) using pyridinium chlorochromate.

(h)

—CH₂COH

O
║

Oxidation of the primary alcohol of (f) using chromic acid. The same product may be formed by similar oxidation of the aldehyde (g). Note how care must be taken to prevent over-oxidation to benzoic acid.

—CH₂CH₂OH → —CH₂COH

K₂Cr₂O₇, H₂SO₄

O
║

(i)

Br Br
| *|
—CHCH₂

Addition of bromine to the carbon-carbon double bond of styrene from part (c). A racemic mixture is produced because the product is chiral.

—CH=CH₂ + Br₂ → —CHCH₂

CCl₄

Br Br
| *|

(j)

—C≡CH

A double dehydrohalogenation of product (i) using sodium amide as the base.

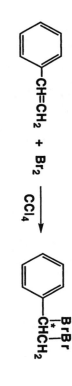

Br Br
| *|
—CHCH₂

1. 3NaNH₂
2. H₂O

→ —C≡CH

(k)

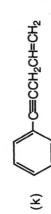

$C \equiv CCH_2CH = CH_2$

The terminal alkyne from (j) is deprotonated with sodium amide to produce the anionic species that reacts with allyl chloride to produce the desired product.

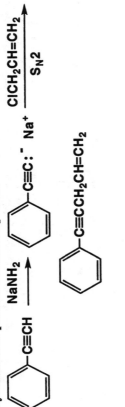

$C \equiv CH \xrightarrow{NaNH_2} C \equiv C:^- \ Na^+ \xrightarrow[S_N2]{ClCH_2CH=CH_2}$

$C \equiv CCH_2CH=CH_2$

(l)

$C \equiv C(CH_2)_5CH_3$

The deprotonated terminal alkyne from (k) reacts with hexyl chloride to produce the desired product.

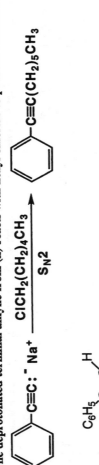

$C \equiv C:^- \ Na^+ \xrightarrow[S_N2]{ClCH_2(CH_2)_4CH_3}$

$C \equiv C(CH_2)_5CH_3$

(m)

$C_6H_5 \overset{H}{\underset{H}{C=C}} (CH_2)_5CH_3$

The alkyne produced in part (l) is reduced with sodium metal in liquid ammonia to produce the desired *trans* alkene.

$C \equiv C(CH_2)_5CH_3 \xrightarrow[NH_3(l)]{2 Na}$

$C_6H_5 \overset{H}{\underset{H}{C=C}} (CH_2)_5CH_3$

(n)

$C_6H_5 \overset{(CH_2)_5CH_3}{\underset{H}{C=C}} H$

The alkyne produced in part (l) is reduced with hydrogen and the Lindlar catalyst to produce the desired *cis* alkene.

$C \equiv C(CH_2)_5CH_3 \xrightarrow[\substack{Lindlar \\ Catalyst}]{H_2}$

$C_6H_5 \overset{(CH_2)_5CH_3}{\underset{H}{C=C}} H$

Problem 20.50 Show how to convert 1-phenylpropane into the following compounds. In addition to this starting material, use any necessary inorganic reagents. Any compound synthesized in one part of this problem may then be used to make any other compound in the problem.

(a)

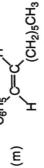

Radical chain bromination of 1-phenylpropane. Bromination is highly regioselective for the benzylic position. A racemic mixture is produced because the product is chiral.

$C_6H_5 \overset{Br}{\underset{*}{\diagup}} + Br_2 \xrightarrow{heat}$

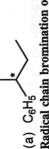

$+ \ HBr$

(b)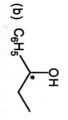

Dehydrohalogenation of product (a) using KOH or other strong base will give an alkene. Acid-catalyzed hydration of the alkene will give the desired alcohol. The reaction is highly regioselective because of the stability of the benzylic carbocation formed by protonation of the alkene. Alternatively, oxymercuration followed by reduction with NaBH₄ forms the same secondary alcohol.

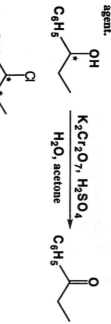

(c) C₆H₅ ...

Oxidation of the secondary alcohol of (b) using chromic acid in aqueous sulfuric acid and the alcohol dissolved in acetone. Alternatively use chromium trioxide in pyridine or pyridinium chlorochromate as the oxidizing agent.

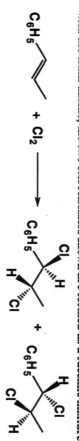

$$K_2Cr_2O_7, H_2SO_4$$
$$H_2O, acetone$$

(d) C₆H₅ ...

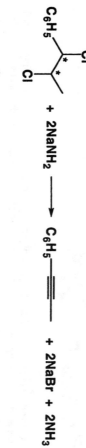

Addition of chlorine by electrophilic addition to the alkene formed in the first reaction of answer (b). Starting with the trans alkene, the two stereoisomers shown are formed as a racemic mixture.

(e) C₆H₅ ...

C₆H₅ — + Cl₂ ——→

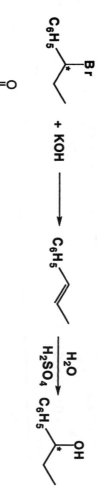

Double dehydrohalogenation of product (d) using sodium amide as the base.

(f) C₆H₅ ...

C₆H₅ + 2NaNH₂ ——→ C₆H₅ + 2NaBr + 2NH₃

Catalytic reduction of the alkyne using the Lindlar catalyst to reduce the alkyne of part (e) to the *cis* alkene.

C₆H₅ ——— + H₂ —Lindlar catalyst→ C₆H₅

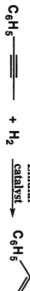

(g) C₆H₅

The trans alkene will be the primary product generated upon dehydrohalogenation of the 1-bromo-1-phenylpropane produced in (a).

Alternatively, the alkyne from part (e) can be treated with Na in ammonia to give the trans alkene.

C₆H₅ ————— + Na $\xrightarrow{NH_3(l)}$ C₆H₅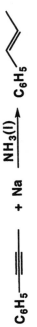

(h) C₆H₅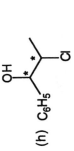

Treatment of the alkene from part (f) or (g) with Cl₂ in H₂O will give the chlorohydrin. Since the stereochemistry of the product is not given, we cannot deduce whether to start with the cis or trans alkene.

C₆H₅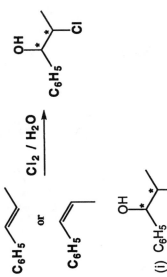

or

C₆H₅

$\xrightarrow{Cl_2 / H_2O}$

(i) C₆H₅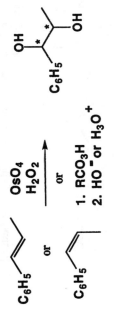

Oxidation of the alkene from part (f) or (g) to a glycol with osmium tetroxide in the presence of hydrogen peroxide. Since the stereochemistry of the diol product is not given, we cannot deduce whether to start with the cis or trans alkene.

C₆H₅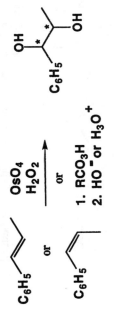

or

C₆H₅

$\xrightarrow[\text{H}_2\text{O}_2]{\text{OsO}_4}$

or

1. RCO₃H
2. HO⁻ or H₃O⁺

Problem 20.51 Carbinoxamine is a histamine antagonist, specifically an H₁-antagonist. The maleic acid salt of the levorotatory isomer is sold as the prescription drug Rotoxamine. Given this retrosynthetic analysis, propose a synthesis of carbinoxamine. (Note: Aryl bromides form Grignard reagents much more readily than aryl chlorides.)

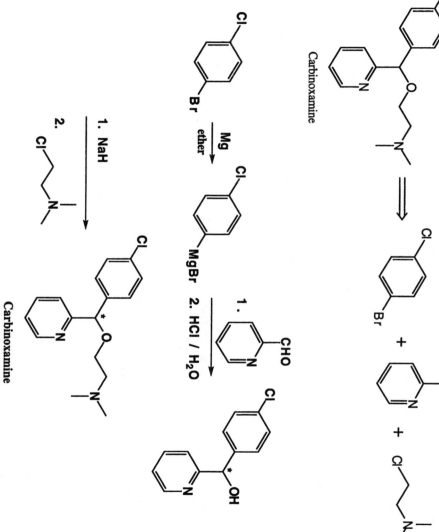

Carbinoxamine

Carbinoxamine

Problem 20.52 Cromolyn sodium, developed in the 1960s, is used to prevent allergic reactions primarily affecting the lungs, as for example exercise-induced emphysema. It is thought to block the release of histamine, which prevents the sequence of events leading to swelling, itching, and constriction of bronchial tubes. Cromolyn sodium is synthesized in the following series of steps. Treatment of one mole of epichlorohydrin with two moles of 2,6-dihydroxyacetophenone in the presence of base gives I. Treatment of I with two moles of diethyl oxalate in the presence of sodium ethoxide gives a diester II. Saponification of the diester with aqueous NaOH gives cromolyn sodium.

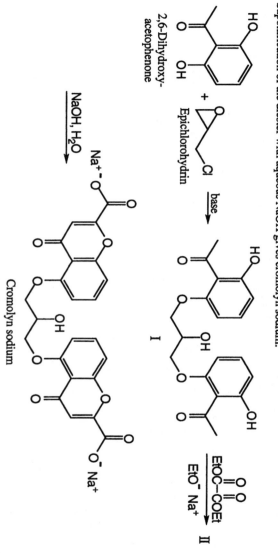

2,6-Dihydroxy-acetophenone

Epichlorohydrin

I

II

Cromolyn sodium

(a) Propose a mechanism for the formation of compound I.

Step 1:

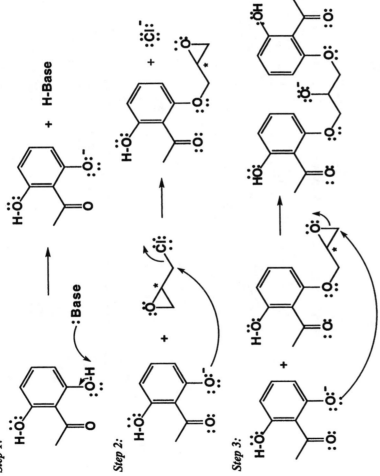

Step 2:

Step 3:

Step 4:

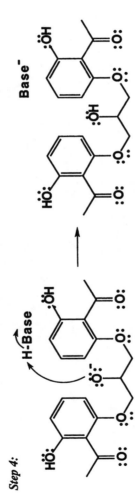

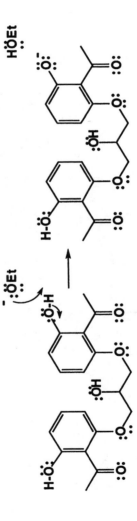

(b) Propose a structural formula for compound II and a mechanism for its formation.

Step 1:

Step 2:

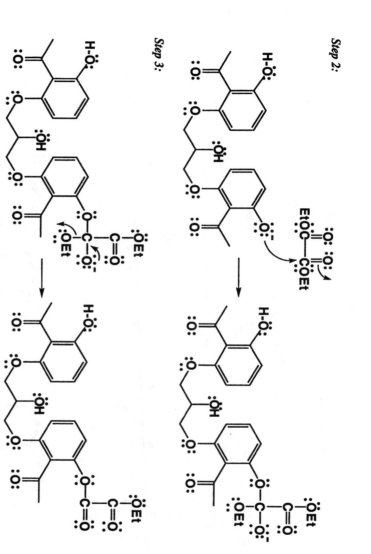

Step 3:

The same sequence occurs on the other phenolic -OH group to give the following intermediate.

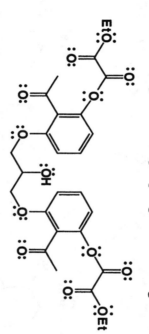

Step 4:

The methyl ketone is deprotonated to give an enolate

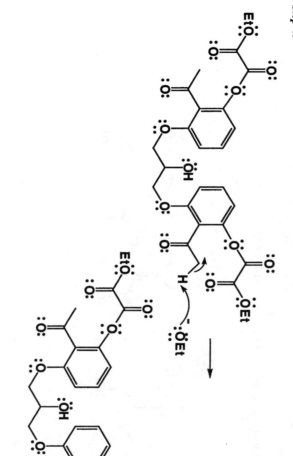

Enolate

The enolate then undergoes an aldol reaction.
Step 5:

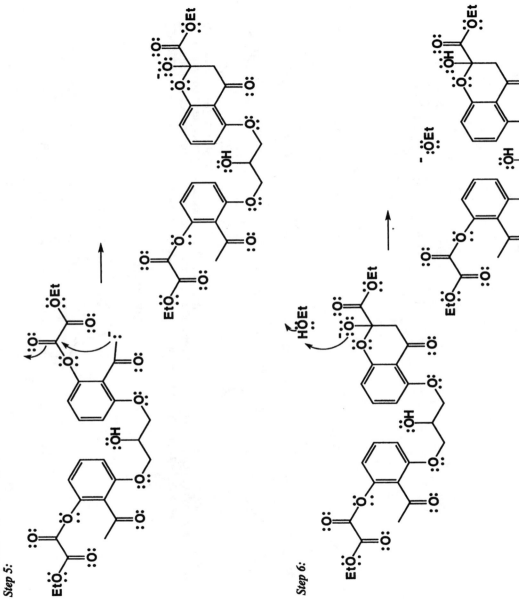

Step 6:

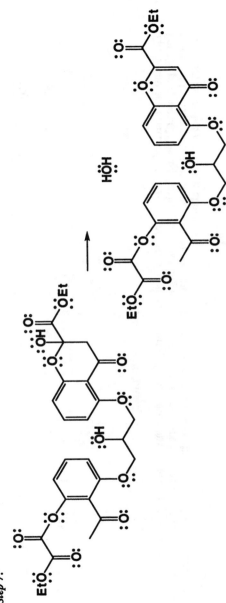

Dehydration completes the process
Step 7:

The same reaction occurs on the other side to give compound II

Compound II

<u>Problem 20.53</u> The following stereospecific synthesis is part of the scheme used by E. J. Corey of Harvard University in the synthesis of erythronolide B, the precursor of the erythromycin antibiotics. In this remarkably simple set of reactions, the relative configurations of five stereocenters are established. [See Corey, E. J., Nicolaou, K. C., and Melvin, L. S. Jr., J. Am. Chem. Soc. 97: 654 (1975)]

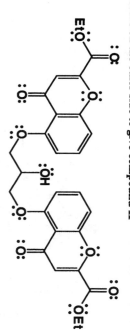

2,4,6-Trimethyl-
phenol

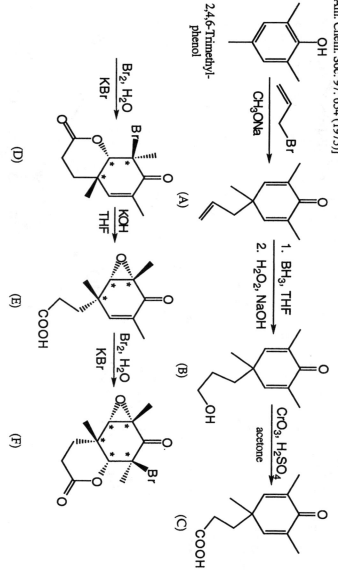

(a) Propose a mechanism for the conversion of 2,4,6-trimethylphenol to compound A.

Step 1:

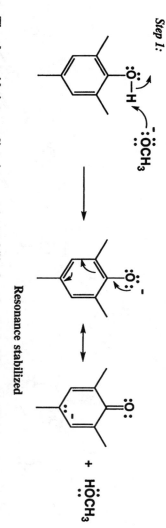

Resonance stabilized

The phenoxide intermediate is resonance stabilized, and a total of four contributing structures can be drawn. For the purposes of this mechanism, the most important is the one shown above in which the negative charge is placed on the ring carbon atom para to the oxygen atom. This partial negative charge is nucleophilic enough to take part in an S$_N$2 reaction as shown.

Step 2:

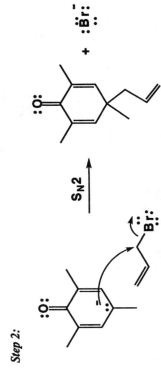

(b) Account for the stereospecificity and regiospecificity of the three steps in the conversion of compound C to compound F.

In the conversion of C to D (as well as E to F), the mechanism involves electrophilic attack of bromine on a double bond of the ring, leading to a bromonium ion intermediate. The bromonium ion intermediate is attacked by the appended carboxyl group to complete the reaction. The bromonium ion intermediate dictates the anti stereochemistry observed, and the length of the carboxylic acid chain leads to the formation of the six-membered ring.

In the conversion of (D) to (E), the mechanism is a more complicated. The carbonyl of the lactone ester is attacked by hydroxide to generate a tetrahedral addition intermediate that collapses by expelling an alkoxide that is set up to attack from the backside of the C-Br bond in an intramolecular S_N2 step to give the observed stereochemistry as shown below.

Step 1:

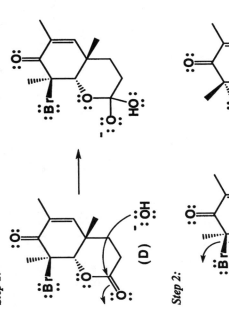

(c) Is compound F produced in this synthesis as a single enantiomer or as a racemic mixture? Explain.

Even though only one structure is drawn in each step, in reality compounds D, E, and F are produced as racemic mixtures. Comound C is achiral, and there are no chiral reagents being used, so all chiral molecules produced subsequently must be made as racemic mixtures. When molecules have this many stereocenters, it often helps to use molecular models when analyzing stereochemistry.

CHAPTER 21
Solutions to the Problems

Problem 21.1 Write the stepwise mechanism for sulfonation of benzene by hot, concentrated sulfuric acid. In this reaction, the electrophile is SO₃ formed as shown in the following equation.

$$H_2SO_4 \rightleftharpoons SO_3 + H_2O$$

In sulfonation of benzene, the electrophile is sulfur trioxide. In Step 1, reaction of benzene with the electrophile yields a resonance-stabilized cation. In Step 2, this intermediate loses a proton to complete the reaction: water is shown as the base accepting the proton.

Step 1:

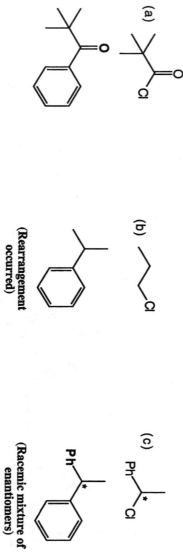

(A resonance-stabilized cation intermediate)

Step 2:

Problem 21.2 Write structural formulas for the products you expect from Friedel-Crafts alkylation or acylation of benzene with

(a)

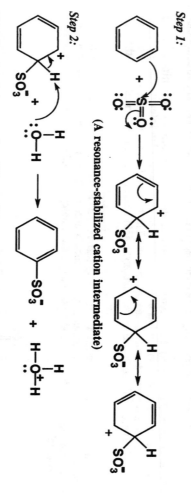

(b)

(c)

(Rearrangement occurred)

(Racemic mixture of enantiomers)

Note that in more strongly acidic conditions, commonly referred to as fuming sulfuric acid, the electrophile can be in the protonated form, namely HSO₃⁺, instead of SO₃.

Problem 21.3 Write a mechanism for the formation of *tert*-butylbenzene from benzene and *tert*-butyl alcohol in the presence of phosphoric acid.

A reasonable mechanism for this reaction involves formation of the *tert*-butyl cation, that then reacts with benzene via electrophilic aromatic substitution.

Step 1:

Step 2:

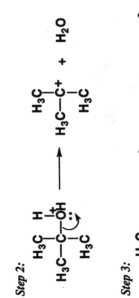

H₃C—C⁺ + H₂O

Step 3:

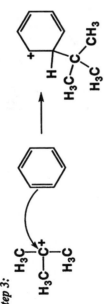

Step 4:

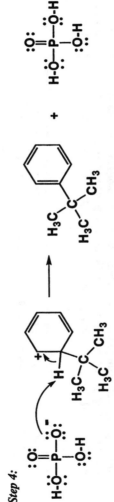

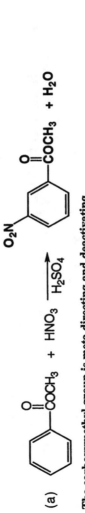

+

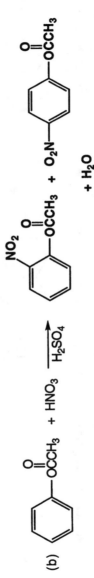

Problem 21.4 Complete the following electrophilic aromatic substitution reactions. Where you predict meta substitution, show only the meta product. Where you predict ortho-para substitution, show both products.

(a)

COCH₃ + HNO₃ $\xrightarrow{\text{H}_2\text{SO}_4}$ O₂N—COCH₃ + H₂O

The carboxymethyl group is meta directing and deactivating.

(b)

OCH₃ + HNO₃ $\xrightarrow{\text{H}_2\text{SO}_4}$ NO₂—OCH₃ + O₂N—OCH₃

+ H₂O

The acetoxy group is ortho/para directing and activating.

Problem 21.5 Predict the major product(s) of each electrophilic aromatic substitution.

When there is more than one substituent on the ring, the predominant product is determined by the more activating (less deactivating) substituent. In addition, steric hindrance precludes reaction between two substituents that are meta with respect to each other.

(a)

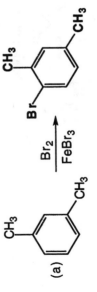

(b)

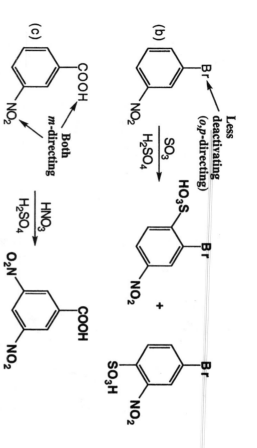

Less deactivating
(*o,p*-directing)

$\xrightarrow[\text{H}_2\text{SO}_4]{\text{SO}_3}$

(c)

Both
m-directing

$\xrightarrow[\text{H}_2\text{SO}_4]{\text{HNO}_3}$

Problem 21.6 In S_N2 reactions of alkyl halides, the order of reactivity is $RI > RBr > RCl > RF$. Alkyl iodides are considerably more reactive than alkyl fluorides, often by factors as great as 10^6. All 1-halo-2,4-dinitrobenzenes, however, react at approximately the same rate in nucleophilic aromatic substitutions. Account for this difference in relative reactivities.

Recall that the overall rate of a reaction is determined by the rate-determining (slow) step. In an S_N2 reaction, departure of the leaving group is involved in the rate-determining step, thus the nature of the leaving group influences the rate of the overall reaction. As shown in the text, the mechanism of nucleophilic substitution with 1-halo-2,4-dinitrobenzenes involves addition to the ring to form a Meisenheimer complex. Since *formation* of this complex is the rate-determining step and departure of the halide is not involved, the nature of the halogen has little influence on the overall rate of the process.

PROBLEMS
Electrophilic Aromatic Substitution: Monosubstitution
Problem 21.7 Write a stepwise mechanism for each of the following reactions. Use curved arrows to show the flow of electrons in each step.

(a)

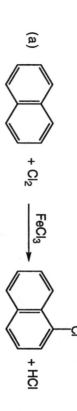

$+ \; Cl_2$

$\xrightarrow{FeCl_3}$

$+ \; HCl$

Chlorination of naphthalene in the presence of aluminum chloride is an example of electrophilic aromatic substitution. It is shown in three steps.

Step 1: Activation of chlorine to form an electrophile.

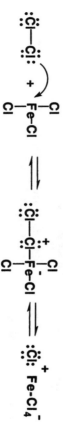

$\overset{..}{\underset{..}{:}Cl}-\overset{..}{\underset{..}{Cl}:} \; + \; \overset{Cl}{\underset{Cl}{Fe}}-Cl \; \Longrightarrow \; \overset{Cl}{\underset{Cl}{:\overset{..}{Cl}-\overset{+}{\overset{..}{Cl}}-\overset{-}{Fe}-Cl}} \; \Longrightarrow \; :\overset{..}{\underset{..}{Cl}}:^{+} \; Fe-Cl_4^{-}$

Step 2: Reaction of the electrophile with the aromatic ring, a nucleophile.

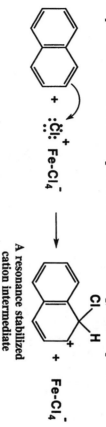

$+ \; :\overset{..}{\underset{..}{Cl}}:^{+} \; Fe-Cl_4^{-} \; \longrightarrow$

A resonance stabilized
cation intermediate

$+ \; Fe-Cl_4^{-}$

Step 3: Proton transfer and reformation of the aromatic ring.

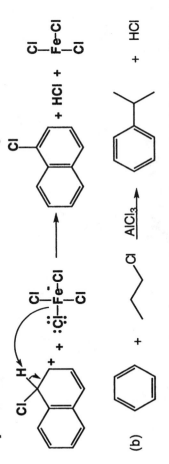

(b)

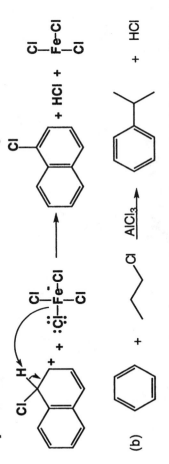

The reaction of benzene with 1-chloropropane in the presence of aluminum chloride involves initial formation of a complex between 1-chloropropane and aluminum chloride, and its rearrangement to an isopropyl cation. This cationic species is the electrophile that undergoes further reaction with benzene.

Step 1: Formation of a complex between 1-chloropropane (a Lewis base) and aluminum chloride (a Lewis acid).

$$CH_3CH_2CH_2\ddot{C}l\!: + Al-Cl \rightleftharpoons CH_3CH_2CH_2\overset{+}{\ddot{C}l}-Al^--Cl + AlCl_4^-$$

Step 2: Rearrangement to form an isopropyl cation.

Step 3: Electrophilic attack on the aromatic ring.

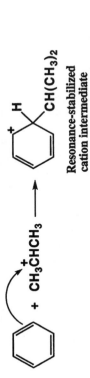

Resonance-stabilized cation intermediate

Step 4: Proton transfer to regenerate the aromatic ring.

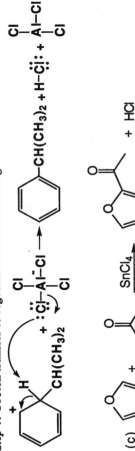

(c)

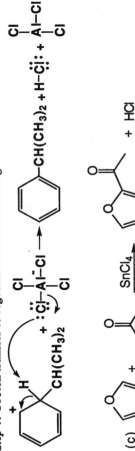

Friedel-Crafts acylation of furan involves electrophilic attack by an acylium ion.
Step 1: Formation of resonance-stabilized acylium ion.

A resonance-stabilized acylium ion

Step 2: Reaction of acylium ion (an electrophile) and furan (a nucleophile) to form a resonance stabilized cation.

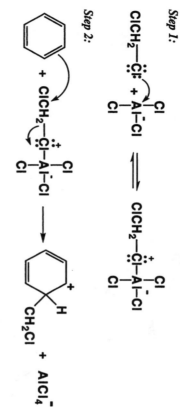

Step 3: Proton transfer and regeneration of aromatic ring.

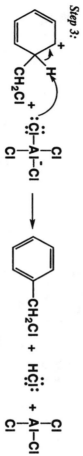

(d) 2 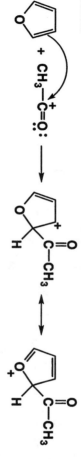 + CH$_2$Cl$_2$ $\xrightarrow{\text{AlCl}_3}$  + 2HCl

Formation of diphenylmethane involves two successive Friedel-Crafts alkylations.

Step 1:

ClCH$_2$–$\ddot{\underset{\cdot\cdot}{Cl}}$: + $\underset{\underset{Cl}{|}}{\overset{\overset{Cl}{|}}{Al}}$–Cl $\rightleftharpoons$ ClCH$_2$–$\overset{+}{\underset{\cdot\cdot}{Cl}}$–$\underset{\underset{Cl}{|}}{\overset{\overset{Cl}{|}}{Al}}$–Cl

Step 2:

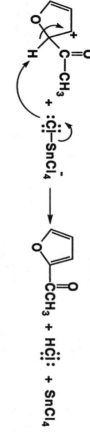 + CH$_2$Cl + AlCl$_4^-$

Step 3:

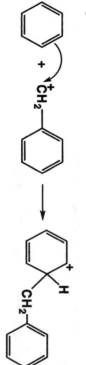 + :$\ddot{\underset{\cdot\cdot}{Cl}}$–$\underset{\underset{Cl}{|}}{\overset{\overset{Cl}{|}}{Al}}$–Cl $\rightarrow$ CH$_2$Cl + H$\ddot{\underset{\cdot\cdot}{Cl}}$: + $\underset{\underset{Cl}{|}}{\overset{\overset{Cl}{|}}{Al}}$–Cl

Step 4:

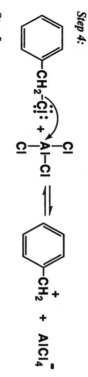 + $\underset{\underset{Cl}{|}}{\overset{\overset{Cl}{|}}{Al}}$–Cl $\rightleftharpoons$ CH$_2^+$ + AlCl$_4^-$

Formation of benzyl chloride completes the first Friedel-Crafts alkylation. This molecule then is a reactant for the second Friedel-Crafts alkylation.

Step 5:

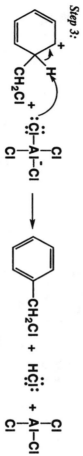

Step 6:

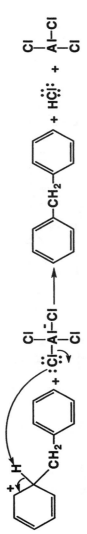

Problem 21.8 Pyridine undergoes electrophilic aromatic substitution preferentially at the 3 position as illustrated by the synthesis of 3-nitropyridine. Note that, under these acidic conditions, the species undergoing nitration is not pyridine, but its conjugate acid.

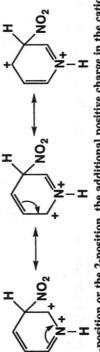

$+ HNO_3 \xrightarrow[300°C]{H_2SO_4}$

$+ H_2O$

Pyridine　　　　　　3-Nitropyridine

Write resonance contributing structures for the intermediate formed by attack of NO_2^+ at the 2, 3, and 4 positions of the conjugate acid of pyridine. From examination of these intermediates, offer an explanation for preferential nitration at the 3 position.

Pyridine is a base, and in the presence of a nitric acid-sulfuric acid mixture, it is protonated. It is the protonated form that must be attacked by the electrophile NO_2^+. For nitration at the 3-position, the additional positive charge in the cation intermediate may be delocalized on three carbon atoms of the pyridine ring. None of the contributing structures, however, places both positive charges on the same atom.

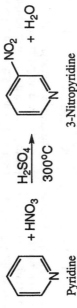

For nitration at the 4-position or the 2-position, the additional positive charge in the cation intermediate is also delocalized on three atoms of the pyridine ring, but one of these contributing structures has a charge of +2 on nitrogen. This situation is thus less stable than that which occurs for nitration at the 3-position.

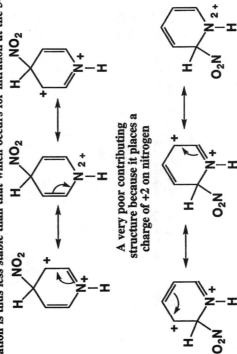

A very poor contributing structure because it places a charge of +2 on nitrogen

A very poor contributing structure because it places a charge of +2 on nitrogen

Problem 21.9 Pyrrole undergoes electrophilic aromatic substitution preferentially at the 2 position as illustrated by the synthesis of 2 nitropyrrole.

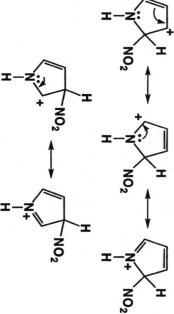

Pyrrole $+ HNO_3$ $\xrightarrow[5°C]{CH_3CO_2H}$ 2-Nitropyrrole $+ H_2O$

Write resonance contributing structures for the intermediate formed by attack of NO_2^+ at the 2 and 3 positions of pyrrole. From examination of these intermediates, offer an explanation for preferential nitration at the 2 position.

Pyrrole is nitrated under considerably milder conditions than pyridine. For nitration at the 2-position, the positive charge on the cation intermediate is delocalized over three atoms of the pyrrole ring whereas for nitration at the 3-position, it is delocalized over only two atoms. The intermediate with the greater degree of delocalization of charge has a lower activation energy for its formation and hence is formed at a faster rate.

Problem 21.10 Addition of *m*-xylene to the strongly acidic solvent HF/SbF5 at –45° gives a new species, which shows ¹H-NMR resonances at δ 2.88 (3H), 3.00 (3H), 4.67 (2H), 7.93 (1 H), 7.83 (1H), and 8.68 (1H). Assign a structure to the species giving this spectrum.

The strong acid results in protonation of the aromatic ring to create a positively charged species as shown below.

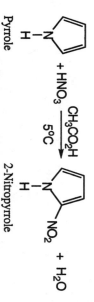

Problem 21.11 Addition of *tert*-butylbenzene to the strongly acidic solvent HF/SbF5 followed by aqueous work-up gives benzene. Propose a mechanism for this dealkylation reaction. What is the other product of the reaction?

A reasonable mechanism for this reaction involves protonation of the aromatic ring, followed by heterolytic bond cleavage and release of the *tert*-butyl cation. After addition of water, *tert*-butyl alcohol will be the other product of the reaction.

Step 1:

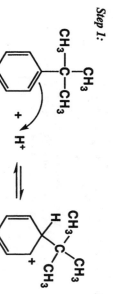

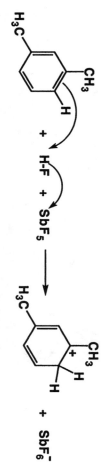

Step 2:

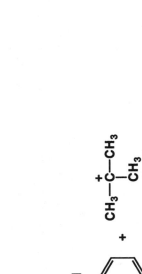

Step 3:

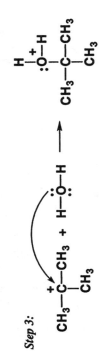

Step 4:

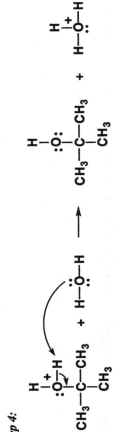

Problem 21.12 What product do you predict from the reaction of SCl_2 with benzene in the presence of $AlCl_3$? What product results if diphenyl ether is treated with SCl_2 and $AlCl_3$?

The Lewis acid, $AlCl_3$, facilitates departure of one of the chlorine atoms from SCl_2, and the resulting electrophile takes part in electrophilic aromatic substitution reaction to create the -SCl derivative that then reacts again to create diphenyl sulfide. The diphenyl sulfide is activated compared to benzene, so this will react further to generate a polymeric product as shown.

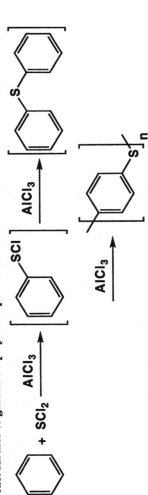

If diphenyl ether were treated in a similar manner, the resulting polymeric species will have alternating ether and thioether functions.

Problem 21.13 Other groups besides H^+ can act as leaving groups in electrophilic aromatic substitution. One of the best leaving groups is the trimethylsilyl group (Me_3Si-). For example, treatment of $Me_3SiC_6H_5$ with CF_3CO_2D rapidly forms DC_6H_5. What are the properties of a silicon-carbon bond that allows you to predict this kind of reactivity?

Based on simple electronegativities, the C-Si bond is polarized such that a partial positive charge is on the Si atom. Furthermore, the heterolytic bond cleavage is facilitated because the $(CH_3)_3Si^+$ cation is so stable.

Disubstitution and Polysubstitution

Problem 21.14 The following groups are ortho-para directors. Draw a contributing structure for the resonance-stabilized aryl cation formed during electrophilic aromatic substitution that shows the role of each group in stabilizing the intermediate by further delocalizing its positive charge.

(a) —OH

(b) —O—CCH₃

(c) —N(CH₃)₂

(d) —NHCCH₃

(e)

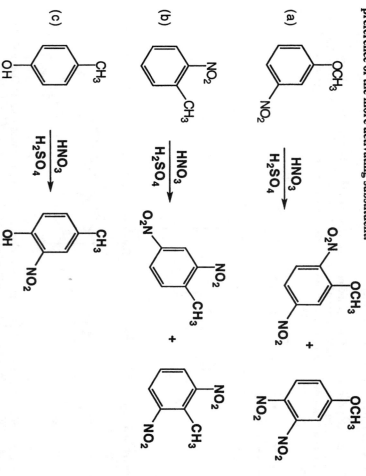

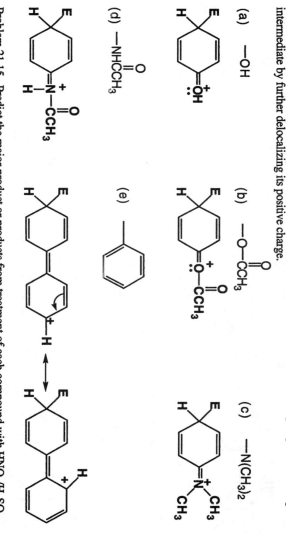

Problem 21.15 Predict the major product or products from treatment of each compound with HNO₃/H₂SO₄.

When there is more than one substituent on a ring, the predominant product is derived from the orientation preference of the more activating substituent.

(a)

(b)

(c)

(d)

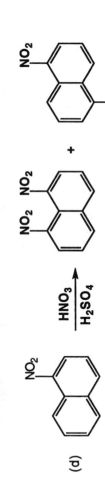

$$\xrightarrow[\text{H}_2\text{SO}_4]{\text{HNO}_3}$$

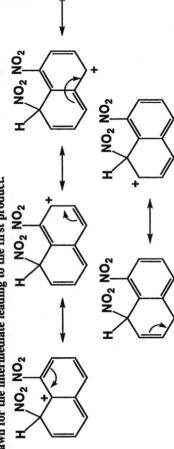

+

Nitration occurs as shown above, because the ring without the nitro group is less deactivated and reacts at the α-positions. The α-positions are more reactive, because in this case, none of the contributing structures place the positive charge adjacent to the existing nitro group. For example, the following five contributing structures can be drawn for the intermediate leading to the first product.

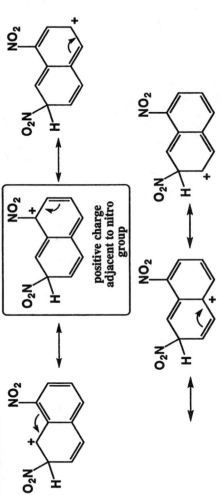

A similar set of contributing structures drawn for reaction at the β-position includes one that places the positive charge adjacent to the existing nitro group. This structure makes a negligible contribution to the resonance hybrid, so the positive charge is less delocalized and this intermediate is less stable than that for α-position substitution.

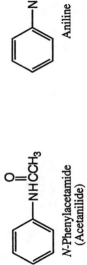

How do you account for the fact that *N*-phenylacetamide (acetanilide) is less reactive toward electrophilic aromatic substitution than aniline?

N-Phenylacetamide
(Acetanilide)

Aniline

The unshared pair of electrons on the nitrogen atom of acetanilide is involved in a resonance interaction with the carbonyl group of the amide, and, therefore, less available for stabilization of an aryl cation intermediate compared to aniline.

Problem 21.17 Propose an explanation for the fact that the trifluoromethyl group is almost exclusively meta directing.

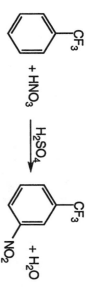

Following are contributing structures for meta and para attack of the electrophile. For meta attack, three contributing structures can be drawn and all make approximately equal contributions to the hybrid. Three contributing structures can also be drawn for ortho/para attack, one of which places a positive charge on the carbon bearing the trifluoromethyl group; this structure makes only a negligible contribution to the hybrid. Thus, for meta attack, the positive charge on the aryl cation intermediate can be delocalized almost equally over three atoms of the ring giving this cation's formation a lower activation energy. For ortho/para attack, the positive charge on the aryl cation intermediate is delocalized over only two carbons of the ring, giving this cation's formation a higher activation energy.

meta attack:

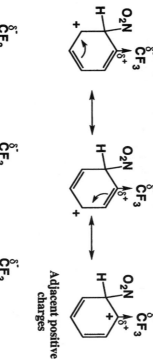

ortho/para attack:

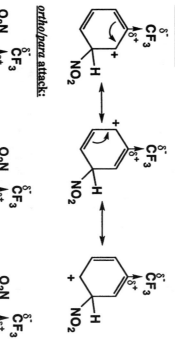

Adjacent positive
charges

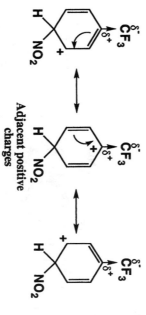

Adjacent positive
charges

Problem 21.18 Suggest a reason why the nitroso group, -N=O, is ortho-para directing although the nitro group, -NO₂, is meta directing.

Like other ortho-para directing groups, the -N=O group has a lone pair of electrons that can stabilize an adjacent positive charge of ortho or para attack via the resonance structures shown below. Without a lone pair of electrons on nitrogen, the nitro group cannot take part in similar stabilization. In fact, an adjacent positive charge is destabilized by the electron-withdrawing nature of the nitro group.

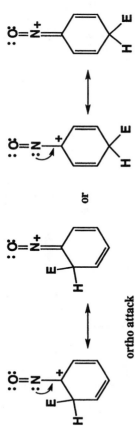

ortho attack   or   para attack

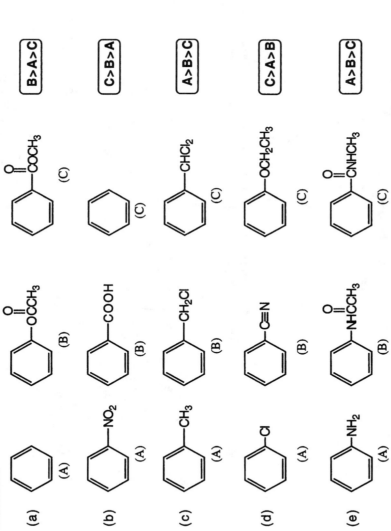

<u>Problem 21.19</u> Arrange the following in order of decreasing reactivity (fastest to slowest) toward electrophilic aromatic substitution.

(a) (A) benzene   (B) OCCH₃   (C) COCH₃   B>A>C

(b) (A) NO₂   (B) COOH   (C) benzene   C>B>A

(c) (A) CH₃   (B) CH₂Cl   (C) CHCl₂   A>B>C

(d) (A) Cl   (B) C≡N   (C) OCH₂CH₃   C>A>B

(e) (A) NH₂   (B) NHCCH₃   (C) CNHCH₃   A>B>C

<u>Problem 21.20</u> For each compound, indicate which group on the ring is the more strongly activating and then draw the structural formula of the major product formed by nitration of the compound.

In the following structures, the more strongly activating group is circled and arrows show the position(s) of nitration. Where both ortho and para nitration is possible, two arrows are shown. A broken arrow shows a product formed in only negligible amounts.

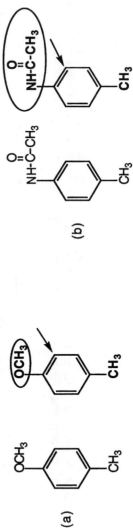

(a)

(b)

Problem 21.21 The following molecules each contain two rings. Which ring in each undergoes electrophilic aromatic substitution more readily? Draw the major product formed on nitration.

(a)

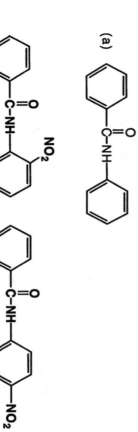

(b)

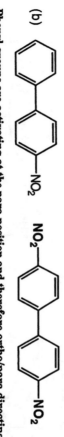

(c)

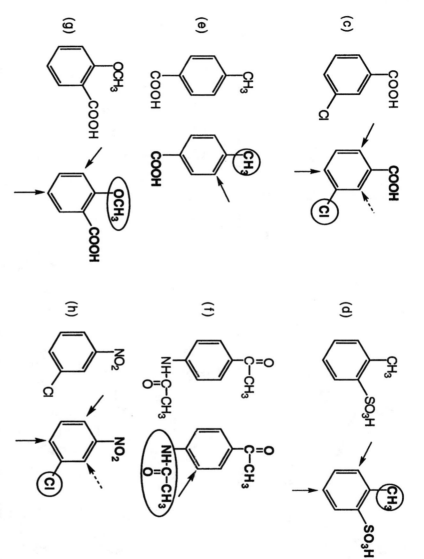

(d)

(e)

(f)

(g)

(h)

The nitrogen side of the amide group is more activating, so nitration produces the ortho/para products on this ring.

Phenyl groups are activating at the para position and therefore ortho/para directing. Note that nitration takes place on the ring that does not already possess the nitro group. Furthermore, nitration does not take place at the ortho position because of steric interactions between the rings.

(c)

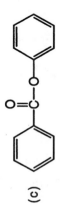

The oxygen side of the ester is more activating, so ortho/para nitration takes place on this ring.

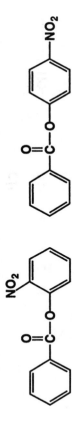

Problem 21.22 Reaction of phenol with acetone in the presence of an acid catalyst gives a compound known as bisphenol A. Bisphenol A is used in the production of epoxy and polycarbonate resins. Propose a mechanism for the formation of bisphenol A.

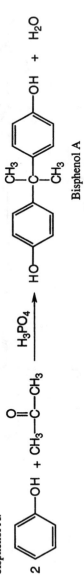

Bisphenol A

The reaction begins by proton transfer from phosphoric acid to acetone to form its conjugate acid which may be written as a hybrid of two contributing structures. The conjugate acid of acetone is an electrophile and reacts with phenol at the para position by electrophilic aromatic substitution to give 2-(4-hydroxyphenyl)-2-propanol. Protonation of the tertiary alcohol in this molecule and departure of water gives a resonance-stabilized cation that reacts with a second molecule of phenol to give bisphenol A.

Step 1:

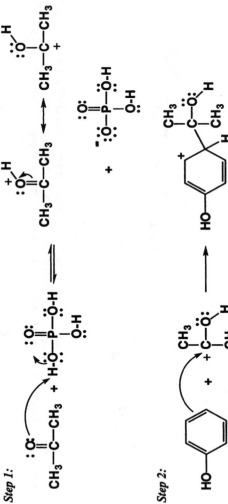

Step 2:

Step 3:

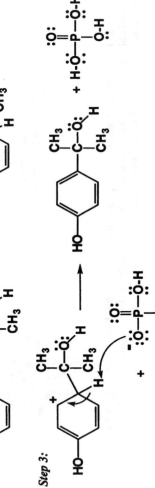

Step 4:

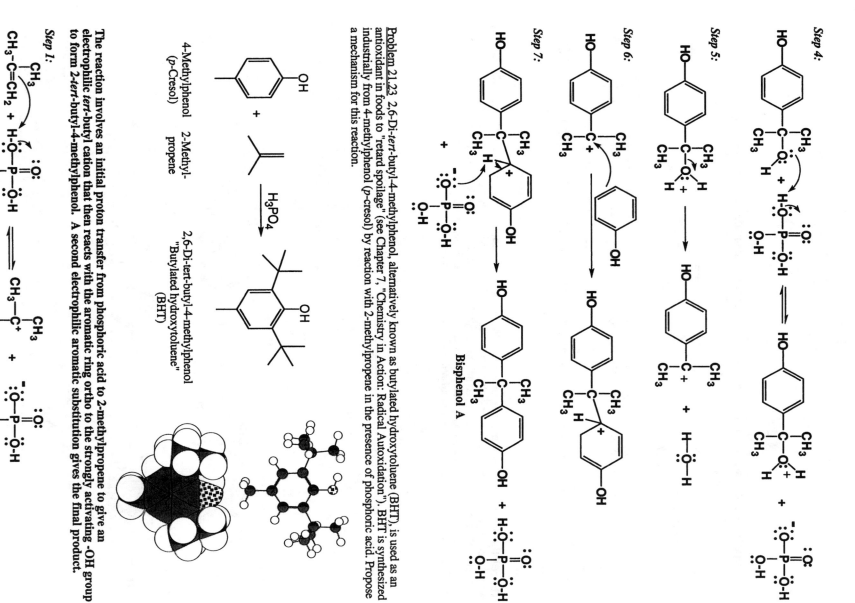

Step 5:

Step 6:

Step 7:

Bisphenol A

Problem 21.23 2,6-Di-*tert*-butyl-4-methylphenol, alternatively known as butylated hydroxytoluene (BHT), is used as an antioxidant in foods to "retard spoilage" (see Chapter 7, "Chemistry in Action: Radical Autoxidation"). BHT is synthesized industrially from 4-methylphenol (*p*-cresol) by reaction with 2-methylpropene in the presence of phosphoric acid. Propose a mechanism for this reaction.

4-Methylphenol   2-Methyl-   2,6-Di-tert-butyl-4-methylphenol
(*p*-Cresol)   propene   "Butylated hydroxytoluene"
                                            (BHT)

H_3PO_4

The reaction involves an initial proton transfer from phosphoric acid to 2-methylpropene to give an electrophilic *tert*-butyl cation that then reacts with the aromatic ring ortho to the strongly activating -OH group to form 2-*tert*-butyl-4-methylphenol. A second electrophilic aromatic substitution gives the final product.

Step 1:

Step 2:

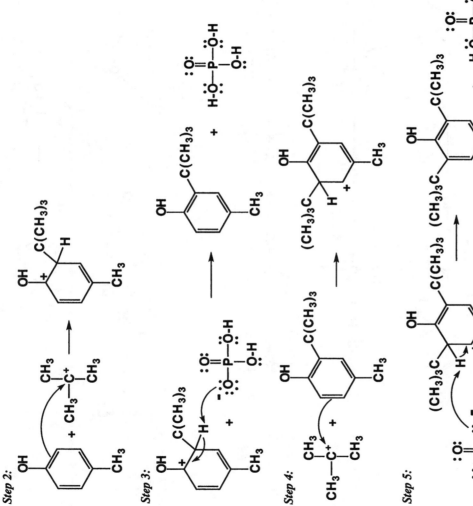

Step 3:

Step 4:

Step 5:

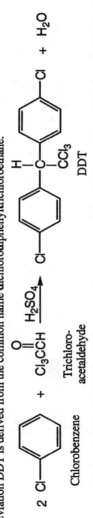

BHT

Problem 21.24 The insecticide DDT is prepared by the following route. Suggest a mechanism for this reaction. The abbreviation DDT is derived from the common name dichlorodiphenyltrichloroethane.

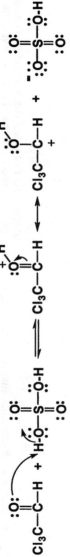

2 Chlorobenzene + Trichloro-acetaldehyde → DDT + H₂O

Propose a mechanism that is highly analogous to that proposed for the formation of Bisphenol A in Problem 21.22 above.

Step 1:

Step 2:

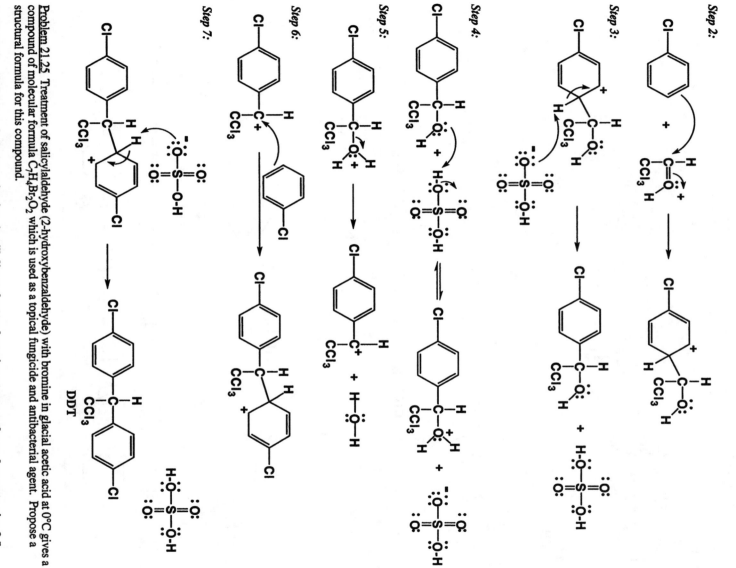

Step 3:

Step 4:

Step 5:

Step 6:

Step 7:

DDT

Problem 21.25 Treatment of salicylaldehyde (2-hydroxybenzaldehyde) with bromine in glacial acetic acid at 0°C gives a compound of molecular formula $C_7H_4Br_2O_2$, which is used as a topical fungicide and antibacterial agent. Propose a structural formula for this compound.

Because the -OH group is more activating, it will direct the two bromine atoms ortho and para to give 3,5-dibromo-2-hydroxybenzaldehyde as the product.

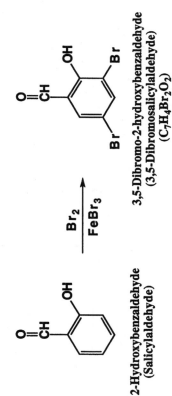

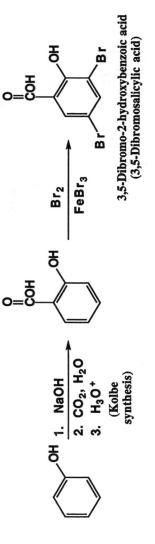

2-Hydroxybenzaldehyde
(Salicylaldehyde)

3,5-Dibromo-2-hydroxybenzaldehyde
(3,5-Dibromosalicylaldehyde)
($C_7H_4Br_2O_2$)

Problem 21.26 Propose a synthesis for 3,5-dibromo-2-hydroxybenzoic acid (3,5-dibromosalicylic acid) from phenol.

A Kolbe reaction followed by bromination leads to 3,5-dibromo-2-hydroxybenzoic acid.

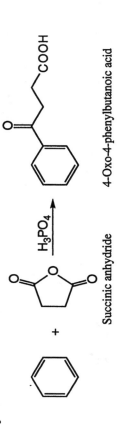
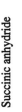

1. NaOH
2. CO_2, H_2O
3. H_3O^+
(Kolbe synthesis)

3,5-Dibromo-2-hydroxybenzoic acid
(3,5-Dibromosalicylic acid)

Problem 21.27 Treatment of benzene with succinic anhydride in the presence of polyphosphoric acid gives the following γ-ketoacid. Propose a mechanism for this reaction.

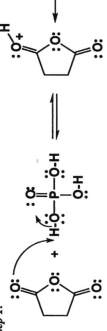

Succinic anhydride

4-Oxo-4-phenylbutanoic acid

Step 1:

Step 2:

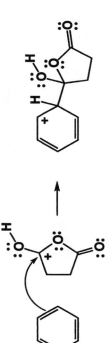

Step 3:

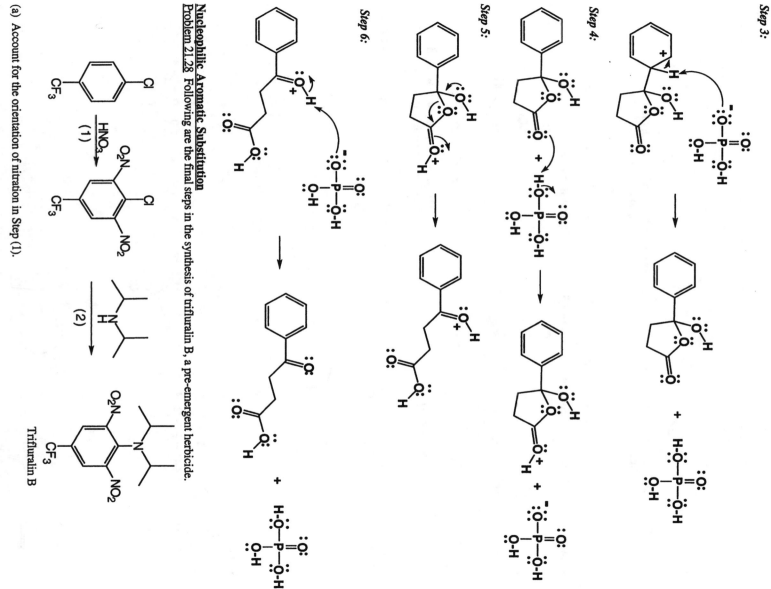

Step 4:

Step 5:

Step 6:

Nucleophilic Aromatic Substitution

Problem 21.28 Following are the final steps in the synthesis of trifluralin B, a pre-emergent herbicide.

Trifluralin B

(a) Account for the orientation of nitration in Step (1).

The trifluoromethyl group is strongly deactivating and meta directing (Problem 21.17). Chlorine is weakly deactivating and ortho/para directing. The combination directs the incoming nitro groups ortho to the chlorine atom.

(b) Propose a mechanism for the substitution reaction in Step (2).

The second step of this transformation is an example of nucleophilic aromatic substitution. In the following structural formulas, the propyl group of dipropylamine is abbreviated R.

Step 1:

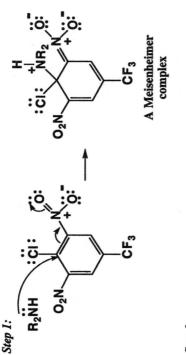

A Meisenheimer complex

Step 2:

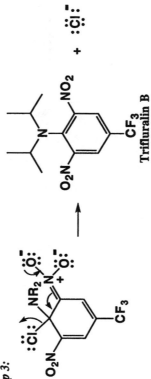

(−H⁺)

Step 3:

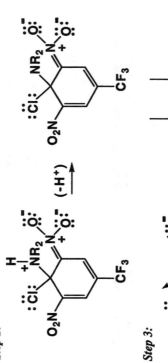

Trifluralin B

+ :Cl:⁻

Problem 21.29 A problem in dyeing fabrics is the degree of fastness of the dye to the fabric. Many of the early dyes were surface dyes; that is, they do not bond to the fabric with the result that they tend to wash off after repeated laundering. Indigo, for example, which gives the blue color to blue jeans, is a surface dye. Color fastness can be obtained by bonding a dye to the fabric. The first such dyes were the so-called reactive dyes, developed in the 1930s for covalent bonding dyes containing -NH₂ groups to cotton, wool, and silk fabrics. In the first stage of the first developed method for reactive dyeing, the dye is treated with cyanuryl chloride, which links the two through the amino group of the dye. The remaining chlorines are then displaced by -OH group groups of cotton (cellulose) or -NH₂ groups of wool or silk (both proteins).

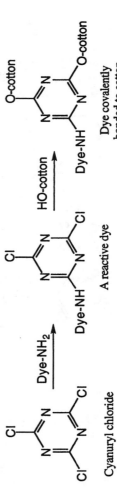

Propose a mechanism for the displacement of a chlorine from cyanuryl chloride by (a) the NH₂ group of a dye and (b) by an -OH group of cotton.

In each reaction, the chlorine atoms are being replaced by nucleophiles, either the amine groups of the dye or the hydroxyl groups of the cotton fibers. The mechanism will be nucleophilic aromatic substitution, analogous to that presented in problem 21.28 above, involving a Meissenheimer complex intermediate. Note that the chloro groups here can be considered the nitrogen analog of an acid chloride.

Syntheses

Problem 21.30 Show how to convert toluene to these compounds.

(a)

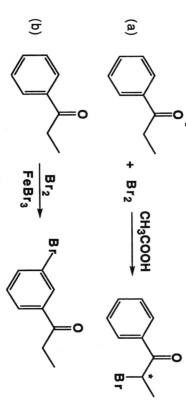

CH₃ →(Br₂, Light or heat)→ CH₂Br + HBr

(b)

CH₃ →(Br₂, FeBr₃)→ Br—⟨⟩—CH₃ + HBr

Problem 21.31 Show how to prepare compounds (a) and (b) from 1-phenyl-1-propanone.

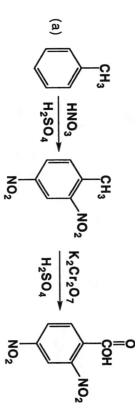

1-Phenyl-1-propanone
(Propiophenone)

(a)

(b)

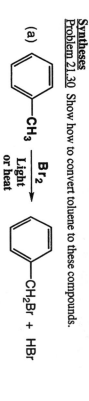

Product (a) is the result of acid catalyzed α-halogenation of a ketone (see chapter 16), while product (b) is the result of electrophilic aromatic substitution. In part (a) a stereocenter is created, so the a racemic mixture of enantiomers will be produced.

(a)

+ Br₂ →(CH₃COOH)→

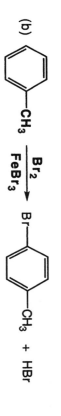

(b)

→(Br₂, FeBr₃)→

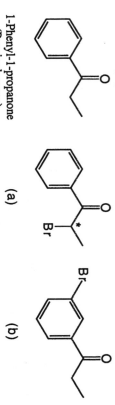

Problem 21.32 Show how to convert toluene to (a) 2,4-dinitrobenzoic acid and (b) 3,5-dinitrobenzoic acid.

Methyl is ortho/para directing. Therefore, toluene can be nitrated twice then oxidized with chromic acid to convert the methyl group into the carboxyl group.

(a)

CH₃ →(HNO₃, H₂SO₄)→ CH₃ (NO₂, NO₂) →(K₂Cr₂O₇, H₂SO₄)→ COOH (NO₂, NO₂)

(b)

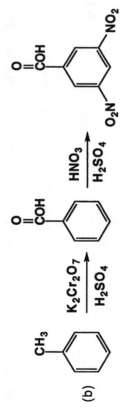

The reaction sequence is very similar to the last one, except now the order of the reactions is reversed because the carboxylic acid group is a meta director.

Problem 21.33 Show reagents and conditions to bring about the following conversions.

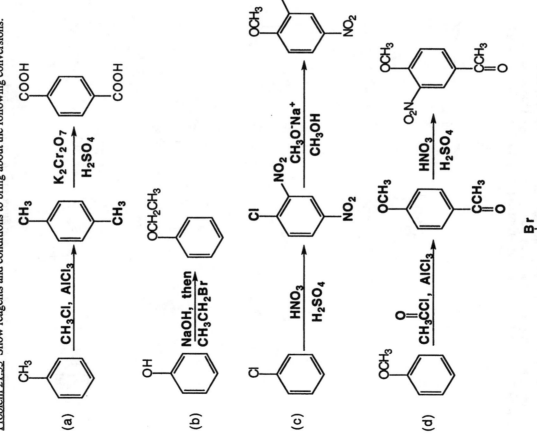

(a)

(b)

(c)

(d)

(e)

480

Solutions

Chapter 21: Aromatics II

Problem 21.34 Propose a synthesis of triphenylmethane from benzene as the only source of aromatic rings, and any other necessary reagents.

Reaction of 3 moles benzene with 1 mole trichloromethane (chloroform) will give triphenylmethane.

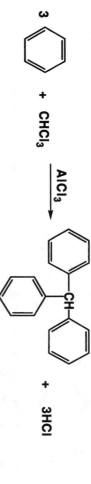

3 [benzene] + CHCl$_3$ $\xrightarrow{\text{AlCl}_3}$ [triphenylmethane] + 3HCl

Problem 21.35 Propose a synthesis for each compound from the given starting material(s). Show all reagents and products in each step of your syntheses.

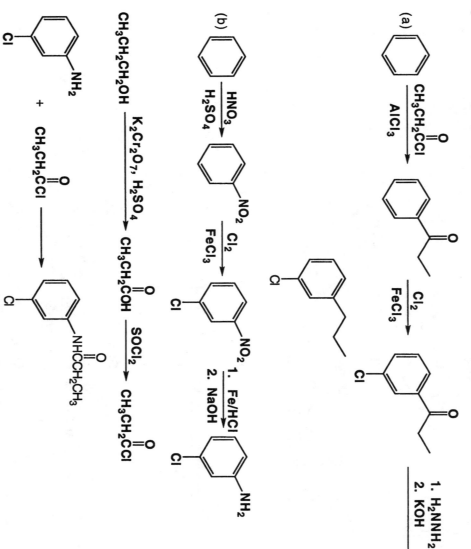

(a) [benzene] $\xrightarrow[\text{AlCl}_3]{\overset{\text{O}}{\text{CH}_3\text{CH}_2\overset{\|}{\text{C}}\text{Cl}}}$ [propiophenone] $\xrightarrow{\text{Cl}_2}{\text{FeCl}_3}$ [product] $\xrightarrow{\text{1. H}_2\text{NNH}_2}{\text{2. KOH}}$

(b) [benzene] $\xrightarrow{\text{HNO}_3}{\text{H}_2\text{SO}_4}$ [nitrobenzene] $\xrightarrow{\text{Cl}_2}{\text{FeCl}_3}$ [product] $\xrightarrow{\text{1. Fe/HCl}}{\text{2. NaOH}}$

CH$_3$CH$_2$CH$_2$OH $\xrightarrow{\text{K}_2\text{Cr}_2\text{O}_7, \text{H}_2\text{SO}_4}$ CH$_3$CH$_2$$\overset{\text{O}}{\overset{\|}{\text{C}}}$OH $\xrightarrow{\text{SOCl}_2}$ CH$_3$CH$_2$$\overset{\text{O}}{\overset{\|}{\text{C}}}$Cl

[3-chloroaniline, Cl ... NH$_2$] + CH$_3$CH$_2$$\overset{\text{O}}{\overset{\|}{\text{C}}}$Cl $\longrightarrow$ [Cl ... NH$\overset{\text{O}}{\overset{\|}{\text{C}}}CH_2CH_3$]

In the above synthesis scheme, the benzene ring is nitrated then chlorinated followed by reduction with Fe/HCl to give the 3-chloroaniline. The 1-propanol is converted to propanoic acid then the acid chloride to get ready for the final step in which the 3-chloroaniline is treated with propanoyl chloride to give the desired amide product.

Problem 21.36 The first widely used herbicide for the control of weeds was 2,4-dichlorophenoxyacetic acid (2,4-D). Show how this compound might be synthesized from phenol and chloroacetic acid by way of the given chlorinated phenol intermediate.

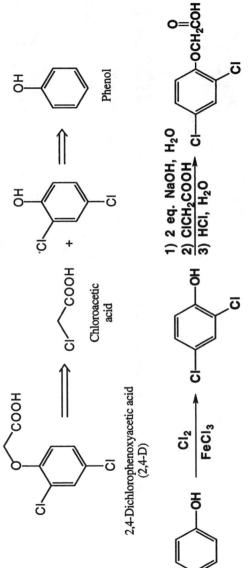

2,4-Dichlorophenoxyacetic acid
(2,4-D)

Chloroacetic
acid

Phenol

Problem 21.37 Phenol is the starting material for the synthesis of 2,3,4,5,6-pentachlorophenol, known alternatively as pentachlorophenol or more simply as penta. At one time, penta was widely used as a wood preservative for decks, siding, and outdoor wood furniture. Draw the structural formula for pentachlorophenol, and describe its synthesis from phenol.

"Penta" is synthesized industrially from phenol. Because the -OH group is strongly activating, reaction of phenol with the first few atoms of chlorine occurs readily. However, after three chlorine atoms are on the ring, it is deactivated enough that harsher conditions are usually required to produce the final product. For this reason, the synthesis of 2,3,4,5,6-pentachlorophenol is usually broken up into to two steps.

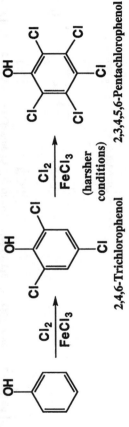

2,4,6-Trichlorophenol

2,3,4,5,6-Pentachlorophenol

Problem 21.38 Starting with benzene, toluene, or phenol as the only sources of aromatic rings, show how to synthesize the following. Assume in all syntheses that mixtures of ortho-para products can be separated into the desired isomer.
(a) 1-Bromo-3-nitrobenzene

Nitro is meta directing; bromine is ortho/para directing. Therefore, to have the two substituents meta to each other, carry out nitration first followed by bromination.

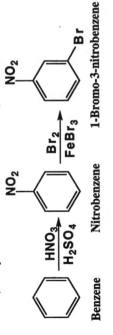

Benzene   Nitrobenzene   1-Bromo-3-nitrobenzene

(b) 1-Bromo-4-nitrobenzene

Reverse the order of steps from part (a). Nitro is meta directing; bromine is ortho-para directing. Therefore, to have the two substituents para to each other, carry out bromination first followed by nitration.

Benzene 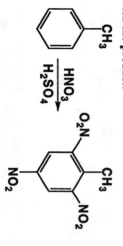 Bromobenzene   1-Bromo-4-nitrobenzene

(c) 2,4,6-Trinitrotoluene (TNT)

The methyl group is ortho/para directing. Therefore, nitrate toluene three successive times then be very careful with the product!

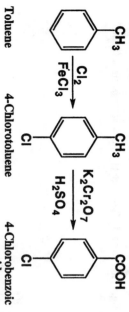

Toluene   2,4,6-Trinitrotoluene

(d) m-Chlorobenzoic acid

The carboxyl group and chlorine atom are meta to each other, an orientation best accomplished by chlorination of benzoic acid (the carboxyl group is meta-directing). Oxidation of toluene with chromic acid gives benzoic acid. Treatment of benzoic acid with chlorine in the presence of ferric chloride gives the desired product.

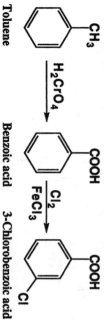

Toluene   Benzoic acid   3-Chlorobenzoic acid

(e) p-Chlorobenzoic acid

Start with toluene. The methyl group is weakly activating and directs chlorination to the ortho/para positions. Separate the desired para isomer and then oxidize the methyl group to a carboxyl group using chromic acid.

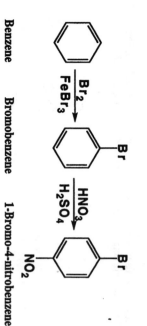

Toluene   4-Chlorotoluene   4-Chlorobenzoic
                                              acid

(f) p-Dichlorobenzene

Treatment of benzene with chlorine in the presence of aluminum chloride gives chlorobenzene. The chlorine atom is ortho/para directing. Treatment with chlorine in the presence of aluminum chloride a second time gives 1,4-dichlorobenzene.

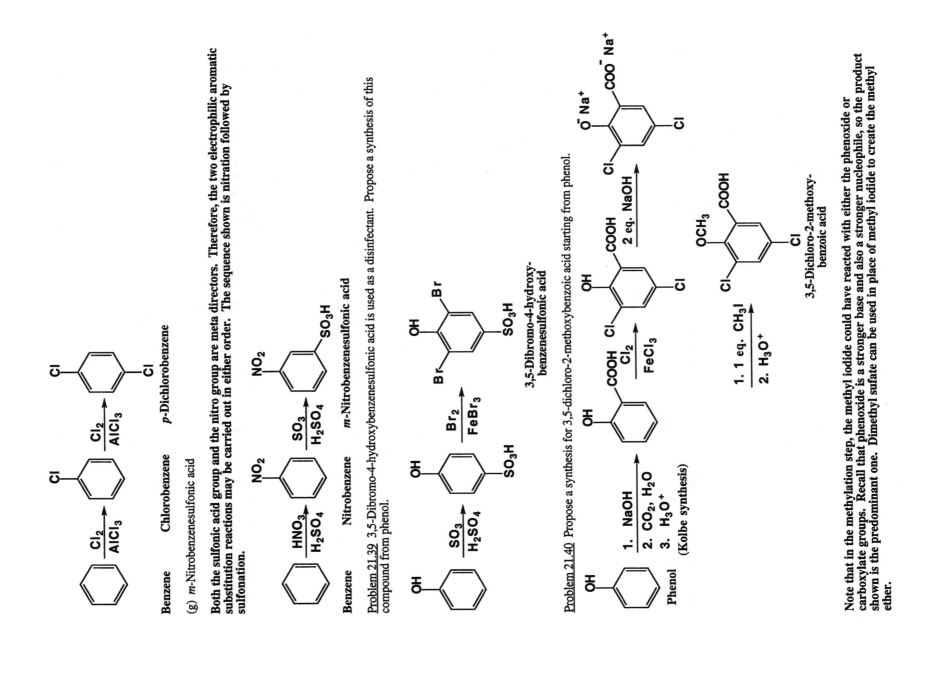

Benzene   Chlorobenzene   *p*-Dichlorobenzene

(g) *m*-Nitrobenzesulfonic acid

Both the sulfonic acid group and the nitro group are meta directors. Therefore, the two electrophilic aromatic substitution reactions may be carried out in either order. The sequence shown is nitration followed by sulfonation.

Benzene   Nitrobenzene   *m*-Nitrobenzenesulfonic acid

Problem 21.39 3,5-Dibromo-4-hydroxybenzenesulfonic acid is used as a disinfectant. Propose a synthesis of this compound from phenol.

3,5-Dibromo-4-hydroxy-benzenesulfonic acid

Problem 21.40 Propose a synthesis for 3,5-dichloro-2-methoxybenzoic acid starting from phenol.

Phenol

3,5-Dichloro-2-methoxy-benzoic acid

Note that in the methylation step, the methyl iodide could have reacted with either the phenoxide or carboxylate groups. Recall that phenoxide is a stronger base and also a stronger nucleophile, so the product shown is the predominant one. Dimethyl sufate can be used in place of methyl iodide to create the methyl ether.

Problem 21.41 The following compound used in perfumery has a violet-like scent. Propose a synthesis of this compound from benzene.

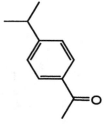

4-Isopropylacetophenone

The isopropyl group is weakly activating and ortho-para directing; the acetyl carbonyl group of the acetyl group is deactivating and meta directing. Therefore, start with benzene, convert it to isopropylbenzene (cumene) and then carry out a Friedel-Crafts acylation using acetyl chloride in the presence of aluminum chloride. Alkylation of benzene can be accomplished using 2-halopropane, 2-propanol, or propene, each in the presence of an appropriate catalyst.

C_6H_6 + CH_3CHCH_3 $\xrightarrow{AlCl_3}$
$\qquad\qquad\quad |$
$\qquad\qquad\quad Cl$

C_6H_6 + CH_3CHCH_3 $\xrightarrow{H_2SO_4}$
$\qquad\qquad\quad |$
$\qquad\qquad\quad OH$

C_6H_6 + $CH_3CH{=}CH_2$ $\xrightarrow{H_3PO_4}$

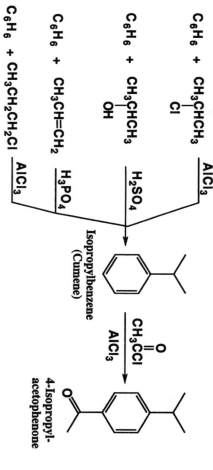

Isopropylbenzene
(Cumene)

$\xrightarrow[\;AlCl_3\;]{\overset{\displaystyle O}{\overset{\|}{CH_3CCl}}}$

4-Isopropyl-
acetophenone

Problem 21.42 Following is the structural formula of musk ambrette, a synthetic musk, essential in perfumes to enhance and retain odor. Propose a synthesis of this compound from m-cresol (3-methylphenol).

C_6H_6 + $CH_3CH_2CH_2Cl$ $\xrightarrow{AlCl_3}$

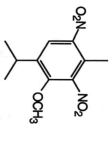

Both methyl and methoxyl groups are ortho/para directing. The methoxyl group is a moderately strong o,p-directing group while the methyl group is only weakly o,p-directing. Introduction of the isopropyl group by acid-catalyzed alkylation gives a mixture of 4-isopropyl-3-methoxytoluene and 2-isopropyl-5-methoxytoluene. Following separation of the desired isomer, nitration both ortho and para to the methoxyl group gives the product.

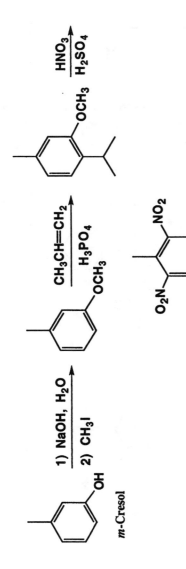

m-Cresol

In the first step, dimethyl sufate can be used in place of methyl iodide to create the methyl ether.

Problem 21.43 Propose a synthesis of this compound starting from toluene and phenol.

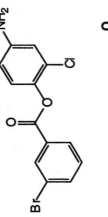

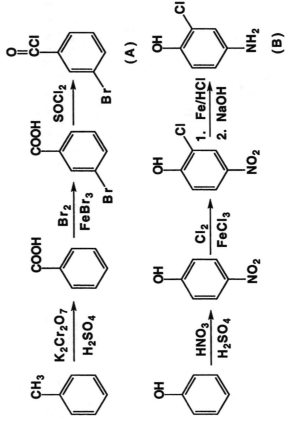

Problem 21.44 When certain aromatic compounds are treated with formaldehyde, CH_2O, and HCl, the CH_2Cl group is introduced onto the ring. This reaction is known as chloromethylation.

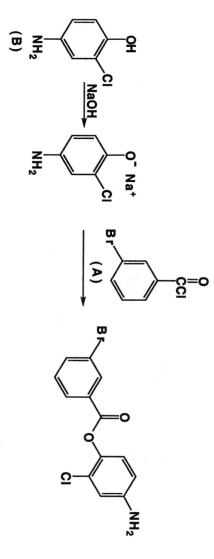

(B)

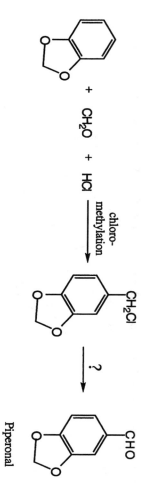

Piperonal

(a) Propose a mechanism for this example of chloromethylation.

The most likely mechanism involves an acid-catalyzed electrophilic aromatic substitution followed by replacement of the resulting -OH group with Cl via a benzyl cation intermediate. The acetal in the bottom of the molecule is likely attacked by the acid, but formaldehyde in the reaction mixture drives the equilibrium to acetal formation.

Step 1:

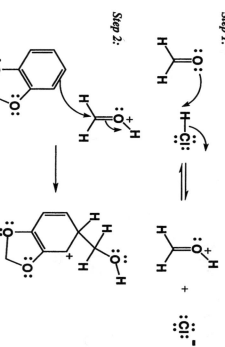

Step 2:

Step 3:

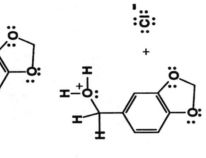

Step 4:

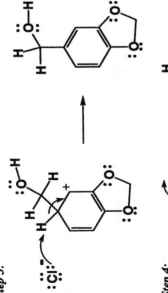

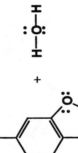

Step 5:

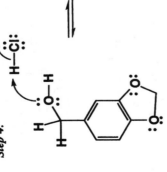

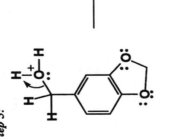

Step 6:

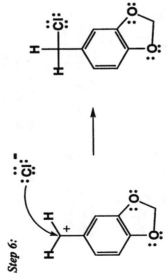

(b) The product of this chloromethylation can be converted to piperonal (see The Merck Index, 12th ed., #7628), which is used in perfumery and in artificial cherry and vanilla flavors. How might the CH$_2$Cl group of the chloromethylation product be converted to a CHO group?

A simple two-step procedure would be to convert the alkyl choride back to the alcohol using NaOH, followed by oxidation with PCC to give the desired aldehyde. Note that the acetal will be stable to these conditions.

Solutions

Problem 21.45 Following is a retrosynthetic analysis for the acaracide (killing mites and ticks) and fungicide dinocap (see The Merck Index, 12th ed., #3340). Given this analysis, propose a synthesis for dinocap from phenol and 1-octene.

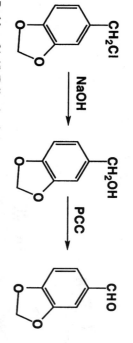

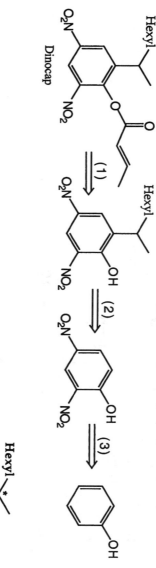

Dinocap

In the above synthesis scheme, the acid catalyzed alkylation step is being carried out on an aromatic ring that contains two nitro groups. Ordinarily, this would render the ring too deactivated to take part in the reaction. However, the presence of the highly activating -OH group adjacent to the site of attack is enough to allow reaction.

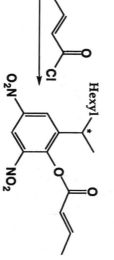

Problem 21.46 Following is the structure of miconazole (see The Merck Index, 12th ed., #6266), an antifungal agent. It is the active ingredient in a number of over-the-counter preparations, including Monistat which is used to treat vaginal yeast infections. One of the compounds needed for the synthesis of miconazole is the trichloro derivative of toluene shown on its right. Show how this derivative can be synthesized from toluene.

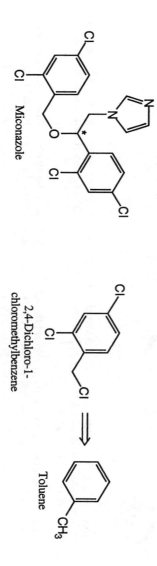

Miconazole

2,4-Dichloro-1-chloromethylbenzene

Toluene

Notice that a chloromethyl group is likely to be weakly deactivating and meta directing due to an inductive effect. Therefore, the best synthetic strategy would be to chlorinate the aromatic ring ortho and para to the methyl group first, then halogenate the benzylic position.

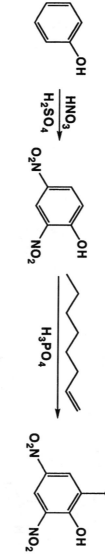

Toluene

$\xrightarrow[\text{FeCl}_3]{\text{Cl}_2}$

$\xrightarrow[\text{heat}]{\text{light or}}$

2,4-Dichloro-1-chloromethylbenzene

Problem 21.47 Following is the structural formula of bupropion, the hydrochloride of which was first marketed in 1985 by Burroughs-Wellcome, now Galaxo-Wellcome, as an antidepressant under the trade name Wellbutrin. During clinical trials, it was discovered that smokers, after one to two weeks on the drug, reported that their craving for tobacco lessened. Further clinical trials confirmed this finding, and the drug was marketed in 1997 under the trade name Zyban as an aid in smoking cessation. Given this retrosynthetic analysis, propose a synthesis for bupropion.

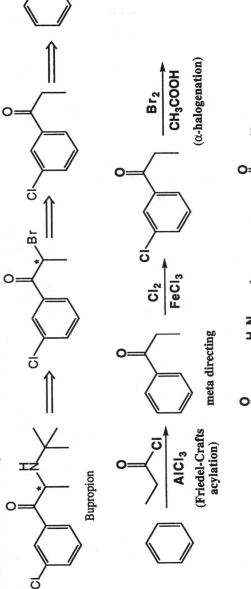

Bupropion

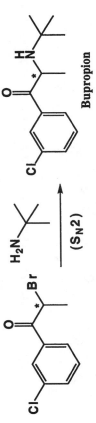

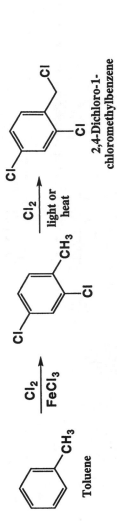

(Friedel-Crafts acylation)

meta directing

$\xrightarrow[\text{CH}_3\text{COOH}]{\text{Br}_2}$

(α-halogenation)

$\xrightarrow[\text{FeCl}_3]{\text{Cl}_2}$

$\xrightarrow[\text{(S}_N2)]{\text{H}_2\text{N}}$

Bupropion

Note that the Friedel-Crafts acylation was carried out first since that produces the desired meta orientation of the chlorine atom and propionyl group. One stereocenter is created in this reaction sequence, so a racemic mixture of enantiomers is produced.

French angelfish *Pomacanthus paru*
Tobago, West Indies

CHAPTER 22
Solutions to the Problems

Problem 22.1 Identify all carbon stereocenters in coniine, nicotine, and cocaine.

(a)

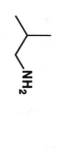

(S)-(+)-Coniine

(b)

(S)-(-)-Nicotine

(c)

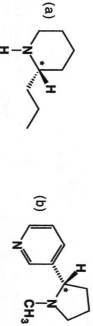

(R)-2-Butanamine

Cocaine

Problem 22.2 Write structural formulas for these amines.
(a) 2-Methyl-1-propanamine   (b) Cyclohexanamine

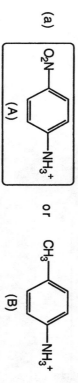

Problem 22.3 Write structural formulas for these amines.
(a) Isobutylamine   (b) Triphenylamine   (c) Diisopropylamine

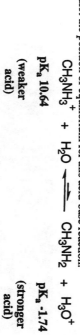

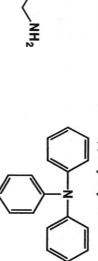

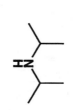

Problem 22.4 Write IUPAC and, where possible, common names for these amines.

(a)

(b)

4-Aminobutanoic acid
(γ-Aminobutyric acid)

(c)

2,2-Dimethylpropanamine
(Neopentylamine)

Problem 22.5 Predict the position of equilibrium for this acid-base reaction.

$$CH_3NH_3^+ + H_2O \rightleftharpoons CH_3NH_2 + H_3O^+$$

pK$_a$ 10.64
(weaker
acid)

pK$_a$ -1.74
(stronger
acid)

Since equilibrium favors formation of the weaker acid, equilibrium will be to the left as shown.

Problem 22.6 Select the stronger acid from each pair of compounds.

(a)

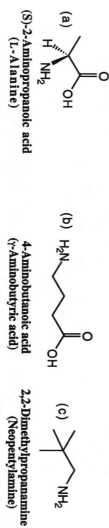

(A)

or

(B)

4-Nitroaniline (pK$_b$ 13.0) is a weaker base than 4-methylaniline (pK$_b$ 8.92). The decreased basicity of 4-nitroaniline is due to the electron-withdrawing effect of the *para* nitro group. Because 4-nitroaniline is the weaker base, its conjugate acid (A) is the stronger acid.

(b)

(C)

or

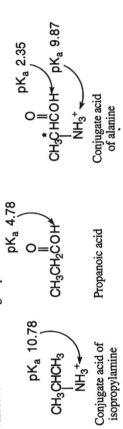

(D)

Pyridine (pK$_b$ 8.75) is a much weaker base than cyclohexanamine (pK$_b$ 3.34). The lone pair of electrons in the sp^2 orbital on the nitrogen atom of pyridine (C) has more s character, so these electrons are less available for bonding to a proton. Because pyridine is a weaker base, its conjugate acid, (C), is the stronger acid

Problem 22.7 Complete each acid-base reaction and name the salt formed.

(a) (CH$_3$CH$_2$)$_3$N + HCl ⟶ **(CH$_3$CH$_2$)$_3$NH$^+$ Cl$^-$**

Triethylammonium chloride

(b)

 + CH$_3$COOH ⟶

Piperidinium acetate

Problem 22.8 Following are structural formulas for propanoic acid and the conjugate acids of isopropylamine and alanine, along with pK$_a$ values for each functional group:

pK$_a$ 10.78

pK$_a$ 4.78

pK$_a$ 2.35

pK$_a$ 9.87

CH$_3$CHCH$_3$
NH$_3^+$

CH$_3$CH$_2$COH
O

CH$_3$CHCOH
O
NH$_3^+$

Conjugate acid
of isopropylamine

Propanoic acid

Conjugate acid
of alanine

(a) How do you account for the fact that the -NH$_3^+$ group of the conjugate acid of alanine is a stronger acid than the -NH$_3^+$ group of the conjugate acid of isopropylamine?

The electron-withdrawing properties of the carboxyl group adjacent to the amine of alanine make the conjugate acid of alanine more acidic than the conjugate acid of isopropylamine.

(b) How do you account for the fact that the -COOH group of the conjugate acid of alanine is a stronger acid than the -COOH group of propanoic acid?

The -NH$_3^+$ group is electron-withdrawing, so the adjacent -COOH group is made more acidic by an inductive effect. This situation is analogous to the electron-withdrawing effects of halogens adjacent to carboxylic acids in molecules such as chloroacetic acid, which has a pK$_a$ of 2.86. In addition, deprotonation of the carboxylic acid function of alanine results in formation of an overall neutral zwitterion. Thus, the carboxylate form of alanine can be thought of as being neutralized by the adjacent positively-charged ammonium ion.

Problem 22.9 In what way(s) might the results of the separation and purification procedure outlined in Example 22.9 be different if:
(a) Aqueous NaOH is used in place of aqueous NaHCO$_3$?

If NaOH is used in place of aqueous NaHCO$_3$, then the phenol will be deprotonated along with the carboxylic acid, so they will be isolated together in fraction A.

(b) The starting mixture contains an aromatic amine, ArNH$_2$, rather than an aliphatic amine, RNH$_2$?

If the starting mixture contains an aromatic amine, ArNH$_2$, rather than an aliphatic amine, RNH$_2$, then the results will be the same. The aromatic amine will still be protonated by the HCl wash, and deprotonated by the NaOH treatment.

Problem 22.10 Show how to bring about each conversion in good yield. In addition to the given starting material, use any other reagents as necessary.

(a)

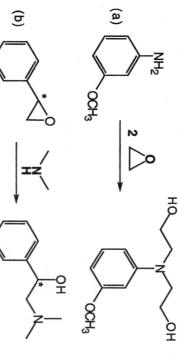

(b)

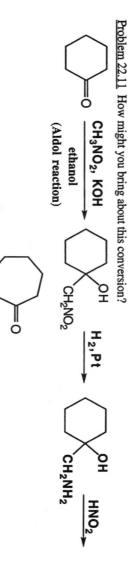

The stereochemistry of the product will be determined by the stereochemistry of the starting material.

Problem 22.11 How might you bring about this conversion?

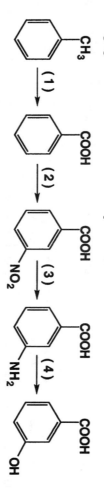

The synthesis begins with an aldol reaction using nitromethane, followed by reduction to give a β-aminoalcohol that then undergoes the ring expansion reaction.

Problem 22.12 Show how to convert toluene to 3-hydroxybenzoic acid using the same set of reactions as in Example 22.12, but changing the order in which some of the steps are carried out.

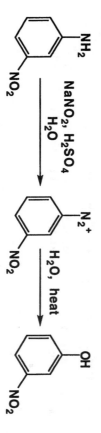

The key to this question is that the methyl group is converted to a meta-directing carboxyl group before the nitration reaction. This leads to the desired product with the hydroxy group in the 3 position.

Step 1: Oxidation at a benzylic carbon (Section 20.6A) can be brought about using chromic acid to give benzoic acid.
Step 2: Nitration of the aromatic ring using HNO₃ in H₂SO₄. The meta-directing carboxyl group gives predominantly the desired 3-nitrobenzoic acid product.
Step 3: Reduction of the nitro group to 3-aminobenzoic acid can be brought about using H₂ in the presence of Ni or other transition metal catalyst. Alternatively, it can be brought about using Zn, Sn, or Fe metal in aqueous HCl.
Step 4: Reaction of the aromatic amine with HNO₂ followed by heating in water gives 3-hydroxybenzoic acid.

Problem 22.13 Starting with 3-nitroaniline, show how to prepare the following compounds.
(a) 3-Nitrophenol

(b) 3-Bromoaniline

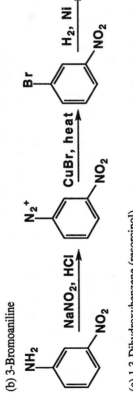

(c) 1,3-Dihydroxybenzene (resorcinol)

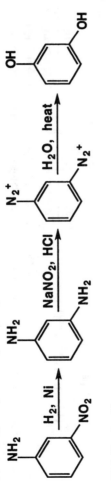

(d) 3-Fluoroaniline

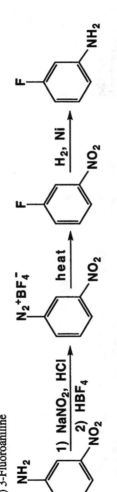

(e) 3-Fluoroaniline

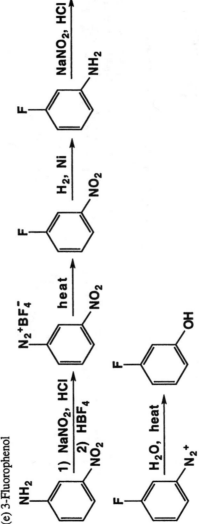

(f) 3-Hydroxybenzonitrile

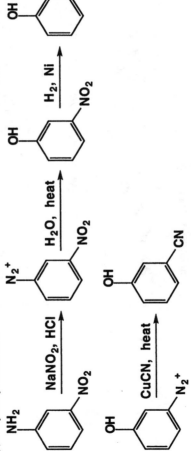

Problem 22.14 The procedure of methylation of amines and thermal decomposition of quaternary ammonium hydroxides was first reported by Hofmann in 1851, but its value as a means of structure determination was not appreciated until 1881 when he published a report of its use in determining the structure of piperidine. Following are the results obtained by Hofmann:

$$C_5H_{11}N \xrightarrow[\substack{\text{2. Ag}_2\text{O, H}_2\text{O} \\ \text{3. heat}}]{\text{1. CH}_3\text{I (excess), K}_2\text{CO}_3} C_7H_{15}N \xrightarrow[\substack{\text{5. Ag}_2\text{O, H}_2\text{O} \\ \text{6. heat}}]{\text{4. CH}_3\text{I (excess), K}_2\text{CO}_3} CH_2=CHCH_2CH=CH_2$$

Piperidine　　　　　　　　　　　　　　　　　　　　(A)　　　　　　　　　　　　　　　1,4-Pentadiene

(a) Show that these results are consistent with the structure of piperidine.

As shown below, the structure of piperidine is consistent with the formulas given, as well as the final product.

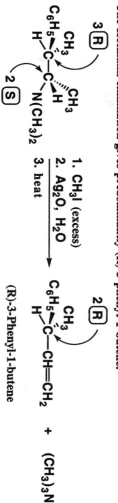

C₅H₁₁N
Piperidine

$$\xrightarrow[\substack{\text{2. Ag}_2\text{O, H}_2\text{O} \\ \text{3. heat}}]{\text{1. CH}_3\text{I (excess)}}$$

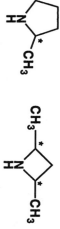

CH₃   CH₃
N
C₇H₁₅N
(A)

$$\xrightarrow[\substack{\text{5. Ag}_2\text{O, H}_2\text{O} \\ \text{6. heat}}]{\text{4. CH}_3\text{I (excess)}}$$

CH₂=CHCH₂CH=CH₂

1,4-Pentadiene

(b) Propose two additional structural formulas (excluding stereoisomers) for $C_5H_{11}N$ that are also consistent with the results obtained by Hofmann.

The following two molecules also have structures that are consistent with the formulas given, as well as the final product. Remember that in Hofmann eliminations, the least substituted alkene is formed predominantly.

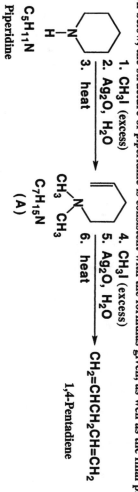

Problem 22.15 In Example 22.15, you considered the product of Cope elimination from the 2R,3S stereoisomer of 2-dimethylamino-3-phenylbutane. What is the product of Hofmann elimination from each of these stereoisomers? What is the product of Cope elimination from each of these stereoisomers?
(a) (2S, 3R) stereoisomer?

The Cope elimination gives predominantly (E)-2-phenyl-2-butene.

$$\xrightarrow[\text{2. heat}]{\text{1. H}_2\text{O}_2}$$

(E)-2-Phenyl-2-butene

+ (CH₃)₂NOH

The Hofmann elimination gives predominantly (R)-3-phenyl-1-butene.

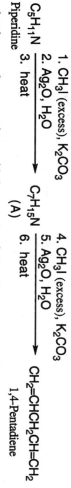

$$\xrightarrow[\substack{\text{2. Ag}_2\text{O, H}_2\text{O} \\ \text{3. heat}}]{\text{1. CH}_3\text{I (excess)}}$$

(R)-3-Phenyl-1-butene

+ (CH₃)₃N

(b) (2S, 3S) stereoisomer?

The Cope elimination gives predominantly (Z)-2-phenyl-2-butene.

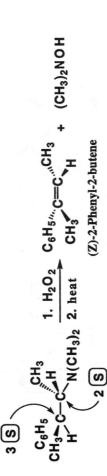

$$\xrightarrow[\text{2. heat}]{\text{1. H}_2\text{O}_2} \quad C_6H_5\cdots C=C\cdots CH_3 \quad + \quad (CH_3)_2NOH$$

(Z)-2-Phenyl-2-butene

The Hofmann elimination gives predominantly (S)-3-phenyl-1-butene.

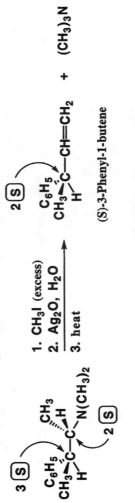

$$\xrightarrow[\text{3. heat}]{\begin{array}{l}\text{1. CH}_3\text{I (excess)}\\ \text{2. Ag}_2\text{O, H}_2\text{O}\end{array}}$$

$$C_6H_5 \quad \overset{}{\underset{H}{C}}-CH=CH_2 \quad + \quad (CH_3)_3N$$

(S)-3-Phenyl-1-butene

PROBLEMS
Structure and Nomenclature
Problem 22.16 Draw structural formulas for these amines and amine derivatives.

(a) *N,N*-Dimethylaniline

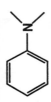

(b) Triethylamine

(c) *tert*-Butylamine

(d) 1,4-Benzenediamine

(e) 4-Aminobutanoic acid

(f) (R)-2-Butanamine

(g) Benzylamine

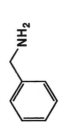

(h) *trans*-2-Aminocyclohexanol

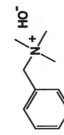

or

(i) 1-Phenyl-2-propanamine
(amphetamine)

(j) Lithium diisopropylamide (LDA)

(k) Benzyltrimethylammonium hydroxide (Triton B)

Problem 22.17 Give an acceptable name for these compounds.

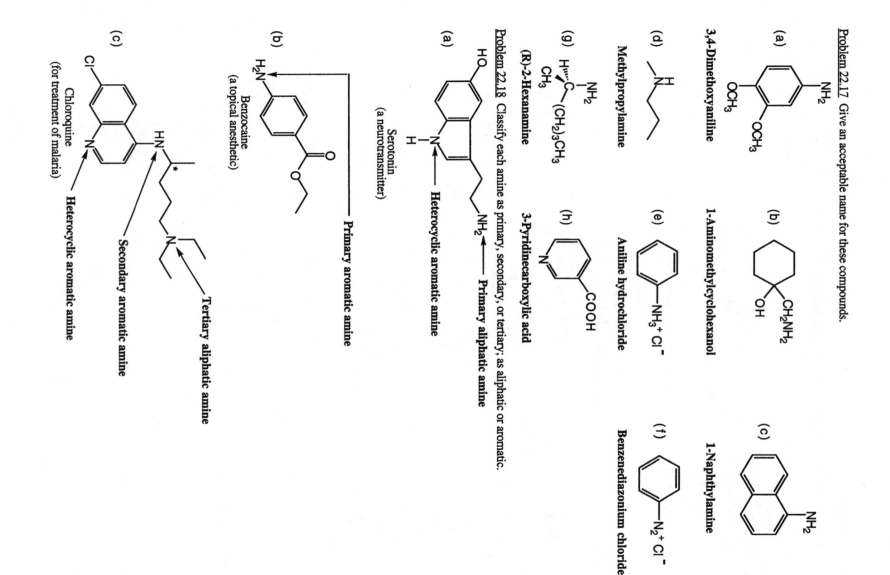

(a)

3,4-Dimethoxyaniline

(b)

1-Aminomethylcyclohexanol

(c)

1-Naphthylamine

(d)

Methylpropylamine

(e)

Aniline hydrochloride

(f)

Benzenediazonium chloride

(g)

(R)-2-Hexanamine

(h)

3-Pyridinecarboxylic acid

Problem 22.18 Classify each amine as primary, secondary, or tertiary; as aliphatic or aromatic.

(a)

Serotonin
(a neurotransmitter)

Heterocyclic aromatic amine

Primary aliphatic amine

(b)

Benzocaine
(a topical anesthetic)

Primary aromatic amine

(c)

Chloroquine
(for treatment of malaria)

Heterocyclic aromatic amine

Secondary aromatic amine

Tertiary aliphatic amine

Problem 22.19 Epinephrine is a hormone secreted by the adrenal medulla. Among its actions, it is a bronchodilator. Albuterol, sold under several trade names, including Proventil and Salbumol, is one of the most effective and widely prescribed antiasthma drugs (see The Merck Index, 12th ed., #217). The R enantiomer of albuterol is 68 times more effective in the treatment of asthma than the S enantiomer.

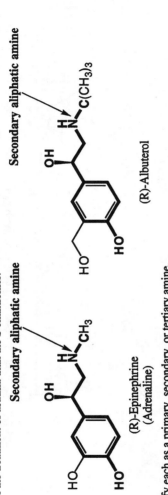

(R)-Epinephrine
(Adrenaline)

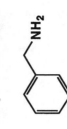

(R)-Albuterol

(a) Classify each as a primary, secondary, or tertiary amine.
(b) Compare the similarities and differences between their structural formulas.

The parts of the molecules that are identical are indicated in bold on the above structures. As far as differences are concerned, epinephrine possesses a second hydroxyl group on the aromatic ring and a methyl group on the amine, while (R)-albuterol has a hydroxymethyl group on the ring and a *tert*-butyl group on the amine.

Problem 22.20 Draw the structural formula for a compound of the given molecular formula.
a) A 2° arylamine, C₇H₉N   (b) A 3° arylamine, C₈H₁₁N   (c) A 1° aliphatic amine, C₇H₉N

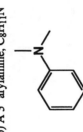

(d) A chiral 1° amine, C₄H₁₁N   (e) A 3° heterocyclic amine, C₆H₁₁N   (f) A trisubstituted 1° arylamine, C₉H₁₃N

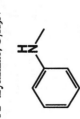

(Other isomers are possible)

(g) A chiral quaternary ammonium salt, C₆H₁₆NCl

There are two answers to this question.

Problem 22.21 Morphine and its *O*-methylated derivative codeine are among the most effective pain killers known. However, they possess two serious drawbacks: they are addictive, and repeated use induces a tolerance to the drug. Increasingly larger doses become necessary; these doses can lead to respiratory arrest. Many morphine analogs have been prepared in an effort to find drugs that are equally effective as pain killers but that have less risk of physical dependence and potential for abuse. Following are several of these. With each name is given its number in The Merck Index.

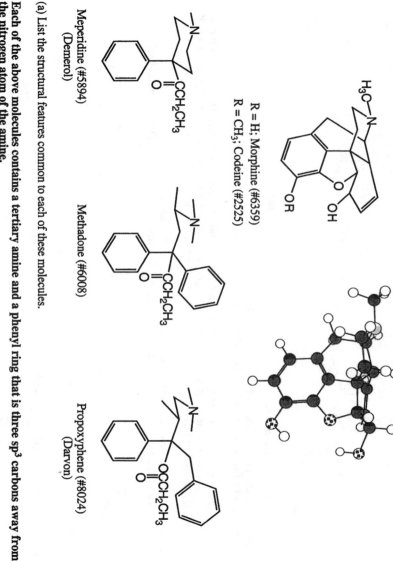

R = H; Morphine (#6359)
R = CH₃; Codeine (#2525)

Meperidine (#5894)
(Demerol)

Methadone (#6008)

Propoxyphene (#8024)
(Darvon)

(a) List the structural features common to each of these molecules.

Each of the above molecules contains a tertiary amine and a phenyl ring that is three sp³ carbons away from the nitrogen atom of the amine.

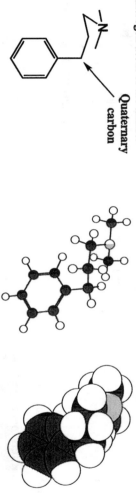

Quaternary
carbon

(b) The Beckett-Casy rules are a set of empirical rules to predict the structure of molecules that bind to morphine receptors and act as analgesics. According to these rules, to provide an effective morphine-like analgesia, a molecule must have (1) an aromatic ring attached to (2) a quaternary carbon and (3) a nitrogen at a distance equal to two carbon-carbon single bond lengths from the quaternary center. Show that these structural requirements are present in this problem.

By inspection of the structures, it can be seen that all of these three structural requirements are present in the molecules mentioned in this problem.

Spectroscopy
Problem 22.22 Account for the formation of the base peaks in these mass spectra.
(a) Isobutylmethylamine, *m/z* 44

Isobutylmethylamine

HN⁺=CH₂
CH₃
m/z = 44

As shown above, the peak at *m/z* = 44 is the result of the characteristic β-cleavage reaction often observed as the base peak in the mass spectra of amines.

(b) Diethylamine, m/z 58

Diethylamine

m/z = 58

As shown above, the peak at *m/z* = 58 is the result of the characteristic β–cleavage reaction often observed as the base peak in the mass spectra of amines.

<u>Problem 22.23</u> Propose a structural formula for compound (A), molecular formula C_5H_{13} N, given its IR and ^{1}H-NMR spectra.

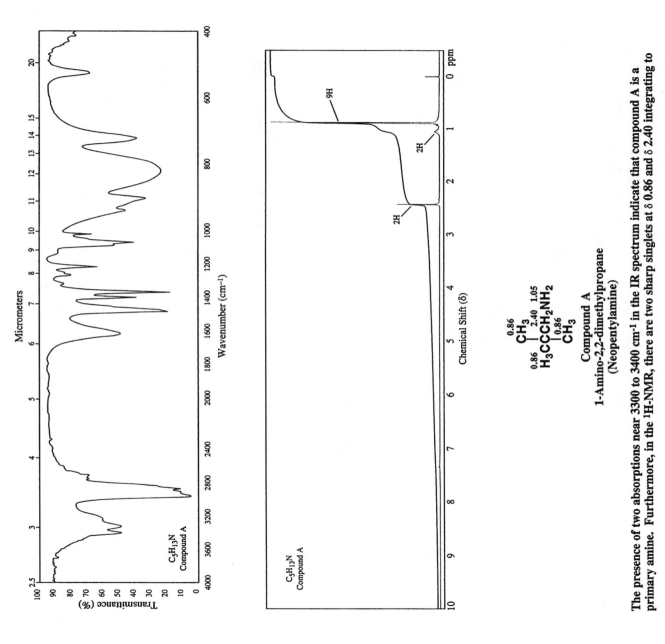

Compound A
1-Amino-2,2-dimethylpropane
(Neopentylamine)

The presence of two absorptions near 3300 to 3400 cm⁻¹ in the IR spectrum indicate that compound A is a primary amine. Furthermore, in the ^{1}H-NMR, there are two sharp singlets at δ 0.86 and δ 2.40 integrating to

9H and 2H, respectively. The above structure is consistent with the index of hydrogen deficiency of zero, since there are no rings or pi bonds.

Basicity of Amines

Problem 22.24 Select the stronger base from each pair of compounds.

(a)

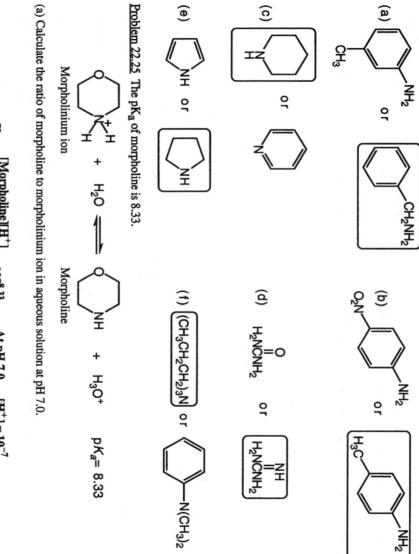

(b)

(c)

(d)

(e)

(f)

Problem 22.25 The pKₐ of morpholine is 8.33.

(a) Calculate the ratio of morpholine to morpholinium ion in aqueous solution at pH 7.0.

$$K_a = \frac{[\text{Morpholine}][H^+]}{[\text{Morpholinium Ion}]} = 10^{-8.33} \quad \text{At pH 7.0} \quad [H^+] = 10^{-7}$$

$$\frac{[\text{Morpholine}]}{[\text{Morpholinium Ion}]} = \frac{10^{-8.33}}{[H^+]} = \frac{10^{-8.33}}{10^{-7.0}} = 10^{-1.33} = \boxed{0.047}$$

(b) At what pH are the concentrations of morpholine and morpholinium ion equal?

The concentrations of morpholine and morpholinium ion will be equal when the pKₐ is equal to the pH, that is, at pH 8.33.

Problem 22.26 Which of the two nitrogens in pyridoxamine (a form of vitamin B₆) is the stronger base? Explain your reasoning.

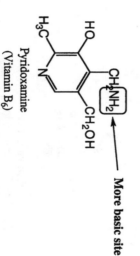

Pyridoxamine
(Vitamin B₆)

→ More basic site

The nitrogen atom of the primary amine is more basic than the pyridine nitrogen atom. This is because the primary amine nitrogen atom is sp³ hybridized, while the pyridine nitrogen is sp² hybridized. The sp² hybridized nitrogen atom has a greater percentage s character, so the electrons are held closer to the nucleus and are less available for interactions with protons.

Problem 22.27 Epibatidine (see The Merck Index, 12th ed., #3647), a colorless oil isolated from the skin of the Ecuadorian poison frog *Epipedobates tricolor* has several times the analgesic potency of morphine. It is the first chlorine-containing, nonopioid (nonmorphine-like in structure) analgesic ever isolated from a natural source.

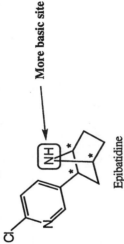

Epibatidine

(a) Which of the two nitrogen atoms of epibatidine is the more basic?

The nitrogen atom of the secondary amine is more basic than the pyridine nitrogen atom. This is because the secondary amine nitrogen atom is sp³ hybridized, while the pyridine nitrogen is sp² hybridized. The sp² hybridized nitrogen atom has a greater percentage s character, so the electrons are held tighter and are less available for interactions with protons.

(b) Mark all stereocenters in this molecule.

The three stereocenters are marked with an asterisk (*).

Problem 22.28 Aniline (conjugate acid pK$_a$ 4.63) is a considerably stronger base than diphenylamine (pK$_a$ 0.79). Account for these marked differences.

$C_6H_5NH_2$ $(C_6H_5)_2NH$

Aniline   Diphenylamine

pK$_a$ 4.63   pK$_a$ 0.79

Conjugate acid pK$_a$

Diphenylamine is a weaker base because the nitrogen lone pair is delocalized by interaction with the pi system of both aromatic rings, as opposed to aniline in which the lone pair on nitrogen is only delocalized by interaction with a single aromatic ring. This delocalization (resonance stabilization) cannot take place when the amine is protonated. Thus, the diphenylammonium ion is a stronger acid because loss of the proton results in greater resonance stabilization (delocalization into two aromatic rings instead of one).

Problem 22.29 Complete the following acid-base reactions and predict the direction of equilibrium (to the right or to the left) for each. Justify your prediction by citing values of pK$_a$ for the stronger and weaker acid in each equilibrium. For values of acid ionization constants, consult Table 22.2 (Acid Strengths, pK$_a$, of the Conjugate Acids of Selected Amines), and Appendix 2 (Acid Ionization Constants for the Major Classes of Organic Acids). Where no ionization constants are given, make the best estimate from the information given in the reference tables and sections.

In all cases, the equilibrium favors formation of the weaker acid (higher pK$_a$) and weaker base (higher pK$_b$). Recall that pK$_a$ + pK$_b$ = 14 for any conjugate acid-base pair.

(a)

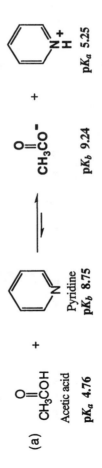

Acetic acid   Pyridine   pK$_b$ 9.24   pK$_a$ 5.25

pK$_a$ 4.76   pK$_b$ 8.75

Equilibrium lies to the right, since the acetate anion and pyridinium ion are the weaker acid and base, respectively.

(b)

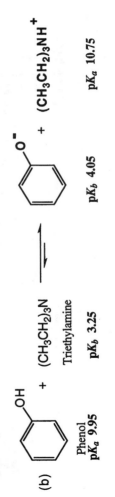

Phenol   Triethylamine   pK$_b$ 4.05   pK$_a$ 10.75

pK$_a$ 9.95   pK$_b$ 3.25

Equilibrium lies to the right, since the phenoxide anion and the triethylammonium species are the weaker base and acid, respectively.

(c) PhC≡CH + NH$_3$ ⇌ PhC≡C$^-$ + NH$_4^+$

Phenylacetylene   Ammonia
pK$_a$ ~25   pK$_b$ 4.74   pK$_b$ ~11   pK$_a$ 9.26

Equilibrium lies to the left, since the alkyne and ammonia are the weaker acid and base, respectively.

(d) PhC≡CH + [(CH$_3$)$_2$CH]$_2$N$^-$ Li$^+$ ⇌ PhC≡C$^-$ Li$^+$ + [(CH$_3$)$_2$CH]$_2$NH

Phenylacetylene   Lithium
                          diisopropylamide
                          (LDA)

pK$_a$ ~25   pK$_b$ ~ -22   pK$_b$ ~11   pK$_a$ ~40

Equilibrium lies to the right, since the alkyne anion and the amine are the weaker base and acid, respectively.

(e) PhCO$^-$Na$^+$ + (CH$_3$CH$_2$)$_3$NH$^+$Cl$^-$ ⇌ PhCOH + (CH$_3$CH$_2$)$_3$N + NaCl

(with C=O shown on PhCO and PhCOH)

Sodium   Triethylammonium
benzoate   chloride
pK$_b$ 9.81   pK$_a$ 10.75   pK$_a$ 4.19   pK$_b$ 3.25

Equilibrium lies to the left, since the carboxylate anion and the triethylammonium ion are the weaker base and acid, respectively.

(f) PhCH$_2$CHNH$_2$ + CH$_3$CHCOH ⇌ PhCH$_2$CHNH$_3^+$ + CH$_3$CHCO$^-$
      |CH$_3$ |HO |=O |CH$_3$(+) |HO |=O

1-Phenyl-2-propanamine   2-Hydroxy-
(Amphetamine)   propanoic acid
                             (Lactic acid)
pK$_b$ ~3   pK$_a$ 3.08   pK$_a$ ~11   pK$_b$ 9.92

Equilibrium lies to the right, since the 1-phenyl-2-propanammonium ion and the carboxylate ion are the weaker acid and base, respectively.

(g) PhCH$_2$CHNH$_3^+$ Cl$^-$ + NaHCO$_3$ ⇌ PhCH$_2$CHNH$_2$ + H$_2$CO$_3$ + Na$^+$ Cl$^-$
      |CH$_3$ |CH$_3$

Amphetamine   Sodium
hydrochloride   bicarbonate
pK$_a$ ~11   pK$_b$ 7.63   pK$_b$ ~3   pK$_a$ 6.36

Equilibrium lies to the left, since the amphetamine ammonium ion and the bicarbonate ion are the weaker acid and base, respectively.

(h) 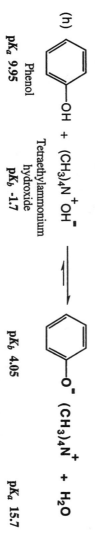 + (CH$_3$)$_4$N$^+$ OH$^-$ ⇌ 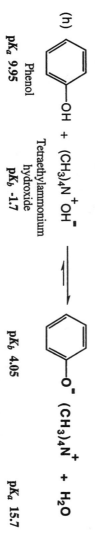O$^-$ (CH$_3$)$_4$N$^+$ + H$_2$O

Phenol   Tetraethylammonium
                    hydroxide
pK$_a$ 9.95   pK$_b$ -1.7   pK$_a$ 15.7

Equilibrium lies to the right, since phenoxide and water are the weaker base and acid, respectively.

Problem 22.30 Quinuclidine and triethylamine are both tertiary amines. Quinuclidine, however, is a considerably stronger base than triethylamine. Stated alternatively, the conjugate acid of quinuclidine is a considerably weaker acid than the conjugate acid of triethylamine. Propose an explanation for these differences in acidity/basicity.

Quinuclidine
(pK_a 10.6)

Triethylamine
(pK_a 8.6)

The protonated form of quinuclidine is more compact than protonated triethylamine, since the alkyl groups of quinuclidine are "tied back" allowing it to be solvated better. For this reason, protonated quinuclidine is a weaker acid, so quinuclidine is a stronger base.

Problem 22.31 Suppose that you have a mixture of these three compounds. Devise a chemical procedure based on their relative acidity or basicity to separate and isolate each in pure form.

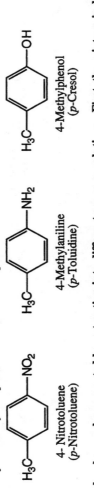

4- Nitrotoluene
(*p*-Nitrotoluene)

4-Methylaniline
(*p*-Toluidine)

4-Methylphenol
(*p*-Cresol)

These molecules can be separated by extraction into different aqueous solutions. First, the mixture is dissolved in an organic solvent such as ether in which all three compounds are soluble. Then, the ether solution is extracted with dilute aqueous HCl. Under these conditions, 4-methylaniline (a weak base) is converted to its protonated form and dissolves in the aqueous solution. The aqueous solution is separated, treated with dilute NaOH, the water-insoluble 4-methylaniline separates, and is recovered. The ether solution containing the other two components is then treated with dilute aqueous NaOH. Under these conditions, 4-methylphenol (a weak acid) is converted to its phenoxide ion and dissolves in the aqueous solution. Acidification of this aqueous solution with dilute HCl forms water-insoluble 4-methylphenol that is then isolated. Evaporation of the remaining ether solution gives the 4-nitrotoluene, which is neither acidic or basic.

Preparation of Amines
Problem 22.32 Propose a synthesis of 1-hexanamine from the following:
(a) A bromoalkane of six carbon atoms

(b) A bromoalkane of five carbon atoms

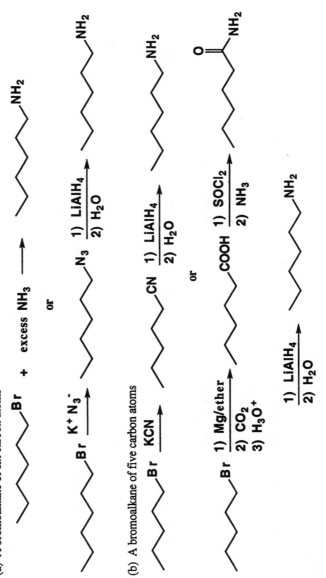

Problem 22.33 Show how to convert each starting material into benzylamine in good yield.

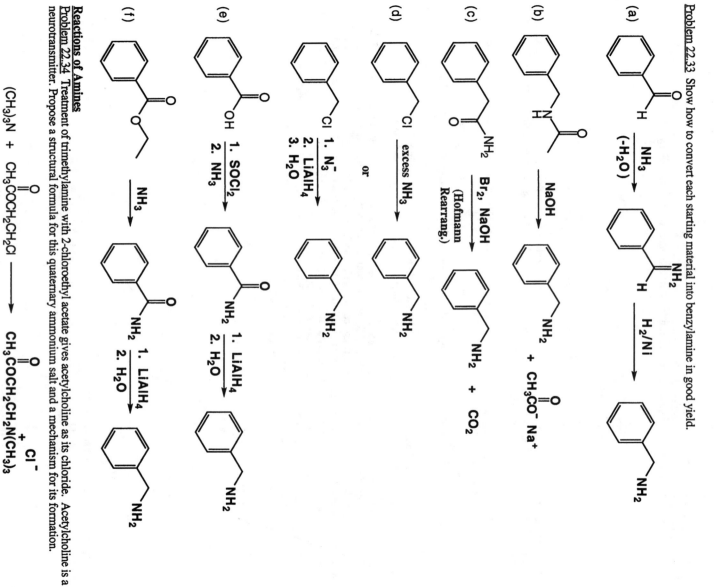

(a)

(b)

(c)

(d)

(e)

(f)

Reactions of Amines

Problem 22.34 Treatment of trimethylamine with 2-chloroethyl acetate gives acetylcholine as its chloride. Acetylcholine is a neurotransmitter. Propose a structural formula for this quaternary ammonium salt and a mechanism for its formation.

$(CH_3)_3N$ + $CH_3COCH_2CH_2Cl$ $\longrightarrow$ $CH_3COCH_2CH_2N(CH_3)_3$ Cl⁻

$C_7H_{16}ClNO_2$
Acetylcholine chloride

Acetylcholine is formed through an S_N2 reaction between trimethylamine and the alkyl chloride, giving the structure shown.

Problem 22.35 *N*-Nitrosamines by themselves are not significant carcinogens. However, they are activated in the liver by a class of iron-containing enzymes (members of the cytochrome P-450 family). Activation involves the oxidation of a C-H bond next to the amine nitrogen to a C-OH group. Show how this hydroxylation product can be transformed into an alkyldiazonium ion, an active carcinogen, in the presence of an acid catalyst.

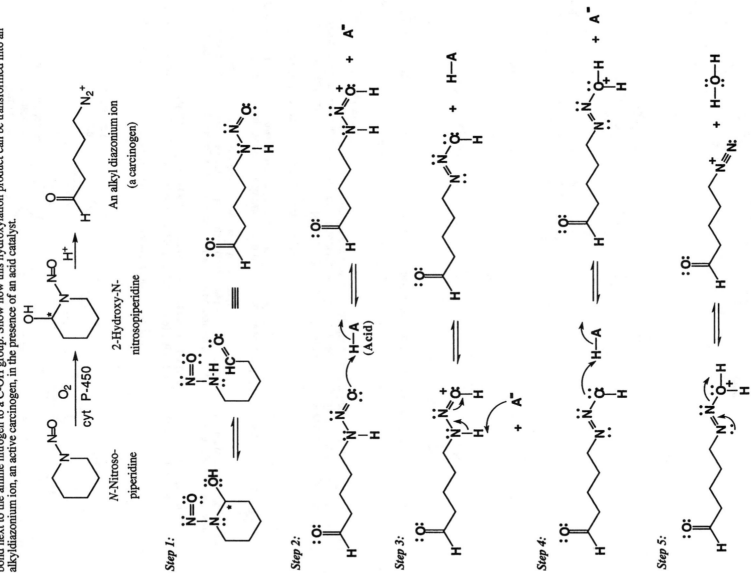

Problem 22.36 Marked similarities exist between the mechanism of nitrous acid deamination of β-aminoalcohols and the pinacol rearrangement. Following are examples of each.

Nitrous acid
deamination of
a β-aminoalcohol:

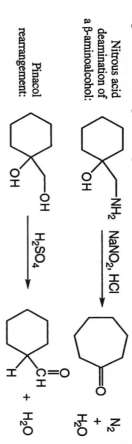

N₂
+
H₂O

Pinacol
rearrangement:

(a) Analyze the mechanism of each rearrangement and list their similarities.

The nitrous acid deamination of a β-aminoalcohol (Section 22.9D) involves formation of a diazonium ion that loses N₂ in concert with a 1,2 shift to create a cation that loses a proton to give the ring expanded ketone.

In the pinacol rearrangement of 1,2 diols (Section 9.7), protonation of one of the alcohol groups leads to departure of water to create a carbocation, followed by migration of an alkyl group to generate a resonance-stabilized cation that loses a proton to give the ketone product.

The similarities between the two mechanisms are that in both cases a 1,2 shift of an alkyl group or hydride produces a cation, that loses a proton to create the final product.

(b) Why does the first reaction give ring expansion but not the second?

In the case of the β-aminoalcohol shown above, the N₂ departs from the primary center not in the ring, requiring a concerted ring expansion for the subsequent alkyl migration step. In the case of the pinacol rearrangement, the tertiary -OH group on the cyclohexane ring departs, precluding the possibility of a ring expanding rearrangement.

(c) Suggest a β-aminoalcohol that would give cyclohexanecarbaldehyde as a product?

1-Hydroxymethylcyclohexanamine would undergo reaction to give cyclohexylcarbaldehyde as shown.

NaNO₂, HCl

1-Hydroxymethyl-
cyclohexanamine

Cyclohexanecarbaldehyde

Problem 22.37 Propose a mechanism for this conversion. Your mechanism must account for the fact that there is retention of configuration at the stereocenter. (S)-Glutamic acid is one of the 20 amino acid building blocks of polypeptides and proteins (Chapter 27).

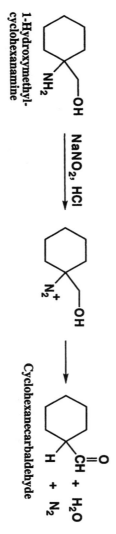

NaNO₂, HCl
0-5°C

+ N₂

(S)-Glutamic acid

Several Steps:

(S)-Glutamic acid

NaNO₂, HCl

The intriguing part of the mechanism is that there is no racemization. This means that it is unlikely that N₂ departs leaving a free cation behind. Such a cation would be sp² hybridized, would be planar and thus would lead to racemization. Therefore, propose the adjacent carbonyl group reacts to form a short-lived three membered ring intermediate that then reacts with the other carboxyl group to make the more stable, five-membered ring intermediate. The stereocenter undergoes double inversion, resulting in net retention of stereochemistry.

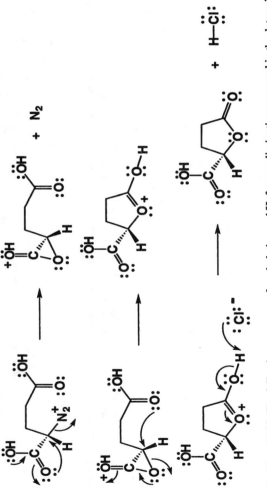

<u>Problem 22.38</u> The following sequence of methylation and Hofmann elimination was used in the determination of the structure of this bicyclic amine.

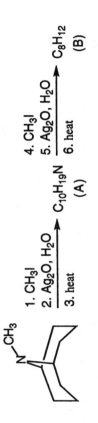

(a) Propose structural formulas for compounds (A) and (B).

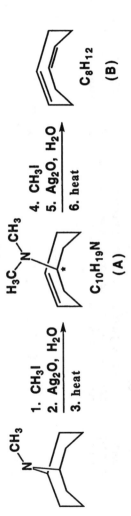

(b) Suppose you were given the structural formula of compound (B) does not unambiguously establish the structure of compound (A), because the amine could be attached at either end of what is a double bond in compound (B). Given this, the starting bicyclic amine cannot be unambiguously identified either, since two different structures are possible. One structure contains the bridging nitrogen atom in a [3.3.1] ring system (shown above), and the other structure contains the bridging nitrogen atom in in a [4.2.1] ring system.

Suppose you were given the structural formula of compound B but only the molecular formulas for compound A and the starting bicyclic amine. Given this information, is it possible, working backward, to arrive at an unambiguous structural formula for compound A? For the bicyclic amine?

Problem 22.39 Propose a structural formula for compound A, $C_{10}H_{16}$, and account for its formation.

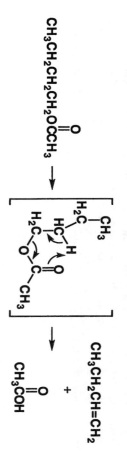

1. CH_3I, 2 moles → 2. H_2O_2 → 3. heat → Compound A $C_{10}H_{16}$

Note how in the following example of a Cope elimination, the syn stereoselectivity of the transition state allows formation of the more substituted alkene, as shown. Note that since the starting material was chiral and a single enantiomer, only a single enantiomer of the product will be created in the reaction sequence.

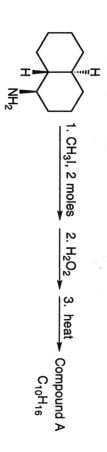

1. CH_3I, 2 moles → 2. H_2O_2 → 3. heat →

Problem 22.40 An amine of unknown structure contains one nitrogen and nine carbon atoms. The ^{13}C-NMR spectrum shows only five signals, all between 20 and 60 ppm. Three cycles of Hofmann elimination sequence [(1) CH_3I; (2) Ag_2O, H_2O; (3) heat] give trimethylamine and 1,4,8-nonatriene. Propose a structural formula for the amine.

The bicyclic amine shown below is the only structure that would explain the five signals in the ^{13}C spectrum as well as the location of the double bonds in 1,4,8-nonatriene.

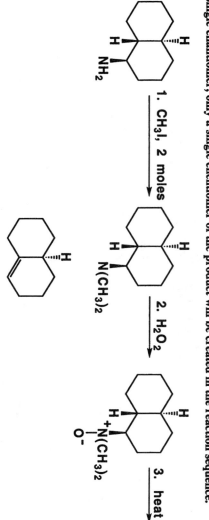

Hofmann
elimination
→

1,4,8-Nonatriene + N(CH₃)₃
Trimethylamine

$C_{10}H_{16}$

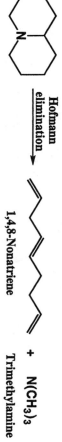

Problem 22.41 The Cope elimination of tertiary amine N-oxides involves a planar transition state and cyclic redistribution of (4n + 2) electrons. The pyrolysis of acetic esters to give an alkene and acetic acid is also thought to involve a planar transition state and cyclic redistribution of (4n + 2) electrons. Propose a mechanism for pyrolysis of the following ester.

$CH_3CH_2CH_2CH_2OCCH_3$ —500°C→ $CH_3CH_2CH=CH_2$ + CH_3COH
Butyl acetate   1-Butene   Acetic acid

A reasonable transition state structure including the likely flow of electrons is shown below:

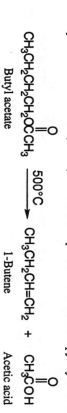

$CH_3CH_2CH_2CH_2OCCH_3$ → → $CH_3CH_2CH=CH_2$ + CH_3COH

Synthesis

Problem 22.42 Propose steps for the following conversions using a reaction of a diazonium salt in at least one step of each conversion.

(a) Toluene to 4-methylphenol (*p*-cresol)

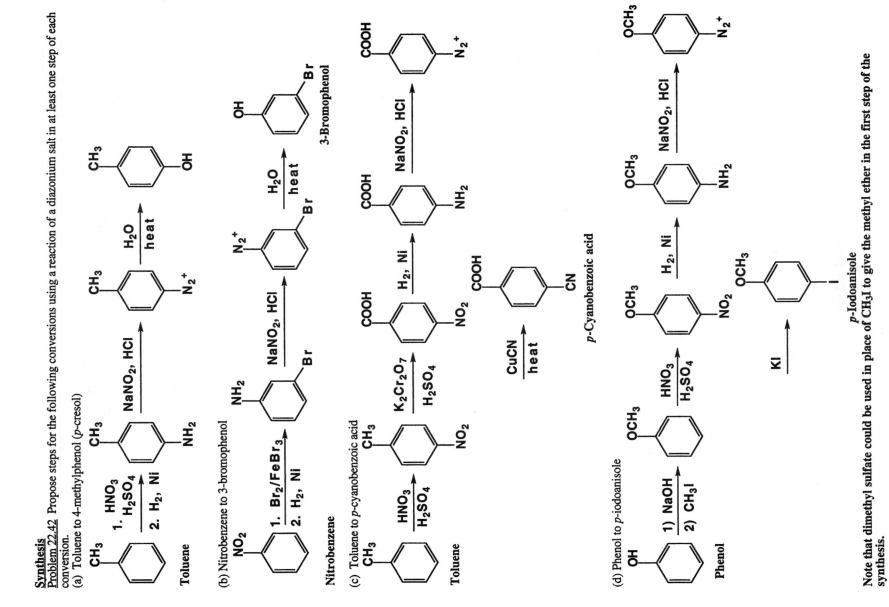

(b) Nitrobenzene to 3-bromophenol

(c) Toluene to *p*-cyanobenzoic acid

(d) Phenol to *p*-iodoanisole

Note that dimethyl sulfate could be used in place of CH₃I to give the methyl ether in the first step of the synthesis.

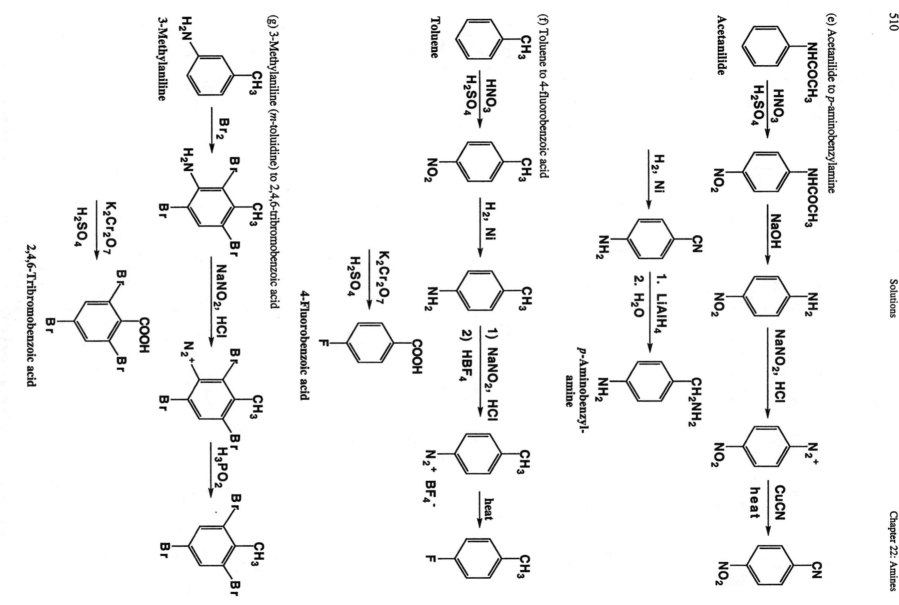

(e) Acetanilide to *p*-aminobenzylamine

Acetanilide

p-Aminobenzyl-amine

(f) Toluene to 4-fluorobenzoic acid

Toluene

4-Fluorobenzoic acid

(g) 3-Methylaniline (*m*-toluidine) to 2,4,6-tribromobenzoic acid

3-Methylaniline

2,4,6-Tribromobenzoic acid

Problem 22.43 Starting materials for the synthesis of the herbicide propranil are benzene and propanoic acid. Show how to bring about this synthesis.

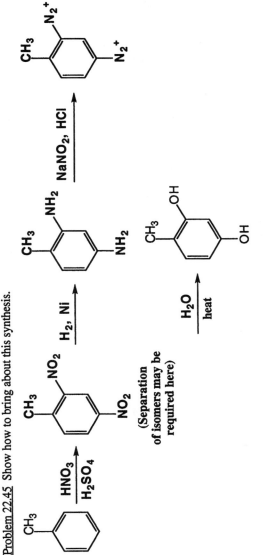

Propanil

The reagents that can be used in each step are listed below:
(1) Aromatic chlorination reaction using Cl₂ and FeCl₃.
(2) Aromatic nitration reaction using HNO₃ and H₂SO₄.
(3) Another aromatic chlorination reaction using Cl₂ and FeCl₃. The new Cl atom is directed to the correct position by the groups already present on the ring.
(4) This reduction can be carried out using H₂ and a transition metal catalyst.
(5) Propanoic acid is first converted to the acid chloride by treatment with SOCl₂ and then reacted with the 3,4-dichloroaniline to give the desired propranil.

Problem 22.44 Show how to bring about each step in the following synthesis.

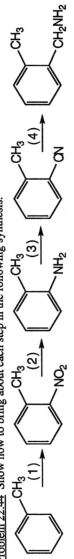

The reagents that can be used in each step are listed below:
(1) Aromatic nitration reaction using HNO₃ and H₂SO₄.
(2) This reduction can be carried out using H₂ and a transition metal catalyst.
(3) This transformation can be accomplished by first turning the amino group into a diazonium salt by reaction with NaNO₂ and HCl, followed by a Sandmeyer reaction using CuCN and heat.
(4) This reduction can be carried out using LiAlH₄ followed by H₂O.

Problem 22.45 Show how to bring about this synthesis.

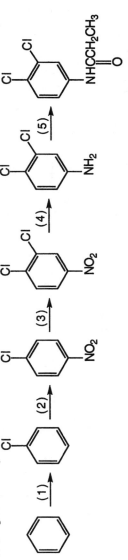

Problem 22.46 Following are steps in a conversion of phenol to 4-methoxybenzylamine. Show how to bring about each step in good yield.

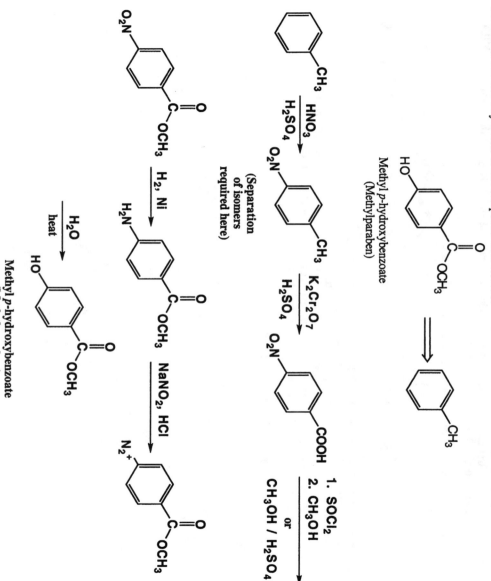

The reagents that can be used in each step are listed below:
(1) Reaction with NaOH to produce the phenoxide, followed by treatment with CH_3I.
(2) Aromatic nitration reaction using HNO_3 and H_2SO_4.
(3) This reduction can be carried out using H_2 and a transition metal catalyst.
(4) This transformation can be accomplished by first turning the amino group into a diazonium salt by reaction with $NaNO_2$ and HCl, followed by a Sandmeyer reaction using CuCN and heat.
(5) This reduction can be carried out using (1) H_2 and a transition metal or (2) $LiAlH_4$ followed by H_2O.

Problem 22.47 Methylparaben (see The Merck Index, 12th ed. #6182) is used as a preservative in foods, beverages, and cosmetics. Provide a synthesis of this compound from toluene.

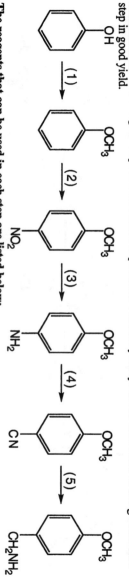

Problem 22.48 Show how to synthesize the following tertiary amine from benzene and any necessary reagents.

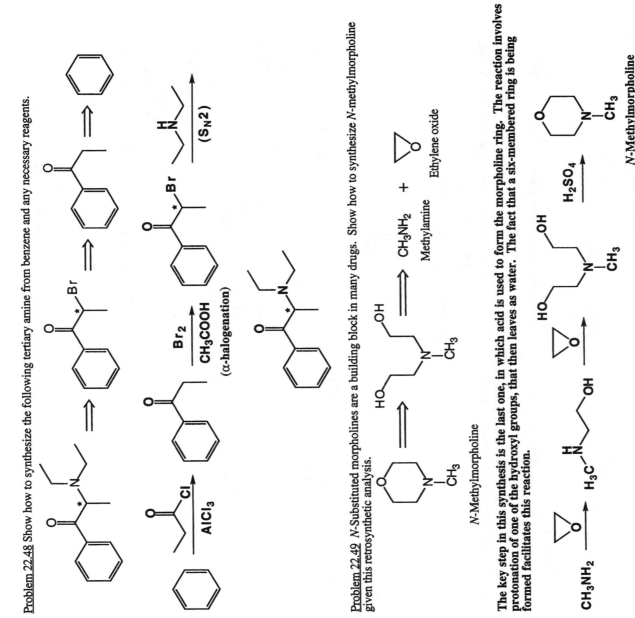

Problem 22.49 *N*-Substituted morpholines are a building block in many drugs. Show how to synthesize *N*-methylmorpholine given this retrosynthetic analysis.

The key step in this synthesis is the last one, in which acid is used to form the morpholine ring. The reaction involves protonation of one of the hydroxyl groups, that then leaves as water. The fact that a six-membered ring is being formed facilitates this reaction.

Problem 22.50 Propose a synthesis for the systemic agricultural fungicide tridemorph (see The Merck Index, 12th ed., #9793) from dodecanoic acid (lauric acid) and propene.

There are two keys to this synthesis. The first is to recognize that the morpholine ring is constructed from an acid-catalyzed reaction between two alcohol groups analagous to that used in problem 22.49. Note that the required diol can be prepared from reaction of an aliphatic amine and the epoxide derived from propene. The second key of the synthesis is to notice that dodecanoic acid must be lengthened by one carbon atom before it is turned into an amine function.

CH₃CH=CH₂ →[RCO₃H]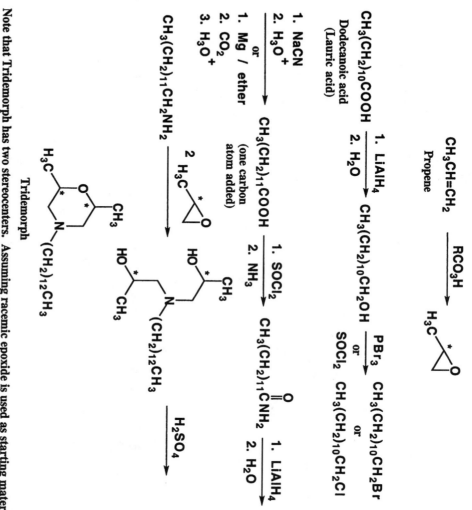
Propene

CH₃(CH₂)₁₀COOH →[1. LiAlH₄ 2. H₂O] CH₃(CH₂)₁₀CH₂OH →[PBr₃ or SOCl₂] CH₃(CH₂)₁₀CH₂Br
Dodecanoic acid   or
(Lauric acid)   CH₃(CH₂)₁₀CH₂Cl

1. NaCN
2. H₃O⁺
   or
1. Mg / ether
2. CO₂
3. H₃O⁺

→ CH₃(CH₂)₁₁COOH →[1. SOCl₂ 2. NH₃] CH₃(CH₂)₁₁C(=O)NH₂ →[1. LiAlH₄ 2. H₂O]
(one carbon
atom added)

CH₃(CH₂)₁₁CH₂NH₂ →[2 H₃C⟨epoxide⟩]

Tridemorph

CH₃(CH₂)₁₁CH₂NH₂ product →[H₂SO₄]

Problem 22.51 The Ritter reaction is especially valuable for the synthesis of 3° alkanamines. In fact, there are few alternative routes to them. This reaction is illustrated by the first step in the following sequence. In the second step, the Ritter product is hydrolyzed to the amine.

(a) Propose a mechanism for the Ritter reaction.

Step 1:

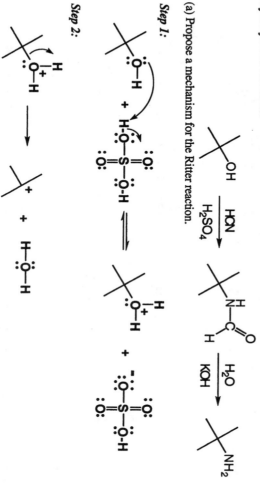

Step 2:

Note that Tridemorph has two stereocenters. Assuming racemic epoxide is used as starting material, a racemic mixture of three stereoisomers (one meso form and two enantiomers) will be formed as the product of the above reaction sequence.

Step 3:

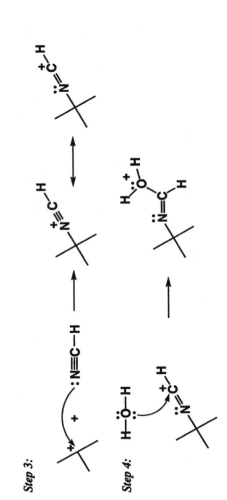

Step 4:

Step 5:

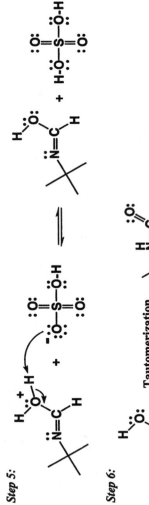

Step 6:

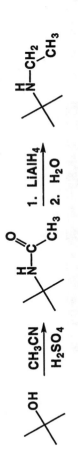

Tautomerization

(b) What is the product of a Ritter reaction using acetonitrile, CH₃CN, followed by reduction of the Ritter product with lithium aluminum hydride?

Problem 22.52 Several diamines are building blocks for the synthesis of pharmaceuticals and agrochemicals. Show how both 1,3-propanediamine and 1,4-butanediamine can be prepared from acrylonitrile

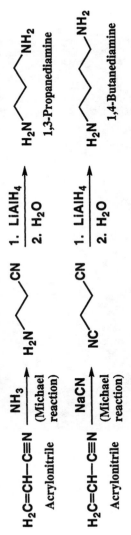

Acrylonitrile is a Michael acceptor that can react with amines (ammonia) or the cyanide anion.

Solutions

Problem 22.53 According to the following retrosynthetic analysis, the intravenous anesthetic 2,6-diisopropylphenol (Propofol, see The Merck Index, 12th ed., #8020) can be synthesized from phenol. Show how this synthesis might be carried out.

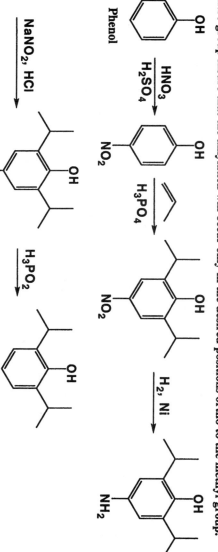

2,6-Diisopropylphenol
(Propofol)

Phenol

In the following synthetic scheme, note how the overall yield of product is increased by having the temporary nitro group insure that the alkylation will occur only in the desired positions ortho to the methyl group.

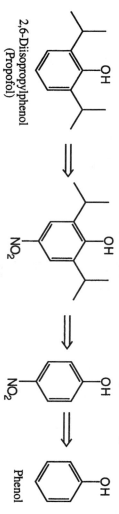

Phenol

2,6-Diisopropylphenol
(Propofol)

Queen angelfish *Holacanthus ciliaris*
Tobago, West Indies

CHAPTER 23
Solutions to the Problems

<u>Problem 23.1</u> Which of these terpenes (Section 5.4) contains conjugated double bonds?

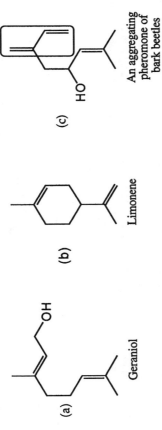

(a)

Geraniol

(b)

Limonene

(c)

HO

An aggregating
pheromone of
bark beetles

Only the aggregating pheromone of bark beetles (c) has conjugated double bonds.

<u>Problem 23.2</u> Estimate the stabilization gained due to conjugation when 1,4-pentadiene is converted to *trans*-1,3-pentadiene. Note that the answer is not as simple as comparing the heats of hydrogenation of 1,4-pentadiene and *trans*-1,3-pentadiene. Although the double bonds are moved from unconjugated to conjugated, the degree of substitution of one of the double bonds is also changed, in this case from a monosubstituted double bond to a trans disubstituted double bond. To answer this question, you must separate the effect due to conjugation from that due to change in degree of substitution.

1,4-Pentadiene   *trans*-1,3-Pentadiene

As stated in the question, there are actually two important issues to be addressed. It is best to isolate the two different interactions by considering analogous systems in Table 23.1. First, *trans*-1,3-pentadiene has an internal double bond while 1,4-pentadiene does not. Internal double bonds are lower in energy since they are more highly substituted. A good estimate of the stabilization energy provided by having an internal versus terminal double bond is 12 kJ/mol, calculated as the difference in energy between 1-butene ($\Delta H° = -127$ kJ/mol) and *trans*-2-butene ($\Delta H° = -115$ kJ/mol).

The overall difference in energy between 1,4-pentadiene ($\Delta H° = -254$ kJ/mol) and *trans*-1,3-pentadiene ($\Delta H° = -226$ kJ/mol) from Table 23.1 is 28 kJ/mol. Corrected by 12 kJ/mol for the presence of the internal double bond in *trans*-1,3-pentadiene leaves:

16 kJ/mol as the stabilization due to conjugation when 1,4-pentadiene is converted to *trans*-1,3-pentadiene.

<u>Problem 23.3</u> Predict the product(s) formed by addition of 1 mole of Br_2 to 2,4-hexadiene.

Predict both 1,2-addition and 1,4-addition. Two new stereocenters are created in each case leading to a racemic mixture of stereoisomers.

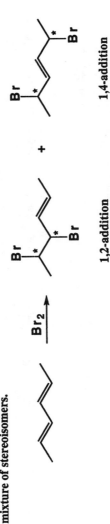

Br_2

Br *   Br *   Br *   Br *

+

1,2-addition   1,4-addition

<u>Problem 23.4</u> What combination of diene and dienophile undergoes Diels-Alder reaction to give each adduct.

(a)

(b)

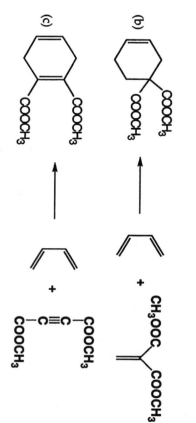

In each part of this problem, the diene is 1,3-butadiene. In (a), the dienophile is one of the double bonds of a second molecule of 1,3-butadiene. In part (b), the dienophile is a 1,1-disubstituted alkene and in part (c), it is a disubstituted alkyne.

(c)

Problem 23.5 Which molecules can function as dienes in Diels-Alder reactions? Explain your reasoning.

(a)    (b)   (c)

To function as a diene, the double bonds must be conjugated and able to assume an s-cis conformation. Both (a) 1,3-cyclohexadiene and (c) 1,3-cyclopentadiene can function as dienes. The double bonds in 1,4-cyclohexadiene are not conjugated and, therefore, this molecule cannot function as a diene.

Problem 23.6 What diene and dienophile might you use to prepare the following Diels-Alder adduct?

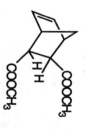

Use cyclopentadiene and the E-alkene. Note that two new stereocenters are formed in the reaction so the product is actually a racemic mixture of the two enantiomers having a trans arrangement of the ester groups.

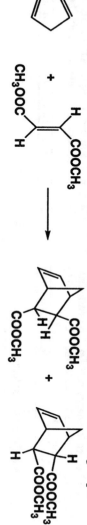

Problem 23.7 Show how to synthesize allyl phenyl ether and 2-butenyl phenyl ether from phenol and appropriate alkenyl halides.

Prepare each by a Williamson ether synthesis. Treat phenol with sodium hydroxide to form sodium phenoxide followed by treatment with the appropriate alkenyl chloride.

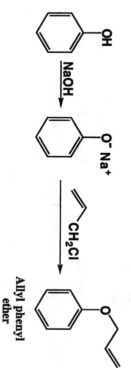

NaOH

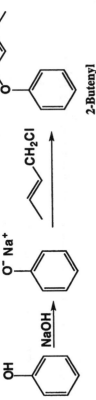

O⁻ Na⁺

CH₂Cl

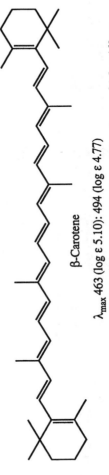

**2-Butenyl
phenyl ether**

<u>Problem 23.8</u> Propose a mechanism for the following Cope rearrangement.

Cope rearrangement gives an enol that then undergoes keto-enol tautomerism (Section 15.11B) to give the observed product.
Step 1:

Cope
rearrangement

Step 2:

keto-enol
tautomerization

320°C

<u>Problem 23.9</u>
Wavelengths in ultraviolet-visible spectroscopy are commonly expressed in nanometers; wavelengths in infrared spectroscopy are sometimes expressed in micrometers. Carry out the following conversions:
(a) 2.5 μm to nanometers. **2.5 μm is equal to 2500 nm.**
(b) 200 nm to micrometers. **200 nm is equal to 0.2 μm.**

<u>Problem 23.10</u> The visible spectrum of β-carotene ($C_{40}H_{56}$; MW 536.89, the orange pigment in carrots) dissolved in hexane shows intense absorption maxima at 463 nm and 494 nm, both in the blue-green region. Because light of these wavelengths is absorbed by β-carotene, we perceive the color of this compound as that of the complement to blue-green, namely, red-orange.

β-Carotene

λ_{max} 463 (log ε 5.10): 494 (log ε 4.77)

Calculate the concentration in milligrams per milliliter of β-carotene that gives an absorbance of 1.8 at 463 nm.

$$c = 1.8/(1.00 \text{ cm})(10^{5.10}) = 1.43 \times 10^{-5} \text{ mol/liter}$$

The molecular weight of β-carotene, $C_{40}H_{56}$, is (12 x 40) + (1 x 56) = 536 g/mol. Concentration in units of milligrams per milliliter is equal to the value of concentration of grams per liter, so:

$$c = (1.43 \times 10^{-5} \text{ mol/liter})(536 \text{ g/mol}) = 7.7 \times 10^{-3} \text{ g/liter}$$

$$\boxed{= 7.7 \times 10^{-3} \text{ milligram / milliliter.}}$$

PROBLEMS

Structure and Stability

Problem 23.11 If an electron is added to 1,3-butadiene, into which molecular orbital does it go? If an electron is removed from 1,3-butadiene, from which molecular orbital is it taken?

When an electron is removed from a conjugated pi system, such as that in 1,3-butadiene, it is taken from the highest occupied molecular orbital (HOMO). When an electron is added to a conjugated pi system, is added to the lowest unoccupied molecular orbital (LUMO).

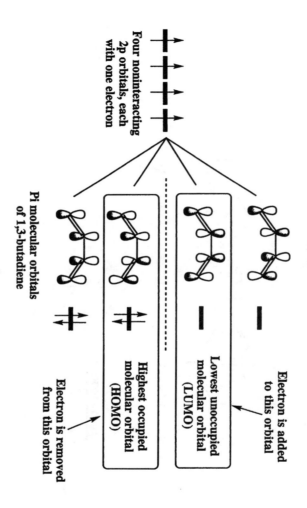

Four noninteracting
2p orbitals, each
with one electron

Pi molecular orbitals
of 1,3-butadiene

Electron is added
to this orbital

Lowest unoccupied
molecular orbital
(LUMO)

Highest occupied
molecular orbital
(HOMO)

Electron is removed
from this orbital

Problem 23.12 Draw an energy diagram (energy versus dihedral angle from 0° to 360°) for rotation about the 2,3 single bond in 1,3-butadiene.

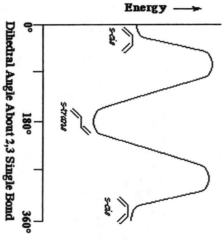

Energy ⟶

s-cis

s-trans

s-cis

0°   180°   360°

Dihedral Angle About 2,3 Single Bond

The s-cis and s-trans conformations are the most stable for 1,3-butadiene, because these provide for an overall planar geometry, which maximizes overlap of the 2p orbitals. The s-trans geometry is more stable than the s-cis geometry because of steric interactions.

Problem 23.13 Draw all important contributing structures for the following allylic carbocations; then rank the structures in order of relative contributions to each resonance hybrid.

(a)

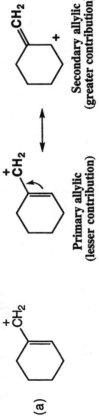

Primary allylic
(lesser contribution)   Secondary allylic
(greater contribution)

The secondary allylic cation makes the greater contribution.

(b) $CH_2=CH-CH=CH-CH_2^+$

$CH_2=CH-CH=CH-CH_2^+ \longleftrightarrow CH_2=CH-CH-CH=CH_2 \longleftrightarrow {}^+CH_2-CH=CH-CH=CH_2$

Primary allylic
(lesser contribution)   Secondary allylic
(greater contribution)   Primary allylic
(lesser contribution)

The secondary allylic cation makes the greater contribution.

(c) $CH_3-\overset{CH_3}{\underset{+}{C}}-CH=CH_2$

$CH_3-\overset{CH_3}{\underset{+}{C}}-CH=CH_2 \longleftrightarrow CH_3-\overset{CH_3}{C}=CH-\overset{+}{C}H_2$

Tertiary allylic
(greater contribution)   Primary allylic
(lesser contribution)

The tertiary allylic cation makes the greater contribution.

Electrophilic Addition to Conjugated Dienes
Problem 23.14 Predict the structure of the major product formed by 1,2-addition of HCl to 2-methyl-1,3-butadiene (isoprene).

A tertiary allylic cation is more stable than a secondary allylic cation. Assume that because the tertiary allylic cation is the more stable of the two, the activation energy is lower for its formation, and accordingly, it is formed at a greater rate than the secondary allylic cation. Therefore, the major product of 1,2-addition is 3-chloro-3-methyl-1-butene.

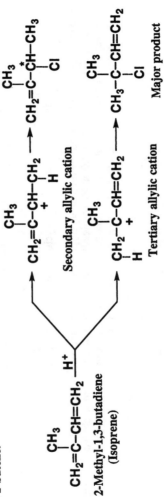

$\overset{CH_3}{\underset{}{CH_2}}=C-CH=CH_2$ $\xrightarrow{H^+}$

2-Methyl-1,3-butadiene
(Isoprene)

$CH_2=C-\overset{CH_3}{\underset{+}{C}}H-CH_2$ → $CH_2=C-\overset{CH_3}{\underset{}{C}H}-CH_2-CH_3$
                    H   Cl

Secondary allylic cation

$CH_2-C=CH-CH_2$ → $CH_3-C-CH=CH_2$
       H   +   CH_3   Cl   CH_3

Tertiary allylic cation   Major product

Problem 23.15 Predict the major product formed by 1,4-addition of HCl to isoprene.

The major product of 1,4-addition is 1-chloro-3-methyl-2-butene.

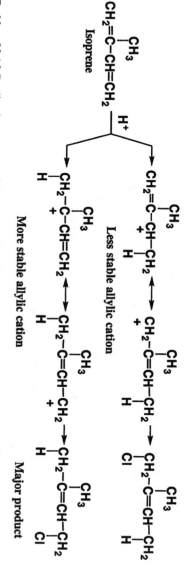

CH₂=C–CH=CH₂
  |
  CH₃

Isoprene

H⁺

CH₂=C–CH–CH₃ ⟷ CH₂–C=CH–CH₃
  | + | +
  CH₃   CH₃

Less stable allylic cation

CH₂=C–CH₂–CH₃ ⟷ CH₂–C=CH–CH₃
    | +   + |
    CH₃   CH₃

More stable allylic cation

H   H   H   H
CH₂–C–CH=CH₂   CH₂–C=CH–CH₂
 | | | | | |
 Cl CH₃   Cl CH₃   Cl

Major product

CH₂–C–CH=CH₂   CH₂–C=CH–CH₂
 + | | | |
     CH₃   Cl CH₃   Cl

Major product

Problem 23.16 Predict the structure of the major 1,2-addition product formed by reaction of 1 mole of Br₂ with isoprene. Also predict the structure of the major 1,4-addition product formed under these conditions.

Following the reasoning developed in the previous problems, predict that the major product of 1,2-addition to isoprene is 3,4-dibromo-3-methyl-1-butene. One new stereocenter is created, so a racemic mixture will be produced.

CH₃
  |
CH₂=C–CH=CH₂ + Br₂ →(1,2-addition)

CH₃
  |
CH₂–C–CH=CH₂
  | |*
  Br Br

Major product

CH₃
  |
CH₂=C–CH=CH₂ + Br₂ →(1,4-addition)

CH₃
  |
CH₂–C=CH–CH₂
  | |
  Br   Br

There is only one 1,4-addition product possible from addition of bromine.

Problem 23.17 Which of the two molecules shown do you expect to be the major product formed by 1,2-addition of HCl to cyclopentadiene? Explain.

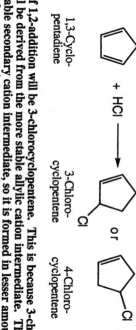

+ HCl ⟶

or

Cl

Cl

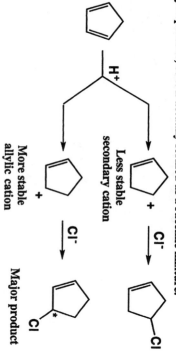

1,3-Cyclo-
pentadiene

3-Chloro-
cyclopentene

4-Chloro-
cyclopentene

The major product of 1,2-addition will be 3-chlorocyclopentene. This is because 3-chlorocyclopentene is the only product that will be derived from the more stable allylic cation intermediate. The 4-chlorocyclopentene derives from a less stable secondary cation intermediate, so it is formed in lesser amounts. A new stereocenter is formed in 3-chlorocyclopentene, so it is actually formed as a racemic mixture.

H⁺

Less stable
secondary cation

+

Cl⁻

More stable
allylic cation

+

Cl⁻

Cl

Cl

Cl*

Major product

Problem 23.18   Predict the major product formed by 1,4-addition of HCl to cyclopentadiene.

The conjugate addition product would also be the 3-chlorocyclopentene. Because of symmetry in the ring, the 1,4-addition product is the same as the predominant 1,2-addition product, both being derived from the allylic cation.

Problem 23.19   Draw structural formulas for the two constitutional isomers of molecular formula $C_5H_6Br_2$ formed by adding one mole of Br_2 to cyclopentadiene.

The two constitutional isomers are the 1,2 and 1,4-addition products. Two new stereocenters are formed, leading to a racemic mixture of stereoisomers.

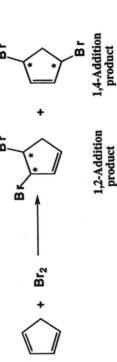

1,2-Addition
product   1,4-Addition
               product

Problem 23.20   What are the expected kinetic and thermodynamic products from addition of one mole of Br_2 to the following dienes

(a)

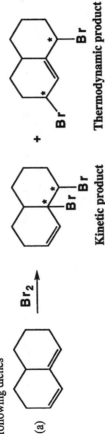

Kinetic product   Thermodynamic product

The kinetic product will be the 1,2-addition product that is derived from the more highly substituted and thus more stable bridged bromonium ion intermediate. The thermodynamic product is the kinetic product. The thermodynamic product is the more stable one, namely the 1,4-addition product, because this has the most highly substituted alkene as shown. Two new stereocenters are created in the molecule as shown, leading to a racemic mixture of stereoisomers.

(b)

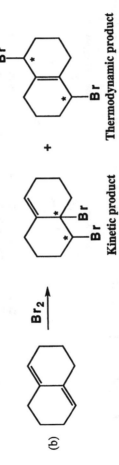

Kinetic product   Thermodynamic product

Because of symmetry in the molecule, both possible bromonium ion intermediates are the same. Thus, there is only one possible 1,2-addition product and this is the kinetic product. The thermodynamic product is the more stable one, namely the 1,4-addition product, because this has the most highly substituted alkene as shown. Once again, two new stereocenters are created in the molecule as shown, leading to a racemic mixture of stereoisomers.

Diels-Alder Reactions
Problem 23.21   Draw structural formulas for the products of reaction of cyclopentadiene with each dienophile.

Note how in both (a) and (b) the predominant endo products are shown. In these to cases a new stereocenter is created, leading to formation of racemic products of the two enantiomers.

(a) $CH_2=CHCl$

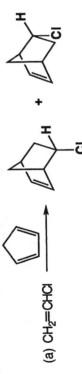

(b) $CH_2=CHOOCH_3$

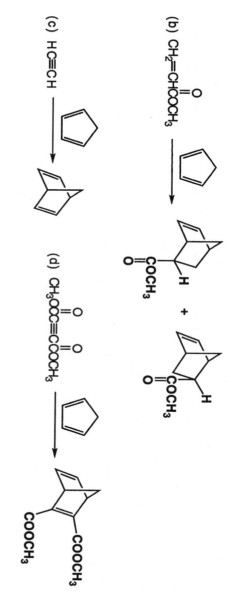

(c) $HC\equiv CH \longrightarrow$

(d) $CH_3OOC\equiv CCOOCH_3$

Problem 23.22 Propose structural formulas for compounds A and B and specify the configuration of compound B.

$+ CH_2=CH_2 \xrightarrow{200°C} C_7H_{10}$

(A)

$\xrightarrow[\text{2) } (CH_3)_2S]{\text{1) } O_3} C_7H_{10}O_2$

(B)

The Diels-Alder adduct is bicyclo[2.2.1]-2-heptene, more commonly known as norbornene. The oxidation product is *cis*-1,3-cyclopentanedicarbaldehyde, a meso compound.

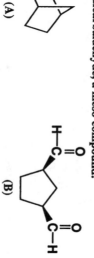

(A)

(B)

Problem 23.23 Under certain conditions, 1,3-butadiene can function both as a diene and a dienophile. Draw a structural formula for the Diels-Alder adduct formed by reaction of 1,3-butadiene with itself.

In the formulas below, one molecule of butadiene is shown as the diene and the other is shown as the dienophile. 4-Vinylcyclohexane has one stereocenter and is formed as a racemic mixture.

Butadiene   Butadiene   4-Vinylcyclohexene

Note that an electron pushing mechanism can be drawn for cyclization of two molecules of butadiene to 1,5-cyclooctadiene. This reaction does not take place because it does not have six electrons moving in a ring. The Diels-Alder reaction, with the relatively low energy arrangement of six electrons moving in a ring takes place instead.

1,5-Cyclooctadiene

Problem 23.24 1,3-Butadiene is a gas at room temperature and a requires gas-handling apparatus to use in a Diels-Alder reaction. Butadiene sulfone is a convenient substitute for gaseous 1,3-butadiene. This sulfone is a solid at room temperature (mp 66°C) and, when heated above its boiling point of 110°C, decomposes by a reverse Diels-Alder reaction to give s-cis-1,3-butadiene and sulfur dioxide. Draw a Lewis structure for butadiene sulfone, and show by curved arrows the path of this reverse Diels-Alder reaction.

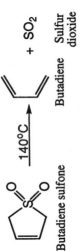

Butadiene sulfone   140°C   Butadiene   + SO₂   Sulfur dioxide

The electron flow in this decomposition is the reverse of that observed in a Diels-Alder reaction.

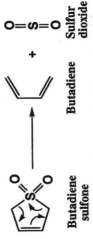

Butadiene sulfone   **Butadiene**   +   O=S=O   **Sulfur dioxide**

Problem 23.25 The following trienone undergoes an intramolecular Diels-Alder reaction to give the product shown. Show how the carbon skeleton of the triene must be coiled to give this product, and show by curved arrows the redistribution of electron pairs that takes place to give the product.

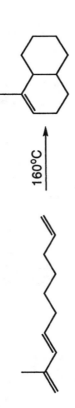

160°C

Locate the position of the carbon-carbon double bond in the product and realize that the two carbons of this double bond were carbons 2 and 3 of the conjugated diene. The other two carbon atoms making up this six-membered ring were from the dienophile. Note that two new stereocenters are formed in the product, and since the starting trienone is not chiral, the product is actually a racemic mixture of the two trans isomers. The trans geometry of the ring fusion (atoms "4" and "9") is a consequence of the concerted nature of the reaction.

this part was the diene

the dienophile was here

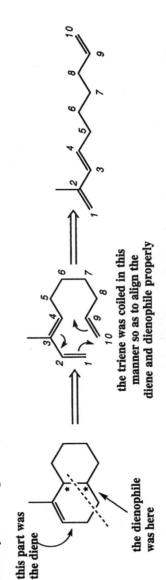

the triene was coiled in this manner so as to align the diene and dienophile properly

Problem 23.26 The following trienone undergoes an intramolecular Diels-Alder reaction to give a bicyclic product. Propose a structural formula for the product. Account for the observation that the Diels-Alder reaction given in this problem takes place under milder conditions (at lower temperature) than the analogous Diels-Alder reaction shown in problem 23.25?

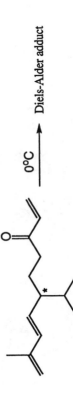

0°C   Diels-Alder adduct

Follow the arrangement of diene and dienophile illustrated in the previous problem.

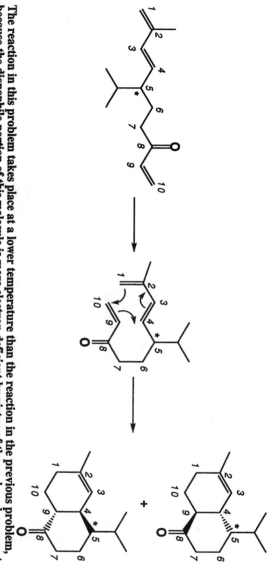

The reaction in this problem takes place at a lower temperature than the reaction in the previous problem, because the dienophile portion of this molecule is more electron deficient by virtue of the carbonyl group. An electron deficient dienophile is a better dienophile, so the reaction is accelerated. Note how the starting material is chiral and presumable racemic, while two new stereocenters are created in the product. The concerted nature of the reaction will lead to a *trans* geometry at the ring fusion (atoms "4" and "9"), and the third stereocenter (atom "5") will lead to a mixture of stereoisomers. If the starting material is racemic, then the product mixture will be racemic as well.

Problem 23.27 The following compound undergoes an intramolecular Diels-Alder reaction to give a bicyclic product. Propose a structural formula for the product.

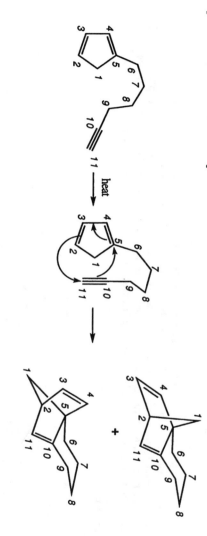

An intramolecular
Diels-Alder adduct

The product has two new stereocenters, but because of the bicyclic structure there are only two possible enantiomers. The different enantiomers arise from reaction on either face of the diene. Only one enantiomer is drawn below as a three-dimensional model.

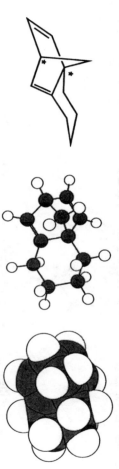

Problem 23.28 Draw a structural formula for the product of this Diels-Alder reaction, including the stereochemistry of the product.

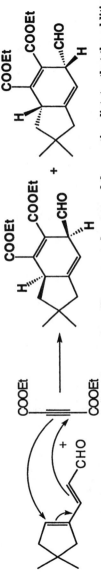

Two new stereocenters are created in the reaction. The concerted nature of the reaction dictates that the addition leads to the two possible cis enantiomers. Note that the products are actually a racemic mixture of the two cis enantiomers, since neither the diene or dienophile has a stereocenter that would favor formation of one enantiomer over the other. The different enantiomers arise from reaction on either face of the diene.

Problem 23.29 Following is a retrosynthetic analysis for the dicarboxylic acid shown on the left.

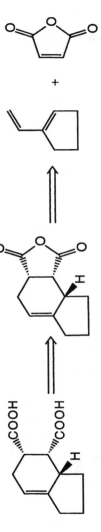

(a) Propose a synthesis of the diene from cyclopentanone and acetylene.

The acetylene is deprotonated, then the resulting alkyne anion is reacted with cyclopentanone to give the cyclopentene derivative and hydrogenation over the Lindlar catalyst. The alcohol is dehydrated to give the cyclopentene derivative and hydrogenation over the Lindlar catalyst completes the synthesis.

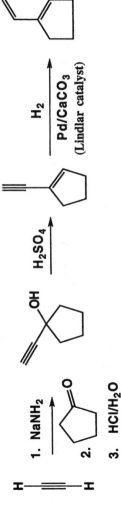

(b) Rationalize the stereochemistry of the target dicarboxylic acid.

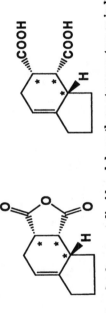

The Diels-Alder adduct and subsequent diacid each have three stereocenters as indicated by asterisks in the above structures. Like all Diels-Alder reactions, the dienophile and diene add in a concerted fashion, fixing the two carbonyls cis to each other as well as fixing the relative configuration of the third stereocenter. Note that the Diels-Alder product and thus the diacid are actually a racemic mixture of enantiomers, since neither the diene or dienophile has a stereocenter that would favor formation of one enantiomer over the other. Only one of the enantiomers was shown in the problem, however, both enantiomers are shown below. The different enantiomers arise from reaction on either face of the diene.

Problem 23.30 One of the published syntheses of warburganal (Problem 5.28) begins with the following Diels-Alder reaction. Propose a structure for Compound A.

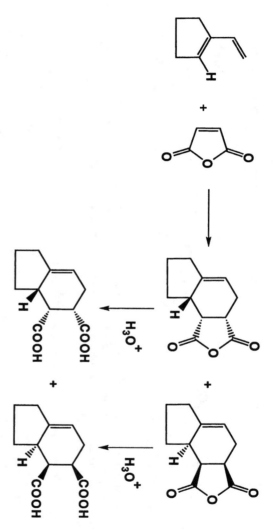

Problem 23.31 The Diels-Alder reaction is not limited to making six-membered rings with only carbon atoms. Predict the products of the following reactions that produce rings with atoms other than carbon in them.

Compound A is a racemic mixture of the two enantiomers resulting from the Diels-Alder reaction as shown. The different enantiomers arise from reaction on either face of the diene.

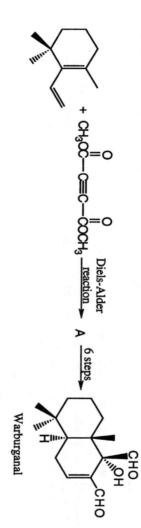

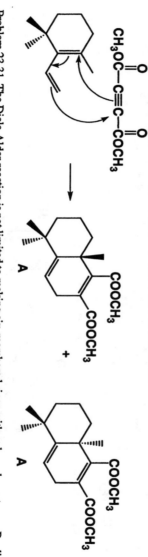

(a)

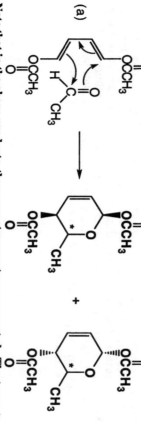

Note that in the above products, three new stereocenters are created. The two ester groups will be cis to each other in all products, but the third stereocenter could be either cis or trans relative to these leading to a complex racemic mixture of stereoisomers.

(b)

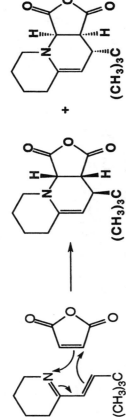

+

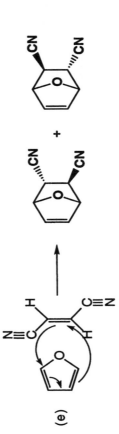

The above products would be formed as a racemic mixture.

(c)

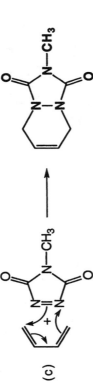

Because of symmetry in the molecule, no new stereocenters are formed.

(d)

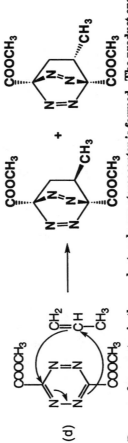

Because of symmetry in the product, only one stereocenter is formed. The product enantiomers will be formed as a racemic mixture.

(e)

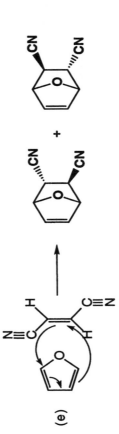

The above products would be formed as a racemic mixture.

Problem 23.32 The first step in a synthesis of dodecahedrane involves a Diels-Alder reaction between the cyclopentadiene derivative (1) and dimethyl acetylenedicarboxylate (2). Show how these two molecules react to form the dodecahedrane synthetic intermediate (3).

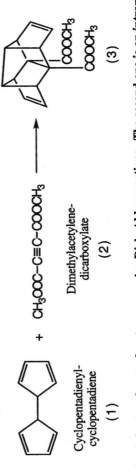

Cyclopentadienyl-
cyclopentadiene
(1)

+ CH₃OOC—C≡C—COOCH₃

Dimethylacetylene-
dicarboxylate
(2)

(3)

This reaction is best understood as two successive Diels-Alder reactions. The second one is an intramolecular reaction that takes place on the product of the first.

Problem 23.33 Bicyclo[2.2.1]-2,5-heptadiene can be prepared in two steps from cyclopentadiene and vinyl chloride. Provide a mechanism for each step.

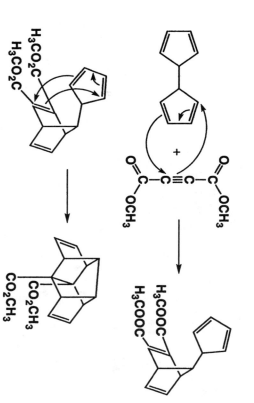

Step 1:

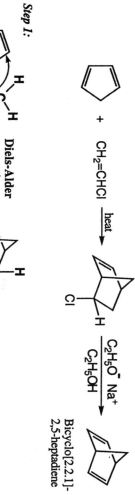

Diels-Alder reaction

$$CH_2=CHCl \xrightarrow{heat}$$

$$\xrightarrow[C_2H_5OH]{C_2H_5O^- \ Na^+}$$

Bicyclo[2.2.1]-
2,5-heptadiene

Step 2:

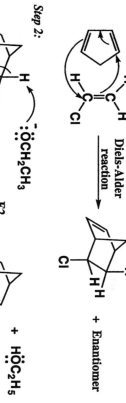

E2

+ HÖC$_2$H$_5$

+ :C̈l:$^-$

+ Enantiomer

Problem 23.34 Treatment of anthranilic acid with nitrous acid gives an intermediate, A, that contains a diazonium ion and a carboxyl group. When this intermediate is heated in the presence of furan, a bicyclic compound is formed. Propose a structural formula for intermediate A and mechanism for formation of the bicyclic product.

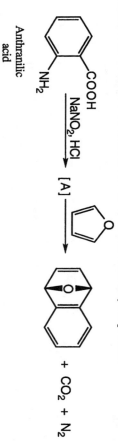

Anthranilic
acid

NaNO$_2$, HCl

[A]

+ CO$_2$ + N$_2$

Several steps:

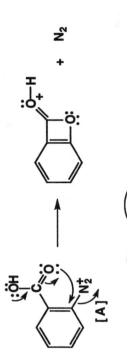

NaNO₂, HCl

[A]

Step 2: This step may or may not be concerted

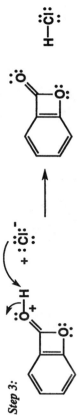

+ N₂

Step 3:

H—Cl̈:

Step 4:

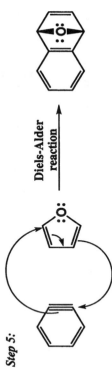

+ :Ö=C=Ö:

Benzyne

Step 5:

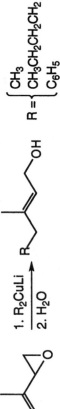

Diels-Alder
reaction

The product is a meso compound.

Problem 23.35 Propose a mechanism for the following reaction, which is called an allylic rearrangement, or, alternatively, a conjugate addition.

1. R₂CuLi
2. H₂O

R = { CH₃, CH₃CH₂CH₂CH₂, C₆H₅ }

Propose that the alkyl group of the Gilman reagent attacks the terminal carbon of the carbon-carbon double bond, bringing about rearrangement of the double bond and opening of the epoxide ring to give a lithium alkoxide. Treatment of this alkoxide with water gives the observed allylic alcohol.

Step 1:

Li(R)Cu—R

Step 2:

Problem 23.36 All attempts to synthesize cyclopentadienone yield only a Diels-Alder adduct. Cycloheptatrienone, however, has been prepared by several methods and is stable.

(a) Draw a structural formula for the Diels-Alder adduct formed by cyclopentadienone.

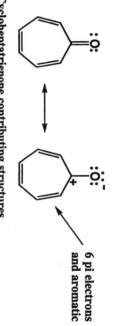

Cyclopentadienone   Cycloheptatrienone

The Diels-Alder adduct is a racemic mixture of two enantiomers as shown.

(b) How do you account for the marked difference in stability of these two ketones?

A major contributing structure for cyclopentadienone has only four pi electrons, an anti-aromatic number, thereby explaining the extreme reactivity of cyclopentadienone.

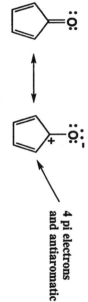

Cyclopentadienone contributing structures

4 pi electrons
and antiaromatic

On the other hand, a major contributing structure of cycloheptatrienone has six pi electrons, a Hückel number, thereby explaining the enhanced stability of cycloheptatrienone and resistance to reaction.

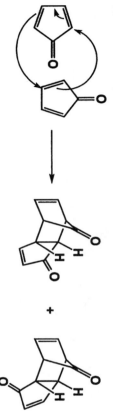

6 pi electrons
and aromatic

Cycloheptatrienone contributing structures

Problem 23.37 Following is a retrosynthetic scheme for the synthesis of the tricyclic diene on the left. Show how to accomplish this synthesis using the indicated carbon sources as starting materials.

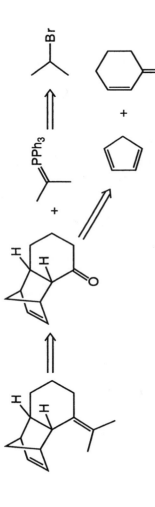

The key elements of this synthesis are a Diels-Alder reaction followed by a Wittig reaction. Note the product is actually a racemic mixture of enantiomers as shown.

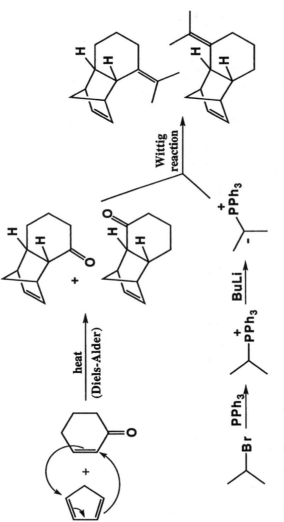

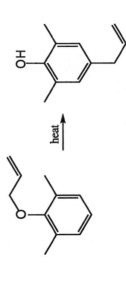

Other pericyclic reactions

Problem 23.32 Claisen rearrangement of an allyl phenyl ether with substituent groups in both ortho positions leads to formation of a para substituted product. Propose a mechanism for the following rearrangement.

Propose rearrangement of the allyl group to the ortho position to form a substituted cyclohexadienone, as in the normal Claisen rearrangement. This intermediate, however, cannot undergo keto-enol tautomerism to become an aromatic compound. A second rearrangement, this time of the allyl group to the para position, gives a substituted cyclohexadienone that can undergo keto-enol tautomerism to give an aromatic compound.

Step 1:

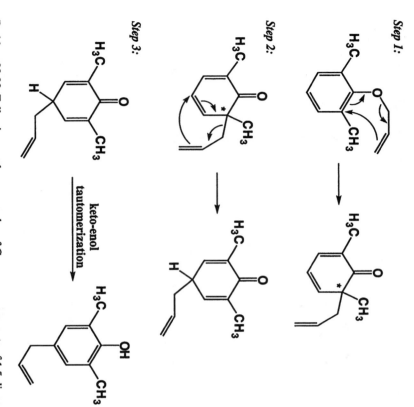

Step 2:

Step 3:

keto-enol
tautomerization

Problem 23.39 Following are three examples of Cope rearrangements of 1,5-dienes. Show that each product can be formed in a single step by a mechanism involving redistribution of six electrons in a cyclic transition state.

Keep in mind as you do these problems that the Cope rearrangement involves flow of three pairs of electrons in a cyclic, six-membered transition state. Bonding in the six carbon atoms participating in the reaction is C=C-C-C-C=C (double-single-single-single-double). Find that combination of carbon atoms and you are well on your way. Numbers have been assigned to the atoms to keep track of the bonds made and broken. It will help to use molecular models when answering these questions.

(a)

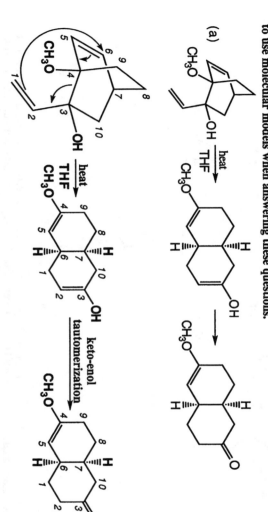

heat
THF

keto-enol
tautomerization

In the above product, two new stereocenters are formed, but only one enantiomer is produced due to the stereochemistry of the starting material. Recall that when a reactant is chiral, it is possible to create a single enantiomer of a product.

(b)

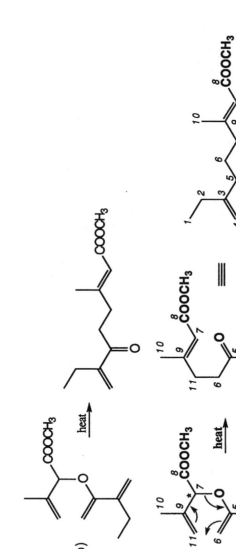

In the above, the starting material is chiral and presumably racemic, but the product has no stereocenters to consider.

(c)

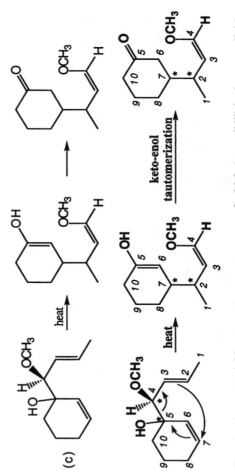

keto-enol
tautomerization

The starting material has two stereocenters, one of which (atom "4") is shown explicitly, while the other (atom "5") is not and thus presumably racemic. As a result, the product of the reaction, which has two new stereocenters created, will be a mixture of stereoisomers.

Problem 23.40 The following transformation is an example of the Carroll reaction, named after the English chemist, M. F. Carroll, who first reported it. Propose a mechanism for this reaction.

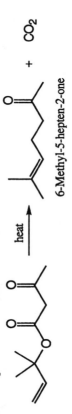

heat

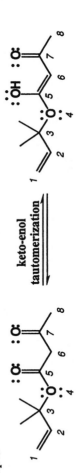

+ CO_2

6-Methyl-5-hepten-2-one

The reaction begins with tautomerization to an enol. Note that this enol is stabilized by conjugation to the adjacent carbonyl.
Step 1:

keto-enol
tautomerization

The rearrangement step involves the familiar six-membered ring transition state with three pairs of electrons moving around the ring.

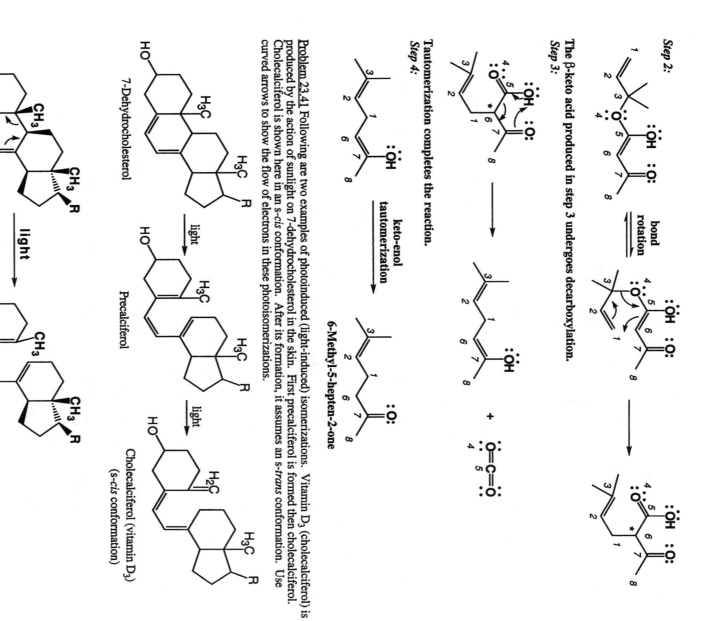

Step 2:

The β-keto acid produced in step 3 undergoes decarboxylation.

Step 3:

bond rotation

Tautomerization completes the reaction.

Step 4:

keto-enol tautomerization

6-Methyl-5-hepten-2-one

Problem 23.41 Following are two examples of photoinduced (light-induced) isomerizations. Vitamin D₃ (cholecalciferol)) is produced by the action of sunlight on 7-dehydrocholesterol in the skin. First precalciferol is formed then cholecalciferol. Cholecalciferol is shown here in an s-cis conformation. After its formation, it assumes an s-trans conformation. Use curved arrows to show the flow of electrons in these photoisomerizations.

7-Dehydrocholesterol

light

Precalciferol

7-Dehydrocholesterol

light

Precalciferol

light

Cholecalciferol (vitamin D₃) (s-cis conformation)

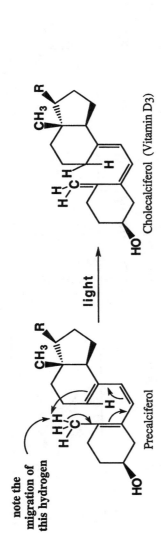

note the
migration of
this hydrogen

Precalciferol

Cholecalciferol (Vitamin D3)

Problem 23.42 Show the product of the following reaction. Include stereochemistry.

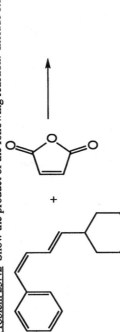

It is helpful to draw the *s-cis* geometry of the diene that is required for reaction. Keeping in mind that four new stereocenters are created in the reaction and the mechanism is concerted, the following two enantiomers will be formed as a racemic mixture. The different enantiomers arise from reaction on either face of the diene.

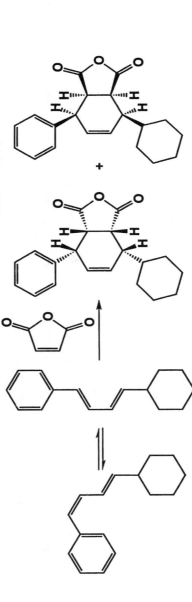

Problem 23.43 Following is a synthesis for the antifungal agent tolciclate (The Merck Index, 12th ed., #9647).

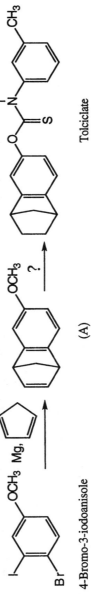

(A)   Tolciclate

4-Bromo-3-iodoanisole

(a) Propose a mechanism for formation of (A).

Step 1:

A benzyne
intermediate

Step 2:

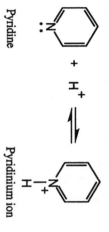

Note that (A) will be formed as a racemic mixture of two enantiomers.

(b) Show how (A) can be converted to toliciate. Use 3-methyl-*N*-methylaniline as the source of the amine nitrogen.

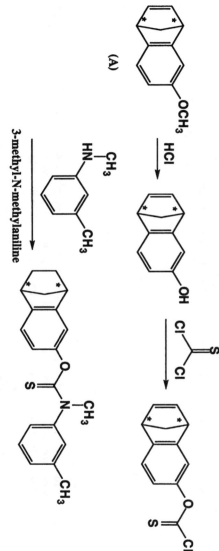

3-methyl-N-methylaniline

Toliciate

Note that each compound involved with this reaction sequence is actually a racemic mixture of enantiomers.

Problem 23.44 Show how to distinguish between 1,3-cyclohexadiene and 1,4-cyclohexadiene by ultraviolet spectroscopy.

1,3-Cyclohexadiene   1,4-Cyclohexadiene

In 1,3-cyclohexadiene, the pi bonds are conjugated so the absorption will occur near 217 nm compared with the 165 nm or so absorption that will be observed for the 1,4-cyclohexadiene that has only isolated (unconjugated) pi bonds.

Problem 23.45 Pyridine exhibits a UV transition of the type n --> p* at 270 nm. In this transition, one of the unshared electrons on nitrogen is promoted from a nonbonding MO to a pi* antibonding MO. What is the effect on this UV peak if pyridine is protonated?

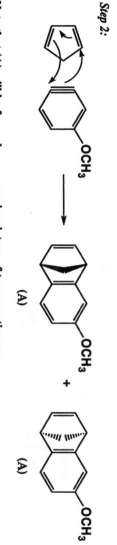

Pyridine   Pyridinium ion

When the pyridium ion is protonated, the lone pair of electrons on nitrogen is tied up in a bond to hydrogen. As a result, it is lower in energy compared with the unprotonated form, so it takes more energy to promote one of these electrons into the π* orbital. Therefore, protonation shifts the absorbance peak to lower wavelength (higher energy).

Problem 23.46 The weight of proteins or nucleic acids in solution is commonly determined by UV spectroscopy using the Beer-Lambert law. For example, the e of double-stranded DNA at 260 nm is 6670 $M^{-1}cm^{-1}$. The formula weight of the repeating unit in DNA (650 Daltons) can be used as the molecular weight. What is the weight of DNA in 2.0 mL of aqueous buffer if the absorbance, measured in a 1-cm cuvette, is 0.75?

According to the Beer-Lambert law:

$$0.75 = (6670)(1)(x)$$

where x equals the unknown concentration of DNA in mol per liter. Rearranging gives:

$$x = 0.75/(6670)(1) = 1.12 \times 10^{-4} \text{ mol/liter}$$

The molecular weight used is that of a single base pair of DNA, namely 650 grams/mol. Furthermore, there is a total of 2.0 mL of solution, so the total weight of double stranded DNA in the sample is:

$$(1.12 \times 10^{-4} \text{ mol/liter})(650 \text{ grams/mol})(2.0 \times 10^{-3} \text{ liter}) =$$

$$\boxed{1.46 \times 10^{-4} \text{ gram}}$$

Problem 23.47 A sample of adenosine triphosphate (ATP) (MW 507, e = 14,700 $M^{-1}cm^{-1}$ at 257 nm) is dissolved in 5.0 mL of buffer. A 250-mL aliquot is removed and placed in a 1-cm cuvette with sufficient buffer to give a total volume of 2.0 mL. The absorbance of the sample at 257 nm is 1.15. Calculate the weight of ATP in the original 5.0-mL sample.

The concentration of sample in the cuvette can be determined by using the Beer-Lambert law as follows:

$$1.15 = (14,700)(1.0 \text{ cm})(x \text{ mol/liter})$$

$$x = 1.15/14,700 = 7.82 \times 10^{-5} \text{ mol/liter}$$

The sample measured in the cuvette is actually diluted 0.250 in 2.0 or 1 in 8 compared to the original unknown sample, so the concentration of ATP in the original unknown sample is:

$$(8)(7.82 \times 10^{-5}) = 6.26 \times 10^{-4} \text{ mol/liter}$$

Since there is a total of 5.0 mL or 5×10^{-3} liter and ATP has a MW 507 grams/mol then the weight of ATP in the original sample can be calculated as:

$$6.26 \times 10^{-4} \text{ mol/liter} = (x \text{ grams})/(506 \text{ grams/mol})(5 \times 10^{-3} \text{ liters})$$

$$x = (6.26 \times 10^{-4})(5 \times 10^{-3} \text{ liters})(506) =$$

$$\boxed{1.59 \times 10^{-3} \text{ grams} = 1.59 \text{ mg}}$$

Pennant coralfish *Heniochus acuminatus*
Rangiroa, French Polynesia

Interchapter
Solutions to the Problems

We do not provide solutions in this Study Guide for the Interchapter Medicinal Chemistry problems. We do, however, provide solutions for them on the instructor's side of our web site. **Your instructor has the option of making them available to you.**

It is our approach in using these problems to concentrate on the types of reactions involved, the types of reagents required, and an analysis of the regiochemistry and stereochemistry in each synthetic sequence. For many steps in many of these problems, there are several reagents that might be used to bring about a desired transformation. We are not so much concerned with which reagent a particular published synthesis uses, but rather that you be able to identify the type of reagent required.

We have found that very often when we suggest a reagent and experimental conditions, students will ask about alternative reagents and experimental conditions. It is difficult for even the most experienced instructors to anticipate in every case what you will ask. It is for this reason that we find it difficult to provide solutions to these problems that will address every question you might raise.

We encourage you to discuss these problems with others in your class, and with your instructors. We hope that through these problems you will come to appreciate how practicing chemists go about developing a synthesis for target molecule, and the beauty and elegance of synthetic organic chemistry.

Stocky hawkfish *Cirrhitus pinnulatus*
Kona coast, Hawaii

CHAPTER 24
Solutions to the Problems

Problem 24.1 Given the following structure, determine the polymer's repeat unit, redraw the structure using the simplified parenthetical notation, and name the polymer.

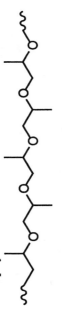

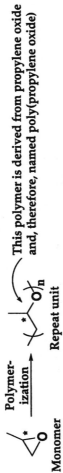

Polymer-
ization
⟶

This polymer is derived from propylene oxide and, therefore, named poly(propylene oxide)

Monomer   Repeat unit

Problem 24.2 Write the repeating unit of the polymer formed from the following reaction and propose a mechanism for its formation.

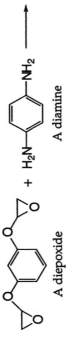

A diepoxide   +   H₂N — ◯ — NH₂   ⟶

A diamine

Following is the structural formula of the repeat unit of this polymer.

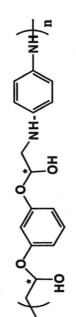

As a mechanism, propose nucleophilic attack of the amine on the less hindered carbon of the epoxide followed by proton transfer from nitrogen to oxygen.

Step 1: Nucleophilic ring opening of the epoxide

Ar-NH₂ +

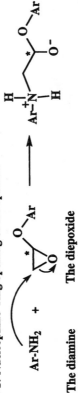

The diamine   The diepoxide

Step 2: Proton transfer from nitrogen to oxygen

Problem 24.3 Show how to prepare polybutadiene that is terminated at both ends with primary alcohol groups.

Treat 1,3-butadiene with two moles of lithium metal to form a dianion followed by addition of monomer units as in Example 24.3 to form a living polymer. Cap the active end groups by treatment of the living polymer with ethylene oxide followed by aqueous acid.

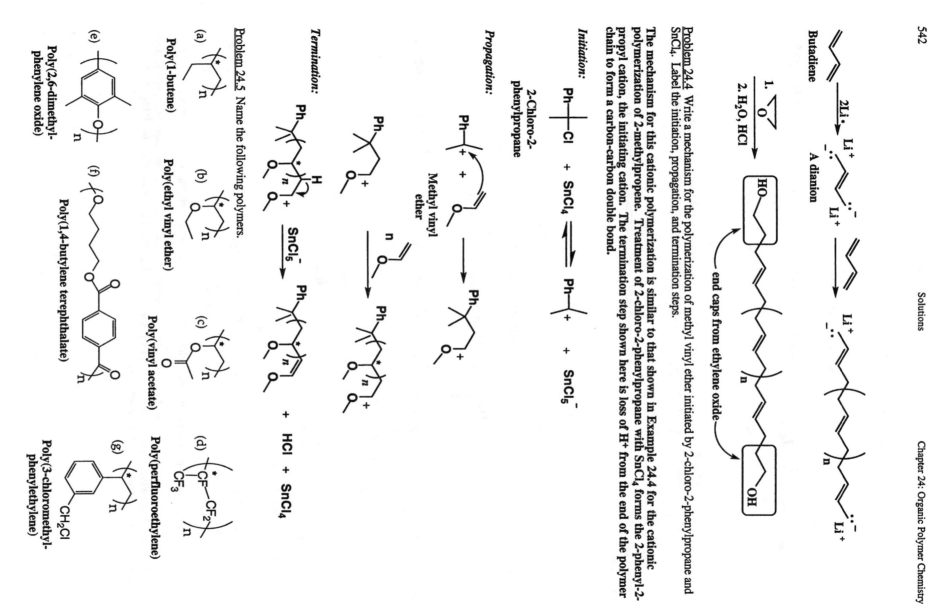

Butadiene

$\xrightarrow{2\text{Li} \cdot}$ **A dianion**

1. $\overset{O}{\triangle}$
2. H_2O, HCl

— end caps from ethylene oxide —

Problem 24.4 Write a mechanism for the polymerization of methyl vinyl ether initiated by 2-chloro-2-phenylpropane and SnCl₄. Label the initiation, propagation, and termination steps.

The mechanism for this cationic polymerization is similar to that shown in Example 24.4 for the cationic polymerization of 2-methylpropene. Treatment of 2-chloro-2-phenylpropane with SnCl₄ forms the 2-phenyl-2-propyl cation, the initiating cation. The termination step shown here is loss of H⁺ from the end of the polymer chain to form a carbon-carbon double bond.

Initiation:

$$Ph\text{—}Cl \;+\; SnCl_4 \;\rightleftharpoons\; Ph\text{—}{+} \;+\; SnCl_5^-$$

2-Chloro-2-phenylpropane

Propagation:

Methyl vinyl ether

Termination:

$$\xrightarrow{SnCl_5^-}$$ $+$ HCl + SnCl₄

Problem 24.5 Name the following polymers.

(a) Poly(1-butene)

(b) Poly(ethyl vinyl ether)

(c) Poly(vinyl acetate)

(d) Poly(perfluoroethylene)

(e) Poly(2,6-dimethyl-phenylene oxide)

(f) Poly(1,4-butylene terephthalate)

(g) Poly(3-chloromethyl-phenylethylene)

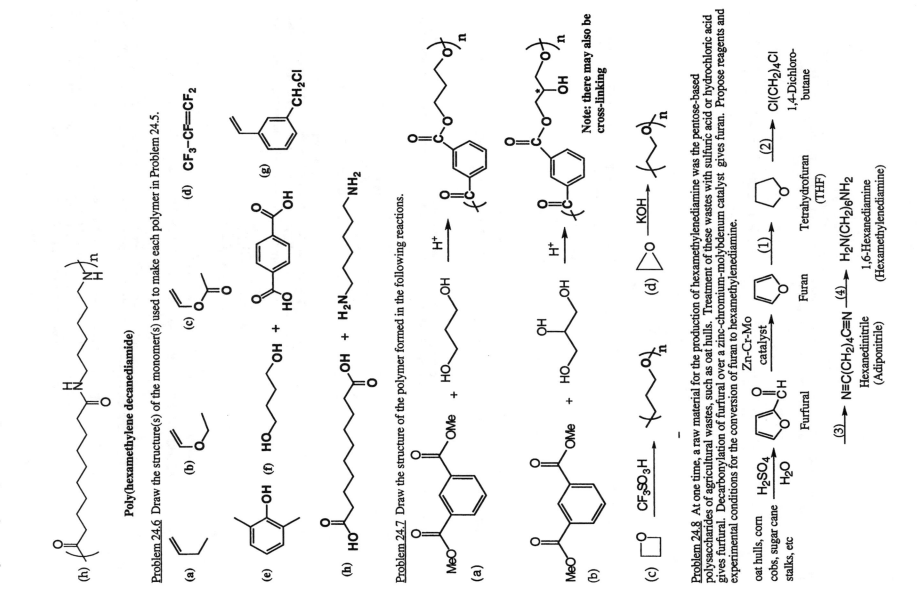

(h)

Poly(hexamethylene decanediamide)

Problem 24.6 Draw the structure(s) of the monomer(s) used to make each polymer in Problem 24.5.

(a)   (b)   (c)   (d) CF₃–CF=CF₂

(e)   (f)   (g)

(h)

Problem 24.7 Draw the structure of the polymer formed in the following reactions.

(a)

(b)

Note: there may also be cross-linking

(c)   CF₃SO₃H

(d)   KOH

Problem 24.8 At one time, a raw material for the production of hexamethylenediamine was the pentose-based polysaccharides of agricultural wastes, such as oat hulls. Treatment of these wastes with sulfuric acid or hydrochloric acid gives furfural. Decarbonylation of furfural over a zinc-chromium-molybdenum catalyst gives furan. Propose reagents and experimental conditions for the conversion of furan to hexamethylenediamine.

oat hulls, corn   H₂SO₄
cobs, sugar cane   ——→
stalks, etc   H₂O

Furfural

Zn-Cr-Mo
catalyst
——(1)——→

Furan

——(2)——→

Tetrahydrofuran
(THF)

Cl(CH₂)₄Cl
1,4-Dichloro-
butane

——(3)——→ N≡C(CH₂)₄C≡N ——(4)——→ H₂N(CH₂)₆NH₂

Hexanedinitrile
(Adiponitrile)

1,6-Hexanediamine
(Hexamethylenediamine)

Step 1: Catalytic hydrogenation using H_2 over a transition metal catalyst.
Step 2: Cleavage of the ether using concentrated HCl at elevated temperature.
Step 3: Treatment of the dihalide with NaCN, by an S_N2 pathway.
Step 4: Catalytic hydrogenation of the cyano groups using H_2 over a transition metal catalyst.

Problem 24.9 Another raw material for the production of hexamethylenediamine is butadiene derived from thermal and catalytic cracking of petroleum. Propose reagents and experimental conditions for the conversion of butadiene to hexamethylenediamine.

$$CH_2=CHCH=CH_2 \xrightarrow{(1)} ClCH_2CH=CHCH_2Cl \xrightarrow{(2)} N\equiv CCH_2CH=CHCH_2C\equiv N \xrightarrow{(3)}$$
Butadiene   1,4-Dichloro-2-butene   3-Hexenedinitrile

$$H_2N(CH_2)_6NH_2$$
1,6-Hexanediamine
(Hexamethylenediamine).

Step 1: 1,4- Addition of Cl_2 to the conjugated diene.
Step 2: Treatment of the dihalide with NaCN, by an S_N2 pathway.
Step 3: Catalytic hydrogenation of the cyano groups and the carbon-carbon double bond using H_2 over a transition metal catalyst.

Problem 24.10 Propose reagents and experimental conditions for the conversion of butadiene to adipic acid.

$$CH_2=CHCH=CH_2 \longrightarrow HOO(CH_2)_4COOH$$
Butadiene   Hexanedioic acid
                                 (Adipic acid)

See Problem 24.9 for the conversion of butadiene to 3-hexenedinitrile. (3) Catalytic hydrogenation of the carbon-carbon double bond in 3-hexenedinitrile followed by (4) hydrolysis of the cyano groups in aqueous acid gives adipic acid.

$$CH_2=CHCH=CH_2 \xrightarrow{(1)} ClCH_2CH=CHCH_2Cl \xrightarrow{(2)} N\equiv CCH_2 CH=CHCH_2C\equiv N \xrightarrow{(3)}$$
Butadiene   1,4-Dichloro-2-butene   3-Hexenedinitrile

$$N\equiv C(CH_2)_4 C\equiv N \xrightarrow{(4)} HOOC(CH_2)_4COOH$$
Hexanedinitrile   Hexanedioic acid
(Adiponitrile)   (Adipic acid)

Problem 24.11 Polymerization of 2-chloro-1,3-butadiene under Ziegler-Natta conditions gives a synthetic elastomer called neoprene. All carbon-carbon double bonds in the polymer chain have the trans configuration. Draw the repeat unit in neoprene.

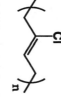

Problem 24.12 Poly(ethylene terephthalate) (PET) can be prepared by this reaction. Propose a mechanism for the step-growth reaction in this polymerization.

$$nCH_3OC \overset{O}{\underset{}{\parallel}} \overline{} COOCH_3 + nHOCH_2CH_2OH \xrightarrow{275^\circ C} (O \overset{O}{\underset{}{\parallel}} C \overline{} COOCH_2CH_2O)_n + 2nCH_3OH$$
Dimethyl terephthalate   Ethylene glycol   Poly(ethylene terephthalate)   Methanol

Propose addition of a hydroxyl group to a carbonyl carbon of dimethyl terephthalate to form a tetrahedral carbonyl addition intermediate, followed by its collapse to give an ester bond of the polymer plus methanol etc. This is an example of transesterification.

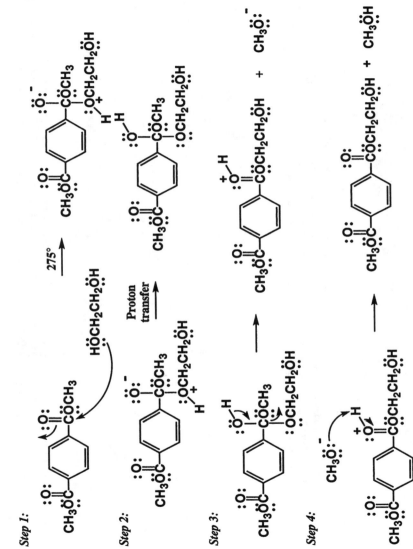

Step 1:

Step 2: Proton transfer

Step 3:

Step 4:

275°

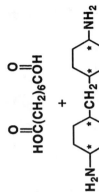

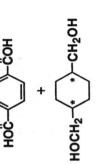

Problem 24.13 Identify the monomers required for the synthesis of these step-growth polymers.

(a)
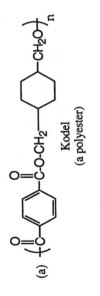
Kodel
(a polyester)

(b)
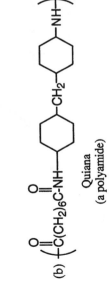
Quiana
(a polyamide)

Solutions

Problem 24.14 Nomex, another aromatic polyamide (aramid) is prepared by polymerization of 1,3-benzenediamine and the diacid chloride of 1,3-benzenedicarboxylic acid. The physical properties of the polymer make it suitable for high strength, high temperature applications such as parachute cords and jet aircraft tires. Draw a structural formula for the repeating unit of Nomex.

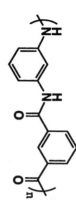

1,3-Benzenediamine

$+$

1,3-Benzene-
dicarbonyl chloride

polymerization $\longrightarrow$ Nomex

Following is the repeat unit in Nomex.

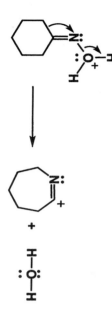

Problem 24.15 Caprolactam, the monomer from which nylon 6 is synthesized, is prepared from cyclohexanone in two steps. In step 1, cyclohexanone is treated with hydroxylamine to form cyclohexanone oxime. Treatment of the oxime with concentrated sulfuric acid in step 2 gives caprolactam by a reaction called a Beckmann rearrangement. Propose a mechanism for the conversion of cyclohexanone oxime to caprolactam.

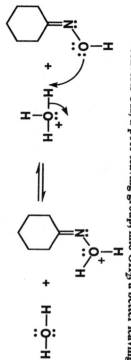

Cyclohexanone

$\xrightarrow{NH_2OH}$

Cyclohexanone
oxime

$\xrightarrow{H_2SO_4}$

Caprolactam

The mechanism is shown divided into six steps.

Step 1: Proton transfer from H_3O^+ to the oxygen atom of the oxime generates an oxonium ion, which converts OH, a poor leaving group, into OH_2, a better leaving group.

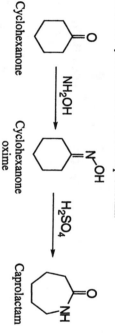

Step 2: Migration of the electron pair of an adjacent carbon-carbon bond to nitrogen accompanies departure of H_2O.

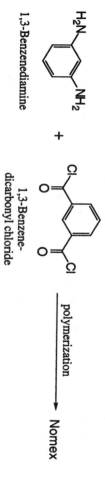

Step 3: Reaction of the carbocation from step 2 with H_2O to give an oxonium ion.

Step 4: Proton transfer to solvent gives the enol of an amide.

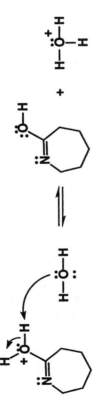

Step 5: Tautomerization of the enol form of the amide gives caprolactam.

keto-enol
tautomerization

<u>Problem 24.16</u> Nylon 6,10 is prepared by polymerization of a diamine and a diacid chloride. Draw the structural formula of each reactant and for the repeat unit in nylon 6,10.

The six-carbon diamine is 1,6-hexanediamine and the ten-carbon diacid chloride is decanedioyl chloride.

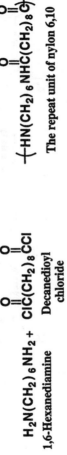

$H_2N(CH_2)_6NH_2$ +
1,6-Hexanediamine

Decanedioyl
chloride

$\{HN(CH_2)_6NHC(CH_2)_8C\}_n$
The repeat unit of nylon 6,10

<u>Problem 24.17</u> Polycarbonates (Section 24.5C) are also formed by using a nucleophilic aromatic substitution route (Section 21.3B) involving aromatic difluoro monomers and carbonate ion. Propose a mechanism for this reaction.

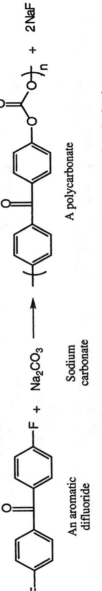

An aromatic
difluoride

Sodium
carbonate

A polycarbonate

+ 2NaF

Review Section 21.3B for the addition-elimination mechanism of nucleophilic aromatic substitution.

Step 1: Nucleophilic addition of carbonate ion to the aromatic ring at the carbon bearing the fluorine atom forms a Meisenheimer complex.

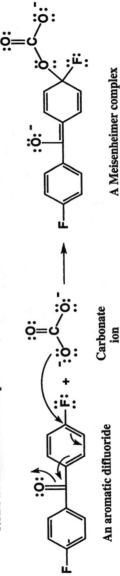

An aromatic difluoride

Carbonate
ion

A Meisenheimer complex

Step 2: Collapse of the Meisenheimer complex with ejection for fluoride ion.

<u>Problem 24.18</u> Propose a mechanism for the formation of this polyphenylurea. To simplify your presentation of the mechanism, consider the reaction of one -NCO group with one -NH₂ group.

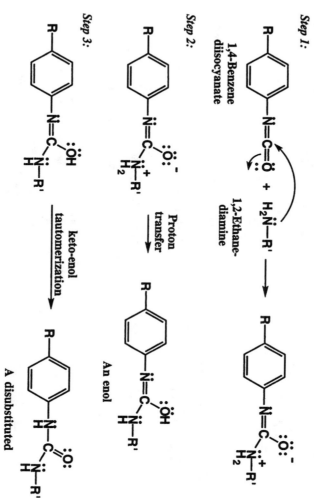

OCN—⬡—NCO   +   H₂N⌒⌒NH₂   ⟶

1,4-Benzenediisocyanate   1,2-Ethanediamine   Poly(ethylene phenylurea)

Propose addition of an amino group to the C=O of the isocyanate group to give an enol, followed by keto-enol tautomerism of the enol to give a disubstituted urea.

Step 1:

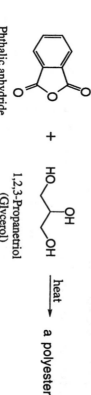

1,2-Ethane-diamine

Step 2:

Proton transfer

An enol

Step 3:

keto-enol tautomerization

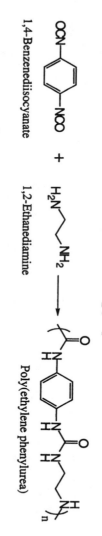

A disubstituted urea

<u>Problem 24.19</u> When equal molar amounts of phthalic anhydride and 1,2,3-propanetriol are heated, they form an amorphous polyester. Under these conditions, polymerization is regioselective for the primary hydroxyl groups of the triol.

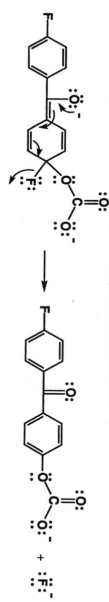

Phthalic anhydride

     +   HO⌒OH, OH

1,2,3-Propanetriol
(Glycerol)

  → heat →   a polyester

(a) Draw a structural formula for the repeat unit of this polyester

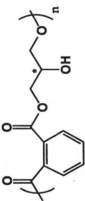

(b) Account for the regioselective reaction with the primary hydroxyl groups only.

The regioselectivity reflects the fact that the 1° hydroxyl groups are more accessible to reaction than the 2° hydroxyl group.

Problem 24.20 The polyester from Problem 24.19 can be mixed with additional phthalic anhydride for each mol of 1,2,3-propanetriol in the original polyester) to form a liquid resin. When this resin is heated, it forms a hard, insoluble, thermosetting polyester called glyptal.
(a) Propose a structure for the repeat unit in glyptal.

The polymer described in Problem 24.19 becomes cross linked as shown.

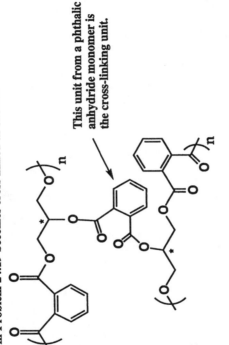

This unit from a phthalic anhydride monomer is the cross-linking unit.

(b) Account for the fact that glyptal is a thermosetting plastic.

Because of the extensive cross linking, the individual polymer chains can no longer be made to flow and, therefore, the polymer cannot be made to assume a liquid state.

Problem 24.21 Propose a mechanism for the formation of the following polymer.

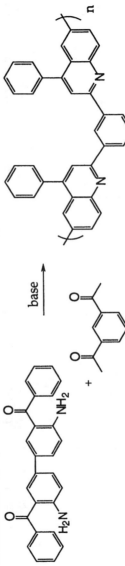

base

One way to attack this problem is to first determine which rings in the product are present in the original monomers, and which are formed during the polymerization. The monomer units are redrawn here to show that new rings are formed during polymerization by aldol reactions followed by dehydration and by imine formation.

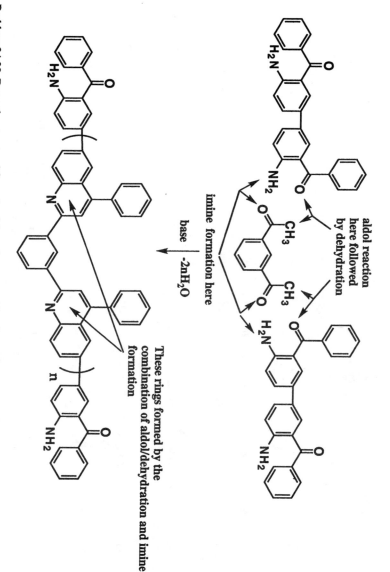

aldol reaction
here followed
by dehydration

These rings formed by the
combination of aldol/dehydration and imine
formation

imine formation here

base | -2nH₂O

Problem 24.22 Draw the structural formula of the polymer resulting from base-catalyzed polymerization of each compound. Would you expect the polymers to be optically active? (S)-(+)-lactide is the dilactone formed from two molecules of (S)-(+)-lactic acid.

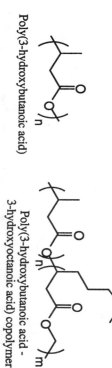

(a)

(S)-(+)-lactide

each stereocenter has the S
configuration; the polymer
is optically active

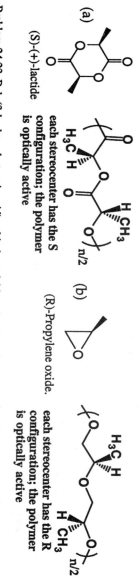

(b)

(R)-Propylene oxide.

each stereocenter has the R
configuration; the polymer
is optically active

Problem 24.23 Poly(3-hydroxybutanoic acid), a biodegradable polyester, is an insoluble, opaque material that is difficult to process into shapes. In contrast, the copolymer of 3-hydroxybutanoic acid and 3-hydroxyoctanoic acid is a transparent polymer that shows good solubility in a number of organic solvents. Explain the difference in properties between these two polymers in terms of their structure.

Poly(3-hydroxybutanoic acid)

Poly(3-hydroxybutanoic acid -
3-hydroxyoctanoic acid) copolymer

The polymer chains of poly(3-hydroxybutanoic acid) can assume a highly ordered arrangement with a high degree of crystallinity, hence its insolubility and its opaque character. In contrast, the polymer chains of the copolymer of 3-hydroxybutanoic acid and 3-hydroxyoctanoic acid have bulky five-carbon chains that effectively prevent polymer chains from assuming any regular ordered structure. As a result, the polymer has little crystalline character, that is, it is an amorphous material with little crystalline character to reflect light.

Problem 24.24 How might you determine experimentally if a particular polymerization is propagating by a step-growth or a chain-growth mechanism?

Analyze the distribution of polymer molecular weights as a function of degree of polymerization. As discussed in the introduction to Section 24.5, high molecular weight polymers are not produced until very late in step-growth polymerization, typically past 99% conversion of monomers to polymers. Given the mechanism of chain-growth polymerization, high molecular weight polymer molecules are produced very early and continuously in the polymerization process.

Problem 24.25 Draw a structural formula for the polymer formed in the following reactions.

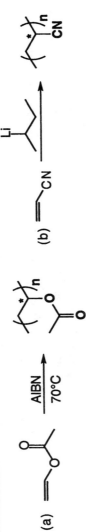

Problem 24.26 Select the monomer in each pair that is more reactive toward cationic polymerization.

The more reactive monomer in each pair is the one forming the more stable carbocation. The first structure in each pair forms the more stable carbocation.

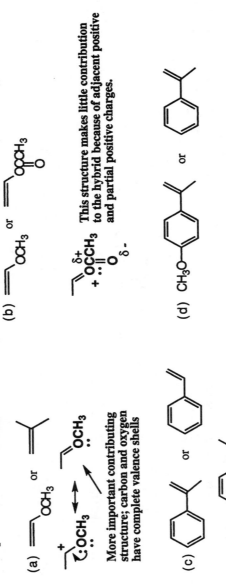

Problem 24.27 Polymerization of vinyl acetate gives poly(vinyl acetate). Hydrolysis of this polymer in aqueous sodium hydroxide gives poly(vinyl alcohol). Draw the repeat units of both poly(vinyl acetate) and poly(vinyl alcohol).

Following are structural formulas for the monomer and repeat unit of each polymer.

Vinyl acetate   Poly(vinyl acetate)   Poly(vinyl alcohol)

Problem 24.28 Benzoquinone can be used to inhibit radical polymerizations. This compound reacts with a radical intermediate, R•, to form a less reactive radical that does not participate in chain propagation steps and, thus, breaks the chain.

Draw a series of contributing structures for this less reactive radical and account for its stability.

This radical can be represented as a hybrid of five contributing structures; three place the single electron on carbon atoms of the ring, and two place it on the oxygen atoms bonded to the ring. This radical is stabilized by the significant degree of delocalization of the single electron.

Problem 24.29 Following is the structural formula of a section of polypropylene derived from three units of propylene monomer.

$$-CH_2CH-CH_2CH-CH_2CH-$$
$$\quad CH_3 \qquad CH_3 \qquad CH_3$$

Draw structural formulas for comparable sections of:

(a) poly(vinyl chloride)

(b) polytetrafluoroethylene

$$-CF_2CF_2-CF_2CF_2-CF_2CF_2-$$

(c) poly(methyl methacrylate) (Plexiglas)

(d) poly(1,1-dichloroethylene)

$$-CH_2CCl_2-CH_2CCl_2-CH_2CCl_2-$$

Problem 24.30 Low-density polyethylene (LDPE) has a higher degree of chain branching than high-density polyethylene (HDPE). Explain the relationship between chain branching and density.

Unbranched polyethylene packs more efficiently into compact structures which have more mass per unit volume than structures formed by packing of branched polyethylene chains. Therefore, unbranched polyethylene has a higher density than branched-chain polyethylene.

Problem 24.31 We saw how intramolecular chain transfer in radical polymerization of ethylene creates a four-carbon branch on a polyethylene chain. What branch is created by a comparable intramolecular chain transfer during radical polymerization of styrene.

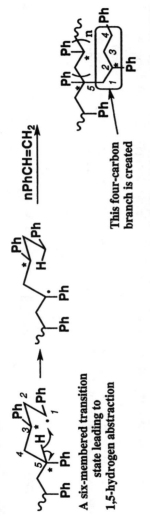

A six-membered transition
state leading to
1,5-hydrogen abstraction

This four-carbon
branch is created

Problem 24.32 Compare the densities of low-density polyethylene (LDPE) and high-density polyethylene (HDPE) with the densities of the liquid alkanes listed in Table 2.4. How might you account for the differences between them?

Given in the table are densities of several liquid alkanes reported in Table 2.4, plus densities for pentadecane, eicosane, and tricosane. As you can see, densities for these unbranched alkanes reach a maximum in the range 0.77-0.79 g/mL, which is significantly less than the density of both LDPE and HDPE. From this data, we conclude that both LDPE and HDPE pack more efficiently (have greater mass per unit volume) than their lower molecular weight counterparts.

Alkane	Formula	Density (g/mL)
Pentane	C_5H_{12}	0.626
Heptane	C_7H_{16}	0.684
Decane	$C_{10}H_{22}$	0.730
Pentadecane	$C_{15}H_{32}$	0.769
Eicosane	$C_{20}H_{42}$	0.789
Tricosane	$C_{30}H_{62}$	0.779
LDPE	$+CH_2+_n$	0.91-0.94
HDPE	$+CH_2+_n$	0.96

Problem 24.33 Natural rubber is the all cis polymer of 2-methyl-1,3-butadiene (isoprene).

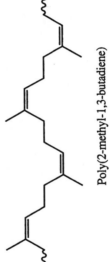

Poly(2-methyl-1,3-butadiene)
(Polyisoprene)

(a) Draw the structural formula for the repeat unit of natural rubber.

(b) Draw the structural formula of the product of oxidation of natural rubber by ozone followed by a workup in the presence of $(CH_3)_2S$. Name each functional group present in this product.

Aldehyde   Ketone

4-Oxopentanal

(c) The smog prevalent in many major metropolitan areas contains oxidizing agents, including ozone. Account for the fact that this type of smog attacks natural rubber (automobile tires, etc.) but does not attack polyethylene or polyvinyl chloride.

Polyethylene and poly(vinyl chloride) do not contain carbon-carbon double bonds, which are susceptible to attack by oxidizing agents such as ozone.

(d) Account for the fact that natural rubber is an elastomer but the synthetic all-trans isomer is not.

The natural cis isomer is kinked by virtue of the cis bond geometry while the all trans isomer has a more uniform staggered polymer chain. The trans synthetic rubber chains can thus pack together better making it more rigid compared to natural rubber.

Problem 24.34 Radical polymerization of styrene gives a linear polymer. Radical polymerization of a mixture of styrene and 1,4-divinylbenzene gives a cross linked network polymer of the type shown in Figure 24.1. Show by drawing structural formulas how incorporation of a few percent 1,4-divinylbenzene in the polymerization mixture gives a cross linked polymer.

Drawn here is a section of the copolymer showing cross linking by one molecule of 1,4-divinylbenzene. Benzene rings derived from, PhCH=CH₂, are shown as Ph.

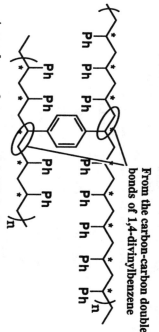

From the carbon-carbon double bonds of 1,4-divinylbenzene

A copolymer of styrene and 1,4-divinylbenzene

Problem 24.35 One common type of cation exchange resin is prepared by polymerization of a mixture containing styrene and 1,4-divinylbenzene (Problem 24.34). The polymer is then treated with concentrated sulfuric acid to sulfonate a majority of the aromatic rings in the polymer.
(a) Show the product of sulfonation of each benzene ring.

The following is a structural formula for a section of the polymer. Structural formulas for only the sulfonated rings are written in full; unsulfonated benzene rings are shown as Ph.

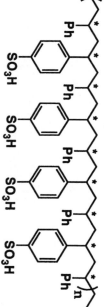

(b) Explain how this sulfonated polymer can act as a cation exchange resin.

The resin is shown in the acid or protonated form. When functioning as a cation exchange resin, cations displace H⁺ and become bound to the negatively charged -SO₃⁻ groups.

Problem 24.36 The most widely used synthetic rubber is a copolymer of styrene and butadiene called SB rubber. Ratios of butadiene to styrene used in polymerization vary depending on the end use of the polymer. The ratio used most commonly in the preparation of SB rubber for use in automobile tires is 1 mol styrene to 3 mol butadiene. Draw a structural formula of a section of the polymer formed from this ratio of reactants. Assume that all carbon-carbon double bonds in the polymer chain are in the cis configuration.

derived from 1,3-butadiene

derived from styrene

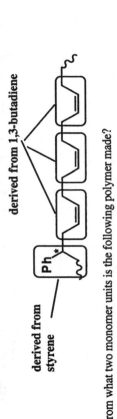

Problem 24.37 From what two monomer units is the following polymer made?

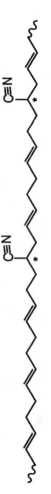

The section of polymer drawn here is derived six 1,3-butadiene monomer units and two acrylonitrile monomer units.

Proboem 24.38 Draw the structure of the polymer formed from ring-opening metathesis polymerization (ROMP) of each monomer.

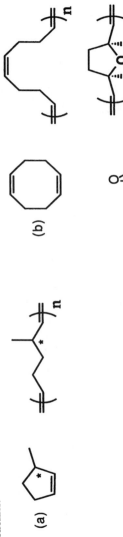

(a)

(b)

(c)

(d)

Reticulated butterflyfish *Chaetodon reticulatus*
Bora Bora, French Polynesia

CHAPTER 25
Solutions to the Problems

Problem 25.1 Draw Fischer projections for all 2-ketopentoses. Which are D-ketopentoses, which are L-ketopentoses, and which are enantiomers? Refer to Table 25.2, and write names of the ketopentoses you have drawn.

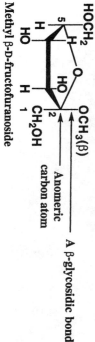

D-Ribulose   L-Ribulose

a pair of enantiomers

D-Xylulose   L-Xylulose

a pair of enantiomers

Problem 25.2 Mannose exists in aqueous solution as a mixture of α-D-mannopyranose and β-D-mannopyranose. Draw Haworth projections for these molecules.

D-Mannose differs in configuration from D-glucose at carbon-2. Therefore, the alpha and beta forms of D-mannopyranose differ from those of alpha and beta D-glucopyranoses only in the orientation of the -OH on carbon-2. Following are Haworth projections for these compounds.

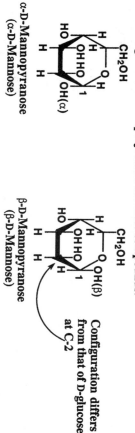

α-D-Mannopyranose
(α-D-Mannose)

β-D-Mannopyranose
(β-D-Mannose)

Configuration differs
from that of D-glucose
at C-2

Problem 25.3 Draw chair conformations for α-D-mannopyranose and β-D-mannopyranose. Label the anomeric carbon atom in each.

D-Mannose differs in configuration from D-glucose at carbon-2. Draw chair conformations for the alpha and beta forms of D-glucopyranose and then invert the configuration of the -OH on carbon-2. For reference, the open chain form of D-mannose is also drawn.

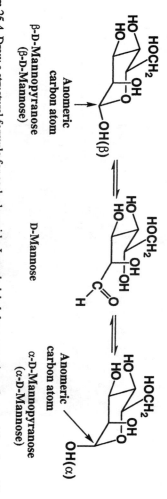

α-D-Mannopyranose
(α-D-Mannose)

β-D-Mannopyranose
(β-D-Mannose)

Anomeric
carbon atom

D-Mannose

Anomeric
carbon atom

Problem 25.4 Draw a structural formula for each glycoside. In each, label the anomeric carbon and the glycosidic bond.
(a) Methyl β-D-fructofuranoside (methyl β-D-fructoside); draw a Haworth projection.

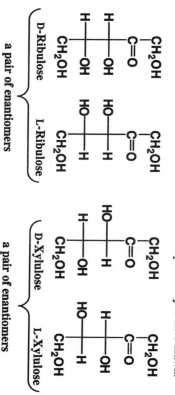

Anomeric
carbon atom

A β-glycosidic bond

Methyl β-D-fructofuranoside

(b) Methyl α-D-mannopyranoside (methyl α-D-mannoside); draw both a Haworth projection and a chair conformation.

Methyl α-D-mannopyranoside
(Chair conformation)

Methyl α-D-mannopyranoside
(Haworth projection)

Problem 25.5 Draw a structural formula for the β-N-glycoside formed between 2-deoxy-D-ribofuranose and adenine.

Following are structural formulas for adenine, the monosaccharide hemiacetal, and the N-glycoside.

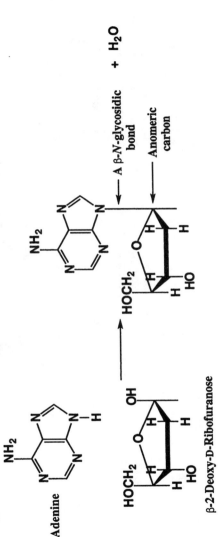

Adenine

β-2-Deoxy-D-Ribofuranose

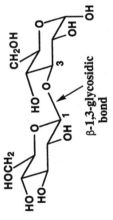

A β-N-glycosidic bond

Anomeric carbon

+ H₂O

Problem 25.6 Draw a chair conformation for the α anomer of a disaccharide in which two units of D-glucopyranose are joined by a β-1,3-glycosidic bond.

β-1,3-glycosidic bond

PROBLEMS

Monosaccharides

Problem 25.7 Explain the meaning of the designations D and L as used to specify the configuration of carbohydrates.

The designations D and L refer to the configuration of the stereocenter farthest from the carbonyl group of the monosaccharide. When a monosaccharide is drawn in a Fischer projection, the reference -OH is on the right in a D-monosaccharide and on the left in an L-monosaccharide. Note that the conventions D and L specify the configuration at one and only one of however many stereocenters there are in a particular monosaccharide.

Problem 25.8 How many stereocenters are present in D-glucose? In D-ribose?

The number of stereocenters present depends on whether you consider the open chain or cyclic forms of the molecules. In each case, an additional stereocenter is present in the cyclic forms due to the anomeric carbon atom. For D-glucose there are 4 (open chain) and 5 stereocenters (cyclic form), while for D-ribose there are 3 (open chain) and 4 (cyclic form) stereocenters.

Problem 25.9 Which carbon of an aldopentose determines whether the pentose has a D or L configuration?

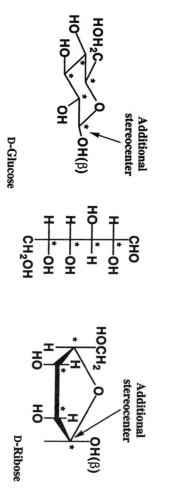

D-Glucose

Additional stereocenter

Additional stereocenter

OH(β)

OH(β)

D-Ribose

The stereochemistry at this carbon determines configuration for aldopentoses

CHO
CHOH
CHOH
CHOH
CH₂OH

The stereochemistry at this carbon determines the configuration for aldopentoses

The reference point for determining D or L configuration is the stereocenter farthest from the carbonyl group. This stereocenter is always the next to the last carbon on the chain.

Problem 25.10 How many aldooctoses are possible? How many D-aldooctoses are possible?

The number of different aldooctoses possible is determined by the number of stereocenters in the open chain form of the molecule. There are 6 stereocenters in the open chain form of aldooctoses so there are $2^6 = 64$ possible. Half of those will have the D configuration, so there are 32 possible D-aldooctoses.

Problem 25.11 Which compounds are D-monosaccharides and which are L-monosaccharides?

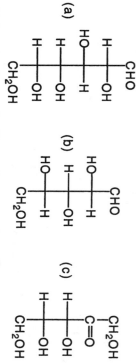

(a)

CHO
H—OH
HO—H
H—OH
H—OH
CH₂OH

(b)

CHO
HO—H
HO—H
HO—H
H—OH
CH₂OH

(c)

CH₂OH
C=O
H—OH
H—OH
CH₂OH

Compounds (a) and (c) are D-monosaccharides, and compound (b) is an L-monosaccharide.

Problem 25.12 Write Fischer projections for L-ribose and L-arabinose.

L-Ribose and L-arabinose are the mirror images of D-ribose and D-arabinose, respectively. The most common error in answering this question is to start with the Fischer projection for the D sugar and then invert the configuration of carbon-4 only. While the monosaccharide thus drawn is an L-sugar, it is not the correct one. All of the stereocenters must be changed to draw the true enantiomers.

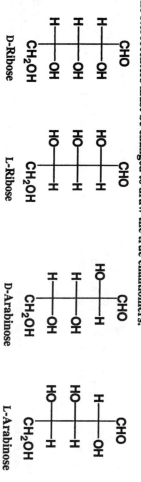

CHO
H—OH
H—OH
H—OH
CH₂OH

D-Ribose

CHO
HO—H
HO—H
HO—H
CH₂OH

L-Ribose

CHO
HO—H
H—OH
H—OH
CH₂OH

D-Arabinose

CHO
H—OH
HO—H
HO—H
CH₂OH

L-Arabinose

Problem 25.13 What is the meaning of the prefix deoxy- as it is used in carbohydrate chemistry?

The prefix deoxy- means "without oxygen".

Problem 25.14 Give L-fucose a name incorporating the prefix deoxy- that shows its relationship to galactose.

A systemic name for L-fucose is 6-deoxy-L-galactose.

Problem 25.15 2,6-Dideoxy-D-altrose, known alternatively as D-digitoxose, is a monosaccharide obtained on hydrolysis of digitoxin, a natural product extracted from foxglove (*Digitalis purpurea*). Digitoxin is widely use in cardiology to reduce pulse rate, regularize heart rhythm, and strengthen heart beat (see The Merck Index, 12th ed., #3206). Draw the structural formula of 2,6-dideoxy-D-altrose.

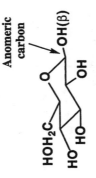

2,6-Dideoxy-D-altose
(D-Digitoxose)

The Cyclic Structure of Monosaccharides
Problem 25.16 Define the term anomeric carbon. In glucose, which carbon is the anomeric carbon?

The new stereocenter created from the carbonyl in forming the cyclic form of a carbohydrate is called the anomeric carbon.

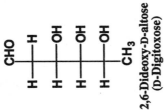

β-D-Glucose

Proboem 25.17 Define (a) pyranose and (b) furanose.

The six-membered hemiacetal ring formed by a carbohydrate is called a pyranose. The five-membered hemiacetal ring formed by a carbohydrate is called a furanose.

Problem 25.18 Which is the anomeric carbon in a 2-ketohexose?

A 2-ketohexose forms a five-membered ring, and the anomeric carbon is the new stereocenter created from the carbonyl.

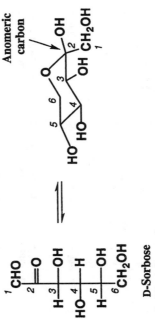

D-Sorbose
(a 2-ketohexose)

Problem 25.19 Are α-D-glucose and β-D-glucose enantiomers? Explain.

They are not enantiomers as only the anomeric carbon atom is different. The other stereocenters are the same, thus α-D-glucose and β-D-glucose are diastereomers.

Problem 25.20 Convert each Haworth projection to an open-chain Fischer projection and name the monosaccharide you have drawn.

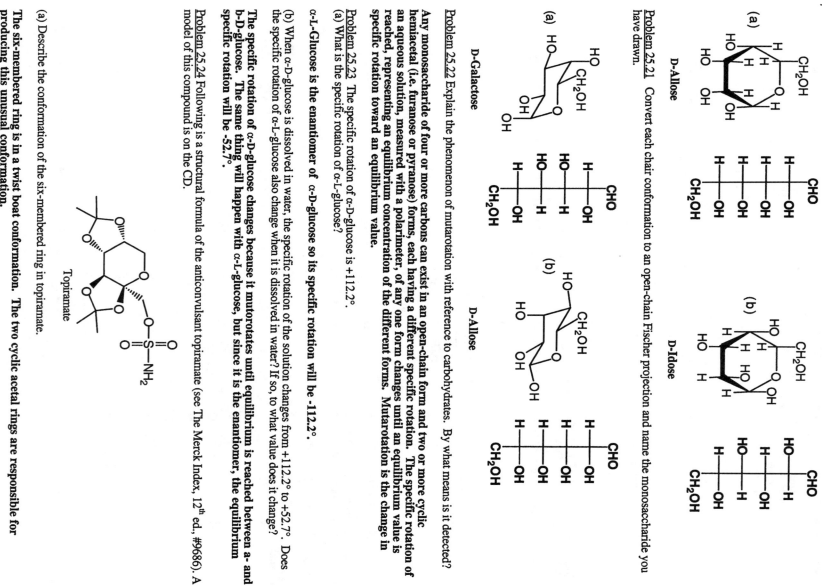

(a)

D-Allose

(b)

D-Idose

Problem 25.21 Convert each chair conformation to an open-chain Fischer projection and name the monosaccharide you have drawn.

(a)

D-Galactose

(b)

D-Allose

Problem 25.22 Explain the phenomenon of mutarotation with reference to carbohydrates. By what means is it detected?

Any monosaccharide of four or more carbons can exist in an open-chain form and two or more cyclic hemiacetal (i.e. furanose or pyranose) forms, each having a different specific rotation. The specific rotation of an aqueous solution, measured with a polarimeter, of any one form changes until an equilibrium value is reached, representing an equilibrium concentration of the different forms. Mutarotation is the change in specific rotation toward an equilibrium value.

Problem 25.23 The specific rotation of α-D-glucose is +112.2°.
(a) What is the specific rotation of α-L-glucose?

α-L-Glucose is the enantiomer of α-D-glucose so its specific rotation will be -112.2°.

(b) When α-D-glucose is dissolved in water, the specific rotation of the solution changes from +112.2° to +52.7°. Does the specific rotation of α-L-glucose also change when it is dissolved in water? If so, to what value does it change?

The specific rotation of α-D-glucose changes because it mutorotates until equilibrium is reached between α- and b-D-glucose. The same thing will happen with α-L-glucose, but since it is the enantiomer, the equilibrium specific rotation will be -52.7°.

Problem 25.24 Following is a structural formula of the anticonvulsant topiramate (see The Merck Index, 12th ed., #9686). A model of this compound is on the CD.

Topiramate

(a) Describe the conformation of the six-membered ring in topiramate.

The six-membered ring is in a twist boat conformation. The two cyclic acetal rings are responsible for producing this unusual conformation.

(b) Draw a structural formula of the cyclic structure remaining after hydrolysis of each amide, ester, and cyclic acetal.
(c) The structure remaining is a monosaccharide. Name the monosaccharide and specify whether it belongs to the D series or the L-series.

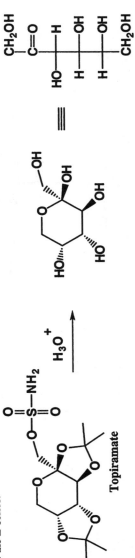

Topiramate

$\xrightarrow{H_3O^+}$

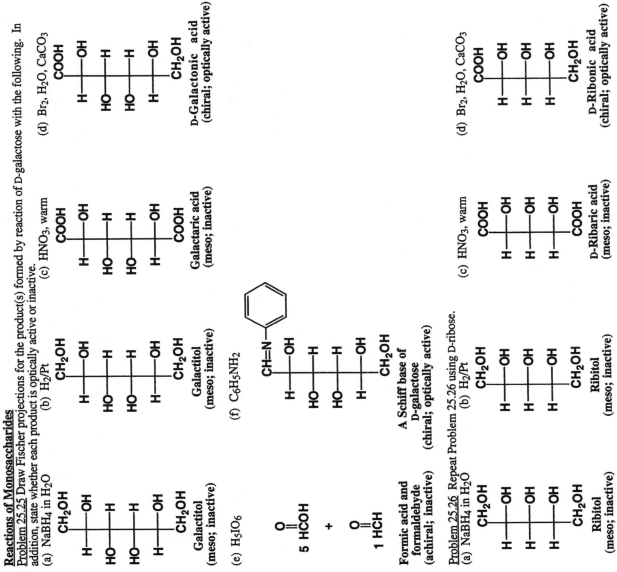

D-Fructose

After complete hydrolysis, D-fructose is the monosaccharide produced.

Reactions of Monosaccharides

Problem 25.25 Draw Fischer projections for the product(s) formed by reaction of D-galactose with the following. In addition, state whether each product is optically active or inactive.

(a) NaBH₄ in H₂O

CH₂OH
H—OH
HO—H
HO—H
H—OH
CH₂OH
Galactitol
(meso; inactive)

(b) H₂/Pt

CH₂OH
H—OH
HO—H
HO—H
H—OH
CH₂OH
Galactitol
(meso; inactive)

(c) HNO₃, warm

COOH
H—OH
HO—H
HO—H
H—OH
COOH
Galactaric acid
(meso; inactive)

(d) Br₂, H₂O, CaCO₃

COOH
H—OH
HO—H
HO—H
H—OH
CH₂OH
D-Galactonic acid
(chiral; optically active)

(e) H₅IO₆

O=HCOH   5
+
O=HCH   1

Formic acid and formaldehyde
(achiral; inactive)

(f) C₆H₅NH₂

CH=N—C₆H₅
H—OH
HO—H
HO—H
H—OH
CH₂OH
A Schiff base of
D-galactose
(chiral; optically active)

Problem 25.26 Repeat Problem 25.25 using D-ribose.

(a) NaBH₄ in H₂O

CH₂OH
H—OH
H—OH
H—OH
CH₂OH
Ribitol
(meso; inactive)

(b) H₂/Pt

CH₂OH
H—OH
H—OH
H—OH
CH₂OH
Ribitol
(meso; inactive)

(c) HNO₃, warm

COOH
H—OH
H—OH
H—OH
COOH
D-Ribaric acid
(meso; inactive)

(d) Br₂, H₂O, CaCO₃

COOH
H—OH
H—OH
H—OH
CH₂OH
D-Ribonic acid
(chiral; optically active)

(e) H_5IO_6

$$\text{4 HCOH} \ (O)$$

$$\text{+}$$

$$\text{1 HCH} \ (O)$$

(f) $C_6H_5NH_2$

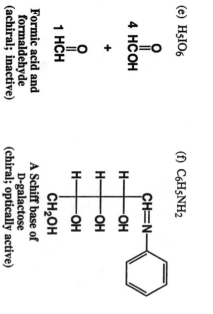

Formic acid and formaldehyde
(achiral; inactive)

A Schiff base of D-galactose
(chiral; optically active)

Problem 25.27 An important technique for establishing relative configurations among isomeric aldoses and ketoses is to convert both terminal carbon atoms to the same functional group. This can be done either by selective oxidation or reduction. As a specific example, nitric acid oxidation of D-erythrose gives meso-tartaric acid (Table 3.1, Section 3.4). Similar oxidation of D-threose gives (2S,3S)-tartaric acid. Given this information and the fact that D-erythrose and D-threose are diastereomers, draw Fischer projections for D-erythrose and D-threose. Check your answers against Table 25.1.

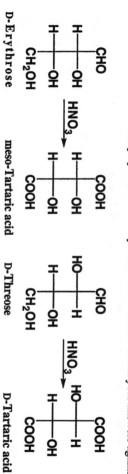

D-Erythrose   meso-Tartaric acid   D-Threose   D-Tartaric acid

Problem 25.28 There are four D-aldopentoses (Table 25.1). If each is reduced with $NaBH_4$, which yield optically active alditols? Which yield optically inactive alditols?

D-Ribose and D-xylose yield different achiral (meso) alditols. D-Arabinose and D-lyxose yield the same chiral alditol.

D-Ribose   Ribitol (meso)

D-Xylose   Xylitol (meso)

D-Arabinose   D-Arabinitol
$[\alpha]_D^{25} = -32°$

D-Lyxose

Problem 25.29 Name the two alditols formed by $NaBH_4$ reduction of D-fructose?

D-Glucitol and D-mannitol. Each differs in configuration only at carbon-2.

Problem 25.30 One pathway for the metabolism of glucose 6-phosphate is its enzyme-catalyzed conversion to fructose 6-phosphate. Show that this transformation can be regarded as two enzyme-catalyzed keto-enol tautomerisms.

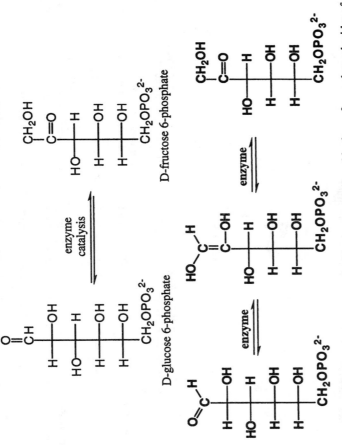

Problem 25.31 L-Fucose, one of several monosaccharides commonly found in the surface polysaccharides of animal cells, is synthesized biochemically from D-mannose in the following eight steps:

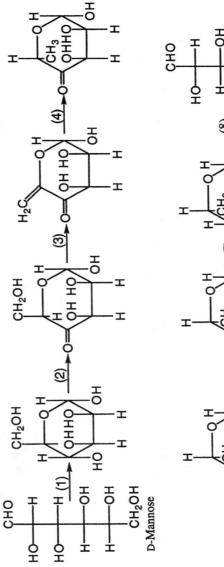

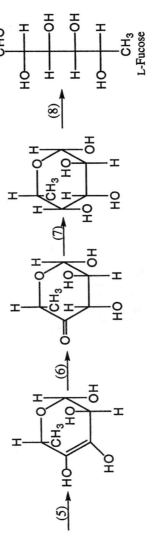

(a) Describe the type of reaction in each step.

Following is the type of reaction in each step.
(1) Formation of a cyclic hemiacetal from a carbonyl group and a secondary alcohol.
(2) A two-electron oxidation of a secondary alcohol to a ketone.
(3) Dehydration of a β-hydroxyketone to an α,β-unsaturated ketone.
(4) A two-electron reduction of a carbon-carbon double bond to a carbon-carbon single bond.
(5) Keto-enol tautomerism of an α-hydroxyketone to form an enediol.

(6) Keto-enol tautomerism of an enediol to form an α-hydroxyketone.

(7) A two-electron reduction of a ketone to a secondary alcohol.

(8) Opening of a cyclic hemiacetal to form an aldehyde and an alcohol.

(b) Explain why it is that this monosaccharide derived from D-mannose now belongs to the L series.

It is the configuration at carbon-5 of this aldohexose that determines whether it is of the D-series or of the L-series. The result of steps 3 and 4 is inversion of configuration at carbon-5 and, therefore, conversion of a D-aldohexose to an L-aldohexose.

Problem 25.32 What is the difference in meaning between the terms glycosidic bond and glucosidic bond?

A glycosidic bond is the bond from the anomeric carbon of a glycoside to an -OR group. A glucosidic bond is a glycosidic bond that yields glucose upon hydrolysis.

Problem 25.33 Treatment of methyl β-D-glucopyranoside with benzaldehyde forms a six-membered cyclic acetal. Draw the most stable conformation of this acetal. Identify each new stereocenter in the acetal.

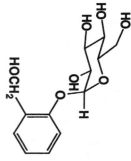

New stereocenter

Problem 25.34 Vanillin, the principal component of vanilla, occurs in vanilla beans and other natural sources as a β-D-glucopyranoside. Draw a structural formula for this glycoside, showing the D-glucose unit as a chair conformation. (See The Merck Index, 12th ed., #10069.)

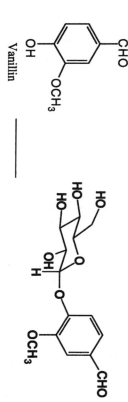

Vanillin

Problem 25.35 Hot water extracts of ground willow and poplar bark are an effective pain reliever. Unfortunately, the liquid is so bitter that most persons refuse it. The pain reliever in these infusions is salicin, a β-glycoside of D-glucopyranose and the phenolic -OH group of 2-(hydroxymethyl)phenol. Draw a structural formula for salicin, showing the glucose ring as a chair conformation. (See The Merck Index, 12th ed., 8476.)

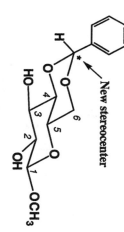

Problem 25.36 Draw structural formulas for the products formed by hydrolysis at pH 7.4 (the pH of blood plasma) of all ester, thioester, amide, anhydride, and glycoside groups in acetyl coenzyme A. Name as many of these compounds as you can.

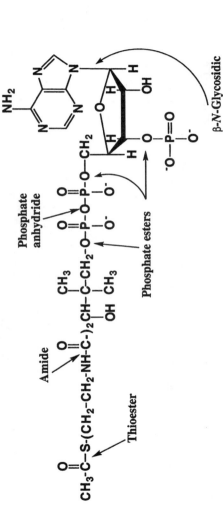

Following are the smaller molecules formed by hydrolysis of each amide, ester, glycosidic, and anhydride bond. They are arranged to correspond roughly to their location from left to right in acetyl CoA.

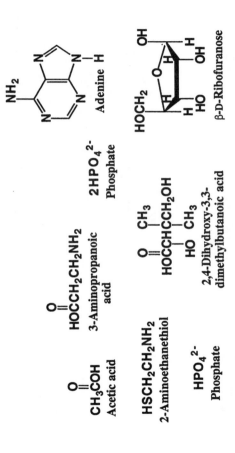

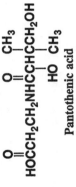

The molecule formed by amide formation between 3-aminopropanoic acid and 2,4-dihydroxy-3,3-dimethylbutanoic acid is given the special name pantothenic acid. Pantothenic acid is a vitamin, most commonly contained in vitamin pills as calcium pantothenate. Its minimum daily requirement (MDR) has not yet been determined.

Disaccharides and Oligosaccharides

Problem 25.37 In making candy or sugar syrups, sucrose is boiled in water with a little acid, such as lemon juice. Why does the product mixture taste sweeter than the starting sucrose solution?

Sucrose is a disaccharide composed of the monosaccharides glucose and fructose linked through a glycosidic bond. The acid catalyzes hydrolysis of the glycosidic bond, and the monomeric glucose and fructose are more soluble than sucrose itself. Because of this, the syrups and candy have higher concentrations of these sugars than is possible with sucrose. As a result, the candy and syrup taste sweeter. Furthermore, fructose actually tastes sweeter than sucrose, having a relative sweetness of 174, compared with 100 for sucrose. Thus, converting the sucrose into fructose increases the sweetness of the mixture.

Problem 25.38 Trehalose, found in young mushrooms and the chief carbohydrate in the blood of certain insects, is a disaccharide consisting of two D-monosaccharide units joined by an α-1,1-glycosidic bond.

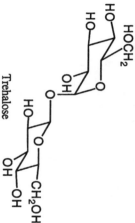

Trehalose

(a) Is trehalose a reducing sugar?

Trehalose is not a reducing sugar because each anomeric carbon is involved in formation of the glycosidic bond.

(b) Does trehalose undergo mutarotation?

It will not undergo mutarotation for the reason given in (a).

(c) Name the two monosaccharide units of which trehalose is composed.

Both monosaccharide units are D-Glucose.

Problem 25.39 The trisaccharide raffinose occurs principally in cottonseed meal (see The Merck Index 12th ed., #8279).

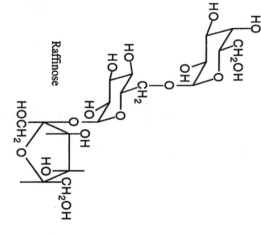

Raffinose

(a) Name the three monosaccharide units in raffinose.

The three monosaccharide units in raffinose, from top to bottom, are D-galactose, D-glucose, and D-fructose.

(b) Describe each glycosidic bond in this trisaccharide.

Reading from left to right, they are D-galactopyranose joined by an α-1,6-glycosidic bond to D-glucopyranose and then D-glucopyranose, in turn, joined by an α-1,2-glycosidic bond to β-D-fructofuranose.

(c) Is raffinose a reducing sugar?

No, it is not a reducing sugar.

(d) With how many mol of periodic acid will raffinose react?

Raffinose will react with 5 mol of HIO₄; 2 mol for the D-galactopyranose ring, 2 mol for the D-glucopyranose ring, and 1 mol for the D-fructofuranose ring.

Problem 25.40 Following is the structural formula of laetrile.

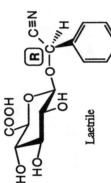

Laetrile

(a) Assign an R or S configuration to the stereocenter bearing the cyano (-CN) group.

The stereocenter has the R-configuration.

(b) Account for the fact that on hydrolysis in warm aqueous acid, laetrile liberates benzaldehyde and HCN.

Hydrolysis of the glycosidic bond in aqueous acid gives D-glucuronic acid and the cyanohydrin of benzaldehyde. This cyanohydrin is in equilibrium with benzaldehyde and HCN.

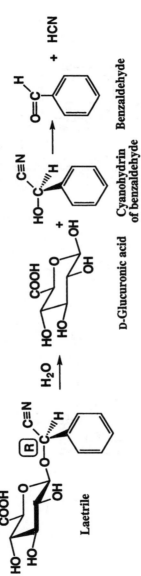

Laetrile

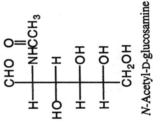

D-Glucuronic acid   Cyanohydrin   Benzaldehyde
                         of benzaldehyde

Polysaccharides
Problem 25.41 What is the difference in structure between oligo- and polysaccharides?

The general term oligosaccharide is often used for carbohydrates that contain from four to ten monosaccharide units. Carbohydrates containing larger numbers of monosaccharide units are called polysaccharides.

Problem 25.42 Why is cellulose insoluble in water?

The polysaccharide chains of cellulose are held together through extensive networks of hydrogen bonds. The chains are held together so tightly that water molecules cannot effectively solvate them.

Problem 25.43 Following is the Fischer projection for N-acetyl-D-glucosamine:

```
        CHO      O
         |       ‖
   H ———|——— NHCCH₃
         |
  HO ———|——— H
         |
   H ———|——— OH
         |
   H ———|——— OH
         |
       CH₂OH
```

N-Acetyl-D-glucosamine

(a) Draw a chair conformation for the α- and β-pyranose forms of this monosaccharide.

To draw the α- and β-pyranose forms, invert configuration at carbon-1.

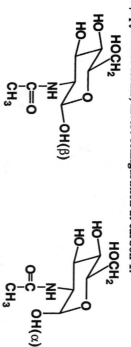

(b) Draw a chair conformation for the disaccharide formed by joining two units of the pyranose form of *N*-acetyl-D-glucosamine by a β-1,4-glycosidic bond. If you drew this correctly, you have the structural formula for the repeating dimer of chitin, the structural polysaccharide component of the shell of lobster and other crustaceans.

Following are Haworth and chair formulas for the β-anomer of this disaccharide.

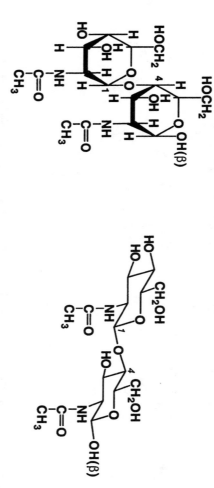

<u>Problem 25.44</u> Propose structural formulas for the following polysaccharides:
(a) Alginic acid, isolated from seaweed, is used as a thickening agent in ice cream and other foods. Alginic acid is a polymer of D-mannuronic acid in the pyranose form joined by β-1,4-glycosidic bonds.

D-Mannuronic acid   D-Galacturonic acid

Following is the chair conformation for repeating disaccharide units of alginic acid.

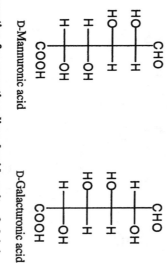

Alginic acid

(b) Pectic acid is the main component of pectin, which is responsible for the formation of jellies from fruits and berries. Pectic acid is a polymer of D-galacturonic acid in the pyranose form joined by α-1,4-glycosidic bonds.

Following is the chair conformation for repeating disaccharide units of pectic acid.

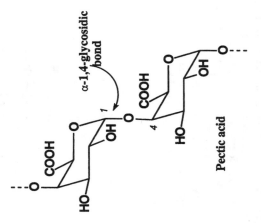

Pectic acid

Problem 25.45 Digitalis (see The Merck Index, 12th ed., 3201) is a preparation made from the dried seeds and leaves of the purple foxglove, *Digitalis purpurea*, a plant native to southern and central Europe and cultivated in the United States. The preparation is a mixture of several active components, including digitalin. Digitalis is used in medicine to increase the force of myocardial contraction and as a conduction depressant to decrease heart rate (the heart pumps more forcefully but less often).

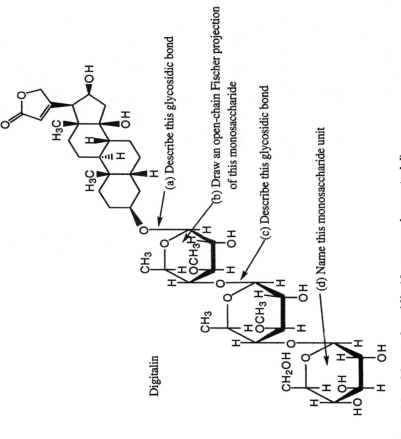

Digitalin

(a) Describe this glycosidic bond

(b) Draw an open-chain Fischer projection of this monosaccharide

(c) Describe this glycosidic bond

(d) Name this monosaccharide unit

a) The indicated bond is a β-glycosidic (the oxygen is equatorial).

b) The first monosaccharide corresponds to the following Fischer projection.

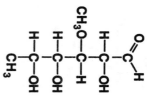

c) This bond is a β-1,4-glycosidic bond.
d) This monosaccharide is glucose.

Problem 25.46 Following is the structural formula of ganglioside GM₂, a macromolecular glycolipid (meaning that it contains lipid and monosaccharide units joined by glycosidic bonds). In normal cells, this and other gangliosides are synthesized continuously and degraded by lysosomes, which are cell organelles containing digestive enzymes. If pathways for the degradation of gangliosides are inhibited, the gangliosides accumulate in the central nervous system causing all sorts of life-threatening consequences. In inherited diseases of ganglioside metabolism, death usually occurs at an early age. Diseases of ganglioside metabolism include Gaucher's disease, Niemann-Pick disease, and Tay-Sachs disease. Tay-Sachs disease is a hereditary defect that is transmitted as an autosomal recessive gene. The concentration of ganglioside GM₂ is abnormally high in this disease because the enzyme responsible for catalyzing the hydrolysis of glycosidic bond (b) is absent.

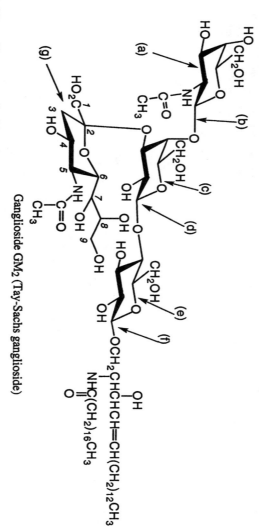

Ganglioside GM₂ (Tay-Sachs ganglioside)

(a) Name this monosaccharide unit.

This monosaccharide is N-acetyl-D-galactosamine

(b) Describe this glycosidic bond (α or β, and between which carbons of each unit).

Because the group is equatorial, this is a β-1,4-glycosidic bond.

(c) Name this monosaccharide unit.

This monosaccharide is D-galactose.

(d) Describe this glycosidic bond.

This is also a β-1,4-glycosidic bond.

(e) Name this monosaccharide unit.

This monosaccharide is D-glucose

(f) Describe this glycosidic bond.

This is a β-glycosidic bond.

(g) This unit is *N*-acetylneuraminic acid, the most abundant member of a family of amino sugars containing nine or more carbons and distributed widely throughout the animal kingdom. Draw the open-chain form of this amino sugar. Do not be concerned with the configuration of the five stereocenters in the open-chain form.

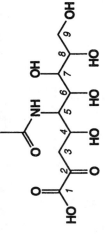

Problem 25.47 Hyaluronic acid acts as a lubricant in the synovial fluid of joints. In rheumatoid arthritis, inflammation breaks hyaluronic acid down to smaller molecules. Under these conditions, what happens to the lubricating power of the synovial fluid?

Smaller molecules are less effective lubricants because they are less able to prevent contact between joint components compared to longer molecules.

Problem 25.48 The anticlotting property of heparin is partly due to the negative charges it carries.
(a) Identify the functional groups that provide the negative charges.

Carboxylates and sulfonates are the functional groups that provide the negative charges of heparin.

(b) Which type of heparin is a better anticoagulant, one with a high or a low degree of polymerization?

Highly polymerized heparin is a better anticoagulant because it is more effective at binding to antithrombin III, its enzyme target.

Problem 25.49 Keratin sulfate is an important component of the cornea of the eye. Following is the repeating unit of this acidic polysaccharide.

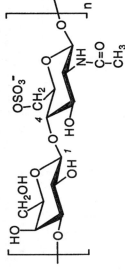

(a) From what monosaccharides or derivatives of monosaccharides is keratin sulfate made?

Keratin sulfate is made from D-Galactose and *N*-acetyl-D-glucosamine.

(b) Describe the glycosidic bond in this repeating disaccharide unit.

This is a β-1,4-glycosidic bond

(c) What is the net charge on this repeating disaccharide unit at pH 7.0?

The repeating disaccharide unit has a single negative charge at pH 7.0 due to the sulfonate group.

CHAPTER 26
Solutions to the Problems

Problem 26.1   How many isomers are possible for a triglyceride containing one molecule each of palmitic, oleic, and stearic acids?

There are three constitutional isomers possible, the difference being which fatty acid is in the middle of the molecule:

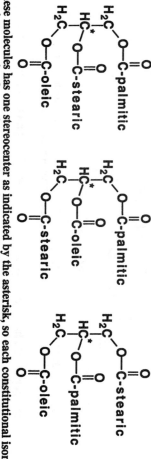

Each of these molecules has one stereocenter as indicated by the asterisk, so each constitutional isomer shown above can exist as a pair of enantiomers. Thus there are 2 x 3 = 6 total isomers possible. Note that for oleic acid, the carbon-carbon double bond is assumed to have the Z (*cis*) configuration only.

(b)   Which of these constitutional isomers are chiral?

They are all chiral because they each have a stereocenter at the middle carbon of the glycerol moiety.

Problems
Problem 26.2   Define the term "hydrophobic."

A good working definition of the term hydrophobic is "dislikes water". Hydrophobic materials are nonpolar, and do not dissolve in water. Because they are not solvated, hydrophobic surfaces of molecules aggregate in polar solvents like water. This aggregation behavior of hydrophobic surfaces is an extremely important phenomenon that underlies a number of important aspects of biological chemistry such as membrane formation, surfactant action, protein folding, and protein-protein recognition.

Problem 26.3   Identify the hydrophobic and the hydrophilic region(s) of a triglyceride.

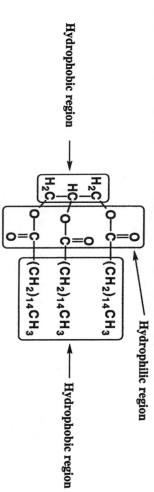

Hydrophobic region

Hydrophilic region

Hydrophobic region

Hydrophobic region

Note that the vast majority of the molecule is hydrophobic, thus explaining why triglycerides are so hydrophobic overall.

Problem 26.4   Explain why the melting points of unsaturated fatty acids are lower than those of saturated fatty acids.

When fatty acids pack together better, the attractive dispersion forces between molecules are stronger, thereby increasing the melting point. Saturated fatty acids can adopt a much more compact structure compared to unsaturated fatty acids that have a kink induced by the *cis* double bond. The more compact saturated fatty acids can pack together better, so their melting points are higher.

Problem 26.5   Which would you expect to have the higher melting point, glyceryl trioleate or glyceryl trilinoleate?

Cis double bonds disrupt chain packing and thus lower melting points. The triglyceride with fewer cis double bonds will have the higher melting point. Each oleic acid unit has only one cis double bond, while each linoleic acid has two (Please see Table 26.1). Glycerol trioleate will have the higher melting point.

Problem 26.6 Draw a structural formula for methyl linoleate. Be certain to show the correct stereochemistry about the carbon–carbon double bonds.

Methyl linoleate

Problem 26.7 Explain why coconut oil is a liquid triglyceride, even though most of its fatty acid components are saturated.

Triglycerides having fatty acids with shorter chains have lower melting points. As can be seen in Table 26.2, coconut oil is 45% lauric acid. Lauric acid is only a C12 fatty acid, so coconut oil has a melting point that is low enough to make it a liquid near room temperature.

Problem 26.8 It is common now to see "contains no tropical oils" on cooking oil labels, meaning that the oil contains no palm or coconut oil. What is the difference between the composition of tropical oils and that of vegetable oils, such as corn oil, soybean oil, and peanut oil?

Tropical oils are considerably richer in low-molecular-weight saturated fatty acids.

Problem 26.9 What is meant by the term "hardening" as applied to fats and oils?

The term "hardening" refers to the process of catalytic hydrogenation using H_2 and a transition metal with polyunsaturated plant oils. By removing the Z double bonds, the reduction reaction allows the fatty acids to pack together better and thus the triacylglycerols become more solid.

Problem 26.10 How many moles of H_2 are used in the catalytic hydrogenation of one mole of a triglyceride derived from glycerol, stearic acid, linoleic acid, and arachidonic acid?

One molecule of H_2 is used per double bond in the triglyceride. Stearic acid does not have any double bonds, linoleic acid has 2 and arachidonic acid has 4 double bonds, respectively. Thus, $2 + 4 = 6$ mole of H_2 will be used per mole of the triglyceride.

Problem 26.11 Characterize the structural features necessary to make a good synthetic detergent.

A good synthetic detergent should have a long hydrocarbon tail and a very polar group at one end. This combination will allow for the production of micelle structures in aqueous solution that will dissolve hydrophobic dirt such as grease and oil. The very polar group should not form insoluble salts with the ions normally found in hard water such as Ca(II), Mg(II), and Fe(III).

Problem 26.12 Following are structural formulas for a cationic detergent and a neutral detergent. Account for the detergent properties of each.

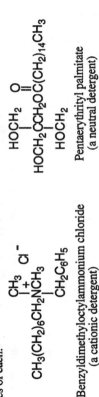

$$CH_3(CH_2)_6CH_2NCH_3 \quad Cl^-$$

Benzyldimethyloctylammonium chloride
(a cationic detergent)

Pentaerythrityl palmitate
(a neutral detergent)

In each case there is a long hydrocarbon tail attached to a very polar group. This combination will allow for the production of micelle structures in aqueous solution that will dissolve nonpolar, hydrophobic dirt such as grease and oil. In the case of benzyldimethyloctylammonium chloride, the polar group is the positively-charged ammonium group, while for the pentaerythrityl palmitate the polar group is composed of the triol functions.

Problem 26.13 Identify some of the detergents used in shampoos and dish washing solutions. Are they primarily anionic, neutral, or cationic detergents.

Most detergents in shampoos and dish washing solutions are anionic detergents such as sodium lauryl sulfate.

$$CH_3(CH_2)_{10}CH_2O-S-O^-Na^+$$

Sodium Lauryl Sulfate

Problem 26.14 Show how to convert palmitic acid (hexadecanoic acid) into the following:

(a) Ethyl palmitate

$$CH_3(CH_2)_{14}\overset{O}{\overset{\|}{C}}OH \ + \ CH_3CH_2OH \ \xrightarrow{H^+} \ CH_3(CH_2)_{14}\overset{O}{\overset{\|}{C}}OCH_2CH_3$$
Ethyl palmitate

(b) Palmitoyl chloride

$$CH_3(CH_2)_{14}\overset{O}{\overset{\|}{C}}OH \ + \ SOCl_2 \ \longrightarrow \ CH_3(CH_2)_{14}\overset{O}{\overset{\|}{C}}Cl$$
Palmitoyl chloride

(c) 1-Hexadecanol (cetyl alcohol)

$$CH_3(CH_2)_{14}\overset{O}{\overset{\|}{C}}OH \ \xrightarrow[2) \ H_2O]{1) \ LiAlH_4, \ ether \ or \ THF} \ CH_3(CH_2)_{14}CH_2OH$$
1-Hexadecanol
(Cetyl alcohol)

(d) 1-Hexadecanamine

$$CH_3(CH_2)_{14}\overset{O}{\overset{\|}{C}}OH \ + \ SOCl_2 \ \longrightarrow \ CH_3(CH_2)_{14}\overset{O}{\overset{\|}{C}}Cl \ \xrightarrow{NH_3} \ CH_3(CH_2)_{14}\overset{O}{\overset{\|}{C}}NH_2$$

$$\xrightarrow[2. \ H_2O]{1. \ LiAlH_4, \ ether \ or \ THF} \ CH_3(CH_2)_{14}CH_2NH_2$$
1-Hexadecanamine

(e) *N,N*-Dimethylhexadecanamide

$$CH_3(CH_2)_{14}\overset{O}{\overset{\|}{C}}OH \ + \ SOCl_2 \ \longrightarrow \ CH_3(CH_2)_{14}\overset{O}{\overset{\|}{C}}Cl \ \xrightarrow{HN(CH_3)_2} \ CH_3(CH_2)_{14}\overset{O}{\overset{\|}{C}}N(CH_3)_2$$
N,N-Dimethylhexadecanamide

Problem 26.15 Palmitic acid (hexadecanoic acid) is the source of the hexadecyl (cetyl) group in the following compounds. Each is a mild surface-acting germicide and fungicide and is used as a topical antiseptic and disinfectant (see The Merck Index, 12th ed., #2074 and #2059). They are examples of quaternary ammonium detergents, commonly called "quats."

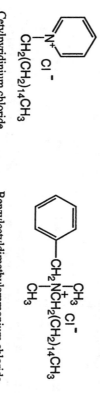

Cetylpyridinium chloride

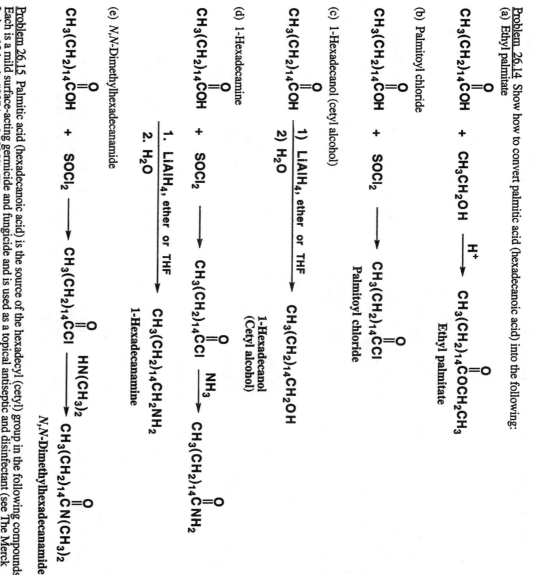

Benzylcetyldimethylammonium chloride

(a) Cetylpyridinium chloride is prepared by treating pyridine with 1-chlorohexadecane (cetyl chloride). Show how to convert palmitic acid to cetyl chloride.

$$CH_3(CH_2)_{14}\overset{O}{\overset{\|}{C}}OH \ \xrightarrow[2) \ H_2O]{1) \ LiAlH_4, \ ether \ or \ THF} \ CH_3(CH_2)_{14}CH_2OH \ \xrightarrow{SOCl_2} \ CH_3(CH_2)_{14}CH_2Cl$$
1-Chlorohexadecane
(Cetyl chloride)

(b) Benzylcetyldimethylammonium chloride is prepared by treating benzyl chloride with N,N-dimethyl-1-hexadecanamine. Show how this tertiary amine can be prepared from palmitic acid.

$$CH_3(CH_2)_{14}\overset{\overset{O}{\|}}{C}OH \quad + \quad SOCl_2 \quad \longrightarrow \quad CH_3(CH_2)_{14}\overset{\overset{O}{\|}}{C}Cl \quad \xrightarrow{HN(CH_3)_2} \quad CH_3(CH_2)_{14}\overset{\overset{O}{\|}}{C}N(CH_3)_2$$

$$\xrightarrow[\text{2) } H_2O]{\text{1) LiAlH}_4, \text{ ether or THF}} \quad CH_3(CH_2)_{14}CH_2N(CH_3)_2$$

N,N-Dimethyl-1-hexadecanamine

Problem 26.16 Lipases are enzymes that catalyze the hydrolysis of esters, especially esters of glycerol. Because enzymes are chiral catalysts, they catalyze the hydrolysis of only one enantiomer of a racemic mixture. For example, porcine pancreatic lipase catalyzes the hydrolysis of only one enantiomer of the following racemic epoxyester. Calculate the number of grams of epoxyalcohol that can be obtained from 100 g of racemic epoxy ester by this method.

A racemic mixture

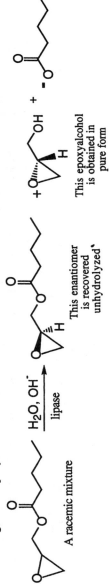

This enantiomer is recovered unhydrolyzed*

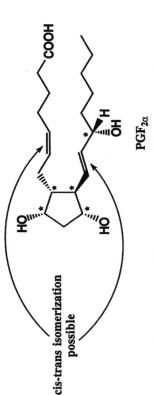

This epoxyalcohol is obtained in pure form

The molecular weight for the epoxyester starting material ($C_8H_{14}O_3$) is 158 g/mol, while the molecular weight of the epoxyalcohol product ($C_3H_6O_2$) is 74 g/mol. Because the starting material is racemic, there is actually only 100 g/2 or 50 g of the starting material epoxyester enantiomer that will be converted to product. As a result, the number of grams of the epoxyalcohol product is calculated as:

$$\frac{(50 \text{ g})(74 \text{ g / mol})}{(158 \text{ g / mol})} = 23.4 \text{ g}$$

Thus, 23.4 g of epoxyalcohol will be produced in this enzyme-catalyzed reaction.

Prostaglandins
Problem 26.17 Examine the structure of $PGF_{2\alpha}$. Identify all stereocenters, all double bonds about which cis-trans isomerism occurs, and state the number of stereoisomers possible for a molecule of this structure.

These double bonds are indicated by the arrows.

cis-trans isomerization possible

$PGF_{2\alpha}$

There are 2^5 stereoisomers possible and 2^2 cis-trans isomers possible for a grand total of $32 \times 4 = 128$ possible stereoisomers.

Problem 26.18 Following is the structure of unoprostone (see The Merck Index, 12th ed., #9984), a compound patterned after the natural prostaglandins (Section 26.3). Rescula, the isopropyl ester of unoprostone, is an antiglaucoma drug used to treat ocular hypertension. Compare the structural formula of this synthetic prostaglandin with that of PGF₂α.

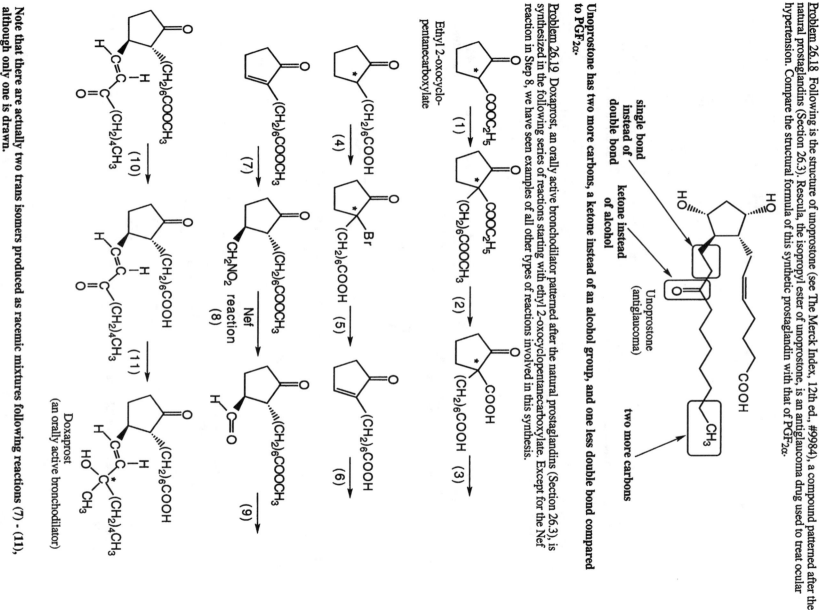

single bond
instead of
double bond

ketone instead
of alcohol

Unoprostone
(antiglaucoma)

two more carbons

Unoprostone has two more carbons, a ketone instead of an alcohol group, and one less double bond compared to PGF₂α.

Problem 26.19 Doxaprost, an orally active bronchodilator patterned after the natural prostaglandins (Section 26.3), is synthesized in the following series of reactions starting with ethyl 2-oxocyclopentanecarboxylate. Except for the Nef reaction in Step 8, we have seen examples of all other types of reactions involved in this synthesis.

Ethyl 2-oxocyclo-
pentanecarboxylate

Doxaprost
(an orally active bronchodilator)

Note that there are actually two trans isomers produced as racemic mixtures following reactions (7) - (11), although only one is drawn.

(a) Propose a set of experimental conditions to bring about the alkylation in Step 1. Account for the regioselectivity of the alkylation, that is, that it takes place on the carbon between the two carbonyl groups rather than on the other side of the ketone carbonyl.

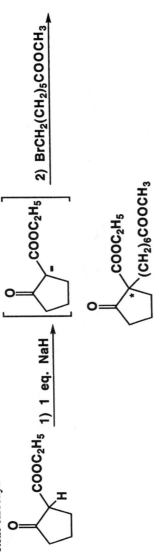

The alkylation reaction occurs at the position shown, because this enolate is the one that is formed predominantly due to the stabilization provided by both of the adjacent carbonyl functions.

(b) Propose experimental conditions to bring about Steps 2 and 3, and propose a mechanism for the loss of carbon dioxide in Step 3.

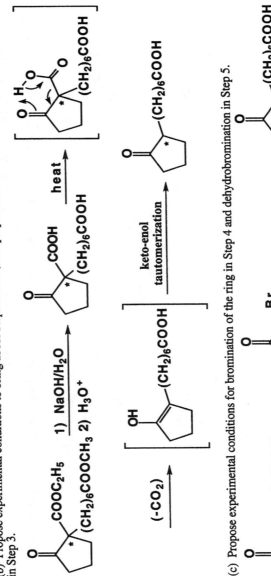

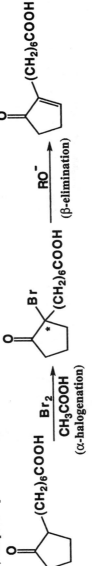

(c) Propose experimental conditions for bromination of the ring in Step 4 and dehydrobromination in Step 5.

(d) Write equations to show that Step 6 can be brought about using either methanol or diazomethane (CH_2N_2) as a source of the -CH_3 in the methyl ester.

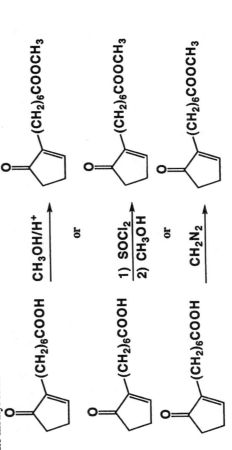

(e) Describe experimental conditions to bring about Step 7 and account for the fact that the trans isomer is formed in this step.

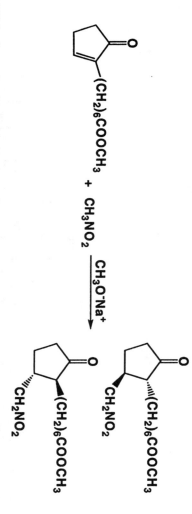

This Michael reaction leads to a trans product because this minimizes nonbonded interaction strain between the nitromethyl and methyl hexanoate groups. Note that two new stereocenters are produced in the reaction, and the product is actually a racemic mixture of the two trans enantiomers.

(f) Step 9 is done by a Wittig reaction. Suggest a structural formula for a Wittig reagent that gives the product shown.

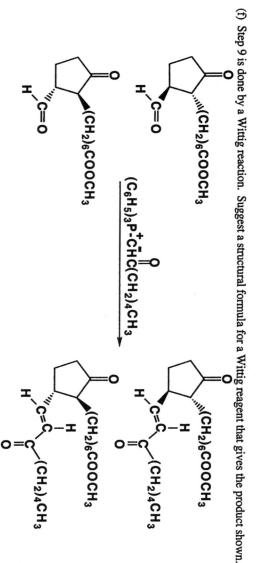

(g) Name the type of reaction involved in Step 10.

Step 10 is an ester hydrolysis reaction.

(h) Step 11 can best be described as a Grignard reaction with methylmagnesium bromide under very carefully controlled conditions. In addition to the observed reaction, what other Grignard reactions might take place in Step 11?

The carboxyl group will be deprotonated by the first equivalent of methylmagnesium bromide. As far as other Grignard reactions go, the cyclopentyl ketone might also take part in a Grignard reaction as shown:

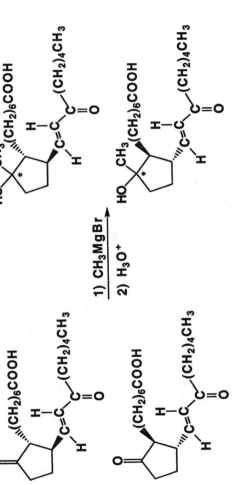

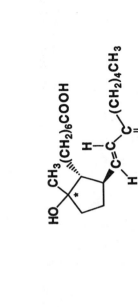

1) **CH₃MgBr**
2) **H₃O⁺**

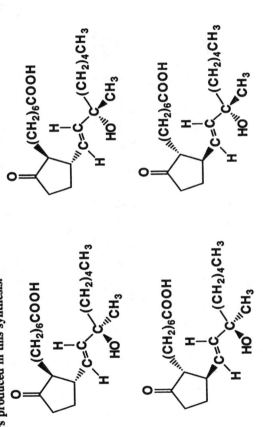

(i) Assuming that the two side chains on the cyclopentanone ring are trans, how many stereoisomers are produced in this synthetic sequence?

There are actually two enantiomers of the trans product produced during the synthesis, although only one is drawn in the problem. The Grignard reaction also gives a new stereocenter, so there are a total of four stereoisomers produced in this synthesis.

Steroids
Problem 26.20 Draw the structural formula for the product formed by treatment of cholesterol with H₂/Pd; with Br₂.

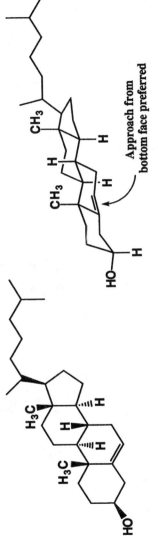

Approach from bottom face preferred

Cholesterol structural formula and conformational formula

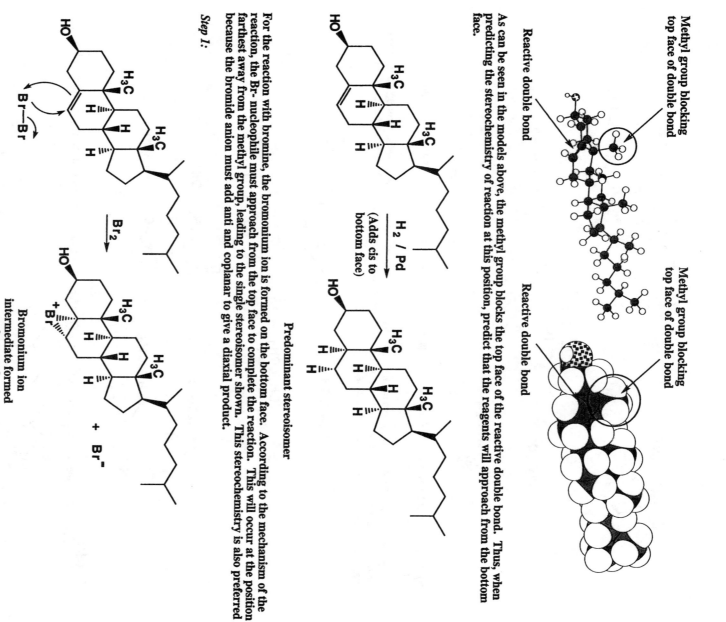

Methyl group blocking
top face of double bond

Methyl group blocking
top face of double bond

Reactive double bond

Reactive double bond

As can be seen in the models above, the methyl group blocks the top face of the reactive double bond. Thus, when predicting the stereochemistry of reaction at this position, predict that the reagents will approach from the bottom face.

Reactive double bond

H_2 / Pd

(Adds cis to
bottom face)

Predominant stereoisomer

For the reaction with bromine, the bromonium ion is formed on the bottom face. According to the mechanism of the reaction, the Br- nucleophile must approach from the top face to complete the reaction. This will occur at the position farthest away from the methyl group, leading to the single stereoisomer shown. This stereochemistry is also preferred because the bromide anion must add anti and coplanar to give a diaxial product.

Step 1:

Br—Br

Br_2

Bromonium ion
intermediate formed
on bottom face

+ Br⁻

Step 2:

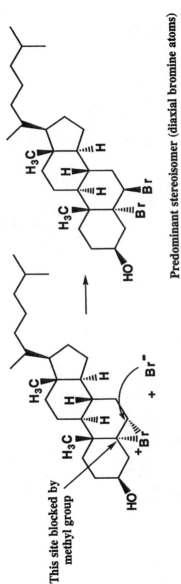

This site blocked by methyl group

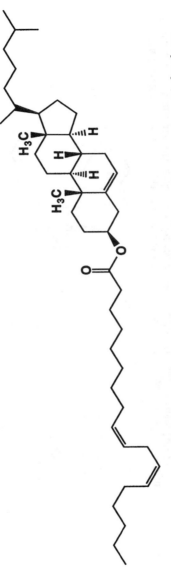

Predominant stereoisomer (diaxial bromine atoms)

Problem 26.21 Both low-density lipoproteins (LDL) and high-density lipoproteins (HDL) consist of a core of triacylglycerols and cholesterol esters surrounded by a single phospholipid layer. Draw the structural formula of cholesteryl linoleate, one of the cholesterol esters found in this core.

Problem 26.22 Examine the structural formulas of testosterone (a male sex hormone) and progesterone (a female sex hormone). What are the similarities in structure between the two? What are the differences?

Overall, these structures are remarkably similar. Both steroids contain the standard four ring steroid structure with the axial methyl groups at C10 and C13. In addition, both structures contain an ene-one group in the A ring. On the other hand, the two structures differ in the nature of the D ring substituent at C17. In testosterone, the substituent is a hydroxyl group and in progesterone it is a ketomethyl group.

Problem 26.23 Examine the model of cholic acid on the CD and account for the ability of this and other bile salts to emulsify fats and oils and thus aid in their digestion.

Cholic acid has the characteristic structure of a soap, so it can emulsify hydrophobic substances. In particular, cholic acid has a large hydrophobic steroid nucleus, and a highly polar carboxylate group.

Problem 26.24 Following is a structural formula for cortisol (hydrocortisone). Draw a stereorepresentation of this molecule showing the conformations of the five- and six-membered rings. See the model of cortisol on the CD.

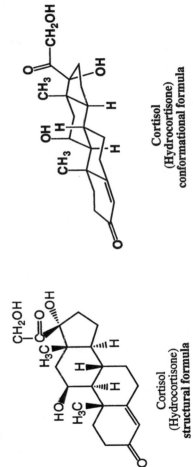

Cortisol (Hydrocortisone) structural formula

Cortisol (Hydrocortisone) conformational formula

Problem 26.25 Much of our understanding of conformational analysis has arisen from studies on the reactions of rigid steroid nuclei. For example, the concept of trans-diaxial ring opening of epoxides was proposed to explain the stereospecific reactions seen with steroidal epoxides. Predict the product when each of the following steroidal epoxides is treated with LiAlH₄;

(a)

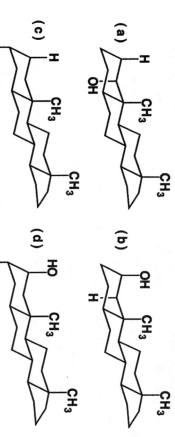

(b)

(c)

(d)

The idea here is that the hydride reagent attacks such that the epoxide opens to give a diaxial product. This dictates that only one product is observed for each reaction, the one that is shown below:

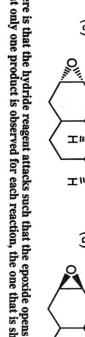

(a)

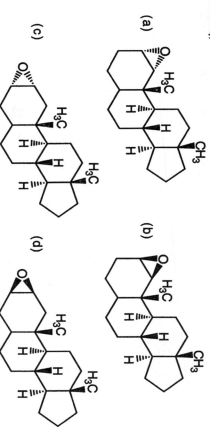

(b)

(c)

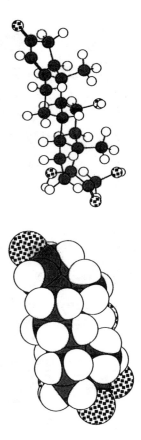

(d)

Phospholipids

Problem 26.26 Draw the structural formula of a lecithin containing one molecule each of palmitic acid and linoleic acid.

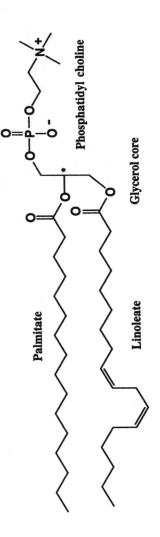

Palmitate

Linoleate

Glycerol core

Phosphatidyl choline

Problem 26.27 Identify the hydrophobic and hydrophilic region(s) of a phospholipid.

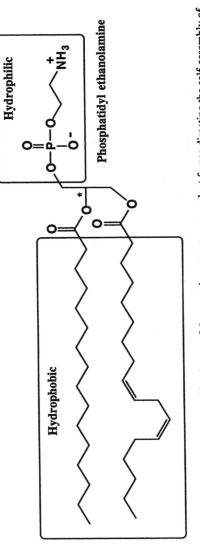

Hydrophilic

Phosphatidyl ethanolamine

Hydrophobic

Problem 26.28 The hydrophobic effect is one of the most important noncovalent forces directing the self-assembly of biomolecules in aqueous solution. The hydrophobic effect arises from tendencies of biomolecules (1) to arrange polar groups so that they interact with the aqueous environment by hydrogen bonding and (2) to arrange nonpolar groups so that they are shielded from the aqueous environment. Show how the hydrophobic effect is involved in directing

(a) Formation of micelles by soaps and detergents.

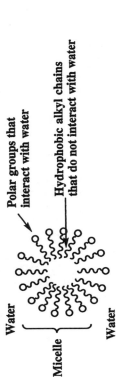

Water

Polar groups that interact with water

Hydrophobic alkyl chains that do not interact with water

Micelle

Water

In micelles, the hydrophobic hydrocarbon tails are associated with each other to form the hydrophobic interior, while the polar groups are associated with each other on the outside surface where they interact with water.

(b) Formation of lipid bilayers by phospholipids.

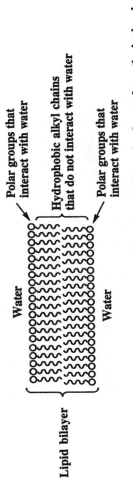

Water

Polar groups that interact with water

Hydrophobic alkyl chains that do not interact with water

Polar groups that interact with water

Lipid bilayer

Water

In lipid bilayers, the hydrophobic hydrocarbon tails are associated with each other to form the hydrophobic inner layer, while the polar head groups are associated with each other on both outside surfaces where they interact with water.

Problem 26.29 How does the presence of unsaturated fatty acids contribute to the fluidity of biological membranes?

Biological membranes must have the proper fluidity to function correctly. Fatty acid components of membranes add the cis double bonds of the unsaturated fatty acids that interfere with chain packing, increasing fluidity of the membrane. Organisms modulate membrane fluidity by controlling the kind of fatty acids in their membranes. For example, organisms living in colder environments tend to have more unsaturated fatty acids to insure membrane fluidity at low temperature.

Problem 26.30 Lecithins can act as emulsifying agents. The lecithin of egg yolk, for example, is used to make mayonnaise. Identify the hydrophobic part(s) and the hydrophilic part(s) of a lecithin. Which parts interact with the oils used in making mayonnaise? Which parts interact with the water?

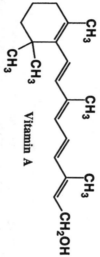

Hydrophilic
Interacts with water

Hydrophobic
Interacts with oil

Fat-soluble Vitamins
Problem 26.31 Examine the structural formula of vitamin A and state the number of cis-trans isomers possible for this molecule.

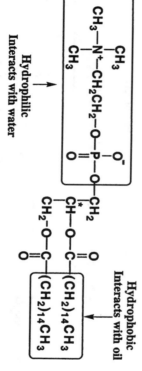

Vitamin A

As shown in the structure above, vitamin A has four double bonds that can be either *cis* or *trans*, thus there are 2⁴ or 16 possible *cis-trans* isomers. Note that the double bond in the ring cannot have *cis-trans* isomers.

Problem 26.32 The form of vitamin A present in many food supplements is vitamin A palmitate. Draw the structural formula of this molecule.

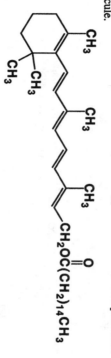

Problem 26.33 Examine the structural formulas of vitamin A, 1,25-dihydroxy-D₃, vitamin E, and vitamin K₁ (Section 26.6). Do you expect them to be more soluble in water or in dichloromethane? Do you expect them to be soluble in blood plasma?

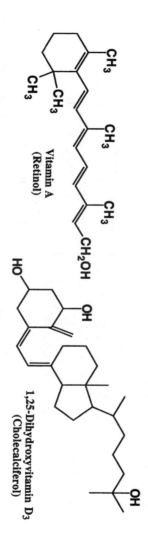

Vitamin A
(Retinol)

1,25-Dihydroxyvitamin D₃
(Cholecalciferol)

Chapter 26: Lipids

Solutions

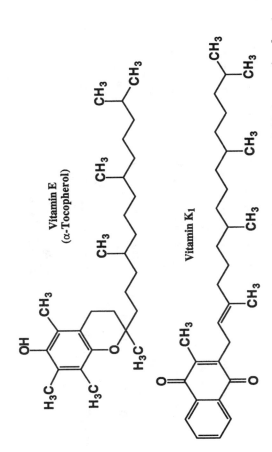

Vitamin E
(α-Tocopherol)

Vitamin K₁

All of these structures are extremely hydrophobic, so they will be more soluble in organic solvents such as dichloromethane than polar solvents such as water. Since blood plasma is an aqueous solution, these vitamins will only be sparingly soluble in blood plasma.

Saddleback butterflyfish *Chaetodon ephippium*
Rangiroa, French Polynesia

CHAPTER 27
Solutions to the Problems

Problem 27.1 Of the 20 protein-derived amino acids shown in Table 27.1, which contain (a) no stereocenter, (b) two stereocenters.

The only amino acid with no stereocenters is glycine (Gly, G). Both isoleucine (Ile, I) and threonine (Thr, T) have two stereocenters as shown with asterisks in the structures below.

$$H_3\overset{+}{N}CH_2COO^-$$

Glycine (Gly, G)

$$CH_3CH_2\overset{*}{C}H\overset{*}{C}HCOO^-$$

Isoleucine (Ile, I)

$$CH_3\overset{*}{C}H\overset{*}{C}HCOO^-$$

Threonine (Thr, T)

Problem 27.2 The isoelectric point of histidine is 7.64. Toward which electrode does histidine migrate on paper electrophoresis at pH 7.0?

An amino acid will have at least a partial positive charge at any pH that is below its isoelectric point. A pH of 7.0 is below the isoelectric point of histidine (7.64), so it will have a partial positive charge. Therefore, at this pH histidine migrates toward the negative electrode.

Problem 27.3 Describe the behavior of a mixture of glutamic acid, arginine, and valine on paper electrophoresis at pH 6.0.

The pI's for glutamic acid, arginine, and valine are 3.08, 10.76, and 6.00, respectively. Therefore, at pH 6.0 glutamic acid is negatively charged, arginine is positively charged, and valine is neutral. Thus, on paper electrophoresis, glutamic acid will migrate toward the positive electrode, arginine will migrate toward the negative electrode, and valine will not move.

Problem 27.4 Draw a structural formula for Lys-Phe-Ala. Label the N-terminal amino acid and the C-terminal amino acid. What is the net charge on this tripeptide at pH 6.0?

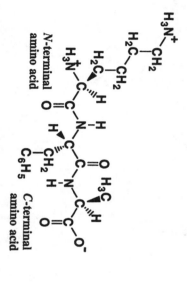

Due to the presence of the basic lysine residue, this tripeptide will have a net positive charge at pH 6.0

Problem 27.5 Which of these tripeptides are hydrolyzed by trypsin? By chymotrypsin?
(a) Tyr-Gln-Val   (b) Thr-Phe-Ser   (c) Thr-Ser-Phe

Based on the substrate specificities listed in Table 27.3, trypsin will not cleave any of these tripeptides because there are no arginine or lysine residues. On the other hand, chymotrypsin will cleave peptides (a) and (b) between the Tyr-Gln and Phe-Ser residues, respectively, but not (c).

<u>Problem 27.6</u> Deduce the amino acid sequence of an undecapeptide (11 amino acids) from the experimental results shown in the accompanying table.

Experimental Procedure	Amino Acid Composition
Undecapeptide	Ala,Arg,Glu,Lys$_2$,Met,Phe,Ser,Thr,Trp,Val
Edman degradation	Ala
Trypsin-Catalyzed Hydrolysis	
Fragment E	Ala,Glu,Arg
Fragment F	Thr,Phe,Lys
Fragment G	Lys
Fragment H	Met,Ser,Trp,Val
Chymotrypsin-Catalyzed Hydrolysis	
Fragment I	Ala,Arg,Glu,Phe,Thr
Fragment J	Lys$_2$,Met,Ser,Trp,Val
Reaction with Cyanogen Bromide	
Fragment K	Ala,Arg,Glu,Lys$_2$,Met,Phe,Thr,Val
Fragment L	Trp,Ser

Based on the Edman degradation result, alanine (Ala) is the *N*-terminal residue of the peptide. Fragment E must have Arg on the *C*-terminal end because it is a peptide produced by trypsin cleavage. Since we know Ala is the *N*-terminal residue, this means fragment E must be of the sequence Ala-Glu-Arg. There must be two lysine residues or an arginine and a lysine residue adjacent to each other based on the appearance of a single lysine residue as Fragment G. Since Fragment J has two lysines and no arginine residues, the two lysine residues must be adjacent to each other. Methionine must be the third to the last residue, because CNBr treatment created fragment L that is only Ser and Trp. In addition, Trp and Ser must be the last two residues. Combining this information with the knowledge that there are two lysine residues adjacent to each other indicates the Fragment J is of the sequence Lys-Lys-Val-Met-Ser-Trp. Note that Val must come after the two Lys residues because of Fragment H. In addition, Trp has to be on the *C*-terminus or a residue would have been cleaved off by chymotrypsin. Phenylalanine must be on the *C* terminus of Fragment I since it results from chymotrypsin cleavage. We already know that Fragment I must start with Ala-Glu-Arg, so the entire sequence of Fragment I must be Ala-Glu-Arg-Thr-Phe.

Putting Fragments I and J together gives the following sequence for the entire peptide:

Ala-Glu-Arg-Thr-Phe-Lys-Lys-Val-Met-Ser-Trp

<u>Problem 27.7</u> At pH 7.4, with what amino acid side chains can the side chain of lysine form salt linkages.

At pH 7.4, the only negatively charged side chains are the carboxylates of glutamic acid and aspartic acid. Therefore, these are the amino acid side chains with which the side chain of lysine can form a salt linkage.

Amino Acids
<u>Problem 27.8</u> What amino acids do these abbreviations stand for?

(a) Phe Phenylalanine	(b) Ser Serine	(c) Asp Aspartic acid	(d) Gln Glutamine
(e) His Histidine	(f) Gly Glycine	(g) Tyr Tyrosine	

<u>Problem 27.9</u> Configuration of the stereocenter in α-amino acids is most commonly specified using the D,L convention. It can also be identified using the R,S convention (Section 3.3). Does the stereocenter in L-serine have the R or the S configuration?

The stereocenter has the S configuration.

Problem 27.10 Assign an R or S configuration to the stereocenter in each amino acid.

(a) L-Phenylalanine

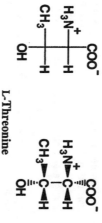

L-Phenylalanine

(b) L-Glutamic acid

L-Glutamic acid

(c) L-Methionine

L-Methionine

All of these L amino acids have the S configuration.

Problem 27.11 The amino acid threonine has two stereocenters. The stereoisomer found in proteins has the configuration 2S,3R. Draw (a) a Fischer projection of this stereoisomer and (b) a three-dimensional representation.

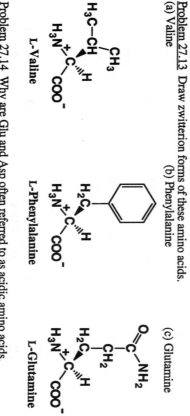

L-Threonine

Problem 27.12 Define the term zwitterion.

A zwitterion refers to the situation in which an acidic and a basic group in the same molecule react with each other to form a dipolar ion or internal salt. A zwitterion has no net charge; it contains one positive charge and one negative charge.

Problem 27.13 Draw zwitterion forms of these amino acids.

(a) Valine

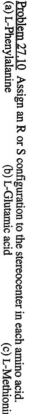

L-Valine

(b) Phenylalanine

L-Phenylalanine

(c) Glutamine

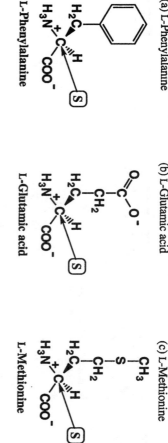

L-Glutamine

Problem 27.14 Why are Glu and Asp often referred to as acidic amino acids.

The side chains of glutamic acid (Glu) and aspartic acid (Asp) have carboxyl groups, so these amino acids are referred to as acidic amino acids.

Problem 27.15 Why is Arg often referred to as a basic amino acid? Which two other amino acids are also basic amino acids?

The guanidino group of arginine (Arg) is strongly basic, so this amino acid is referred to as a basic amino acid. Note that this means arginine is positively charged at neutral pH. Lysine (Lys) and histidine (His) are also referred to as basic amino acids because their side chains contain a primary amine and imidazole functions, respectively.

Problem 27.16 What is the meaning of the alpha as it is used in α-amino acid?

The alpha in α-amino acids indicates that the amino group is on the carbon atom that is α to the carboxyl group.

Problem 27.17 Several β-amino acids exist. There is a unit of β-alanine, for example, contained within the structure of coenzyme A (Problem 25.41). Write the structural formula of β-alanine.

$$\overset{+}{H_3}NCH_2CH_2COO^-$$

β-Alanine

Problem 27.18 Although only L-amino acids occur in proteins, D-amino acids are often a part of the metabolism of lower organisms. The antibiotic actinomycin D, for example, contains a unit of D-valine, and the antibiotic bacitracin A contains units of D-asparagine and D-glutamic acid. Draw Fischer projections and three-dimensional representations for these three D-amino acids.

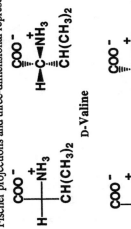

D-Valine

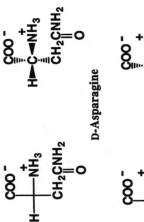

D-Valine

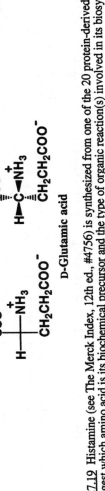

D-Asparagine

D-Glutamic acid

Problem 27.19 Histamine (see The Merck Index, 12th ed., #4756) is synthesized from one of the 20 protein-derived amino acids. Suggest which amino acid is its biochemical precursor and the type of organic reaction(s) involved in its biosynthesis (for example, oxidation, reduction, decarboxylation, nucleophilic substitution)

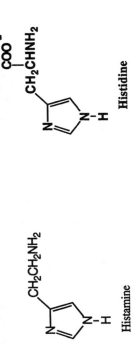

Histamine

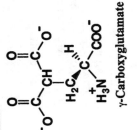

Histidine

Histamine is derived from the amino acid histidine and is the result of a biosynthetic decarboxylation reaction. Note how both histamine and histidine are drawn in the form present at basic pH.

Problem 27.20 As discussed in the "Chemistry in Action: Vitamin K, Blood Clotting, and Basicity," (Chapter 26), vitamin K participates in carboxylation of glutamic acid residues of the blood-clotting protein prothrombin.
(a) Write a structural formula for γ-carboxyglutamate.

γ-Carboxyglutamate

(b) Account for the fact that the presence of γ-carboxyglutamate escaped detection for many years; on routine amino acid analyses, only glutamate was detected.

This amino acid was not detected because it is a β-dicarboxylic acid and therefore easily decarboxylated under conditions of routine amino acid analysis.

Problem 27.21 Both norepinephrine (see The Merck Index, 12th ed., #6788) and epinephrine (see The Merck Index, 12th ed., #3656) are synthesized from the same protein-derived amino acid. From which amino acid are they synthesized and what types of reactions are involved in their biosynthesis?

(a)

Norepinephrine

(b)

Epinephrine
(Adrenaline)

L-Tyrosine

Norepinephrine and epinephrine are derived from the amino acid tyrosine. In both cases, biosynthesis of these molecules involves decarboxylation, aromatic hydroxylation (a two-electron oxidation) ortho to the original aromatic -OH group, and hydroxylation (a second two-electron oxidation) of the benzylic methylene group. Epinephrine is also methylated on the α-amino group. Note how all of the molecules in the problem are drawn in the form present at basic pH.

Problem 27.22 From which amino acid are serotonin (see The Merck Index, 12th ed., #8607) and melatonin (see The Merck Index, 12th ed., #5857) synthesized and what types of reactions are involved in their biosynthesis?

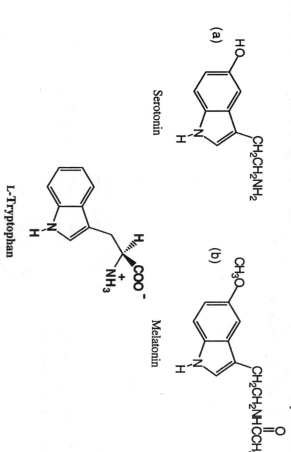

(a)

Serotonin

(b)

Melatonin

L-Tryptophan

Serotonin and melatonin are derived from the amino acid tryptophan. In both cases, biosynthesis of these molecules involves decarboxylation. In the case of serotonin there is also an aromatic hydroxylation (a two-electron oxidation). For melatonin the phenolic -OH group of serotonin is methylated and the primary amino group is acetylated. Note how all of the molecules in the problem are drawn in the form present at basic pH.

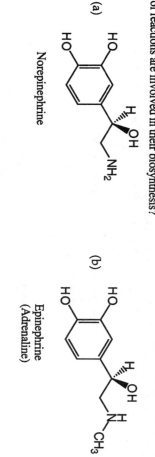

Acid-Base Behavior of Amino Acids

Problem 27.23 Draw the structural formula for the form of each amino acid most prevalent at pH 1.0.

(a) Threonine

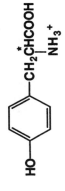

$$CH_3\overset{*}{C}HCHCOOH$$
$$\underset{NH_3^+}{|}$$

(b) Arginine

$$H_2N\overset{NH_2^+}{\overset{||}{C}}NHCH_2CH_2CH_2\overset{*}{C}HCOOH$$
$$\underset{NH_3^+}{|}$$

(c) Methionine

$$CH_3SCH_2CH_2\overset{*}{C}HCOOH$$
$$\underset{NH_3^+}{|}$$

(d) Tyrosine

$$HO— \overset{*}{C}H_2CHCOOH$$
$$\underset{NH_3^+}{|}$$

Problem 27.24 Draw the structural formula for the form of each amino most prevalent at pH 10.0.

(a) Leucine

$$(CH_3)_2CHCH_2\overset{*}{C}HCOO^-$$
$$\underset{NH_2}{|}$$

(b) Valine

$$(CH_3)_2CH\overset{*}{C}HCOO^-$$
$$\underset{NH_2}{|}$$

(c) Proline

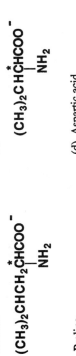

(d) Aspartic acid

$$^-OOCCH_2\overset{*}{C}HCOO^-$$
$$\underset{NH_2}{|}$$

Problem 27.25 Write the zwitterion form of alanine and show its reaction with:

(a) 1 mole NaOH

$$CH_3\overset{*}{C}HCOO^- \ + \ 1 \ mole \ \ NaOH \longrightarrow \ CH_3\overset{*}{C}HCOO^-$$
$$\underset{NH_3^+}{|} \underset{NH_2}{|}$$

(b) 1 mole HCl

$$CH_3\overset{*}{C}HCOO^- \ + \ 1 \ mole \ \ HCl \longrightarrow \ CH_3\overset{*}{C}HCOOH$$
$$\underset{NH_3^+}{|} \underset{NH_3^+}{|}$$

Problem 27.26 Write the form of lysine most prevalent at pH 1.0 and then show its reaction with the following. Consult Table 27.2 for pKₐ values of the ionizable groups in lysine.

At pH 1.0, the most prevalent form of lysine has both amino groups as well as the carboxylic acid group protonated and a total charge of +2 as shown in the following structure.

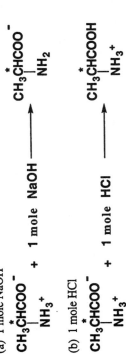

$$\overset{+}{H_3}NCH_2CH_2CH_2CH_2\overset{*}{C}HCOOH \xrightarrow{\ \ \ } pK_a \ 2.10$$
$$\underset{NH_3^+}{|} \ \ pK_a \ 9.82$$
$$pK_a \ 9.95$$

(a) 1 mole NaOH

$$\overset{+}{H_3}NCH_2CH_2CH_2CH_2\overset{*}{C}HCOO^-$$
$$\underset{NH_3^+}{|}$$

(b) 2 moles NaOH

$$H_3NCH_2CH_2CH_2CH_2\overset{*}{C}HCOO^-$$
$$\underset{NH_2}{|}$$

(c) 3 moles NaOH

$$H_2NCH_2CH_2CH_2CH_2\overset{*}{C}HCOO^-$$
$$\underset{NH_2}{|}$$

Problem 27.27 Write the form of aspartic acid most prevalent at pH 1.0 and then show its reaction with the following. Consult Table 27.2 for pK_a values of the ionizable groups in aspartic acid.

At pH 1.0, the most prevalent form of aspartic acid has both carboxylic acid groups as well as the amino group protonated and a total charge of +1 as shown in the following structure.

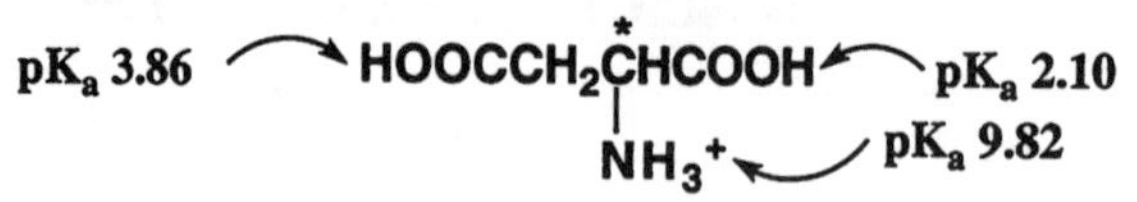
pKa 3.86
HOOCCH2CHCOOH
NH3+
pKa 2.10
pKa 9.82

(a) 1 mole NaOH

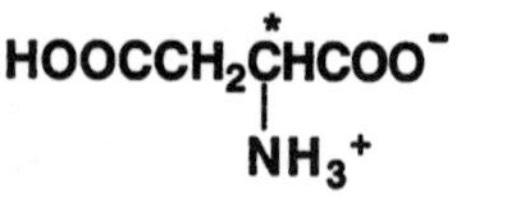
HOOCCH2CHCOO-
NH3+

(b) 2 mole NaOH

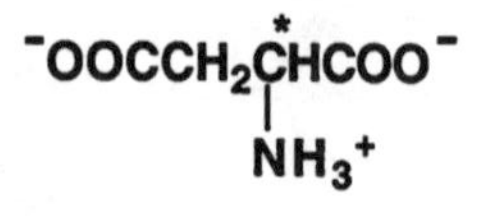
-OOCCH2CHCOO-
NH3+

(c) 3 mole NaOH

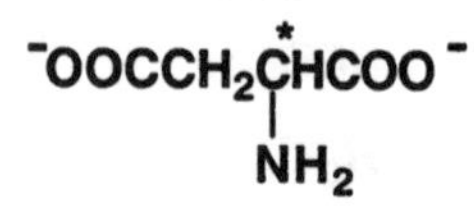
-OOCCH2CHCOO-
NH2

Problem 27.28 Given pK_a values for ionizable groups in Table 27.2, sketch curves for the titration of (a) glutamic acid with NaOH, and (b) histidine with NaOH.

Glutamic acid has pK_a values of 2.10, 4.07, and 9.47 so the titration curve would look something like the following:

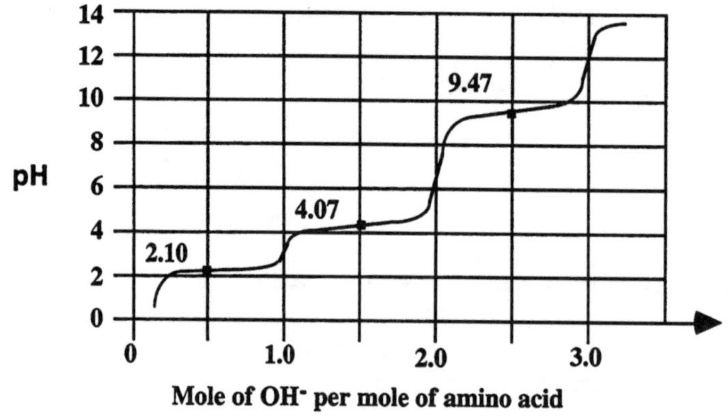
pH
14
12
10
8
6
4
2
0
2.10
4.07
9.47
0
1.0
2.0
3.0
Mole of OH- per mole of amino acid

Histidine has pK_a values of 1.77, 6.10, and 9.18 so the titration curve would look something like the following:

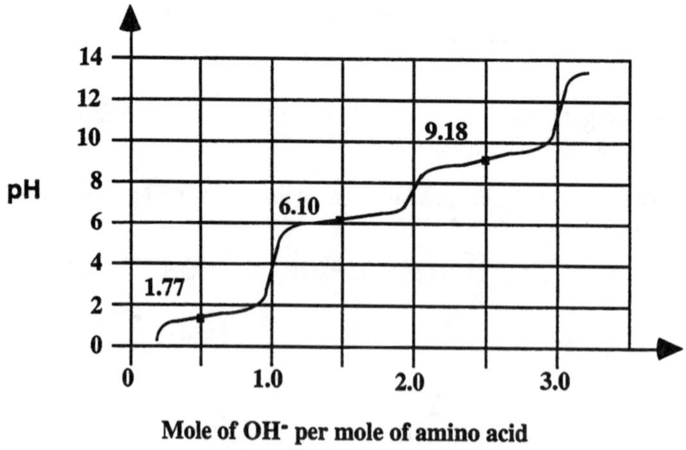
pH
14
12
10
8
6
4
2
0
1.77
6.10
9.18
0
1.0
2.0
3.0
Mole of OH- per mole of amino acid

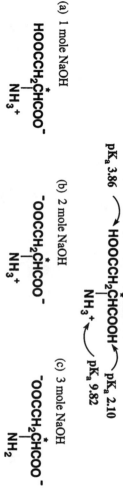

Problem 27.29 Draw a structural formula for the product formed when alanine is treated with the following reagents.

(a) Aqueous NaOH

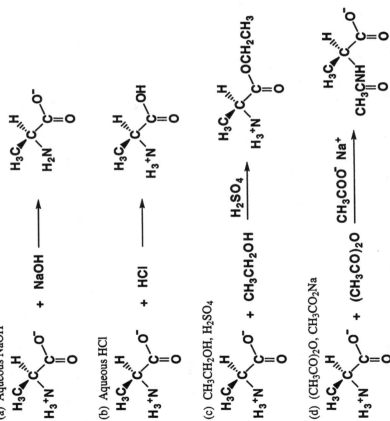

(b) Aqueous HCl

(c) CH₃CH₂OH, H₂SO₄

(d) (CH₃CO)₂O, CH₃CO₂Na

Problem 27.30 For lysine and arginine, the isoelectric point, pI, occurs at a pH where the net charge on the nitrogen-containing groups is +1 and balances the charge of -1 on the α-carboxyl group. Calculate pI for these amino acids.

The pI will occur when the nitrogen-containing groups have a total charge of +1, and this occurs halfway between their respective pKa values.
For lysine:

$$pI = \frac{1}{2}\ (pK_a\ \alpha\text{-NH}_3^+ + pK_a\ \text{side chain-NH}_3^+) = \frac{8.95 + 10.53}{2} = \boxed{9.74}$$

For arginine:

$$pI = \frac{1}{2}\ (pK_a\ \alpha\text{-NH}_3^+ + pK_a\ \text{side chain guanidium}) = \frac{9.04 + 12.48}{2} = \boxed{10.76}$$

Problem 27.31 For aspartic and glutamic acids, the isoelectric point occurs at a pH where the net charge of +1 on the α-amino group. Calculate pI for these amino acids.

The pI will occur when the two acid groups have a total charge of -1, and this occurs halfway between their respective pKa values.
For aspartic acid:

$$pI = \frac{1}{2}\ (pK_a\ \alpha\text{-COOH} + pK_a\ \text{side chain-COOH}) = \frac{2.10 + 3.86}{2} = \boxed{2.98}$$

For glutamic acid:

$$pI = \frac{1}{2}\ (pK_a\ \alpha\text{-COOH} + pK_a\ \text{side chain-COOH}) = \frac{2.10 + 4.07}{2} = \boxed{3.08}$$

Problem 27.32 Account for the fact that the isoelectric point of glutamine (pI 5.65) is higher than the isoelectric point of glutamic acid (pI 3.08).

Amino acids have no net charge at their pI. For this to happen with glutamic acid, the net charge on the α-carboxyl and side chain carboxyl groups must be -1 to balance the +1 charge of the α-amino group. This will occur at a pI = (1/2)(2.10 + 4.07) = 3.08. The amide side chain of glutamine is already neutral near neutral pH, so the pI of the amino acid is determined by the values for the only ionizable groups, namely the α-carboxyl group and α-amino groups, according to the equation pI = (1/2)(2.17 + 9.03) = 5.6. This value is near that of the other amino acids with non-ionizable functional groups on their side chains.

Problem 27.33 Enzyme-catalyzed decarboxylation of glutamic acid gives 4-aminobutanoic acid (Section 27.1D). Estimate the pI of 4-aminobutanoic acid.

There is little if any inductive effect operating between the amino and carboxyl groups of 4-aminobutanoic acid because there are three methylene groups between them. Thus, the pK_a of the amino group of 4-aminobutanoic acid is like that of a simple amino group, near 10.0. Similarly, the pK_a of the carboxyl group is like that of a simple carboxyl group, near 4.5. Given these estimates for the pK_a values, the pI would be:

$$pI = \frac{1}{2}\,(pK_a\,\alpha\text{-COOH} + pK_a\,\alpha\text{-NH}_2) = \frac{4.5 + 10.0}{2} = \boxed{7.25}$$

Problem 27.34 Guanidine and the guanidino group present in arginine are two of the strongest organic bases known. Account for this basicity.

The guanidino group is strongly basic because of resonance stabilization of the protonated guanidinium ion as shown below:

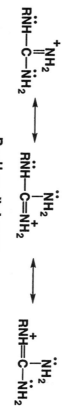

R = H or alkyl group

Problem 27.35 At pH 7.4, the pH of blood plasma, do the majority of protein-derived amino acids bear a net negative charge or a net positive charge?

The majority of amino acids have a pI near 5 or 6, so they will bear a net negative charge at pH 7.4.

Problem 27.36 Do the following molecules migrate to the cathode or to the anode on electrophoresis at the specified pH?

The key to determining which way the molecules migrate is to estimate the net charge on the molecules at the given pH. Molecules with a net positive charge will migrate toward the negative electrode and molecules with a net negative charge will migrate toward the positive electrode. Molecules at a pH above their isoelectric point have a net negative charge, and molecules at a pH below their isoelectric point have a net positive charge, and molecules at a pH that equals their isoelectric point have no net charge.

(a) Histidine at pH 6.8

pI = 7.64, so at pH 6.8 histidine has a net positive charge and migrates toward the negative electrode (cathode).

(b) Lysine at pH 6.8

pI = 9.74, so at pH 6.8 lysine has a net positive charge and migrates toward the negative electrode (cathode).

(c) Glutamic acid at pH 4.0

pI = 3.08, so at pH 4.0 glutamic acid has a net negative charge and migrates toward the positive electrode (anode).

(d) Glutamine at pH 4.0

pI = 5.65, so at pH 4.0 glutamine has a net positive charge and migrates toward the negative electrode (cathode).

(e) Glu-Ile-Val at pH 6.0

The glutamic acid residue has a carboxyl group that will be largely deprotonated at pH 6.0, so the overall molecule will have a net negative charge and will migrate toward the positive electrode (anode).

(f) Lys-Gln-Tyr at pH 6.0

The lysine residue has an amino group on its side chain that will be protonated at pH 6.0, so the molecule will have a net positive charge and will migrate toward the negative electrode (cathode).

Problem 27.37 At what pH would you carry out an electrophoresis to separate the amino acids in each mixture of amino acids?

Recall that an amino acid below its isoelectric point will have some degree of positive charge, an amino acid above its isoelectric point will have some degree of negative charge, and an amino acid at its isoelectric point will have no net charge.

(a) Ala, His, Lys

Electrophoresis could be carried out at pH 7.64, the isoelectric point of histidine (His). At this pH, the histidine is neutral and would not move, the lysine (Lys) will be positively charged and will move toward the negative electrode, and the alanine (Ala) will be slightly negatively charged and will move toward the positive electrode.

(b) Glu, Gln, Asp

Electrophoresis could be carried out at pH 3.08, the isoelectric point of glutamic acid (Glu). At this pH, the glutamic acid is neutral and would not move, the glutamine (Gln) will be positively charged and will move toward the negative electrode, and the aspartic acid (Asp) will be slightly negatively charged and will move toward the positive electrode.

(c) Lys, Leu, Tyr

Electrophoresis could be carried out at pH 6.04, the isoelectric point of leucine (Leu). At this pH, the leucine is neutral and would not move, the lysine (Lys) will be positively charged and will move toward the negative electrode, and the tyrosine (Tyr) will be slightly negatively charged and will move toward the positive electrode.

Problem 27.38 Examine the amino acid sequence of human insulin (Figure 27.16), and list each Asp, Glu, His, Lys, and Arg in this molecule. Do you expect human insulin to have an isoelectric point nearer that of the acidic amino acids (pI 2.0 - 3.0), the neutral amino acids (pI 5.5 - 6.5), or the basic amino acids (pI 9.5 - 11.0)?

A listing of the amino acids present are shown below:

aspartic acid (Asp)	0
glutamic acid (Glu)	4
histidine (His)	2
lysine (Lys)	1
arginine (Arg)	1

The charge will only be neutral when there are four positively charged residues to neutralize the four negative charges of the carboxylates from the four Glu residues. For this to happen, the Lys, Arg, and both His residues must be positively charged. Since the imidazole of His is not protonated until the pH is below 6, the entire molecule will only be neutral around this pH. Thus, insulin is expected to have an isoelectric point nearer to that of the neutral amino acids. Its measured isoelectric point is 5.30 to 5.35.

Problem 27.39 A chemically modified guanidino group is present in cimetidine (Tagamet), a widely prescribed drug for the control of gastric acidity and peptic ulcers. Cimetidine reduces gastric acid secretion by inhibiting the interaction of histamine with gastric H2 receptors. In the development of this drug, a cyano group was added to the substituted guanidino group to alter its basicity. Do you expect this modified guanidino group to be more basic or less basic than the guanidino group of arginine? Explain.

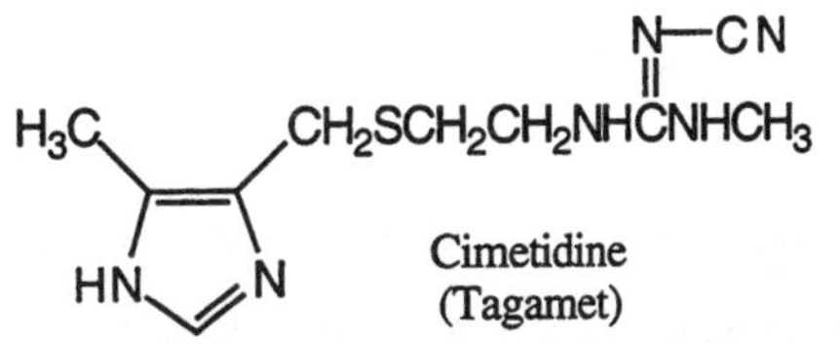

Cimetidine
(Tagamet)

A cyano group is electron-withdrawing. As a result, the guanidino function will have less electron density available to interact with a proton, and will be less basic than a similar guanidino group without the cyano group.

Problem 27.40 Draw a structural formula for the product formed when alanine is treated with the following reagents.

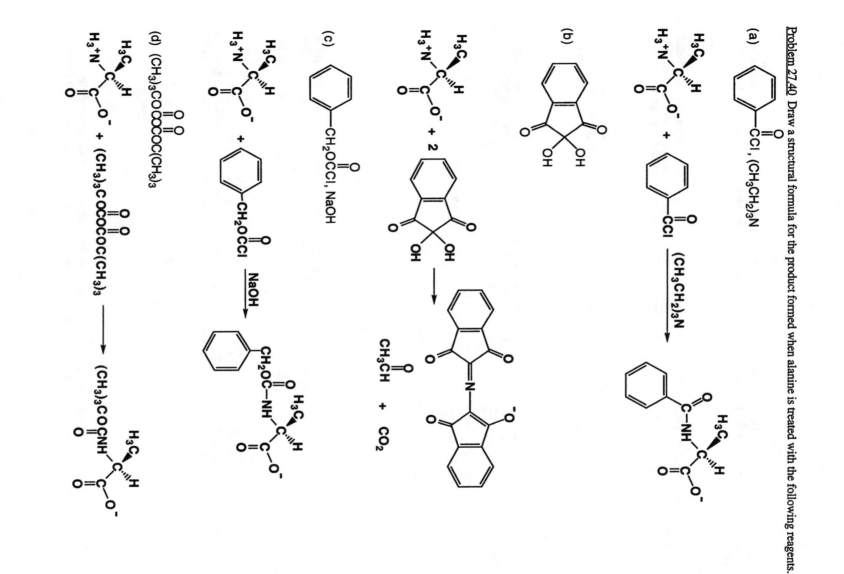

(e) Product (c) + L-alanine ethyl ester + DCC

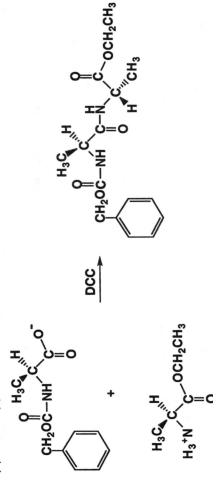

(f) Product (d) + L-alanine ethyl ester + DCC

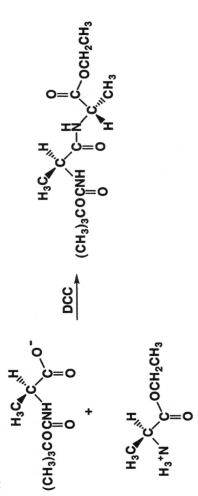

Primary Structure of Polypeptides and Proteins

Problem 27.41   If a protein contains four different SH groups, how many different disulfide bonds are possible if only a single disulfide bond is formed? How many different disulfides are possible if two disulfide bonds are formed?

If only one disulfide bond were to be formed from the four different cysteine residues, then there are a total of 6 different disulfide bonds that can be formed. There are three possibilities if two disulfide bonds are to be formed.

Problem 27.42   How many different tetrapeptides can be made under the following conditions?
(a) The tetrapeptide contains one unit each of Asp, Glu, Pro, and Phe.

There could be any of the four residues in the first position, any of the remaining three amino acids in the second position and so on. Thus, there are 4 x 3 x 2 x 1 = 24 possible tetrapeptides.

(b) All 20 amino acids can be used, but each only once.

Using the same logic as in (a), there are 20 x 19 x 18 x 17 = 116,280 possible tetrapeptides.

Problem 27.43   A decapeptide has the following amino acid composition:

Ala₂, Arg, Cys, Glu, Gly, Leu, Lys, Phe, Val

Partial hydrolysis yields the following tripeptides:

Cys-Glu-Leu + Gly-Arg-Cys + Leu-Ala-Ala + Lys-Val-Phe + Val-Phe-Gly

One round of Edman degradation yields a lysine phenylthiohydantoin. From this information, deduce the primary structure of this decapeptide.

Due to the Edman degradation result, the Lys residue must be at the *N*-terminus. Given this information, the rest of the peptide sequence is deduced because of overlap among the tripeptide sequences as shown below.

The complete peptide is:

Lys-Val-Phe-Gly-Arg-Cys-Glu-Leu-Ala-Ala

The peptides fit as follows:

Lys-Val-Phe
Val-Phe-Gly
Gly-Arg-Cys
Cys-Glu-Leu
Leu-Ala-Ala

Problem 27.44 Following is the primary structure of glucagon a polypeptide hormone of 29 amino acids. Glucagon is produced in the α-cells of the pancreas and helps maintain blood glucose levels in a normal concentration range.

1   5   10   15
His-Ser-Glu-Gly-Thr-Phe-Thr-Ser-Asp-Tyr-Ser-Lys-Tyr-Leu-Asp-Ser-Arg-Arg-

            20   25   29
Ala-Gln-Asp-Phe-Val-Gln-Trp-Leu-Met-Asn-Thr

(a) Phenyl isothiocyanate

Which peptide bonds are hydrolyzed when this polypeptide is treated with each reagent?

This reagent only hydrolyzes the *N*-terminal amino acid, so the His-Ser bond would be hydrolyzed. The site of cleavage is indicated by the j.

1   5   10   15
His∥Ser-Glu-Gly-Thr-Phe-Thr-Ser-Asp-Tyr-Ser-Lys-Tyr-Leu-Asp-Ser-Arg-

            20   25   29
Arg-Ala-Gln-Asp-Phe-Val-Gln-Trp-Leu-Met-Asn-Thr

(b) Chymotrypsin

Chymotrypsin catalyzes the hydrolysis of the peptide bonds that are located on the carboxyl side of phenylalanine, tyrosine, and tryptophan residues. The sites of cleavage are indicated by the j.

1   5   10   15
His-Ser-Glu-Gly-Thr-Phe∥Thr-Ser-Asp-Tyr∥Ser-Lys-Tyr∥Leu-Asp-Ser-

            20   25   29
Arg-Arg-Ala-Gln-Asp-Phe∥Val-Gln-Trp∥Leu-Met-Asn-Thr

(c) Trypsin

Trypsin catalyzes the hydrolysis of the peptide bonds that are located on the carboxyl side of arginine and lysine residues. The sites of cleavage are indicated by the j.

1   5   10   15
His-Ser-Glu-Gly-Thr-Phe-Thr-Ser-Asp-Tyr-Ser-Lys∥Tyr-Leu-Asp-Ser-

            20   25   29
Arg∥Arg∥Ala-Gln-Asp-Phe-Val-Gln-Trp-Leu-Met-Asn-Thr

(d) Br-CN

Cyanogen bromide cleaves on the *C*-terminal side of methionine residues. The site of cleavage is indicated by the j.

1   5   10   15
His-Ser-Glu-Gly-Thr-Phe-Thr-Ser-Asp-Tyr-Ser-Lys-Tyr-Leu-Asp-Ser-

            20   25   29
Arg-Arg-Ala-Gln-Asp-Phe-Val-Gln-Trp-Leu-Met∥Asn-Thr

Problem 27.45 A tetradecapeptide (14 amino acid residues) gives the following peptide fragments on partial hydrolysis. From this information, deduce the primary structure of this polypeptide. Fragments are grouped according to size.

Pentapeptide Fragments	Tetrapeptide Fragments
Phe-Val-Asn-Gln-His	Gln-His-Leu-Cys
His-Leu-Cys-Gly-Ser	His-Leu-Val-Glu
Gly-Ser-His-Leu-Val	Leu-Val-Glu-Ala

The complete peptide is:
   Phe-Val-Asn-Gln-His-Leu-Cys-Gly-Ser-His-Leu-Val-Glu-Ala

The peptides fit as follows:
   Phe-Val-Asn-Gln-His
                Gln-His-Leu-Cys
                       His-Leu-Cys-Gly-Ser
                                   Gly-Ser-His-Leu-Val
                                             His-Leu-Val-Glu
                                                    Leu-Val-Glu-Ala

Problem 27.46 Draw a structural formula of these tripeptides. Mark each peptide bond, the *N*-terminal amino acid, and the *C*-terminal amino acid.
(a) Phe-Val-Asn

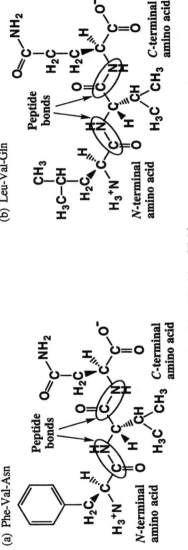

(b) Leu-Val-Gln

Problem 27.47 Estimate the pI of each tripeptide on Problem 27.46.

These pI values can be estimated by using the pK_a of the amino group for the *N*-terminal amino acid, and the pK_a of the carboxylic acid group for the *C*-terminal amino acid. Using the values for the appropriate amino groups and carboxylic acid groups listed in table 27.2 leads to the values of pI = 1/2(9.24 + 2.02) = 5.63 and pI = 1/2(9.76 + 2.17) = 5.96 for (a) Phe-Val-Asn and (b) Leu-Val-Gln, respectively.

Problem 27.48 Glutathione (G-SH, see The Merck Index, 12th ed., #4483), one of the most common tripeptides in animals, plants, and bacteria, is a scavenger of oxidizing agents. In reacting with oxidizing agents, glutathione is converted to G-S-S-G.

Glutathione

(a) Name the amino acids in this tripeptide.

The amino acids in glutathione are glutamic acid (Glu), cysteine (Cys), and glycine (Gly).

(b) What is unusual about the peptide bond formed by the *N*-terminal amino acid?

The *N*-terminal glutamic acid is linked to the next residue by an amide bond between the carboxyl group of the side chain, not the α-carboxyl group.

(c) Write a balanced half-reaction for the reaction of two molecules of glutathione to form a disulfide bond. Is glutathione a biological oxidizing agent or a biological reducing agent?

$$2 \text{ G-SH} \longrightarrow \text{G-S-S-G} + 2\text{H}^+ + 2\text{e}^-$$

The glutathione is oxidized in this process, so it is a biological reducing agent.

(d) Write a balanced equation for reaction of glutathione with molecular oxygen, O_2, to form G-S-S-G and H_2O. Is molecular oxygen oxidized or reduced in this process?

$$4 \text{ G-SH} + O_2 \longrightarrow 2 \text{ G-S-S-G} + 2 \text{ H}_2\text{O}$$

Molecular oxygen is reduced in this process.

Problem 27.49 Following is a structural formula and ball-and-stick model for the artificial sweetener aspartame (see The Merck Index, 12th ed., #874). Each amino acid has the L configuration.

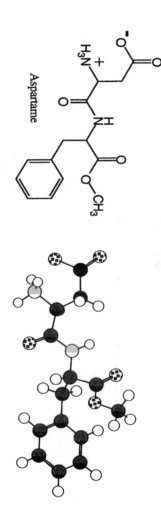

Aspartame

(a) Name the two amino acids in this molecule.

Aspartame is composed of aspartic acid (Asp) attached via a peptide bond to the methyl ester of phenylalanine (Phe).

(b) Estimate the isoelectric point of aspartame?

These pI value can be estimated by using the pK$_a$ of the amino group for the *N*-terminal amino acid, and the pK$_a$ of the carboxylic acid group for aspartic acid residue. Using the values listed in table 27.2 leads to the values of pI = 1/2(9.82 + 3.86) = 6.84.

(c) Draw structural formulas for the products of hydrolysis of aspartame in 1 M HCl.

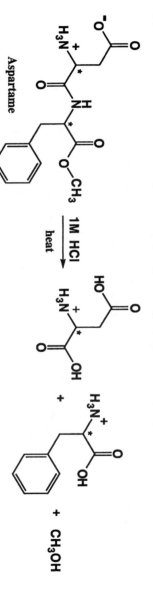

Aspartame

Problem 27.50 2,4-Dinitrofluorobenzene, very often known as Sanger's reagent after the English chemist Frederick Sanger who popularized its use, reacts selectively with the N-terminal amino group of a polypeptide chain. Sanger was awarded the 1958 Nobel Prize for chemistry for his work in determining the primary structure of bovine insulin. One of the few persons to be awarded two Nobel Prizes, he also shared the 1980 award in chemistry with American chemists, Paul Berg and Walter Gilbert, for the development of chemical and biological analyses of DNAs.

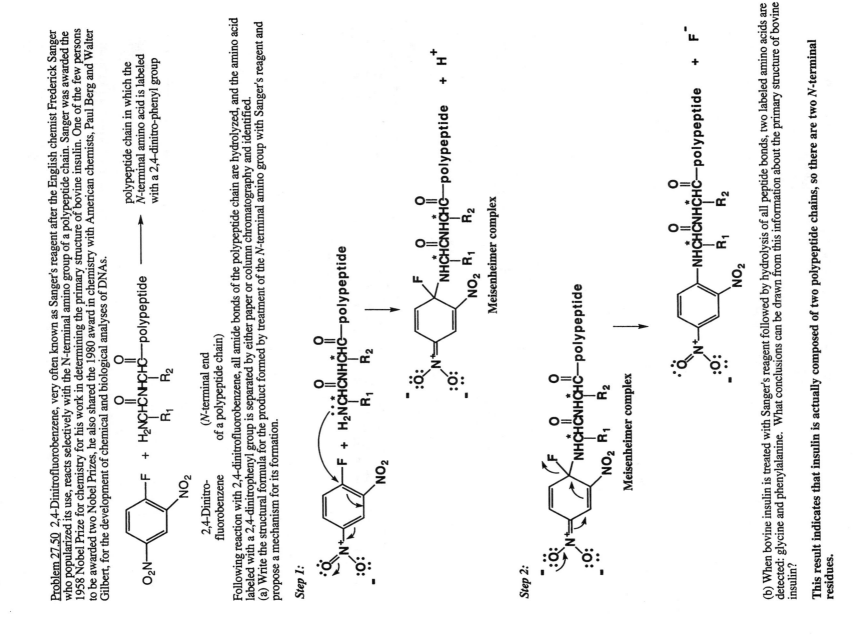

Following reaction with 2,4-dinitrofluorobenzene, all amide bonds of the polypeptide chain are hydrolyzed, and the amino acid labeled with a 2,4-dinitrophenyl group is separated by either paper or column chromatography and identified.
(a) Write the structural formula for the product formed by treatment of the N-terminal amino group with Sanger's reagent and propose a mechanism for its formation.

(b) When bovine insulin is treated with Sanger's reagent followed by hydrolysis of all peptide bonds, two labeled amino acids are detected: glycine and phenylalanine. What conclusions can be drawn from this information about the primary structure of bovine insulin?

This result indicates that insulin is actually composed of two polypeptide chains, so there are two N-terminal residues.

(c) Compare and contrast the structural information that can be obtained from use of Sanger's reagent with that from use of the Edman degradation.

Sanger's reagent only allows identification of the *N*-terminal amino acid, while the Edman degradation can sequentially determine the sequence of amino acids at the *N*-terminus.

Synthesis of Polypeptides

Problem 27.51 In a variation of the Merrifield solid-phase peptide synthesis, the amino group is protected by a fluorenylmethoxycarbonyl (FMOC) group. This protecting group is removed by treatment with a weak base such as the secondary amine, piperidine. Write a balanced equation and propose a mechanism for this deprotection.

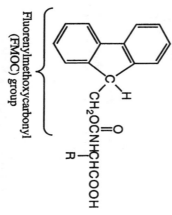

Fluorenylmethoxycarbonyl
(FMOC) group

The key to the mechanism is that the FMOC group is unusually acidic for a hydrocarbon, since deprotonation generates a relatively stable dibenzocyclopentadienyl anion. The stability of this anion is analogous to cyclopentadiene itself. The following is a reasonable mechanism for the removal of FMOC protecting groups in the presence of a weak base such as piperidine.

Step 1:

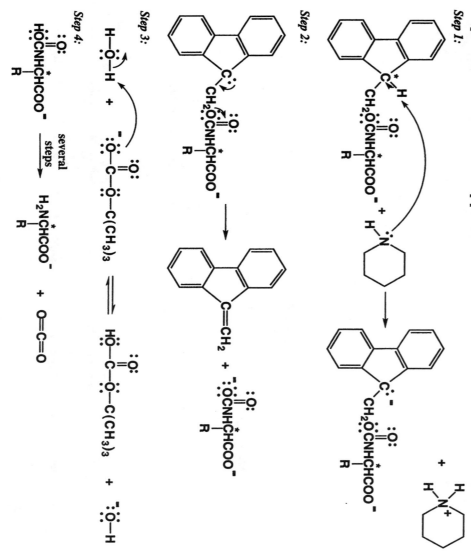

Step 2:

Step 3:

Step 4:

Problem 27.52 The BOC-protecting group may be added by treatment of an amino acid with di-*tert*-butyl dicarbonate as shown in the following reaction sequence. Propose a mechanism to account for formation of these products.

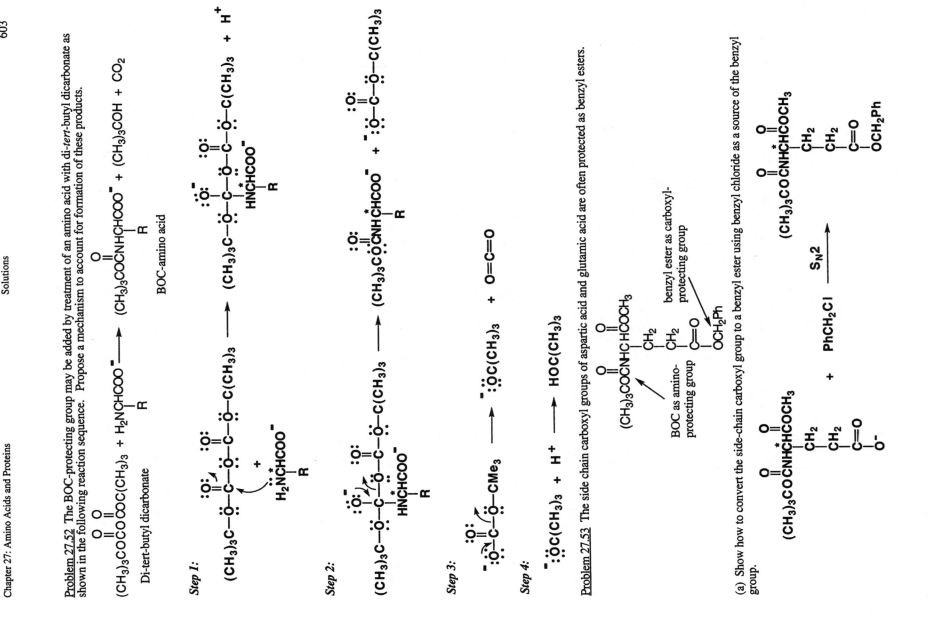

Di-tert-butyl dicarbonate

BOC-amino acid

Step 1:

Step 2:

Step 3:

Step 4:

Problem 27.53 The side chain carboxyl groups of aspartic acid and glutamic acid are often protected as benzyl esters.

BOC as amino-protecting group

benzyl ester as carboxyl-protecting group

(a) Show how to convert the side-chain carboxyl group to a benzyl ester using benzyl chloride as a source of the benzyl group.

(b) How do you deprotect the side-chain carboxyl under mild conditions without removing the BOC protecting group at the same time?

The benzyl esters can be removed under very mild conditions by hydrogenolysis using hydrogen in the presence of a transition metal catalyst.

Three-Dimensional Shapes of Polypeptides and Proteins

Problem 27.54 Examine the α-helix conformation. Are amino acid side chains arranged all inside the helix, all outside the helix, or is their arrangement random?

All of the amino acid side chains extend outside the helix.

Problem 27.55 Distinguish between intermolecular and intramolecular hydrogen bonding between the backbone groups on polypeptide chains. In what type of secondary structure do you find intermolecular hydrogen bonds? In what type do you find intramolecular hydrogen bonding?

Substantial intermolecular hydrogen bonding is possible with β-sheet secondary structures, while only minimal intermolecular hydrogen bonding is possible with α-helix secondary structures. This is because the entire edges of β-sheets have backbone hydrogen bonding groups exposed, while α-helices have only the hydrogen bonding groups exposed on the ends. Both β-sheets and α-helices are stabilized by extensive intramolecular hydrogen bonding.

Problem 27.56 Many plasma proteins found in an aqueous environment are globular in shape. Which amino acid side chains would you expect to find on the surface of a globular protein and in contact with the aqueous environment? Which would you expect to find inside, shielded from the aqueous environment? Explain.

(a) Leu   (b) Arg   (c) Ser   (d) Lys   (e) Phe

In general, charged or hydrophilic amino acids are exposed to the aqueous solution on the surface of a globular protein. Thus, (b) Arg, (c) Ser, and (d) Lys will be on the surface. The hydrophobic amino acids (a) Leu and (e) Phe will generally be inside the protein, shielded from the aqueous environment.

Problem 27.57 Denaturation of a protein is a physical change, the most readily observable result of which is loss of biological activity. Denaturation stems from changes in secondary, tertiary, and quaternary structure through disruption of noncovalent interactions including hydrogen bonding and hydrophobic interactions. Two common denaturing agents are sodium dodecyl sulfate (SDS) and urea. What kinds of noncovalent interactions might each reagent disrupt?

The SDS disrupts the hydrophobic forces that keep the non-polar residues in the interior of the structure away from the aqueous solvent and the urea disrupts the hydrogen bonds of the protein that stabilize higher order structure.

Dusky batfish *Platax pinnatus*
Moorea, French Polynesia

CHAPTER 28
Solutions to the Problems

Problem 28.1 Draw structural formulas for these compounds.
(a) 2'-Deoxythymidine 5'-monophosphate

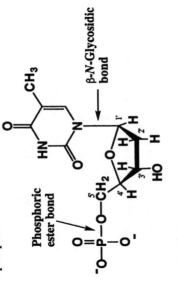

Phosphoric ester bond

β-*N*-Glycosidic bond

(b) 2'-Deoxythymidine 3'-monophosphate

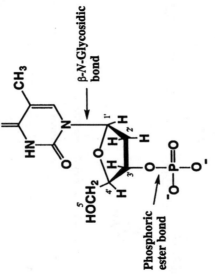

β-*N*-Glycosidic bond

Phosphoric ester bond

Problem 28.2 Write the structural formula for the section of DNA that contains the base sequence CTG and is phosphorylated on the 3' end only.

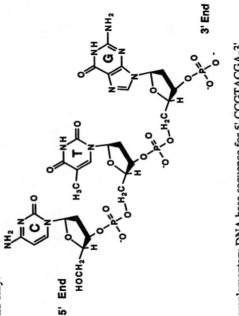

5' End

3' End

Problem 28.3 Write the complementary DNA base sequence for 5'-CCGTACGA-3'.

The complementary sequence would be 3'-GGCATGCT-5'

Problem 28.4 Here is a portion of the nucleotide sequence in phenylalanine tRNA.

3'-ACCACCUGCUCAGGCCUU-5'

Write the nucleotide sequence of its DNA complement.

Remember that the base uracil (U) in RNA is complementary to adenine (A) in DNA. The complement DNA sequence of the above RNA sequence would be:

5'-TGGTGGACGAGTCCGGAA-3'.

Problem 28.5 The following section of DNA codes for oxytocin, a polypeptide hormone.

3'-ACG-ATA-TAA-GTT-TTA-ACG-GGA-GAA-CCA-ACT-5'

(a) Write the base sequence of the mRNA synthesized from this section of DNA.

The base sequence of the mRNA synthesized from this section of DNA would be:

5'-UGC-UAU-AUU-CAA-AAU-UGC-CCU-CUU-GGU-UGA-3'

(b) Given the sequence of bases in part (a), write the primary structure of oxytocin.

The primary sequence of oxytocin would be:

Amino terminus- Cys-Tyr-Ile-Gln-Asn-Cys-Pro-Leu-Gly -Carboxyl terminus

Note how the last codon, UGA, does not code for an amino acid, but rather is the stop signal.

Problem 28.6 The following is another section of the bovine rhodopsin gene. Which of the endonucleases given in Example 28.6 will catalyze cleavage of this section.

SacI

HpaII

5'-ACGTCGGGTCGTCGTCCTCTCGGGGTGGTGAGTCTTCCGGCTCTTCT-3'

The *SacI* and *HpaII* cleavage sites are shown on the sequence above.

PROBLEMS

Nucleosides and Nucleotides

Problem 28.7 A pioneer of designing and synthesizing antimetabolites that could destroy cancer cells was George Hitchings at Burroughs Wellcome Co. In 1942 he initiated a program to discover DNA antimetabolites, and in 1948 he and Gertrude Elion synthesized 6-mercaptopurine (The Merck Index, 12th ed., #5919), a successful drug for treating acute leukemia. Another DNA antimetabolite synthesized by Hitchings and Elion was 6-thioguanine (see The Merck Index, 12th ed., #9473). Hitchings, Elion along with Sir James W. Black won the 1988 Nobel Prize for Physiology or Medicine for their discoveries of "important principles of drug treatment". In each drug, the oxygen at carbon 6 of the parent molecule is replaced by divalent sulfur. Draw structural formulas for the enethiol (the sulfur equivalent of an enol)) forms of 6-mercaptopurine and 6-thioguanine.

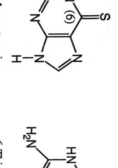

6-Mercaptopurine

6-Thioguanine

The enethiol forms are shown below:

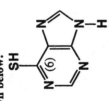

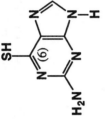

Problem 28.8 Following are structural formulas for cytosine and thymine. Draw two additional tautomeric forms for cytosine and three additional tautomeric forms for thymine.

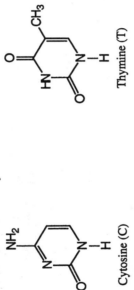

Cytosine (C)   Thymine (T)

Three additional tautomeric forms for cytosine are:

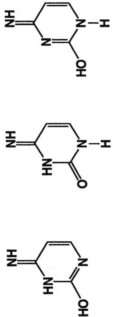

Four additional tautomeric forms for thymine are:

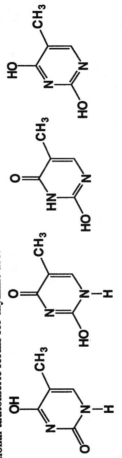

Problem 28.9 Draw structural formulas for a nucleoside composed of
(a) β-D-Ribose and adenine   (b) β-Deoxy-D-ribose and cytosine

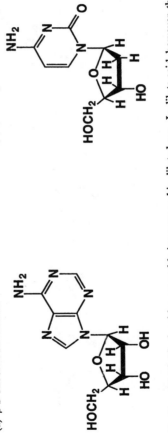

Problem 28.10 Nucleosides are stable in water and in dilute base. In dilute acid, however, the glycosidic bond of a nucleoside undergoes hydrolysis to give a pentose and a heterocyclic aromatic amine base. Propose a mechanism for this acid-catalyzed hydrolysis.

Acid-catalyzed glycosidic bond hydrolysis is most pronounced for purine nucleosides. A reasonable mechanism involves protonation of the heterocyclic base to create a good leaving group that is displaced by water to produce the product pentose and free base. The reaction of guanosine in acid (H-A) is shown below.

Step 1:

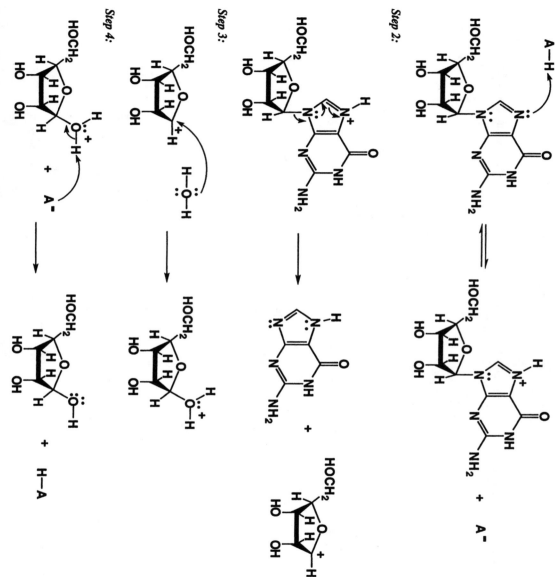

Step 2:

Step 3:

Step 4:

Problem 28.11 Explain the difference in structure between a nucleoside and a nucleotide.

A nucleoside consists of a D-ribose or 2'-deoxy-D-ribose bonded to an heterocyclic aromatic base by a β-N-glycosidic bond. A nucleotide is a nucleoside that has one or more molecules of phosphoric acid esterified at an -OH group of the monosaccharide, usually at the 3' and/or 5' -OH group.

Problem 28.12   Write the structural formula for each nucleotide and estimate its net charge at pH 7.4, the pH of blood plasma.
(a) 2'-Deoxyadenosine 5'-triphosphate (dATP)

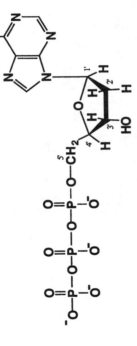
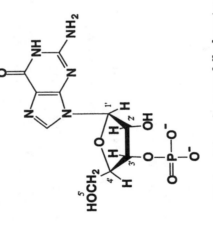

The values for the first three pK$_a$'s of dATP are all below 5.0, so these are fully deprotonated at pH 7.4. The fourth pK$_a$ of dATP is 7.0, so that at pH 7.4 there is approximately a 70:30 ratio of species with a net charge of -4 or -3, respectively. This ratio was determined using the Henderson-Hasselbalch equation.

(b) Guanosine 3'-monophosphate (GMP)

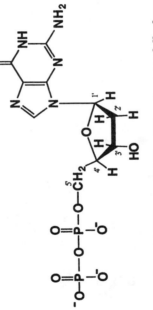

The two pK$_a$ values for GMP are well below 7.4, so these are fully deprotonated, leading to an overall charge of -2.

(c) 2'-Deoxyguanosine 5'-diphosphate (dGDP)

The values for the first two pK$_a$'s of dGDP are all below 5.0, so these are fully deprotonated at pH 7.4. The third pK$_a$ of dATP is 6.7, so that at pH 7.4 there is approximately a 83:17 ratio of species with a net charge of -3 or -2, respectively. This ratio was determined using the Henderson-Hasselbalch equation.

Problem 28.13 Cyclic-AMP, first isolated in 1959, is involved in many diverse biological processes as a regulator of metabolic and physiological activity. In it, a single phosphate group is esterified with both the 3' and 5' hydroxyls of adenosine. Draw the structural formula of cyclic-AMP.

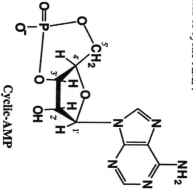

Cyclic-AMP

The Structure of DNA

Problem 28.14 Why are deoxyribonucleic acids called acids? What are the acidic groups in their structure?

Deoxyribonucleic acids are called acids because the phosphodiester groups of the backbone are acidic. At neutral pH, they are fully deprotonated, leading to the anionic nature of DNA.

Problem 28.15 Human DNA contains approximately 30.4% A. Estimate the percentages of G, C, and T and compare them with the values presented in Table 28.1.

The A residues must be paired with T residues, so estimate that there is also 30.4% T. A and T must therefore account for 30.4% + 30.4% = 60.8% of the bases. That leaves (100% - 60.8%) / 2 = 39.2% / 2 = 19.6% each for G and C. In Table 28.1, there is actually slightly less T than expected, so there is also slightly more G and C than expected.

Problem 28.16 Draw the structural formula of the DNA tetranucleotide 5'-A-G-C-T-3'. Estimate the net charge on this tetranucleotide at pH = 7.0. What is the complementary tetranucleotide to this sequence?

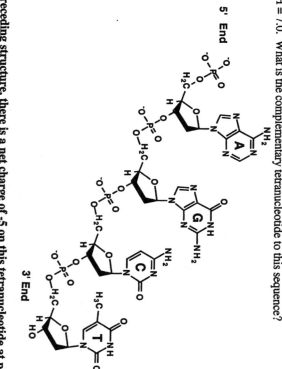

5' End

3' End

Problem 28.17 List the postulates of the Watson-Crick model of DNA secondary structure.

Major postulates of the Watson-Crick model are that:
1) A molecule of DNA consists of two antiparallel strands coiled in a right handed manner about the same axis, thereby creating a double helix.
2) The bases project inward toward the helix axis.

As shown in the preceding structure, there is a net charge of -5 on this tetranucleotide at pH 7.0. This oligonucleotide is self-complementary, that is the complementary oligonucleotide also has the sequence 5'-A-G-C-T-3'.

3) **The bases are paired through hydrogen bonding, with a purine paired to a pyrimidine so that each base pair is of the same size and shape.**

4) **In particular A pairs with T and G pairs with C.**

5) **The paired bases are stacked one on top of another in the interior of the double helix.**

6) **There is a distance of 0.34 nm between adjacent stacked paired bases.**

7) **There are ten paired bases per turn of the helix, and these are slightly offset from each other. The slight offset provides two grooves of different dimensions along the helix, the so-called major and minor grooves.**

Problem 28.18 The Watson-Crick model is based on certain experimental observations of base composition and molecular dimensions. Describe these observations and show how the Watson-Crick model accounts for each.

Chargaff found that in different organisms, the amount of A always equals the amount of T and the amount of G always equals the amount of C, even though different organisms have different ratios of A to G. The base-pairing postulates of the Watson-Crick model fully explain the observed ratios of bases. The geometry of the Watson-Crick model also accounts perfectly for the periodicity observed in the X-ray diffraction data.

Problem 28.19 If you read J. D. Watson's account of the discovery of the structure of DNA, The Double Helix, you will find that for a time in their model-building studies, he and Crick were using alternative (and incorrect, at least in terms of their final model of the double helix) tautomeric structures for some of the heterocyclic bases.

(a) Write at least one alternative tautomeric structure for adenine.

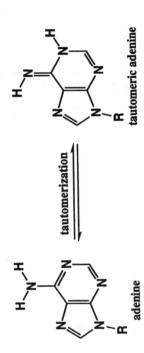

adenine   *tautomerization*   tautomeric adenine

(b) Would this structure still base-pair with thymine, or would it now base-pair more efficiently with a different base and if so, with what base?

This tautomeric form of adenine would not be able to base pair with thymine, but it would be able to form a reasonable base pair with cytosine, as shown below:

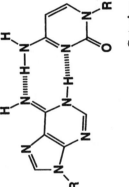

Cytosine

Problem 28.20 Compare the α-helix of proteins and the double helix of DNA in these ways.

(a) The units that repeat in the backbone of the polymer chain.

The α-helix of a protein is composed of amino acids, so the repeating unit of the backbone is a carboxyl group bonded to a tetrahedral carbon atom and a nitrogen atom. Of course, the carboxyl group and nitrogen atoms are linked in amide bonds. The repeating unit of the double helix in DNA is a 2′-deoxy-D-ribose unit linked in 3′-5′ phosphodiester bonds.

(b) The projection in space of substituents along the backbone (the R groups in the case of amino acids; purine and pyrimidine bases in the case of double-stranded DNA) relative to the axis of the helix.

The α-helix of a protein has the R groups pointed out away from the helix axis. The DNA bases of the double helix are pointed inward, toward the helix axis.

Problem 28.21 Discuss the role of the hydrophobic interactions in stabilizing:

(a) Double-stranded DNA

In the DNA double helix, the relatively hydrophobic bases are stacked on the inside, surrounded by the relatively hydrophilic sugar-phosphate backbone that is on the outside of the structure. The stacking of the hydrophobic bases minimizes contact with water.

(b) Lipid bilayers

In lipid bilayers, the hydrophobic hydrocarbon tails are associated with each other to form the hydrophobic inner layer, while the polar head groups are associated with each other on both outside surfaces.

(c) Soap micelles

In micelles, the hydrophobic hydrocarbon tails are associated with each other to form the hydrophobic interior, while the polar groups are associated with each other on the outside surface.

Problem 28.22 Name the type of covalent bond(s) joining monomers in these biopolymers.

(a) Polysaccharides   (b) Polypeptides   (c) Nucleic acids

Polysaccharides have glycosidic linkages, polypeptides have amide linkages and nucleic acids have phosphodiester linkages between the monomers, respectively.

Problem 28.23 In terms of hydrogen bonding, which is more stable, an A-T base pair or a G-C base pair?

A G-C base pair is held together by 3 hydrogen bonds, while an A-T base pair is held together by only two hydrogen bonds. Thus, in terms of hydrogen bonds alone, a G-C base pair is more stable than an A-T base pair.

Problem 28.24 At elevated temperatures, nucleic acids become denatured, that is, they unwind into single-stranded DNA. Account for the observation that the higher the G-C content of a nucleic acid, the higher the temperature required for its thermal denaturation.

G-C base pairs have three hydrogen bonds between them, while A-T base pairs have only two. Thus, the G-C base pairs are held together with stronger overall attractive forces and require higher temperatures to denature.

Problem 28.25 Write the DNA complement for 5'-ACCGTTAAT-3'. Be certain to label which is the 5' end and which is the 3' end of the complement strand.

The complementary sequence is 3'-TGGCAATTA-5'

Problem 28.26 Write the DNA complement for 5'-TCAACGAT-3'.

The complementary sequence is 3'-AGTTGCTA-5'

Ribonucleic Acids (RNA)

Problem 28.27 Compare the degree of hydrogen bonding in the base pair A-T found in DNA with that in the base pair A-U found in RNA.

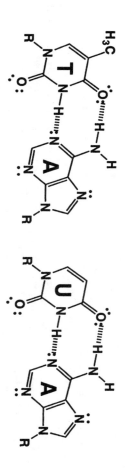

The only difference between uracil (U) and thymine (T) is the presence of a methyl group at the 5 position of thymine, that is absent in uracil. As can be seen in the structures, the presence or absence of this methyl group has very little influence on hydrogen bonding.

Problem 28.28 Compare DNA and RNA in these ways:

(a) Monosaccharide units

DNA contains 2'-deoxy-D-ribose units, while RNA contains D-ribose units.

(b) Principal purine and pyrimidine bases

DNA		RNA	
Purines	Pyrimidines	Purines	Pyrimidines
Adenine	Thymine	Adenine	Uracil
Guanine	Cytosine	Guanine	Cytosine

(c) Primary structure

The monosaccharide unit in DNA is 2'-deoxy-D-ribose, the monosaccharide in RNA is D-ribose. The bases are the same between the two types of nucleic acids, except thymine is found in DNA while uracil is found in RNA. DNA is usually double stranded and RNA is primarily single stranded. In both DNA and RNA, the primary sequence consists of linear chains of the nucleic acids linked by phosphodiester bonds involving the 3' and 5' hydroxyl groups of the monosaccharide units.

(d) Location in the cell

DNA is found in cell nuclei, while the bulk of RNA occurs as ribosome particles in the cytoplasm.

(e) Function in the cell

DNA serves to store and transmit genetic information, and RNA is primarily involved with the transcription and translation of that genetic information during the production of proteins.

Problem 28.29   What type of RNA has the shortest lifetime in cells?

Messenger RNA has the shortest lifetime in cells, usually on the order of a few minutes or less. This short lifetime is thought to allow for very tight control over how much protein is produced in a cell at any one time.

Problem 28.30   Write the mRNA complement for 5'-ACCGTTAAT-3'. Be certain to label which is the 5' end and which is the 3' end of the mRNA strand.

The mRNA complement would be 3'-UGGCAAUUA-5'

Problem 28.31   Write the mRNA complement for 5'-TCAACGAT-3'.

The mRNA complement would be 3'-AGUUGCUA-5'

The Genetic Code
Problem 28.32   What does it mean to say that the genetic code is degenerate?

The genetic code is referred to as degenerate because more than one codon can code for the same amino acid. This is because there are 64 different codons, but only twenty amino acids and a stop signal for which coding is needed.

Problem 28.33   Write the mRNA codons for
(a) Valine   GUU, GUC, GUA, GUG   (b)   Histidine   CAU, CAC
(c) Glycine   GGU, GGC, GGA, GGG

Problem 28.34   Aspartic acid and glutamic acid have carboxyl groups on their side chains and are called acidic amino acids. Compare the codons for these two amino acids.

All of the codons for these two acidic amino acids begin with GA. The codons for aspartic acid are GAU and GAC, while the codons for glutamic acid are GAA and GAG.

Problem 28.35   Compare the structural formulas of the amino acids phenylalanine and tyrosine. Compare also the codons for these two amino acids.

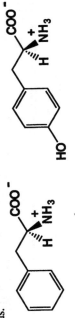

L-Phenylalanine   L-Tyrosine

Phenylalanine has a phenyl group while tyrosine has a phenol group. The mRNA codons for phenylalanine are UUU and UUC, while the mRNA codons for tyrosine are UAU and UAC.

Problem 28.36 Glycine, alanine, and valine are classified as nonpolar amino acids. Compare the codons for these three amino acids. What similarities do you find? What differences do you find?

Glycine	Alanine	Valine
GGU	GCU	GUU
GGC	GCC	GUC
GGA	GCA	GUA
GGG	GCG	GUG

All of these amino acids have four mRNA codons, all codons start with G, and in each case, the first two bases of the codon are identical for a given amino acid. This makes the last base irrelevant.

Problem 28.37 Codons in the set CUU, CUC, CUA, and CUG all code for the amino acid leucine. In this set, the first and second bases are identical, and the identity of the third base is irrelevant. For what other sets of codons is the third base also irrelevant, and for what amino acid(s) does each set code?

The third base is also irrelevant for GUX (valine), GCX (alanine), GGX (glycine), ACX (threonine), CCX (proline), CGX (arginine), and UCX (serine). In the preceding codons, X stands for any of the bases.

Problem 28.38 Compare the codons with a pyrimidine, either U or C, as the second base. Do the majority of the amino acids specified by these codons have hydrophobic or hydrophilic side chains?

The majority of amino acids with a pyrimidine in the second position of their codons are hydrophobic. This set contains phenylalanine, leucine, isoleucine, methionine, valine, proline, and alanine. Only serine and threonine have a pyrimidine in the second position and also have a somewhat hydrophilic side chain.

Problem 28.39 Compare the codons with a purine, either A or G, as the second base. Do the majority of the amino acids specified by these codons have hydrophilic or hydrophobic side chains?

The majority of amino acids with a purine in the second position of their codons are hydrophilic. This set contains histidine, glutamine, asparagine, lysine, aspartic acid, glutamic acid, arginine, cysteine, and serine. Only glycine and tryptophan are not hydrophilic, while tyrosine is a special case that is aromatic with a polar group.

Problem 28.40 What polypeptide is coded for by this mRNA sequence?

5'-GCU-GAA-GUC-GAG-GUG-UGG-3'

This mRNA codes for the following polypeptide:

Amino terminus- Ala-Glu-Val-Glu-Val-Trp -Carboxyl terminus.

Problem 28.41 The alpha chain of human hemoglobin has 141 amino acids in a single polypeptide chain. Calculate the minimum number of bases on DNA necessary to code for the alpha chain. Include in your calculation the bases necessary for specifying termination of polypeptide synthesis.

The minimum number of bases needed for the alpha chain of human hemoglobin must code for the 141 amino acids as well as three extra bases for the stop codon. Therefore, the minimum number of bases that will be required is (3 x 141) + (1 x 3) = 426 bases.

Problem 28.42 In HbS, the human hemoglobin found in individuals with sickle-cell anemia, glutamic acid at position 6 in the beta chain is replaced by valine.
(a) List the two codons for glutamic acid and the four codons for valine.

The two mRNA codons for glutamic acid are GAA and GAG, while the four mRNA codons for valine are GUU, GUC, GUA, and GUG.

(b) Show that one of the glutamic acid codons can be converted to a valine codon by a single substitution mutation, that is, by changing one letter in the codon.

Both of the glutamic acid codons can be converted to valine by replacing the central A with a U residue.

CHAPTER 29
Solutions to the Problems

Problem 29.1 Under anaerobic (without oxygen) conditions, glucose is converted to lactate by a metabolic pathway called anaerobic glycolysis or, alternatively, lactate fermentation. Is anaerobic glycolysis a net oxidation, a net reduction, or neither?

$$C_6H_{12}O_6 \xrightarrow[\text{glycolysis}]{\text{Anaerobic}} 2\ CH_3\overset{\overset{\displaystyle OH}{|}}{C}HCO_2^- + 2\ H^+$$

Glucose   Lactate

The overall process of anaerobic glycolysis that converts glucose to lactate is neither an oxidation or a reduction. No electrons are involved in the balanced half-reaction.

Problem 29.2 Does lactate fermentation result in an increase or decrease in blood pH?

Lactate fermentation leads to an increase of the H⁺ concentration in the bloodstream, therefore the bloodstream pH decreases.

Problem 29.3 Write structural formulas for palmitic, oleic, and stearic acids, the three most abundant fatty acids.

Palmitic and stearic acids are fully saturated, having 16 and 18 carbons in their chains, respectively. Oleic acid has 18 carbons and a single *cis* double bond.

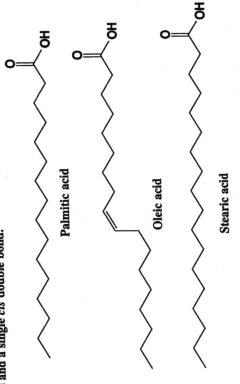

Palmitic acid

Oleic acid

Stearic acid

Problem 29.4 A fatty acid must be activated before it can be metabolized in cells. Write a balanced equation for the activation of palmitic acid.

Activation of a fatty acid involves formation of a thioester with coenzyme A. The proton is derived from the thiol group of CoA-SH.

$$CH_3(CH_2)_{14}\overset{\overset{\displaystyle O}{\|}}{C}O^- + CoA\text{-}SH + ATP \longrightarrow CH_3(CH_2)_{14}\overset{\overset{\displaystyle O}{\|}}{C}SCoA$$

$$+ AMP + P_2O_7{}^{4-} + H^+$$

Palmitic acid

Problem 29.5 Name three coenzymes necessary for β-oxidation of fatty acids. From what vitamin is each derived?

The three coenzymes needed for β-oxidation are:
1) Coenzyme A (CoA-SH) derived from the vitamin pantothenic acid.
2) Nicotine adenine dinucleotide (NAD⁺) derived from the vitamin niacin.
3) Flavin adenine dinucleotide (FAD) derived from the vitamin riboflavin (vitamin B₂).
All three coenzymes contain a molecule of adenosine

Problem 29.6 We have examined β-oxidation of saturated fatty acids, such as palmitic acid and stearic acid. Oleic acid, an unsaturated fatty acid, is also a common component of dietary fats and oils. This unsaturated fatty acid is degraded by β-oxidation but, at one stage in its degradation, requires an additional enzyme named enoyl-CoA isomerase. Why is this enzyme necessary, and what isomerization does it catalyze?

If you count the carbon atoms in oleic acid carefully, you will see that after three rounds of β-oxidation you are left with the following fragment that is then isomerized by enoyl-CoA isomerase to the *trans*-enoyl-CoA derivative needed for the next step of β-oxidation.

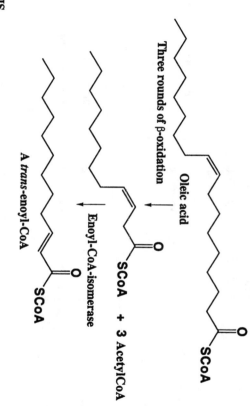

Oleic acid

Three rounds of β-oxidation

SCoA + 3 AcetylCoA

Enoyl-CoA-isomerase

SCoA

A *trans*-enoyl-CoA

GLYCOLYSIS

Problem 29.7 Name one coenzyme required for glycolysis. From what vitamin is it derived?

The one coenzyme required for glycolysis is NAD⁺, which is derived from the vitamin niacin.

Problem 29.8 Number the carbon atoms of glucose 1 through 6 and show from which carbon atom the carboxyl group of each molecule of pyruvate is derived.

By numbering the carbon atoms of glucose and following the different atoms through the pathway it can be seen that the carboxyl group carbon atoms are derived from carbon atoms 3 and 4 of glucose.

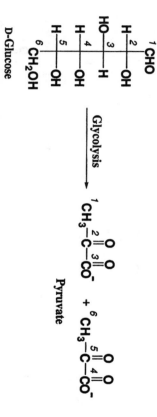

D-Glucose

Glycolysis →

Pyruvate

Problem 29.9 How many moles of lactate are produced from three moles of glucose?

During anaerobic glycolysis, 2 moles of lactate are produced for each mole of glucose used. 6 Moles of lactate will be produced from 3 moles of glucose.

Problem 29.10 Although glucose is the principal source of carbohydrates for glycolysis, fructose and galactose are also metabolized for energy.
(a) What is the main dietary source of fructose? of galactose?

The main dietary source of D-fructose is in the disaccharide sucrose, or table sugar, in which D-fructose is combined with D-glucose. The main dietary source of D-galactose is the disaccharide lactose, from milk, in which D-galactose is combined with D-glucose.

(b) Propose a series of reactions by which fructose might enter glycolysis.

Fructose could be converted to fructose 6-phosphate, and enter glycolysis at reaction 3, where it will be converted to fructose 1,6-bisphosphate.

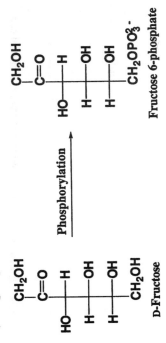

D-Fructose

Phosphorylation →

Fructose 6-phosphate

(c) Propose a series of reactions by which galactose might enter glycolysis.

D-Galactose can be isomerized at C-4 to produce D-glucose and thereby enter glycolysis at the beginning.

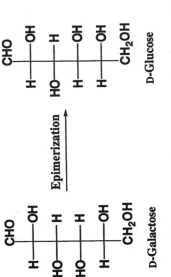

D-Galactose

Epimerization →

D-Glucose

Problem 29.11　How many moles of ethanol are produced per mole of sucrose through the reactions of glycolysis and alcoholic fermentation? How many moles of CO_2 are produced?

A total of 4 moles of ethanol and 4 moles of carbon dioxide are produced from 1 mole of sucrose. This can be seen be remembering that 1 mole of the disaccharide sucrose is first hydrolyzed to 1 mole of glucose and 1 mole of fructose. Each of these 6-carbon monosaccharides enters glycolysis to give 2 moles of pyruvate, so a total of 4 moles of pyruvate are produced for each mole of sucrose used. Each mole of pyruvate is converted to 1 mole of ethanol and 1 mole of carbon dioxide, so a total of 4 moles of ethanol and 4 moles of carbon dioxide are produced for each mole of sucrose.

Problem 29.12　Glycerol derived from hydrolysis of triglycerides and phospholipids is also metabolized for energy. Propose a series of reactions by which the carbon skeleton of glycerol might enter glycolysis and be oxidized to pyruvate.

Glycerol enters glycolysis through the following enzyme catalyzed steps that lead to glyceraldehyde 3-phosphate, which is converted into pyruvate according to the normal glycolysis pathway.

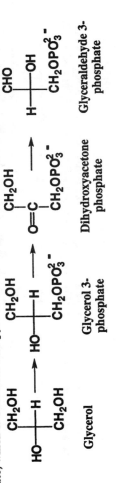

Glycerol

Glycerol 3-
phosphate

Dihydroxyacetone
phosphate

Glyceraldehyde 3-
phosphate

Problem 29.13　Ethanol is oxidized in the liver to acetate ion by NAD^+.
(a) Write a balanced equation for this oxidation.

The production of acetate ion from ethanol is an overall 4 electron oxidation, so two moles of NAD^+ are required for every mole of ethanol. In addition, 2 protons are produced along with the proton that will dissociate from acetic acid to give acetate.

$$CH_3CH_2OH \ + \ 2 \ NAD^+ \ \longrightarrow \ CH_3\overset{\overset{\displaystyle O}{\|}}{C}O^- \ + \ 2 \ NADH \ + \ 3 \ H^+$$

(b) Do you expect the pH of blood plasma to increase, decrease, or remain the same as a result of metabolism of a significant amount of ethanol?

The pH of blood plasma will drop due to the protons produced as the result of metabolism of a significant amount of ethanol.

Problem 29.14 Write a mechanism to show the role of NADH in the reduction of acetaldehyde to ethanol.

For this reaction, NADH delivers a hydride equivalent, and a group on the enzyme (denoted as A) delivers a proton to the oxygen atom.

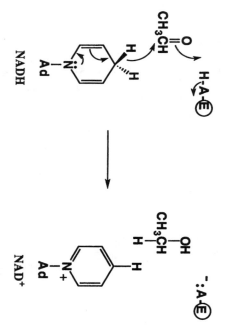

Problem 29.15 When pyruvate is reduced to lactate by NADH, two hydrogens are added to pyruvate; one to the carbonyl carbon, the other to the carbonyl oxygen. Which of these hydrogens is derived from NADH?

As can be seen in the mechanism given in the answer to Problem 29.14, the NADH delivers a hydride equivalent, H⁻. This species is highly nucleophilic and reacts with the electrophilic carbonyl carbon atom.

Problem 29.16 Review the oxidation reactions of glycolysis and β-oxidation and compare the types of functional groups oxidized by NAD⁺ with those oxidized by FAD.

NAD⁺ oxidizes a secondary alcohol to a ketone (reaction 3 of β-oxidation) as well as an aldehyde to a carboxylic acid derivative (reaction 6 of glycolysis). FAD oxidizes a carbon-carbon single bond to a carbon-carbon double bond (reaction 1 of β-oxidation).

Problem 29.17 Why is glycolysis called an anaerobic pathway?

Glycolysis is called an anaerobic pathway because no oxygen is used. Glycolysis probably first evolved in organisms that appeared before there was oxygen in the environment.

Problem 29.18 Which carbons of glucose appear as CO_2 as a result of alcoholic fermentation?

As shown in the answer to Problem 29.8, it is carbons 3 and 4 of D-glucose that end up as the carboxylic acid carbons of pyruvate. These same two carbon atoms, carbons 3 and 4, end up as CO_2 as a result of alcoholic fermentation.